INCLUSION PHENOMENA IN INORGANIC, ORGANIC, AND ORGANOMETALLIC HOSTS

Advances in Inclusion Science

Inclusion Phenomena in Inorganic, Organic, and Organometallic Hosts

Proceedings of the Fourth International Symposium on Inclusion Phenomena and the Third International Symposium on Cyclodextrins Lancaster, U.K., 20–25 July 1986

Edited by

JERRY L. ATWOOD
Department of Chemistry, University of Alabama, U.S.A.

and

J. ERIC D. DAVIES
Department of Chemistry, University of Lancaster, U.K.

Reprinted from
Journal of Inclusion Phenomena, Vol. 5, Nos. 1, 2 and 4 (1987)

D. Reidel Publishing Company

A MEMBER OF THE KLUWER ACADEMIC PUBLISHERS GROUP

Dordrecht / Boston / Lancaster / Tokyo

Library of Congress Cataloging-in-Publication Data

CIP

International Symposium on Inclusion Phenomena
  (4th: 1986: Lancaster, Lancashire)
  Inclusion phenomena in inorganic, organic, and
organometallic hosts.

  (Advances in inclusion science)
  "Reprinted from Journal of inclusion phenomena, vol. 5, nos. 1, 2, and 4 (1987)."
  Includes bibliographies and indexes.
  1. Clathrate compounds—Congresses. 2. Cyclodextrins—Congresses. I. Atwood, J. L. II. Davies, J. E. D. III. International Symposium on Cyclodextrins (3rd: 1986: Lancaster, Lancashire) IV. Journal of inclusion phenomena. V. Title. VI. Series.

QD474.I59 1986 541.2'2 87-28778
ISBN-13:978-94-010-8269-3 e-ISBN-13:978-94-009-0387-5
DOI: 10.1007/978-94-009-0387-5

Published by D. Reidel Publishing Company,
P.O. Box 17, 3300 AA Dordrecht, Holland.

Sold and distributed in the U.S.A. and Canada
by Kluwer Academic Publishers,
101 Philip Drive, Norwell, MA 02061, U.S.A.

In all other countries, sold and distributed
by Kluwer Academic Publishers Group,
P.O. Box 322, 3300 AH Dordrecht, Holland.

All Rights Reserved
© 1987 by D. Reidel Publishing Company, Dordrecht, Holland
Softcover reprint of the hardcover 1st edition 1987

No part of the material protected by this copyright notice may be reproduced or
utilized in any form or by any means, electronic or mechanical
including photocopying, recording or by any information storage and
retrieval system, without written permission from the copyright owner

Table of Contents

Due to the contents of this volume having appeared previously in various issues of the *Journal of Inclusion Phenomena*, and the original pagination being retained, the page numbering here is not always consecutive. The reader's attention is directed to the contents list below. The additional *italic* page numbering in the Poster Session section is intended to make the use of this volume more convenient.

INTRODUCTION	xi
PREFACE	1
THE GLEANER / The Conference Daily Newspaper	3

CYCLODEXTRINS

TSUNEJI NAGAI / Developments in Cyclodextrin Applications in Drug Formulations	29
M. KATA and B. SELMECZI / Increasing the Solubility of Drugs through Cyclodextrin Complexation	39
S. P. JONES and G. D. PARR / Ability of the Acetotoluides to Form Cyclodextrin Inclusion Complexes	45
STEPHEN, F. LINCOLN, JOHN H. COATES, BRUCE G. DODDRIDGE, and ANDREA M. HOUNSLOW / The Inclusion of the Drug Diflunisal by Alpha- and Beta-Cyclodextrins. A Nuclear Magnetic Resonance and Ultraviolet Spectroscopic Study	49
Y. INOUE, M. KITAGAWA, H. HOSHI, M. SAKURAI, and R. CHÛJÔ / Geometry of α-Cyclodextrin Inclusion Complex with *m*-Nitrophenol Deduced from Quantum Chemical Analysis of Carbon-13 Chemical Shifts	55
ROBERT L. SCHILLER, STEPHEN F. LINCOLN, and JOHN H. COATES / The Inclusion of Pyronine B and Pyronine Y by Beta- and Gamma-Cyclodextrins. A Kinetic and Equilibrium Study	59
NIGEL J. CLAYDEN, CHRISTOPHER M. DOBSON, STEPHEN J. HEYES, and PHILIP J. WISEMAN / ^{2}H NMR Studies of Metallocenes in Host Lattices	65
TOMASZ KOŚCIELSKI, DANUTA SYBILSKA, and JANUSZ JURCZAK / Separation Processes in Gas-Liquid Chromatography Based on Formation of α-Cyclodextrin - Chiral Hydrocarbons Inclusion Complexes	69
KENJIRO HATTORI and KEIKO TAKAHASHI / Novel HPLC Adsorbents by Immobilization of Modified Cyclodextrins	73
G. TSOUCARIS, G. LE BAS, N. RYSANEK, and F. VILLAIN / Conformational and Enantiomeric Discrimination in Cyclodextrin Inclusion Compounds	77

ENZYME MODELS AND HYDROPHOBIC CAGES

CHUL-JOONG YOON, HIROSHI IKEDA, RYOICHI KOJIN, TSUKASA IKEDA, and FUJIO TODA / Reaction of Cyclodextrin-Nicotinamide as a NADH Coenzyme Model	85

TSUKASA IKEDA, RYOICHI KOJIN, CHUL-JOONG YOON, HIROSHI IKEDA, MASAO IIJIMA, and FUJIO TODA / Catalytic Activity of β-Cyclodextrin-Histamine   93

M. PRUDHOMME, G. DAUPHIN, J. GUYOT, G. JEMINET, and N. GRESH / Modifications of Benzoxazole Ring Substituents in A.23187 (Calcimycin). Effect on Cation Carrier Properties   99

TOSHIMI SHIMIZU, YOSHIO TANAKA, and KEISHIRO TSUDA / Ion Transport, Ion Extraction, and Ion Binding by Synthetic Cyclic Octapeptide   103

NALIN PANT, MICHAEL MANN, and ANDREW D. HAMILTON / Synthetic Analogs of Peptide-Binding Antibiotics   109

J. C. LOCKHART and H. GREY / Molecular Graphics in the Study of the Calcium-Binding Sites of Carp Parvalbumin and Other Proteins   113

CYNTHIA J. BURROWS and RICHARD A. SAUTER / Synthesis and Conformational Studies of a New Host System Based on Cholic Acid   117

G. D. ANDREETTI, G. CALESTANI, F. UGOZZOLI, A. ARDUINI, E. GHIDINI, A. POCHINI, and R. UNGARO / Solid State Studies on $p.t$-Butyl-calix[6]arene Derivatives   123

F. SEVERCAN / Topology of N-Ethylmaleimide in Normal Human Erythrocyte Membrane   127

MACROCYCLIC COMPLEXES AND IONOPHORES

AMIRA ABOU-HAMDAN, IAN M. BRERETON, ANDREA M. HOUNSLOW, STEPHEN F. LINCOLN, and THOMAS M. SPOTSWOOD / An Equilibrium and Kinetic Study of the Complexation of Lithium and Sodium Ions by the Cryptand 4,7,13-Trioxa-1,10-diazabicyclo-[8.5.5]-eicosane ($C21C_5$)   137

D. E. FENTON, B. P. MURPHY, R. PRICE, P. A. TASKER, and D. J. WINTER / Metal-Free Macrocycles via Template Method: a Starting Point for Selective Complexation Studies   143

THOMAS W. BELL, ALBERT FIRESTONE, FRIEDA GUZZO, and LAIN-YEN HU / Torands: Planar Polyazamacrocyclic Ligands for Metal Ions   149

SEBASTIANO PAPPALARDO, FRANCESCO BOTTINO, PAOLO FINOCCHIARO, ANTONINO MAMO, and FRANK R. FRONCZEK / Synthesis of Symmetrical N-Tosyldiazamacrocycles and Complexation Properties of their Derivatives   153

GUY LEPROPRE and JACQUES FASTREZ / Size and Charge Dependence of Binding by Azacyclophanes   157

FRANZ P. SCHMIDTCHEN / Multiple Recognition in Polytopic Anion Hosts   161

FRANCO BENETOLLO, GABRIELLA BOMBIERI, and MARY R. TRUTER / Crystal Structures of 1:1 Complexes Between Urea and Two Crown Ether Derivatives of Phthalic Acid   165

M. N. BELL, A. J. BLAKE, R. O. GOULD, A. J. HOLDER, T. I. HYDE, A. J. LAVERY, G. REID, and M. SCHRÖDER / Transition Metal Complexes of Homoleptic Polythia Crowns   169

J. VECIANA and A. DURÁN / Isolation, Properties and Association Phenomena of Alkaline Salts and their Crown Complexes of a Radical Anion   173

MAREK PIETRASZKIEWICZ / Synthesis and Complexing Properties of a Chiral Macrocyclic Molecular Receptor with Convergent Binding Sites   177

I. GOLDBERG, H. SHINAR, G. NAVON, and W. KLAUI / Organometallic Ionophore for Alkali Metal Cations   181

TABLE OF CONTENTS

INCLUSION COMPOUNDS

Yu. A. DYADIN, G. N. CHEKHOVA, and N. P. SOKOLOVA / Solid Clathrate Solutions   187

Yu. A. DYADIN, V. R. BELOSLUDOV, G. N. CHEKHOVA, and M. Yu. LAVRENTIEV / Clathrate Thermodynamics for the Unstable Host Framework   195

Yu. A. DYADIN, F. V. ZHURKO, E. Ya. ALADKO, Yu. M. ZELENIN, and L. A. GAPONENKO / Clathrate Formation in Water-Tetraalkyl Ammonium Iodide Systems at High Pressure   203

F. H. HERBSTEIN, M. KAPON, and G. M. REISNER / Catenated and Non-Catenated Inclusion Complexes of Trimesic Acid   211

H. L. WIENER, L. ILARDI, P. LIBERATI, L. DENGLER, S. A. JEFFAS, S. SABA, and N. O. SMITH / Selectivity of the Host $Ni(4\text{-mepy})_4(NCS)_2$ Towards Aromatic Guests   215

D. W. DAVIDSON, M. A. DESANDO, S. R. GOUGH, Y. P. HANDA, C. I. RATCLIFFE, J. A. RIPMEESTER, and J. S. TSE / Some Physical and Thermophysical Properties of Clathrate Hydrates   219

TOSCHITAKE IWAMOTO, SHIN-ICHI NISHIKIORI, and TAI HASEGAWA / Three-Dimensional Metal Complex Hosts Built of α,ω-(Long-Carbon-Chain)-diaminoalkane Ligand Bridging Two-Dimensional Cyanometal Complex Network: Hofmann-Diaminoalkane-Type Clathrates   225

ROGER BISHOP, IAN G. DANCE, STEPHEN C. HAWKINS, and MARCIA L. SCUDDER / Molecular Determinants of a New Family of Helical Tubuland Host Diols   229

D. D. MacNICOL, P. R. MALLINSON, A. MURPHY, and C. ROBERTSON / Synthesis and Structure of Hexakis(p-hydroxyphenyloxy)-benzene: a Versatile Analogue of the Hydrogen-Bonded Hexameric Unit of β-Hydroquinone   233

J. VECIANA, J. CARILLA, C. MIRAVITLLES, and E. MOLINS / Free Radicals as Host Molecules   241

L. PANG and E. A. C. LUCKEN / ^{35}Cl Nuclear Resonance Studies of CCl_4 as a Guest Molecule in Various Clathrates   245

MIKIJI MIYATA, FUSAHARU NOMA, KEN OKANISHI, HIROMORI TSUTSUMI, and KIICHI TAKEMOTO / Inclusion Polymerization of Diene and Diacetylene Monomers in Deoxycholic Acid and Apocholic Acid Canals   249

INTERCALATES

J. H. CHOY, C. E. KIM, K. W. HYUNG, and J. C. PARK / Correlation Between Layer Charge and Activation Energy of Thermally Induced Deintercalation in Organo-Layer Silicates   253

S. AKYÜZ and T. AKYÜZ / Infrared Spectral Investigations of Adsorption and Oxidation of N,N-Dimethylaniline by Sepiolite, Loughlinite and Diatomite   259

A. V. MISCHENKO, Yu. V. MORONOV, P. P. SAMOJLOV, and V. E. FEDOROV / Lithium Intercalation Cluster Compounds   263

K. F. GADD / Metal-Containing Cellulose: Some Novel Materials   265

ZEOLITES

G. CELESTANI, V. SANGERMANO, C. RIZZOLI, G. BACCA, and G. D. ANDREETTI / Generation and Management of Three-Dimensional Structural Diagrams for Zeolites on Standard Graphical Support of an IBM-PC   269

TABLE OF CONTENTS

MARK D. HOLLINGSWORTH, KENNETH D. M. HARRIS, WILLIAM JONES, and JOHN M. THOMAS / ESR and X-ray Diffraction of Diacyl Peroxides in Urea and Aluminosilicate Hosts   273

W. DEPMEIER / Aluminate Sodalites – a Family of Inclusion Compounds with Strong Host-Guest Interactions   279

H. GIES / Synthesis, Crystallographic, and Thermal Properties of a New Porous Silica   283

POSTER SESSION

L. ANDĚRA and E. SMOLKOVÁ-KEULEMANSOVÁ / The Effect of Water Vapour on the Cyclodextrin-Solute Interaction in Gas-Solid Chromatography   397 [289]

NEVIN CELEBI, OSAMU SHIRAKURA, YOSHIHARU MACHIDA, and TSUNEJI NAGAI / The Inclusion Complex of Piromidic Acid with Dimethyl-β-cyclodextrin in Aqueous Solution and in the Solid State   407 [299]

ANDREA GERLÓCZY, LAJOS SZENTE, JÓZSEF SZEJTLI, and ANNA FÓNAGY / Improvement of Fat Digestion in Rats by Dimethyl-β-cyclodextrin   415 [307]

K. BUJTÁS, T. CSERHÁTI, and J. SZEJTLI / Reduction of Phytotoxicity of Nonionic Tensides by Cyclodextrins   421 [313]

J. SZEMÁN, E. FENYVESI, J. SZEJTLI, H. UEDA, Y. MACHIDA, and T. NAGAI / Water Soluble Cyclodextrin Polymers: Their Interaction with Drugs   427 [319]

M. SZÖGYI, T. CSERHÁTI, and J. SZEJTLI / Cyclodextrins Lessen the Membrane Damaging Effect of Nonionic Tensides   433 [325]

L. SZENTE, J. SZEJTLI, and LE TUNG CHAU / Effect of Cyclodextrin Complexation on the Reduction of Menthone and Iso-Menthone   439 [331]

DAO-DAO ZHANG, NAI-JU HUANG, LING XUE, and YONG-MING HUANG / β-Cyclodextrin-Catalyzed Effects on the Hydrolysis of Esters of Aromatic Acids   443 [335]

HIDETAKE SAKURABA, HIROKAZU ISHIZAKI, YOSHIO TANAKA, and TOSHIMI SHIMIZU / Asymmetric Halogenation and Hydrohalogenation of Styrene in Crystalline Cyclodextrin Complexes   449 [341]

MIYOKO SUZUKI, YOSHIO SASAKI, JÓZSEF SZEJTLI, and ÉVA FENYVESI / ^{13}C Nuclear Magnetic Resonance Spectra of Cyclodextrin Monomers, Derivatives and their Complexes with Methyl Orange   459 [351]

TSUYOSHI KIJIMA, SATOSHI TAKENOUCHI, and YOSHIHISA MATSUI / Complexes of Na-, Ca-, and Zn-Montmorillonites with an Aminated Cyclodextrin   469 [361]

YOSHIAKI FUKUSHIMA and SHINJI INAGAKI / Synthesis of an Intercalated Compound of Montmorillonite and 6-Polyamide   473 [365]

J. H. CHOY, Y. J. SHIN, G. DEMAZEAU, and P. HAGENMULLER / Isomorphous Substitution Effects on the Thermally Induced Interlayer Reaction in N-Hexylammonium Layered Aluminosilicates   483 [375]

ARZU SUNGUR, SEVIM AKYÜZ, and J. ERIC D. DAVIES / Vibrational Spectroscopic Studies of 4,4'-Bipyridyl Metal(II) Tetracyanonickelate Complexes and Their Clathrates   491 [383]

PAUL D. BEER and ANTHONY D. KEEFE / The Synthesis of Metallocene Calix[4]arenes   499 [391]

TABLE OF CONTENTS

J. REBIZANT, M. R. SPIRLET, P. P. BARTHÉLEMY, and J. F. DESREUX / Solid State and Solution Structures of the Lanthanide Complexes with Cryptand (2.2.1): Crystallographic and NMR Studies of a Dimeric Praseodymium (2.2.1) Cryptate Containing Two μ-Hydroxo Bridges   505 [397]

RICHARD A. BARTSCH, DAVID A. BABB, and BRIAN E. KNUDSEN / Synthesis and Alkali Metal Cation Complexation of N-Aryl [3.2.2] Cryptands   515 [407]

PAUL D. BEER, CHRISTOPHER J. JONES, JON A. McCLEVERTY, and SITHY S. SALAM / Redox Responsive Metal Complexes Containing Cation Binding Sites   521 [413]

YASUKO TAKAHASHI / Binding Properties of Alginic Acid and Chitin   525 [417]

KAZUO AMAYA / The Molecular Anvil Model of an Enzyme Taking into Consideration the Flexibility of Enzyme Molecules   535 [427]

JANUSZ LIPKOWSKI and TOSCHITAKE IWAMOTO / Display of Cross-Sections Showing Packing in Inclusion Compounds   545 [437]

Erratum:
S. F. LINCOLN, J. H. COATES, B. G. DODDRIDGE, and A. M. HOUNSLOW / The Inclusion of the Drug Diflunisal by Alpha- and Beta- Cyclodextrins. A Nuclear Magnetic Resonance and Ultraviolet Spectroscopic Study, *J. Incl. Phenom.* **5**, 49–53 (1987).   551 [443]

AUTHOR INDEX   445

SUBJECT INDEX   453

Introduction

The contents of this volume originate from the joint Inclusion Phenomena/Cyclodextrins Symposium held at Lancaster in July 1986. Consisting of 50 extended abstracts and 21 original contributions, the reader will find an up-to-date survey of the current state of research into, and applications of, inclusion compounds.

Topics covered range from cyclodextrin complexes and their use as media for selective chemical reagents and their applications in chromatography and in the pharmaceutical and agricultural areas; the synthesis of new hosts, particularly those containing hydrophobic cavities; the characterisation of inclusion compounds using crystallographic and spectroscopic techniques; the use of inclusion compounds as enzyme models; macrocyclic complexes and ionophores; to intercalates and zeolites.

The Symposium was extremely successful, being attended by some 250 delegates drawn from 23 nations. It is hoped that the reader will recapture the flavour of the meeting from reading this volume.

Preface

The joint meeting comprising the 4th International Symposium on Inclusion Phenomena and the 3rd International Symposium on Cyclodextrins was held on 20 - 25 July, 1986 at the University of Lancaster, Great Britain, and followed on from the previous joint meeting held in Tokyo in July, 1984. The meeting was sponsored by the Royal Society of Chemistry.

The action packed week consisted of the presentation of 14 Plenary Lectures and 19 Invited Lectures; a Poster Session at which 100 posters were on display (the participants were fortified during this four hour event by a liberal supply of free food and drink!) and two informal Evening Sessions where liquid refreshments were available to lubricate the vocal chords of delegates. The 220 participants from 23 countries were thus kept completely occupied throughout the meeting.

One extremely useful feature of the meeting was the preparation of a daily report summarising the highlights of the previous day's events. This report - The Daily Gleaner - was prepared by a group of postgraduate students from Imperial College, London and the Universities of Lancaster and Sheffield, under the guidance of Fraser Stoddart (Sheffield) and Dick Wife (Shell, Amsterdam).

A tremendous amount of work, frequently stretching into the early hours of the morning, was required of the group to ensure that The Daily Gleaner was ready by the start of the morning session. After a few days delegates were literally queuing up to collect their copy hot off the photocopier. As a tribute to the group, and to give a flavour of the interdisciplinary nature of the oral presentations, we reproduce the entire copy in this issue.

The proceedings of the meeting will be published in three issues of Volume 5 - Numbers 1, 2 and 4. Because of the high quality of the material presented in the Poster Session, the Editorial Board members present at the meeting thought it wise to try and encourage all the poster authors to contribute to the Proceedings Issues and thus ensure a comprehensive record of the exciting material presented at the Poster Session.

It was appreciated that every author would not be able to submit an original MS due to other publication commitments. In this event authors were encouraged to submit an 'extended abstract' which thus avoids the problem of duplicate publication.

Volume 5 Numbers 1 and 2 thus contain 49 extended abstracts and Number 4 will contain the original papers. This issue contains the extended abstracts of posters dealing with cyclodextrins, enzyme models, and hydrophobic cage design.

Two separate meetings will be held in 1988. The 4th Cyclodextrin meeting will be held in Munich in April 1988 and the 5th Inclusion meeting will be held in Alabama in September 1988. Further details of these meetings can be obtained from Professor Szejtli (Chinoin Ltd, Budapest, Hungary) and from Professor Atwood (University of Alabama, Alabama 35486, USA) respectively.

Eric Davies

The Sunday Gleaner

Omnibus Edition   **Sunday   27 July 1986**

4th Int. Symposium Inclusion Phenomena/3rd Int. Symposium Cyclodextrins
Lancaster 20-25 July 1986

"The experiment has been set up, the place is Lancaster and the theme is how one molecule recognises another...."

A collection of the reports which appeared in "The Daily Gleaner".

- No. 1 LEHN
- No. 2 SUTHERLAND/GOKEL/KOGA/NEWKOME
         THOMAS/MAXWELL
- No. 3 BRESLOW/TABUSHI/NAGAI/UEKAMA/HARATA
         STEZOWSKI/COLLET/WEBER/HART/SAENGER
- No. 4 DALE/HAMILTON/JURCZAK/RIPMEESTER
- No. 5 BELL/BUSCH/FENTON/REINHOUDT
         WILLIAMS(DH)/WILLIAMS(DJ)/SAUNDERS/MURRAY-RUST
- No. 6 VÖGTLE
         SHANZER/WHITLOCK/SAUVAGE

POSTERS

EVENING SESSIONS - SZEJTLI & WIFE

OVERVIEW - CRAMER & LANGLEY

Some Statistics

That was the week that was!

STOP PRESS : - LANCASTER 26 JULY
              HOSTS RECOGNISE RECEPTIVE GUESTS : COMPLEX IDEAS WITH
              EXCELLENT YIELDS.

The Daily Gleaner

No. 1   **Monday 21 July 1986**

A report on yesterday's proceedings at the 4th Inclusion Phenomena/ 3rd Cyclodextrin Symposium.

In Today!s Issue — **LEHN** : "It's the concepts that pull you"

Following the opening of the Symposium and the welcome to the participants by the Vice-Chancellor of the University of Lancaster, the theme of the Meeting was set by **Fraser Stoddart** (Sheffield). In keeping with the previous Symposia, this Meeting will continue to be innovative and to promote discussion.

The opening Plenary Lecture by **Jean-Marie Lehn** (Strasbourg and Paris) illustrated how different molecular architectures can be conceived within a single philosophy.

The practical and effective synthesis of photoactive lanthanide cryptates (1), capable of effective light conversion, has found application in the design of novel fluorescent immunoassay systems. The macrobicyclic bipyridyl cryptands prevent the normal solvation quenching of these cations. Remarkably effective DNA cleavage was accomplished using intercalators (2) based on 2,7-diazapyrenium cations.

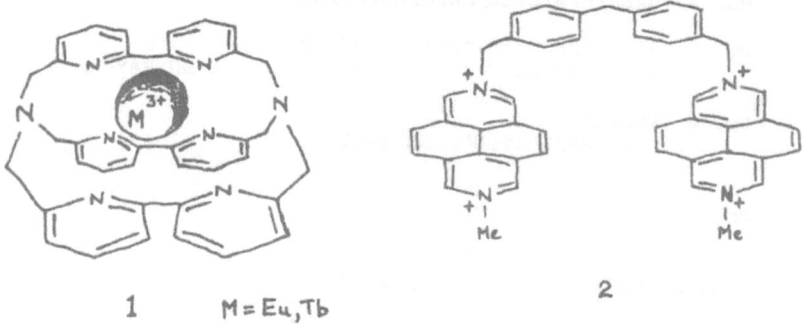

The pH-selective transport of K^+ and Ca^{2+} ions across a liquid membrane was achieved using bis-tartaro-18-crown-6 derivatives. The incorporation of caroviologens (3) into vesicle bilayer membranes promises to be a viable entry into molecular wires.

The discussion ranged from querying leaky vesicles to comparing the merits of inorganic and organic brains. It appears that what's between our ears is ahead.

The Daily Gleaner

No. 2 Tuesday 22 July 1986

A report on yesterday's proceedings at the 4th Inclusion Phenomena/ 3rd Cyclodextrin Symposium.

In Today's Issue —
- **SUTHERLAND** : The hole lecture
- **GOKEL** : More than "half a lecture"
- **KOGA** : What amAZAing hosts!
- **NEWKOME** : Sensing the "two dollar bit"
- **THOMAS** : Shapes of things to come
- **MAXWELL** : Star-Wars catalysts

The development of selectivity in synthetic receptors was discussed by **Ian Sutherland** (Liverpool). The known affinity of crown ethers for alkylammonium ions was developed to produce recognition enhancement by use of tricyclic systems (1), incorporating two diaza crown ether rings bridged by aromatic sub-units. Complexation studies on alkyl- and dialkyl-ammonium ions employing NMR spectroscopy were described. As far as the stereochemistry of binding of organic cations is concerned, small (12- and 15-) rings were shown to be better than large (18-) ones. The use of rigid spacer groups (**a,b,c**) between the crown moieties maintained a well-defined cavity size producing substrate selectivity, which was described as "coarse tuning". Furthermore, changing the size of the diaza crown rings allowed "fine tuning" of this selectivity by controlling the penetration of the ammonium group into the ring. Replacement of one of the rigid aromatic bridges by a polyether chain produced an adjustable tricyclic system with somewhat broader substrate recognition.

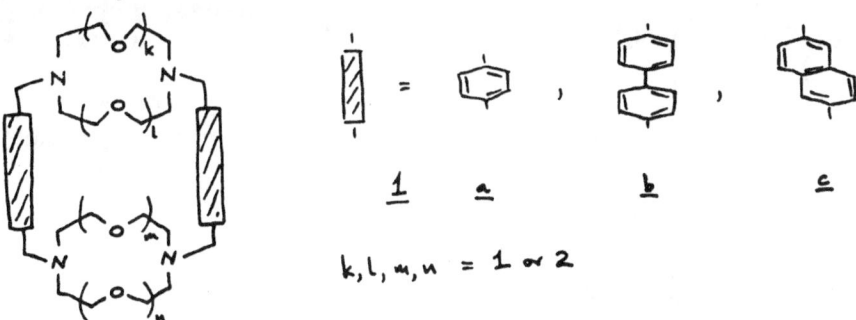

George Gokel (Miami) gave a vibrant presentation on the metal ion inclusion complexes of lariat ethers. These synthetic receptors are being designed to parallel the properties of natural hosts such as valinomycin. In addition, the investigation reveals the diverse potential of such systems. A brief resume of the properties of the simpler lariat ethers, copiously illustrated by X-ray crystal structures, preceded the development of new so-called bracchial lariat ethers. The diaza macrocycles (**2a,2b**) were demonstrated to have variable complexation geometries dependant on the metal cation and the group R. Potassium ion binding to the peptide lariats (3) utilises the amide carbonyl function on each side-chain. The analogy was drawn to the valinomycin binding.

		Na⁺	K⁺
2a	R = H	syn	syn
2b	R = Me	syn	anti

The subsequent debate entertained further applications and potential studies. Specifically, introduction of radical anions into the lariat could produce molecules suitable for electrochemical transport. An extension to the complexation of lanthanide ions echoed the concept alluded to by Jean-Marie Lehn the previous evening.

Cyclophanes incorporating the diphenylmethane skeleton (**4a**) with rigid hydrophobic cavities that complex aromatic guests such as durene and 2,7-dihydroxynaphthalene were described by **Kenji Koga** (Tokyo). By increasing the distance between the $(CH_2)_n$ spacers from 3.5 to 6.0Å, the host (**4b**) can be rendered suitable for complexation with large aliphatic guests such as deoxycholate.

4a X = ⌬ ; d = 3.5Å

4b X = naphthyl ; d = 6.0Å

Also, the incorporation of chirality into cyclophanes afforded units (**5**) which have been used in the asymmetric reduction (NaBH$_4$) of 1-naphthylglyoxylic acid. Although the selectivity is rather low (10% ee), the potential of chiral cyclophanes in asymmetric synthesis is evident.

George Newkome (Miami) demonstrated the commercial application of inclusion complexes in electronic sensors. The macrocyclic hexalactam (**6**), which can complex chloroform, was coated on to the surface of a piezo-electric crystal. Complexation of chloroform produced changes in its vibration frequency with respect to a non-coated crystal, dependent upon the substrate concentration. Inexpensive devices ($2) with sensitivities of parts per 10^{15} have been developed. They are being used in the detection of trace amounts of chloroform as an environmental pollutant.

6

$R = CH_2Ph$

John Thomas (Cambridge) opened his lecture by giving a brief introduction on the structure of zeolites. Examples of selective synthesis caused by Brönsted catalysis within the interlamellar space were given. The characteristics of ABC-6 type zeolites, composed of labelled 6- and 8-membered rings, were described and their structures were established by real space imaging and optical diffractometry resulting in a zeolite "signature". Structural studies of zeolites involved the use of spectroscopic techniques, including ESR, ^{13}C, ^{29}Si, ^{1}H and ^{27}Al NMR, together with neutron scattering and electron microscopy. Elegant use of computer graphics illustrated very clearly the position of the guests in the recently solved structures of pyridine and benzene in Zeolite L. Structures were suggested for some zeolites which have not as yet been proven to exist.

The day finished with a lecture by **Ian Maxwell** (Shell, Amsterdam) on the engineering of zeolites to produce highly selective catalysts. By reducing the Al content (acidity) in zeolites, the hydrocarbon cracking efficiency is decreased while the octane number of the products is increased, giving a commercially more valuable product. Zeolites can be used to select certain molecules based on their shapes because of their small pore size (5-7Å). para-Xylene can be produced from a mixture of ortho-, meta- and para-isomers in a 1:2:97 ratio using phosphorus-modified ZSM-5 Zeolite.
Combining two or more zeolites in series leads to multifunctional catalysts. Two zeolites are used commercially to dewax gasoline; the first isomerises straight-chain hydrocarbons to the thermodynamic mixture of straight- and branched-isomers while the second is used for separation of the branched hydrocarbons. The straight-chain isomers are then recycled and re-isomerised.
It appears now that zeolites can indeed be engineered to achieve selective catalysis on a large industrial scale.

The Daily Gleaner

No. 3   Wednesday 23 July 1986

A report on yesterday's proceedings at the 4th Inclusion Phenomena/3rd Cyclodextrin Symposium.

In Today's Issue —
- **BRESLOW**   : That's the limit!
- **TABUSHI**   : AB BA CA DA BRA
- **NAGAI**   : Short stay guests
- **UEKAMA**   : Ins and outs...
- **HARATA**   : ...ups and downs
- **STEZOWSKI**: Models give ideas not answers
- **COLLET**   : All that was vacant was the name
- **WEBER**   : Household Hosts
- **HART**   : Have wheels will travel
- **SAENGER**   : "It's neat chemistry, huh?"

"Somebody will have to check that I'm not making all this up" was one of **Ron Breslow's** (Columbia) comments during his high-speed presentation on a novel catalytic process he has developed. The remark was justified by his report of catalytic turnover numbers in the order of 10^{12}. Specific 9-chlorination of a steroid nicotinate ester (1) in >90% yield in 5 min. was achieved using metal/template systems like 2 to direct the reaction.

A mechanism which does <u>not</u> defy the diffusion limit was proposed and compared to "a double-barrelled shotgun where we reload both barrels at once". The astounding potential of this system was reinforced by his very latest results which reveal that he can now place an ester, thiocyanate and fluoride group with the same precision but as yet lower catalytic turnovers. Attempts to build some more complexity into the system are obviously under way since he mentioned that he is trying to incorporate a cyclodextrin. The reason behind this might be found in his closing comment- "I hesitate to call this an artificial enzyme because it's such a simple system".

The contribution from **Iwao Tabushi** (Kyoto) was based on his work with chemically-modified cyclodextrins as enzyme models. The asymmetrically bifunctionalised cyclodextrins (3a,3b) were synthesised in order to mimic the aminotransferase activity of Vitamin B_6. The A-B regioisomer (3a) was used to effect the transformation of keto-acids into L-amino acids with 96% ee. The elegance of the system was demonstrated by the fact that the B-A regioisomer (3b) performs the same reaction on keto-acids to give the corresponding D-amino acids acids with identical enantiomeric excesses.

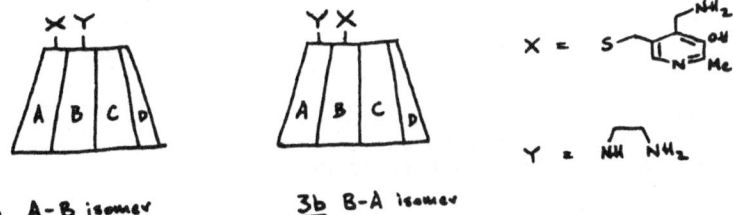

3a A-B isomer   3b B-A isomer

Tsuneji Nagai (Tokyo) reported some developments in the use of cyclodextrins (CDs) in drug formulation. The stability and dissolution rate of Cinnarizine (4) was greatly enhanced by adduct formation with βCD. However, the co-administration of a competing agent such as DL-phenylalanine was found to be necessary to enhance the bioavailability of the drug. The percutaneous penetration of the antifungal agent Tolnaftate (5) was increased by complexation with DMβCD.

4   5   6

The pharmaceutical theme was maintained by **Kaneto Uekama** (Kumamoto) who concentrated on the use of chemically-modified cyclodextrins. He described how di- and trimethyl-α- and β-cyclodextrins could be used to improve the stability, solubility, dissolution rate and bioavailability of different drugs such as HCFU (6), Prednisolone and Vitamins K1, K2 and K3. Looking to the future, the applications of hydroxypropylated, ethylated and maltosylated cyclodextrins were described.

Kazuaki Harata (Ibaraki) gave a well-illustrated lecture on host-guest interactions in cyclodextrins. His talk emphasised the indispensible role of X-ray single crystal structure determinations in the detailed study of the many varied host-guest interactions of these complexes. His many excellent slides clearly demonstrated how substitution at the hydroxy centres alter not only the size and shape of the free host cavity, but also changed the chiral recognition of the cyclodextrin. An interesting example of this was the take up in different fashions of D- and L-mandelic acid with TMβCD. Several carefully chosen examples of how the mode of complexation alters with substitution were given. For example, the mode of complexation of benzaldehyde in TM-α-CD was different compared to that found in native α-CD.

The post-prandial lecture was delivered by **John Stezowski** (Stuttgart) on the use of synthetic oligonucleotides as models for drug-DNA interactions. Studies on the interactions of anthracyclines with hepta- and octa-nucleotides gave similar results to those obtained with natural DNA, thus proving the usefulness of these as models. The nucleotides were then employed to study the behaviour of new alkyl-acridines as these drugs are known to intercalate with DNA. It was shown that, by alkylation of the amino nitrogen, the active form of the drug was now a tautomer of the free compound (**7**). This work should also allow the study of whether or not these drugs exert their toxic effects by cross-linking the DNA strands.

Preliminary work showed that hepta- and octa-mers do not form double strands (i.e. stable duplexes), but that nona- and pentadeca-mers do. Duplexes of alkylated oligonucleotides were also synthesised in order to study the effect of reducing the number of hydrogen bonds between the two strands. Thus, these synthetic oligonucleotides offer viable and simple models to explore the effects of chemically-modifying the nucleic acid structure in DNA.

The cryptophanes, lucidly described by **André Collet** (Paris), are a family of rigid, lipophilic hosts which display elegant inclusion properties and tremendous aesthetic appeal. High selectivities and strengths were demonstrated for the complexation of the halomethanes within the context of unravelling the thermodynamics and kinetics of the inclusion process. In particular, cryptophane (**8**) exhibited high selectivity for chloroform over dichloromethane in $CDCl_2CDCl_2$. The conceptual extension to the water-soluble cryptophane (**9**), which scavenges trace amounts of the halomethanes from water, was equally successful regarding selectivity. Free energy measurements were cited to support the contention that the hydrophobic effect was greater for chloroform than for dichloromethane using (**9**) in D_2O. This calculation correlates with expectations based on the molecular volumes of the two guest species, a theme reiterated during the lecture. Enantiomeric differentiation is yet another elegant property displayed by these chiral systems. Furthermore, during the subsequent debate it was proposed that a kinetic resolution of racemic halomethanes could be effected using chiral cryptophanes with intrinsically slower exchange rates.

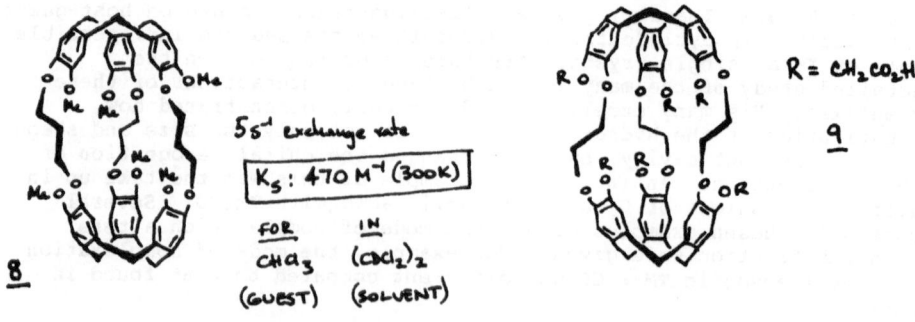

Edwin Weber (Bonn) introduced us to the term "coordinatoclathrate" which describes a species whose structure lies somewhere between a coordination complex and a true clathrate. 1,1'-Binapthyl-2,2'-dicarboxylic acid (BNDA)(**10**) forms 1:1 and 1:2 crystalline inclusion compounds with more than 50 polar and apolar guests. The polarity of the guest determines the nature of the channels formed in the crystal. Of the two extremes (host-host hydrogen bonds or host-guest hydrogen bonds) the ring size of the resultant hydrogen bonded complex varies from the 7-membered dimer ring observed in the 1:2 complex of BNDA with DMF, to the 24-membered ring comprising an open channel structure as observed with ethylene glycol. The hydrocarbon scissor host with the functional groups removed, is capable of forming complexes with nonaliphatic hydrocarbons. Structural modifications include changes in the nature of the linkage, the functional groups and also in the basic structure of the molecule. Modification of the sensor group results in a change in the hydrogen bond functionality; in the methyl ester no inclusion behaviour is observed. The bridging linkage must be rigid and any change in position or length results in a loss of inclusion activity. Spiro-type hosts (**11**) were described, and also form 2:1 and 1:1 inclusion complexes with a large number of solvents. Roof-type hosts include FADA (**12**) and its maleic analogue MADA, which do not include the smaller alcohols but do include some larger molecules. A suggestion for future work involved the use of these hosts as active site models and an analogy was drawn between the BNDA complex with imidazole and water, and the protolytic enzyme Streptomyles Griseus Protease A.

Harold Hart (Michigan) continued in the same vein as the previous speaker, but with more emphasis on the crystallographic aspects of the work. Compound (**13**) is a "wheels and axle" molecule, with the long molecular axis as the axle, and the sp^3 carbons at each end as the wheels. The host molecules line up forming channels with the bulky end groups acting as "spacers", preventing close packing of the hosts. These channels are continuous throughout the crystal, with various guest molecules therein included. Work was done on replacing the hydroxy groups by aryl substituents, and on varying the nature of the axle. Again, in many cases, a variety of inclusion compounds were formed.

The N,N'-ditritylurea (DTU) system posesses potential hydrogen bonding sites, as well as "spacers". Again many molecular complexes were formed, and the X-ray structures performed. Interesting is the strong correlation in crystallographic parameters; 6 out of the 10 complexes studied all had the same space group, which had, within narrow limits, similar cell parameters. However, despite this apparent similarity, the complexes exhibited completely different modes of complexation, although again the molecules had the long molecular axis lined up with the long crystallographic axis, with the guests running parallel. Further work was again carried out on variously substituted derivatives of DTU, including asymmetric compounds.

Wolfram Saenger (Berlin), with very little notice, gave a talk on the enzyme Ribonuclease T1 that is effectively both a catalyst and a host for guanosine-2'-phosphate (2'-GMP). A crystal structure determination clearly showed that the enzyme included the 2'-GMP. The negative charge was shown to be concentrated on the outside of the complex, with the positive charge in a band at the centre. The 2'-GMP was shown to be completely included, showing how one phosphate ester was in an exposed positon from one face, enabling a facile removal.

PRELOG TELEX

```
Professor V Prelog
Laboratorium für Organische Chemie
Eidgenossischen Technischen Hochschule
Zürich
Switzerland
```

WE WISH YOU A MOST HAPPY DAY. WE TOO ENJOY YOUR WORK AND THIS WEEK

WE SHALL CERTAINLY SLAUGHTER A PIG.

J Fraser Stoddart

On behalf of the Organising Committee and the Delegates from the
Fourth International Symposium on Inclusion Phenomena.

Jack Dunitz in this month's Chemistry in Britain (p606)........
On Prelog at 80:-

"If you want to be happy for an hour, buy a bottle of wine;

 if you want to be happy for a week, slaughter a pig;

 if you want to be happy for a year, get married;

 if you want to be happy for your life, enjoy your work."

                                                    V Prelog

The Daily Gleaner

No. 4 Thursday 24 July 1986

A report on yesterday's proceedings at the 4th Inclusion Phenomena/3rd Cyclodextrin Symposium.

In Today's Issue -   **DALE**   : Got it all cornered
                     HAMILTON   : Killer Pac-Man
                     JURCZAK   : Success under great pressure!
                     RIPMEESTER : Listening to the guests

A comparison of conformations in various free and complexed crown ethers and aza-crowns using dynamic ^{13}C NMR spectroscopy was presented by **Johannes Dale** (Oslo). Crown ethers derived from the basic 12-crown-4 unit were examined since these exhibit the largest chemical shift changes when conformational changes occur on going from the free to complexed form. Simple complexes [e.g. (**1**)] of the ligands with alkali metals were examined in order to obtain information on the thermodynamics and kinetics of ligand exchange. These principles were then applied to more complex systems such as the bicyclic (**2**) and tricyclic (**3**) hosts. In all such systems, it was shown that conformational changes, such as those between so-called "corner-side" carbon and vicinal and geminal hydrogens involve a sequence of steps. For example, the process of geminal H-H exchange in derivatives of (**2**) involves both C-C rotation and nitrogen-inversion.

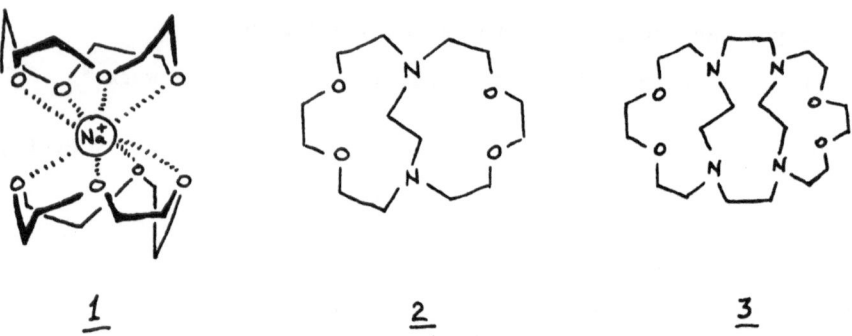

<u>1</u>   <u>2</u>   <u>3</u>

We were very fortunate that **Andrew Hamilton** (Princeton) agreed to present his lecture at extremely short notice.
Vancomycin (**4**), a natural heptapeptide antibiotic of intricate structure, was described by him as "almost a molecular Pac-Man". Vancomycin functions by highly selective binding of <u>D</u>-Ala-<u>D</u>-Ala residues, so interfering with the biosynthesis of bacterial cell walls. The strategy involves mimicking the biological construction of vancomycin in order to retain the antibiotic function as well as reducing the intrinsic toxicity. The binding site interaction involves six hydrogen bonds from the vancomycin to the <u>D</u>-Ala-<u>D</u>-Ala residues. The synthesis of analogues has been designed to incorporate the identified active groups without recourse to a total synthesis.

Preparations of analogous cyclic tripeptides (**5a,5b**) together with preliminary binding studies were reported. Tetramethylammonium acetate binds only weakly and this observation was ascribed to the lack of ammonium ion functionality which is present biologically. Studies using NMR spectroscopy (DMSO-d_6) suggest that acylated D-Ala-D-Ala binding is occurring; however, as yet, the evidence remains equivocal. Future studies will be concentrated upon the characterisation of the binding and in addition will involve synthesis of analogues more closely related to the natural systems.

An alternative approach to an old problem was demonstrated by **Janusz Jurczak (Warsaw)** in his use of high pressure rather than high dilution conditions to direct intramolecular reactions in the synthesis of diazamacrobicycles. Under reaction pressures of >5 kbar, the Menshutkin quaternisation was employed in the clean, high-yielding syntheses of chiral aza-crown ethers such as (**6**) and (**7**), the latter being produced in only one diastereoisomeric form. The novelty of the work described in the lecture provoked a lengthy and wide-rangeing discussion.

John Ripmeester (Ottawa) has used CP/MAS ^{13}C NMR to study the orientation of the guests in the clathrates of Dianin's compound, ß-quinol, α-cyclodextrin and urea. This gives useful information on the presence of different isomeric forms of the guest. Clathrate hydrates can also be characterised using an approach that utilises the NMR properties of heavier nuclei such as ^{129}Xe, where the spectrum depends both on cage shape as determined by the anisotropic shift and the cage size as determined by the chemical shift.

The Daily Gleaner

No. 5   Friday 25 July 1986

A report on yesterday morning and afternoon's proceedings at the 4th Inclusion Phenomena/3rd Cyclodextrin Symposium.

In Today's Issue -   BELL   : The holy of holies
                     BUSCH   : ConTEMPLATING life
                     FENTON   : Molecular heavy breathing
                     REINHOUDT   : Chemistry beyond the catalogue
                     WILLIAMS (DH) : It's not so tough
                     WILLIAMS (DJ) : X-Rated picture show
                     SAUNDERS   : Motion pictures
                     MURRAY-RUST   : Cost-effective chemistry

Tom Bell (New York) presented a structured and lucid exposition on the design and synthesis of rigid, planar ionophores. The rigid/flexible dichotomy, the important contribution of dipoles in the cavity and problems of solubility were all taken into account in the design. The preparation of the hexaazakekulene derivative (1) was achieved via one of the proposed synthetic routes, although characterisation of the final product proved troublesome until it was realised that the compound was in fact the calcium adduct. The macrocycle's remarkable affinity for calcium ions was highlighted by the fact that all of the metal was sequestered from trace sources.

The fact that **Daryle Busch** (Ohio) "hadn't heard about a transition metal doing anything this week" must have prompted him to detail us with some of the systems he has developed as mimics for the cytochrome P-450 oxidation system. The cavity size of lacunar cyclidenes (2), controlled by varying the R groups, was shown to affect markedly some Fe oxidation reactions. The measurement of ^{1}H NMR relaxation times in the presence of bound paramagnetic Cu^{2+} very elegantly determined the position and orientation of guests such as ^{n}BuOH included in vaulted cyclidenes (3). Evidence for the simultaneous binding of alcohols and dioxygen in a Co^{2+} vaulted cyclidene to give a ternary complex was obtained in the same way, although the results elicited the remark- "If we were to believe these numbers... and I am not prepared to believe them!"

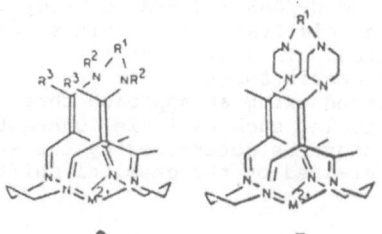

After quoting Lewis Carroll and telling us that "words mean what I want them to mean", **David Fenton** (Sheffield) discussed his studies of binucleating ligands. Simple synthetic routes, usually <u>via</u> a template method, provided both acyclic and cyclic Schiff bases. These molecules ligate two metal ions which are separated by a bridging donor species. X-Ray crystallography was the main tool of structural elucidation and provided some interesting and often slightly surprising results. The potential of these systems to complex anions was demonstrated by a crystallographic structure showing a perchlorate ion tetrahedrally surrounded by four bicopper macrocycles. This result asked a "chicken and egg" type question- did the perchlorate ion template the cavity or move into a cavity already created? Again the problem of rigidity versus flexibility in guest molecules was mentioned and these systems' flexibility was shown to be necessary for effective modelling of bimetallobiosites.

David Reinhoudt (Twente) gave a typically positive presentation on the complexation of neutral guests. The "Flexible Synthesis of Rigid Molecules" described his recent efforts on the design and synthesis of pyridohemispherands (**4**), including ^{13}C NMR T_1 measurements of their adducts with malononitrile. A study of the acidity of intra-annular functional groups enabled him to "have arrived at the ideal receptor molecule for urea", the carboxylic acid crown ether (**5**) which solubilises urea in $CHCl_3$. The success of incorporating an electrophilic group into the molecule led him to the design and recently completed synthesis of the bireceptor (**6**) containing bound Ni^{2+} as the electrophilic centre. The obvious potential of this system and the elegant approach justify his final comment- "This is the future; and it is starting in our hands."

Dudley Williams (Cambridge) described his work in determining the active site of the vancomycin group of antibiotics in an easy and enjoyable way. These antibiotics prevent the cross-linking of the glycoprotein strands in the cell wall of Gram positive bacteria. This occurs by binding the <u>D</u>-Ala-<u>D</u>-Ala terminus of the protein. Acetyl-<u>D</u>-Ala-<u>D</u>-Ala was used as a model compound to study the binding of the protein to the antibiotic. Nuclear Overhauser effect NMR experiments carried out on this model with a number of these antibiotics (ristocetin and teicoplanin) allowed the characterisation of the binding site. This is basically a hydrophobic pocket into which the <u>D</u>-Ala-<u>D</u>-Ala residue fits and is held by hydrogen bonds between the carboxylate groups of the alanine residues and three amine groups of the amino acid backbone of the antibiotic. A number of modifications to the binding site allowed the determination of the of the other groups which form the hydrophobic pocket. Consideration was given as to how this antibiotic could be formed by modifications in the conformation of a linear heptapeptide. This lecture illustrated how important molecular recognition is in nature and how non-covalent forces can produce very selective complexes.

It was with great personal pleasure that many of our reporters listened to **David Williams** (London) discuss the role that X-ray crystallography does, and must, play in the design of receptor molecules. In a talk ranging from the second sphere coordination of transition metals to the encapsulation of the bipyridinium herbicides, David showed us how solid state crystal structures of inclusion complexes have been used to optimise the critical balance between H-bonding, charge transfer, and electrostatic interactions in molecular receptor design. Furthermore, comparisons of the X-ray structures of inclusion complexes with those of the free hosts has led to the design and synthesis of new receptor molecules requiring only the smallest possible conformational changes to accommodate specific guest species. And, as he promised not to mention the occasions when the crystals we supplied did not contain exactly what they should have, we in turn shall not mention that leaving crystals in their mother liquor on a radiator tends to diminish their solid state character!

A kaleidoscopic display of molecular graphics was given both by **Martin Saunders** (Smith, Kline and French) and by **Peter Murray-Rust** (Glaxo). Martin Saunders is clearly not of like mind with "people who think HCN is a large molecule". The lecture illustrated both the present uses and the future developments of simple and realistic electrostatic potential calculations to model molecular interactions. Applications include fascinating molecular-scale calcalations of the dynamic trajectories of biologically-important molecules. Successful studies indicate the utility of electrostatic potentials in modelling interactions of a chiral hplc stationary phase with substrate molecules. In harmony with the previous lecture, the conformational changes during complexation of dibenzo-30-crown-10 with diquat was discussed. Calculations reinforce the solid state evidence that the total encapsulation of the diquat by the crown ether in a horse-shoe conformation is the most favourable conformation. However, the correlation of physical measurements with the empirical calculations should be treated with some caution according to Peter Murray-Rust. His lecture concentrated on the problems of molecular modelling and on methods of their refinement. Commonality in structure and functional group interactions between molecules are investigated using the Cambridge Crystallographic Data File. Qualitative predictions can thus be made if a sufficiently large sample size is taken. As examples, the directional H-bonding of alcohols to ketones, epoxides, and ethers were cited. In addition, the erudite conformational studies on the cyclic decapeptide Tyrocidine-A were also reported.

The Daily Gleaner

No. 6　　　　　　　　　　　　　　　　　　　　**Friday   25 July 1986**

A report on the proceedings not previously covered in other issues at the 4th Inclusion Phenomena/3rd Cyclodextrin Symposium.

In Today's Issue :- **VÖGTLE**　:　Gripping stuff
　　　　　　　　　　SHANZER　:　Let's twist again
　　　　　　　　　　WHITLOCK　:　The house that Howard built
　　　　　　　　　　SAUVAGE　:　A tricycle made for two

In a lecture of noteworthy clarity, **Fritz Vögtle** (Bonn) described his development of a series of novel synthetic receptors. Molecules were designed with large three dimensional cavities in which steric effects, hydrophobic binding, and more specific binding sites were all carefully considered. These concepts were condensed together in a family of receptor molecules containing donor groups held apart by hydrophobic spacer groups.

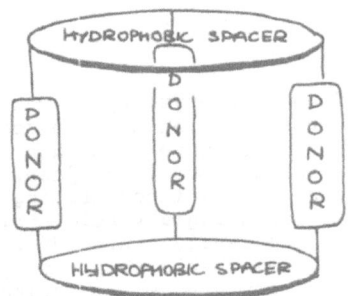

The first compound with three catechol residues and 1,3,5-substituted benzene rings as spacer groups demonstrated a remarkable affinity for Fe^{3+}, whilst a subsequent compound featuring bipyridyl donor groups showed an equally impressive ability to complex with Fe^{2+}. Another type of host featuring a 4,4',4''-triphenylmethyl group as a large spacer unit and carboxy-based groups as donors was described. These species were found to form complexes with simple aromatic guests. The synthesis of all of these polymacrocycles were effected elegantly and in yields not normally achieved with such complicated systems. He finished by touching on the wealth of potential chemistry promised by this series of compounds based on their novel selectivities and high complexing power. These included complexing anions and even ion pairs not to mention the possibility of carrying out and catalysing reactions in the cavity.

The opening lecture on the last day was given by **Abraham Shanzer** (Rehovot). Expanding upon a topic briefly mentioned the previous evening by Fritz Vogtle, he explained how his search for a flexible ionophore for Fe^{3+} transport led him to synthesise analogues of enterobactin (1). This exhibits an extremely high binding constant (10^{52} M^{-1}) for Fe^{3+} which cannot be explained by the coordination of the iron alone. There must also be some extra contribution and this is due to the formation of a helical arrangement by the three sidearms when coordinating the iron. He then outlined his efforts to reproduce this helical arrangement synthetically using acyclic polyamides.

"Substance is nothing, form is everything!" were the opening remarks of **Howard Whitlock** (Wisconsin) during his provocative discussion on the concept of molecular architecture - the use of molecular building blocks to make molecules for a specific purpose. He has designed a series of cyclophanes (e.g. 2 & 3) incorporating naphthalene walls separated by spacer units and a rigidly held pyridinium group to provide a "stickiness" to the cavity. Already these new receptors have been shown to exhibit remarkable selectivity for the binding of phenols at the expense of everything else, including anilines and benzoic acids.

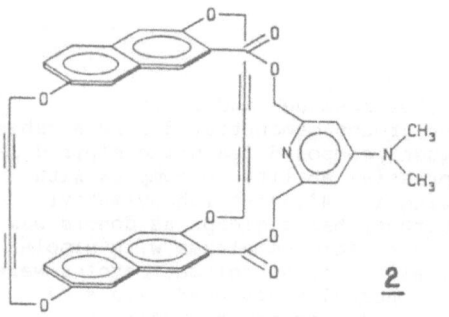

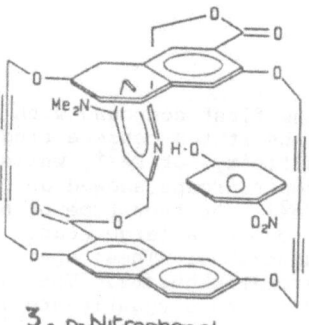

3. p-Nitrophenol

"As everyone knows, transition metals are magic" and **Jean-Pierre Sauvage** certainly used them as wands to weave together his threads. With Cu(I) as a template to hold them in the correct orientation, extended phenanthroline units were stitched together in high yields to give interlocking macrocycles named catenates (**4**). Demetallation with cyanide ion afforded the free catenand, which was shown to form a Ni(I) catenate, stabilising the Ni(I) towards oxidation with a topological factor of 10^5 relative to the $(dmp)_2Ni^+$ complex. The synthesis of [3]-catenands consisting of three interlocked macrocyclic rings was also described.

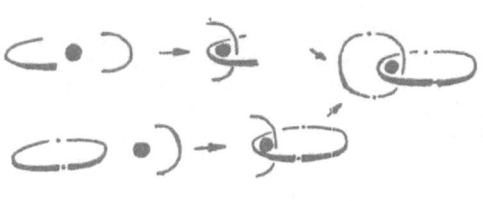

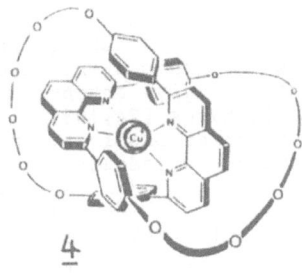

4

POSTERS

A review of some of the posters presented; poster numbers (as they appear in the abstracts) are included in brackets [ ].

The **POSTER OSCAR** (£150, donated by Royal Dutch Shell, Amsterdam) was won by **John Thomas'** group for their contribution [100] entitled - **The Inclusion of Benzene and Pyridine in the Tunnels of a Zeolite**. The runner's-up prize (£75, Royal Dutch Shell, Amsterdam) was won by **David Reinhoudt's** group for their poster [63] entitled - **The Use of Electrophiles in the Complexation of Urea by Macrocyclic Host Molecules**.

The posters on the topic of cyclodextrins were an interesting cross-section through the range of chemistry of these compounds. Most of the posters described work with natural cyclodextrins, but increasingly, attention is being focused on chemically-modified derivatives [24,108], especially the methylated cyclodextrins which show great promise in many applications. Several contributors studied cyclodextrin complexation at a fundamental level by many methods including solubility, filtration cell studies, thermal analysis, ^{1}H, ^{2}H, ^{13}C and ^{19}F NMR spectroscopy in solution and in the solid state, spectrophotometry, circular dichroism and X-ray crystallography [1,3,5,6,8,9,23,26,27]. The use of cyclodextrins to influence reactions selectively and/or catalytically was also described [2,11]. Reactions of cyclodextrin adducts in the solid state showed enhanced selectivities, and reactions in solution with possible industrial applications were described [12], as well as their uses in modelling biological reactions [21,22]. There was a lot of information on the potential pharmaceutical benefits of using cyclodextrins as drug-delivery systems [7,17,18], and in one case as a bile substitute to solubilise fat in vivo [13]. In an agrochemical application, cyclodextrins were found useful in plant protection because of their ability to complex non-ionic tenzides [19,20]. The increasing use of cyclodextrins to effect chromatographic separations also received deserving attention [14,15,16,109], and their potentially powerful applications as stationary phases are evident.

A study of open-chain and macrocyclic receptors based on bis-cholic acid in complexation with neutral polar guests was presented [36] in a clear and concise form by **Cynthia Burrows** (NY State). The open chain host (1) can exist in several conformations, of which the two extremes are the "open" and "closed" forms. Using ^{1}H NMR spectroscopy, it was shown that, in the latter case, the hydroxyl functions are directed towards a central cavity space and the conformation is stabilised by internal hydrogen bonding. When polar guests such as methanol are complexed by this host, a similar hydrogen bonding network between host and guest is observed.

The challenge to complex uncharged lipophilic molecules was elegantly
described [37] by **Israel Goldberg** (Tel Aviv). The requirement for
well defined cavities of suitable properties for inclusion was
emphasised. This work can be seen as part of a new trend to impart
rigidity to well characterised (solution state) hosts normally
associated with the solid state.

Among the redox active macropolycyclic molecules described [45] by
Paul Beer (Birmingham), were the metallo-calix-[4]-arenes (2) which
include neutral substrates. With ^{t}butylamine as substrate (**S**) the
oxidation potential of the iron atom in the ferrocene complex is
increased, but unfortunately the process is irreversible.

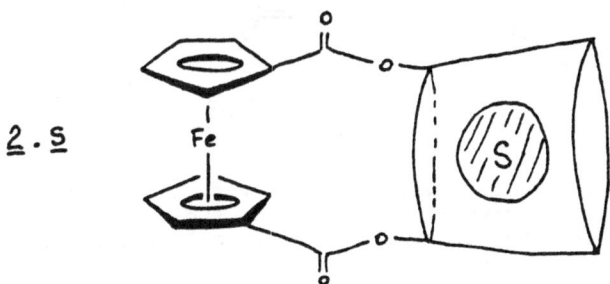

Potential developments in cancer therapy must include the
utilisation of 'targetted macrocycles' for selective radionucleotide
delivery. It is necessary to ensure rapid and complete
radionucleotide complexation under physiological conditions to give
a kinetically-stable complex. This goal has been achieved [49]
using the water-soluble macrocycles (3) and (4). Targetting is
achieved using monoclonal antibodies. While this ensures
selectivity, it also requires specific monoclonal antibodies which
do not cause radionucleotide accumulation in the biological system.
The applicability is clearly reliant on the limitations of
monoclonal antibody techniques.

To many people the phrase "liquid clathrates" may appear at first to
be a contradiction in terms, but **Jerry Atwood** (Alabama) has shown us
[61] that the phrase can be used effectively to describe a new
mesomorphic state, incorporating many of the features of solid state
clathrates and also solution inclusion phenomena. The ability of
these liquid clathrate phases to extract hydrocarbons from solutions
in model studies represents an exciting advance in the possibility
of producing cheap coal-derived petroleum.

In general, crown ethers do not form strong inclusion complexes with neutral molecules, thus much work on binding urea has instead focused on binding the more strongly held uronium cation. **David Reinhoudt** (Enschede) has solved many of the problems associated with this work [63] through the use of crown ethers possessing readily ionisable groups (5) and, more recently, bireceptor molecules (6) capable of stabilising a urea complex through metal cation co-complexation.

With David's work in mind, it was therefore very interesting to see a poster [85] reporting crystal structures of stoicheiometric 1:1 complexes of urea (NOT the uronium ion) with small crown ethers containing ionisable groups.

Andrew Hamilton (Princeton) reported [late submission] the recent synthesis and characterisation of a macrocycle (7), designed specifically to control the environment around a complexed metal in order to influence its catalytic properties. The macrocycle was synthesised from a bipyridyl unit and two valine residues. It was designed to provide (a) ligands capable of stabilising a range of metal oxidation states, (b) chiral groups which are close enough to influence the approach to the metal, (c) a cavity in order to effect substrate specificity and (d) sufficient stability to withstand oxidation. Complexation with a number of transition metals has been successful and preliminary studies on the Fe^{3+} complex show that in the presence of iodosylbenzene in acetonitrile, it will epoxidise cyclooctene in 55% yield.

Of the many other posters on the subjects of Macrocycles, Clathrates, Intercalates, Surface Absorption and Zeolites, many deserve mention but lack of space permits only the Oscar winner to be described. The contribution [100] of the group of **John Thomas** (Cambridge) showed how the positions of small organic molecules such as benzene and pyridine included within zeolites could be determined. The poster was particularly notable for its spectacular use of colour graphics special effects.

FOURTH INTERNATIONAL SYMPOSIUM INCLUSION PHENOMENA
THIRD INTERNATIONAL SYMPOSIUM CYCLODEXTRINS

Lancaster 20-25 July 1986

EVENING SESSIONS

The **Evening Sessions** were an experiment. A unique mix of scientific debate and liquid refreshments were floated at a meeting for the first time to our knowledge. The idea was to promote informal discussion and argument, catalysed by short talks from some of the poster presenters. The activities were sustained and controlled by **Dick Wife** (Shell, Amsterdam) and **József Szejtli** (Chinoin, Budapest). The **First Evening Session**, nominally on solid state phenomena, was to some extent an acclimatisation. People were getting used to the idea and the format. **David Wernick** (Tel Aviv) discussed **Catalysis by Cyclodextrins in the Solid State**, **Roger Bishop** (New South Wales) talked at length on **Helical Tubuland Hosts** and **Ric Zarzycki** (Sheffield) spoke briefly on **Transition Metal Adducts with Cyclodextrins**. The response to the talks themselves was rather limited, but other topics surfaced in between the presentations and several impromptu discussions did ensue. By the end of the session, people were beginning to warm to the idea and this was one of the factors which led to the success of the **Second Evening Session**, nominally dealing with the solution state. The presentations were given by **Cynthia Burrows** (New York State) on **Chiral Hydrophobic Cavities from Cholic Acid**, **David Parker** (Durham) on **Macrocycles/Antibodies as Anti-Tumour Agents**, **John Toner** (Kodak, Rochester, USA) on **Specific Ionophores** and **Steve Lincoln** (Adelaide) on **Double Guests in Cyclodextrins**. There was a good deal of discussion and debate throughout the session, not only on the topics presented, but also on subjects introduced by speakers from the floor. Everyone present appeared to enjoy themselves and the general opinion voiced afterwards was that the experiment had been sufficient of a success to warrant repetition.

FOURTH INTERNATIONAL SYMPOSIUM INCLUSION PHENOMENA
THIRD INTERNATIONAL SYMPOSIUM CYCLODEXTRINS
Lancaster 20-25 July 1986

OVERVIEW

It was entirely appropriate that the fine statesman of science, **Friedrich Cramer** (Göttingen) had the last word today. In his own inimitable way he provided for us an overview of the whole area of molecular recognition illustrated simply by Emil Fischer's "Lock and Key" concept.

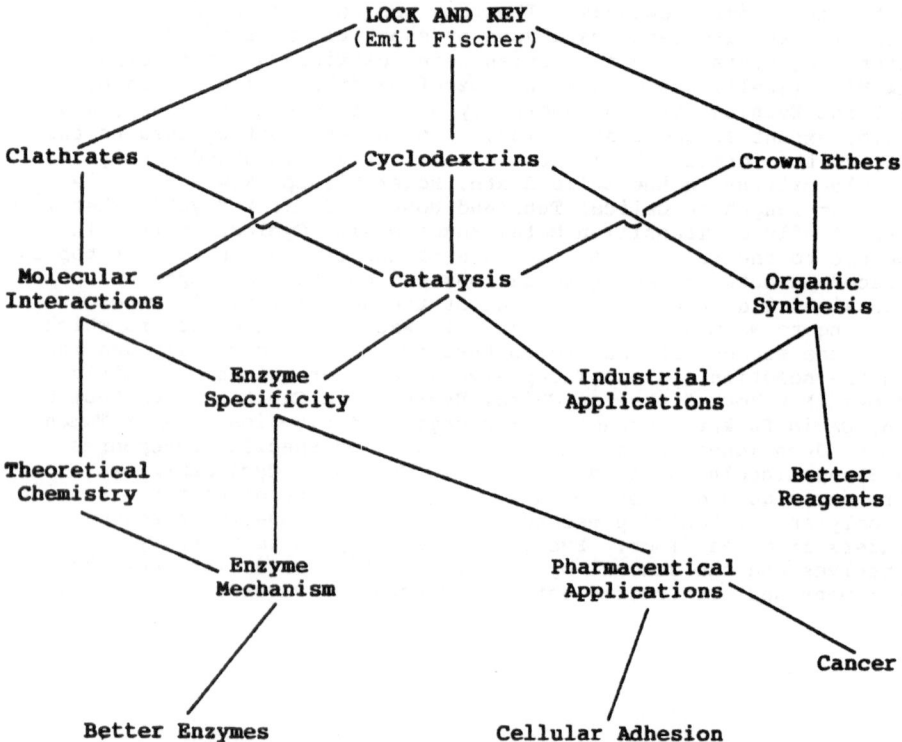

By leading us through the different processes which result in molecular recognition, he painted a picture of how important this whole area is in physical, chemical and biological sciences. Beginning with the simple example of pyrophosphate hydrolysis by cyclodextrins and building up to how cell walls in mammalian tissue recognise sugar residues, he succeeded in completing a picture to which many individual strokes had been added during the course of the week. One was left with the feeling of having been led through a park full of beautiful and exciting sights by a well-loved and knowledgeable teacher.

The Overview Session was chaired in an extremely lively manner by **Bernard Langley** (ICI, Academic Relations). The rapidity with which he spoke was matched only by the quickness of his wit and it was fitting that the meeting concluded on such a high note.

FOURTH INTERNATIONAL SYMPOSIUM INCLUSION PHENOMENA

THIRD INTERNATIONAL SYMPOSIUM CYCLODEXTRINS

Lancaster 20-25 July 1986

That was the week that was

The Lancaster meeting was a friendly meeting, a hard-working experiment. It was also a week filled with the many excellent presentations in lectures and posters, and discussions stretching into the early hours of the morning, some serious and some not so serious. We appreciate the effort and the energy of each delegate throughout this meeting. In the two years before the next meeting in Alabama, we wish you all success in your own experiments, and to seeing you as hosts or as guests.

SESSION CHAIRMEN

Fraser Stoddart (Sheffield), **Giovanni Andreetti** (Parma), **Wolfram Saenger** (Berlin), **David MacNicol** (Glasgow), **Eric Davies** (Lancaster), **Jerry Atwood** (Alabama), **Mary Truter** (London), **David Ollis** (Sheffield), **Bernard Langley** (ICI Academic Relations).

EVENING SESSION CHAIRMEN

Dick Wife (Shell, Amsterdam), **József Szejtli** (Chinoin, Budapest).

LOCAL ORGANISING COMMITTEE

Convener: **Eric Davies** (Lancaster) and **John Gibson** (Royal Society of Chemistry), **David MacNicol** (Glasgow), **Fraser Stoddart** (Sheffield), **Dick Wife** (Shell, Amsterdam), **David Williams** (London).

The REPORTERS and STEWARDS who toiled deep into the night to produce **The Daily Gleaner**.

Imperial College of Science and Technology
Alexandra Slawin

Lancaster University
Vivian Knott, Mark Houghton, Andrew Haise (Projectionist).

Sheffield University
David Berrisford, Paola Bonaccorsi, Surinder Chana, Lawrence Cullen, Mark Garner, Colin Gemmell, Franz Kohnke, David Leigh, John Mathias, David Mulligan, Mark Reddington, Neil Spencer, Michael Wali, Ric Zarzycki.

DEVELOPMENTS IN CYCLODEXTRIN APPLICATIONS IN DRUG FORMULATIONS

Tsuneji Nagai
Department of Pharmaceutics, Hoshi University
Ebara, Shinagawa-ku, Tokyo 142
Japan

ABSTRACT. With regard to recent developments in cyclodextrin (CD) applications in drug formulations, here will be described on the basis mainly of (a) novel preparative methods of CD inclusion complexes, (b) effects of CDs on bioavailability and disposion of drugs and (c) absorption enhancement by CD derivatives in transdermal application. (a) When inclusion complex of cinnarizine (CN) with β-CD was prepared by a spray-drying method, it was very stable under heating and highly humid conditions. (b-1) CDs gave influence on hypnotic potency and disposition of barbiturates in intravenous and intraperitoneal administrations. (b-2) The bioavailability of CN on oral administration of the complex, which was comparable with that of CN alone, was enhanced by simultaneous administration of competing agents, such as DL-phenylalanine. (c) When tolnaftate (TOL), antifungal drug, was administered percutaneously in the form of the complex with dimethyl-β-CD and water-soluble β-CD polymer, it was absorbed in the skin, and the concentration was kept high compared with the case of TOL alone.

1. INTRODUCTION

It may be said that recent developments in cyclodextrin (CD) applications in drug formulations are concerned with: stabilization; enhancement of solubility; novel preparative methods of inclusion complexes; enhancement of bioavailability; reduction of topical or hemolytic side effects on administration; absorption enhancement in transdermal application.
 As it is well known, the advantage of using CDs mainly comes from the inclusion complex formation.
 When it is considered to use some excipients in pharmaceutical preparations, the safety of the materials upon the administration should be guaranteed. Regarding this point, CDs are confirmed to have no problem (1-3). At the present, α- and β-CD are included in Japanese Standards for Ingredients of Drugs, a kind of Japanese national formulary.

β-CD is the most practical to use, but its complexes are usually slightly soluble. Therefore, various derivatives of CDs are under development (4-6). Alkylation of hydroxyl function of CDs has resulted in an enhanced water-solubility and these derivative have been found to be effective in binding guest molecules. An application of water-soluble CD polymers to pharmaceutical preparations is under investigation (7). Anyway, it may take a time to guarantee the safety upon the administration of the derivatives which are not yet officially registered, and thus it takes time to get some products of these derivative on the market.

Already in Japan, there are commercial products of prostaglandins with α- and β-CD on market which have been developed by Ono Pharmaceutical Company (8). Therefore, CDs often play a very important role in materializing new drugs.

From the methodological point of view to a development of CD application in pharmaceutical formulation, the way of approach seems to have been almost established. For example, in stablization or solubilization, usually the effect of the addition of CDs is examined first, and then a further investigation and evaluation of the preparation is done in comparison with the preparations without CDs or with the existing formulations.

Here will be described some new approaches to CD applications in pharmaceutical formulations on the basis of our recent trials. They are concerned with (a) preparative methods of CD inclusion complexes (9); (b) effects of CDs on bioavailability and drug disposition (10); and (c) absorption enhancement by CD derivatives in transdermal application (11).

2. SPRAY-DRYING METHOD TO PREPARE CD INCLUSION COMPLEXES

2.1. Preparative Methods of CD Inclusion Complexes in Solid State

Most common preparative methods of includion complexes in solid state are a coprecipitation method based on phase solubility, and kneading. However, these methods may be suitable in preparation of the complexes in laboratory scale, but not in manufacturing scale. The present situation that the materialized products of CD complexes are not so many may be partly due to a difficulty in preparation of them in manufacturing scale.

More than ten years ago, when it was at the early stage of CD applications in pharmaceutical formulations in Japan, the authors developed a preparative method of inclusion complexes by freeze-drying (12). Recently, the authors tried to develop a spray-drying method to prepare the inclusion complex, because spray-drying method generally can be extended to production on a large scale (9).

2.2. Inclusion Complex of Cinnarizine (CN) with β-CD by Spray-drying Method

CN is widely used orally for a treatment of cerebral diseases. It is physico-chemically a weak base and its solubility in water is very poor, and its bioavailability is variable depending on the products and the condition of administration. Clinically, preparations of CN are usually administered to old patients. Many of these patients are in achlorhydria or anacidity. Therefore, it is necessary to develop a preparation of CN which dissolves well. In a series, the authors confirmed the inclusion complex formation of CN with β-CD (11).

The process of the spray-drying method was: the solution, in which CN was dissolved at pH 11.00 with 1 N NaoH, was made pH 7.0 with 1 N HCl, and then spray-dried.

In comparison with this method, usual coprecipitation methods take 7 days to obtain the product. Moreover, the filtration process is necessary. There is, therefore, a difficulty in preparing the preparation on a large scale. The spray-drying method gives a good yield in a short time and may be suitable for an extension to a manufacturing scale. Between the preparations by coprecipitation and spray-drying, there was no remarkable difference in powder X-ray patterns and dissolution behavior.

Concerning the stability of CN, it was confirmed that no degradation took place in the process in the spray-drying method. CN is rather stabilized with β-CD. When the preparations were examined under heating and high humid conditions, they were found to be quite stable. The dissolution rate of the samples stored under high humid conditions was rapid compared with the initial value. This result may be due to the increase of the water content in the sample, as it gives no serious problem in final formulation design, but rather favorable.

As a result, it may be possible that a spray-drying method affords a promising means for preparations of solid inclusion complexes on a manufacturing scale, if the drug is fitting to this method.

3. INFLUENCE OF CD ON HYPNOTIC POTENCY AND DISPOSITION OF BARBITURATES IN INTRAVENOUS AND INTRAPERITONEAL ADMINISTRATIONS

3.1. Complex Formation of Barbiturates with CDs

Barbiturates are known to form stable inclusion complexes with CDs in aqueous solution. When the stability constants of inclusion complexes of hexobarbital (HBA), thiopental (TPA) and pentobarbital (PBA) with CDs were determined, those with β-CD were the largest, then with γ-CD next, and with α-CD the smallest for these three drugs (10a).

3.2. Effect on Sleeping Time in Intravenous Administration in Mice

In the dose of equimolar/kg, the sleeping time was significantly shorter in administration of drugs with β-CD, γ-CD and α-CD in the order of the stability constant than the respective drugs alone, while there was no difference between the cases with Pullulan and dextran and that of HBA (10a), as shown in Table I.

Table I. Effect of Additives on Sleeping Time of HBA, TPA and PBA in Mice after Intravenous Administration

Additives	Sleeping time (min)		
	HBA	TPA	PBA
None	13.88 ± 1.43	24.67 ± 2.16	44.05 ± 4.40
α-CD	5.05 ± 0.82	19.57 ± 2.33	39.29 ± 3.01
β-CD	2.60 ± 0.33	14.97 ± 1.08	26.00 ± 3.70
γ-CD	4.68 ± 0.63	25.25 ± 2.57	28.22 ± 3.60
PUL	14.15 ± 1.62	14.81 ± 2.55	45.15 ± 4.85
DEX	12.00 ± 1.25	17.08 ± 2.12	40.02 ± 2.03

Each value represents the mean ± S.E. of more than 7 determinations.

Regarding the effect of concentration of the additives on sleeping time of HBA after intravenous administration in mice, a linear relationship was found between the sleeping time in logarithmic scale and the molar ratio of additives to HBA. The sleeping time decreased with an increase in concentration of CDs, and was almost constant for the cases of Pulluran and dextran.

3.3. Effect on Sleeping Lag in Intraperitoneal Administration in Mice

In the case of intraperitoneal administration, it was possible to determine the sleeping lag of barbiturates, though it was impossible in intravenous one because the onset of the effect was so rapid.

As shown in Figure 1 (10b), the sleeping lag was significantly larger in the presence of β-CD than that of corresponding drugs alone. On the other hand, the sleeping time was significantly shorter in the presence of β-CD than that of corresponding drugs alone, as it was similar to the case of intravenous administration.

The difference in hypnotic potency between the two routes of administration might come from the reason that barbiturates and CDs, which were administered intraperitoneally, were not rapidly diluted by systemic circulation and also had to take a process of the penetration into the vein.

Therefore, in general it is possible that CDs give influence on the permeation of drugs. This situation gets predominant in the case of injections.

In case of oral administration, the enhancement of dissolution

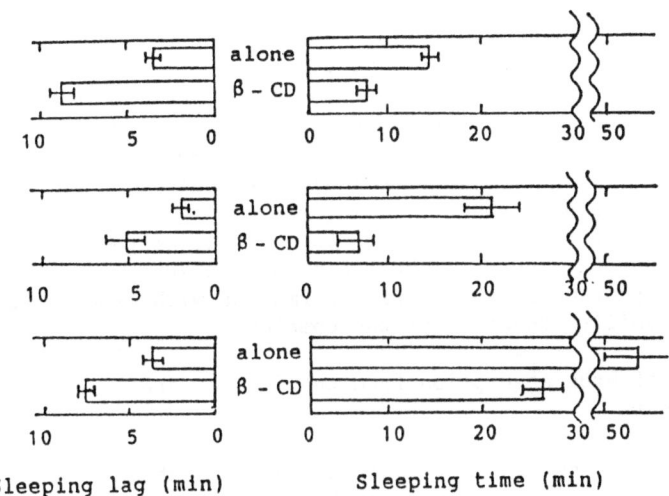

Figure 1. Effect of β-CD on Sleeping Lag (left) and Sleeping Time (right) of HBA (top), TPA (middle) and PBA (bottom) in Mice after Intraperitoneal Administration.
Drug (215.3/kg) with and without β-CD (215.3μmol/kg) was simaultaneously administered to mice.
Each bar represents the mean ± s.e. of more than 5 determinations.

rate of drugs by complex formation with CDs overcomes the inhibitory effect of the complex formation on the permeation of drug molecules, if the stability constant is not so high. However, in such a case as CN with β-CD which will be described next, the inhibitory effect of the complex formation on the permeation through intestinal membrane may not be negligible.

4. ENHANCEMENT OF BIOAVAILABILITY OF β-CD COMPLEX ON ORAL ADMINISTRATION WITH A COMPETING AGENT.

4.1. Complex Formation of CN with β-CD and Its Effect on the Bioavailability

As mentioned already, the inclusion complex formation of CN with β-CD was confirmed by the solubility method, X-ray diffractometry and IR spectrophotometry. Then, it was found that the dissolution rate of CN was enhanced 30 time or more by the complex formation with β-CD compared with that of intact CN at pH 5.0, while the bioavailability of CN in beagle dogs was not enhanced in oral administration of the inclusion complex.

Usually, an inclusion complex formation of a drug with β-CD brings about an enhancement of solubility, dissolution rate of the drug, and as the result, the bioavailability is enhanced. However,

the bioavailability is not always enhanced. The absorption rate
through membrane itself may generally be lowered by the complex
formation, because only free drug molecules can be absorbed.
Therefore, if the stability constant is large, the complex formation
is not so effective in an enhancement of the bioavailability. The
inclusion complex of CN with β-CD may be an example of this case.

The stability constant of the inclusion complex of CN with β-CD
which was determined by the solubility method was 6,200/Mol (13).
This is a very large one, as in many cases it is of the order of
100/Mol. Therefore, the fact that there was no enhancement of
bioavailability of CN by the complex formation with β-CD was due to
this large stability constant of the complex.

4.2. Process of Drug Absorption from the Complex and Role of the
Competing Agent

Th authors attempted to enhance the bioavailability of CN by
administering its β-CD complex together with another compound which
competes with the β-CD in complex formation in aqueous solution, as
this additive is called competing agent (10d).

As shown in Figure 2, in the case of the complex alone, the

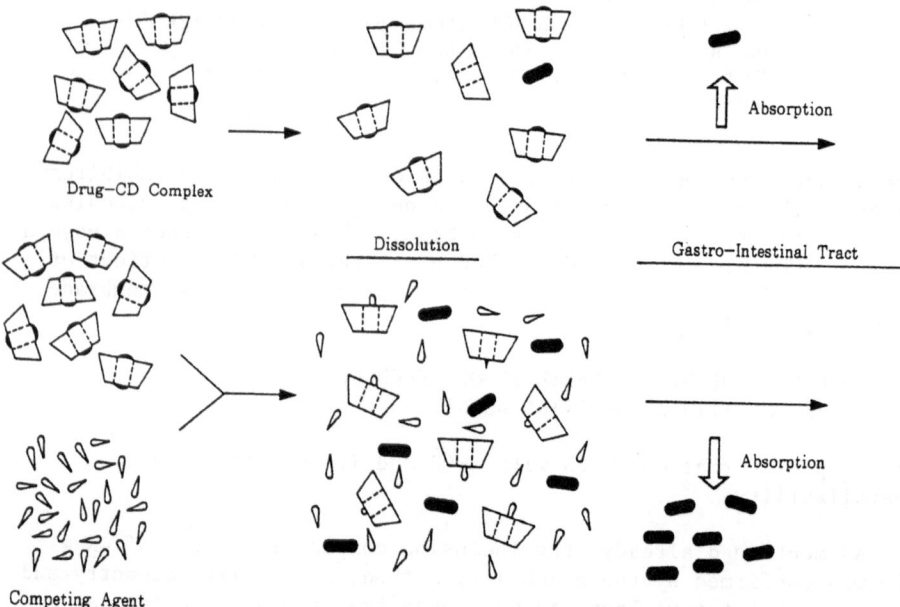

Figure 2. Process of Drug Absorption from Cyclodextrin Complex and Role of a Competing Agent

concentration of free drug is not high. On the other hand, when the
complex is administered together with the competing agent, both the
complex and the competing agent dissolve, and the concentration of

free drug that is available for absorption may increase.

As the competing agent, DL-phenylalanine was chosen, because it has been reported that L-phenylalanine forms an inclusion complex with β-CD with a stability constant of 1000/Mol. Moreover, DL-phenylalanine seems to be a pharmaceutically acceptable excipient.

The plasma level-time curves in comparison between CN alone and its combination with DL-phenylalanine showed that DL-phenylalanine does not act as an absorption promoter to CN directly, while those in comparison between the complex alone and its combination with DL-phenylalanine showed that the administration of the complex with DL-phenylalanine brings about a clear increase in plasma level and AUC. Therefore, it can be said that DL-phenylalanine acts as the competing agent.

4.2. Enhancement of Bioavailability of CN with an Increase of the Amount of DL-Phenylalanine as the Competing Agent

The bioavailability parameters in the combination of the complex with DL-Phenylalanine, C_{max} and AUC, increased with the dose of DL-phenylalanine. There was observed no change in T_{max} in the simultaneous administration with DL-phenylalanine, though the reason is not clear because a bioavailability phenomenon is a quite complicated one.

When C_{max} and AUC were plotted against the dose of DL-phenylalanine, there was a linear relationship between C_{max} and the dose. On the other hand, AUC increased hyperbolically with the dose of DL-phenylalanine. It may be difficult to explain the difference in profile between C_{max} and AUC. Anyway, it can be said that the increase in the concentration of free CN by the competing agent enhanced C_{max} and AUC.

4.3. Evaluation of Competing Effect by <u>In Vitro</u> Method

In order to investigate the effect of the competing agent on the <u>in vitro</u> membrane permeability of CN, the apparent penetration rate constant was determined using a Sartorious Absorption Simulator (10d). The experiment was done also in comparison with L-isoleucine and L-leucine.

As shown in Figure 3, the penetration rate of CN decreased with the addition of β-CD (No. 3). This may be due to a decrease in the concentration of free drug by the inclusion complex formation with β-CD. When DL-phenylalanine was added, the penetration rate constant was restored (No. 5). This means an increase in free drug concentration by the competing action of DL-phenylalanine. DL-phenylalanine did not affect the penetration rate of CN in the case without β-CD (No. 2).

When L-leucine and L-isoleucine were added, the penetration rate constant was restored (No. 6 and 7, respectively). Therefore, L-leucine and L-isoleucine also have some competing action, and L-isoleucine is stronger than L-leucine with respect to this activity.

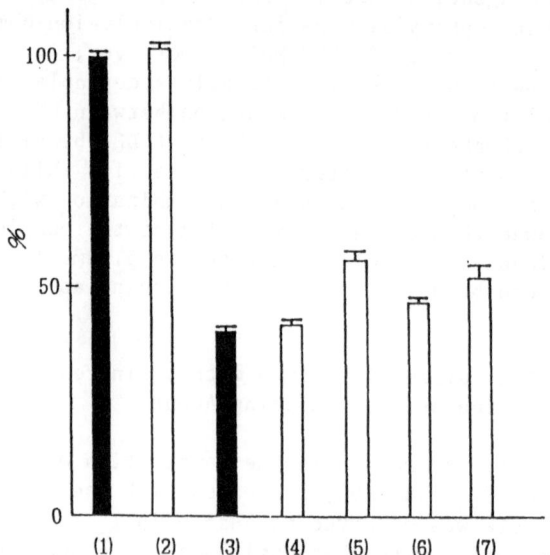

Figure 3. Apparent Penetration Rate Constant of CN from the Solutions containing the Additives in Comaparison with that of CN alone (%)

(1) CN alone (200mg); (2) CN (200mg) + DL-Phe;
(3) CN (200mg) + βCD (1232mg);
(4) CN (200mg) + βCD (1232mg) + DL-Phe (897mg)
(5) CN (200mg) + βCD (1232mg) + DL-Phe (1424mg)
(6) CN (200mg) + βCD (1232mg) + L-Leu (1424mg)
(7) CN (200mg) + βCD (1232mg) + L-Ile (1424mg)

4.4. Competing Effect of L-Isoleucine

As L-isoleucine showed a competing effect with β-CD in <u>in vitro</u> penetration study in Figure 3, it was examined in absorption study in beagle dogs (10e). C_{max} increased linearly with the dose of L-isoleucine in the combination of the complex L-isoleucine. AUC also increased in a similar way.

5. ABSORPTION ENHANCEMENT IN PERCUTANEOUS ADMINISTRATION OF DRUGS

5.1. Effects of CDs on Skin Barrier

In the pharmaceutical field, development of transdermal therapeutic systems (TTS) is booming, and thus skin has become a matter of interest because of its potency as the route of systemic administration. However, skin is essentially a barrier against penetration and permeation of external substances including drugs. One of the available method to improve the transdermal absorption of drugs is to reduce this barrier function of skin by the aid of enhancers.

Sezaki and his group (14) investigated the effects of β-CD and di-O-methyl-β-CD on percutaneous penetration of butylparaben, indomethacin and sulfanilic acid through the skin obtained from a guinea pig, and gave a proposal as: the complex of drug with CD does slightly penetrate the skin; the drug penetration is decreased as a consequence of the decrease of the free drug fraction by the complex formation with CD; CD enhances the percutaneous penetration of hydrophilic molecule, such as salicylic acid, by varying the skin barrier function; the effect of cyclodextrin on the percutaneous drug absorption can be regulated to some extent by means of its chemical modification such as methylation. Here, the effect of CDs on the skin barrier may be most important in absorption enhancement and it seems to be related to extraction of its components, such as cholesterol and triglyceride, by the complex formation.

5.2. Effects of Dimethyl-β-CD and Water-Soluble β-CD Polymer on Percutaneous Absorption of Tolnaftate

Tolnaftate (TOL) is an antifungal drug and is one of the practically insoluble drugs. Actually, this substance results in a poor percutaneous absorption, and so it is worth trying to enhance the topical bioavailability.

As an in vitro result, the solubility of TOL increased with an increase of dimethyl-β-CD (11). There was a difference in x-ray diffraction pattern between the coprecipitate and the physical mixture of TOL with dimethyl-β-CD. The thermogram by differential scanning calorimetry also gave a similar result. Dissolution profile of the coprecipitate was also different from the physical mixture.

The powder preparations for external use of TOL alone, the physical mixture and the coprecipitate mentioned above were applied to the shaved back of mice in the form of slurry, and the plasma and skin concentrations were determined by HPLC method.

As the result, the skin concentration of TOL was the highest in the case of coprecipitate with dimethyl-β-CD. The physical mixture also gave a high skin concentration compared with TOL alone.

When water-soluble β-CD polymer for the percutaneous absorption of TOL, in the same way as the above case of dimethyl-β-CD. The complex formation of drugs with this CD polymer will be published (7). The skin concentration of TOL was the highest in the case of coprecipitate. The physical mixture also gave a high skin concentration compared with TOL alone. Additionally, when the plasma concentration of TOL was determine, the tendency was similar to the skin concentration.

These results suggested that the powder dosage form composed of the coprecipitate of TOL with dimethyl-β-CD or water-soluble cyclodextrin polymer may afford a useful means for percutaneous absorption of drugs.

References

(1) M. Makita, N. Kojima, Y. Hashimoto, H. Ida, M. Tsuji, and Y. Fujisaki, *Oyo Yakuri*, <u>10</u>, 449 (1975).
(2) FDSC Report No. 55-093-1,2 (1980): private communication from Niohon Shokuhin Kako Co., Ltd., 3-4-1, Marunouchi, Tokyo, Japan.
(3) D.W. Frank, J.E. Gray, and R.N. Weaver, *Am. J. Pathol.*, <u>83</u>, 367 (1976).
(4) I. Tabushi, *4th International Symposium on Inclusion Phenomena*, Lancaster, U.K., July, 1986.
(5) K. Uekama, *4th International Symposium on Inclusion Phenomena*, Lancaster, U.K., July, 1986.
(6) K. Harata, *4th International Symposium on Inclusion Phenomena*, Lancaster, U.K., July, 1986.
(7) J. Szeman, H. Ueda, E. Fenyvesi, J. Szejtli, Y. Machida and T. Nagai, *Chem. Pharm. Bull.*, accepted for publication.
(8) Ono Pharm. Co., Ltd., *J.P.* 47-39057 (1972); *J.P.* 50-3362 (1975); *J.P.* 50-35319 (1975); *J.P.* 50-35324 (1975).
(9) T. Tokumura, Y. Tsushima, K. Tatsuishi, M. Kayano, Y. Machida, and T. Nagai, *Yakuzaigaku*, <u>45</u>, 1 (1985).
(10) a) S. Shirakura, N. Nambu and T. Nagai, *Yakuzaigaku*, <u>45</u>, 27 (1985); b) *Idem.*, *Chem. Pharm. Bull.*, <u>33</u>, 3521 (1985); c) *Idem.*, *J. Incl. Phenom.*, <u>2</u>, 613 (1984); d) T. Tokumura, M. Namba, Y. Tsushima, K. Tatsuishi, M. Kayano, Y. Machida, and T. Nagai, *J. Pharm. Sci.*, <u>75</u>, 391 (1986); e) *Idem.*, *Chem. Pharm. Bull.*, <u>34</u>, 1275 (1986).
(11) J. Szeman, H. Ueda, Y. Watanabe, E. Fenyvesi, J. Szejtli, Y. Machida and T. Nagai, *Drug Design and Delivery*, submitted.
(12) M. Kurozumi, N. Nambu, and T. Nagai, *Chem. Pharm. Bull.*, <u>23</u>, 3062 (1975).
(13) T. Tokumura, H. Ueda, Y. Tsushima, M. Kasai, M. Kayano, I. Amada, and T. Nagai, *Chem. Pharm. Bull.*, <u>32</u>, 4179 (1984).
(14) H. Okamoto, H. Komatsu, M. Hashida and H. Sezaki, *Int. J. Pharm.*, <u>30</u>, 35 (1986).

INCREASING THE SOLUBILITY OF DRUGS THROUGH CYCLODEXTRIN COMPLEXATION

M. Kata and B. Selmeczi
Department of Pharmaceutical Technology, University Medical School
P. O. Box 121
H-6701 Szeged
Hungary

ABSTRACT. The bioavailability of pharmaca which dissolve in water only with difficulty is very limited. The cyclodextrins /CDs/, and primarily β-CD and γ-CD, were successfully applied to increase the dissolution characteristics and hence the bioavailability of drugs: furosemide, hydrochlorothiazide, mebendazole, metronidazole, spironolactone, tofisopam, vinpocetine base, etc. From these pharmaca, products were made by mixing, kneading, grinding, freeze-drying, spray-embedding and precipitation.
   The more important factors on which the dissolution and bioavailability of the drugs depend are concluded.

Key words: cyclodextrin complexation, solubility characteristics of drugs, increase of bioavailability of pharmaca.

1. INTRODUCTION

Certain non-polar pharmaca dissolve in water only slightly and slowly. The dissolution characteristics of some of these active ingredients can be increased by CD complexation /1-6/. Chiefly those drugs can be taken into consideration which satisfy the following requirements:
- the guest pharmacon molecule should be partly or totally non-polar,
- its relative molecular mass should be between 100 and about 500,
- its chemical structure should permit the formation of inclusion complexes, and
- its single dose should not be more than 30-50 mg.
   From such pharmaca, products were made by different methods and were subsequently investigated. It was established how the dissolution characteristics of the drugs change as a function of the CD derivatives, of the % compositions and of the methods of preparation of the products.

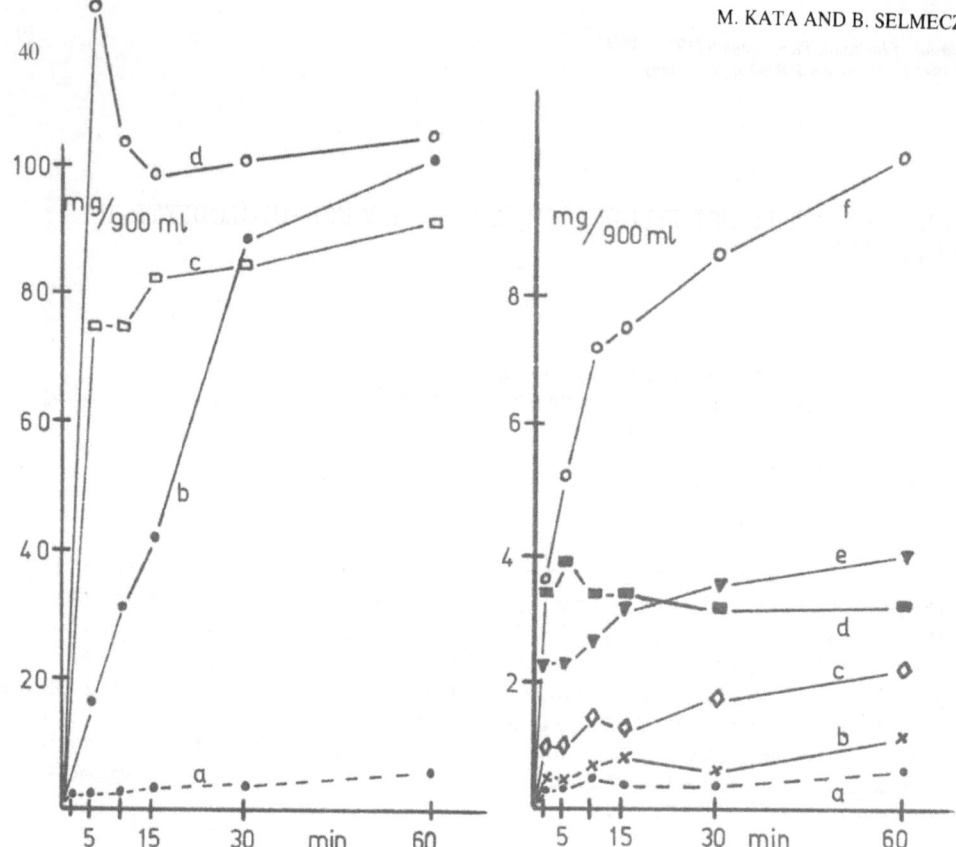

Fig. 1: Spironolactone Fig. 2: Mebendazole

2. MATERIALS AND METHODS

The investigated pharmaca meet the pharmacopoeial specifications, and the α-, β-, γ- and dimethyl-β-CD /DMβ-CD/ are products of Chinoin Chemical and Pharmaceutical Works Ltd. /Budapest, Hungary/.

The physical mixes were made by mixing, the spray embeddings with a NIRO-Atomizer apparatus, the inclusion complexes by precipitation, the kneaded products by kneading and the freeze-dried products with a Leybold GT 2 apparatus.

Dissolution tests were performed with the rotation basket method, with 900 ml of 37 ^{o}C water at 50-100 rpm, according to USP XX and XXI. /7/. The dissolved pharmacon content was measured spectrophotometrically /Spektromom 195 instrument/ and the CD content was measured in solution with concentrated sulphuric acid and anthrone at 625 nm /8/.

The products and inclusion phenomena with CD were studied by means of X-ray diffractometry, DTA and electron microscopy /9/.

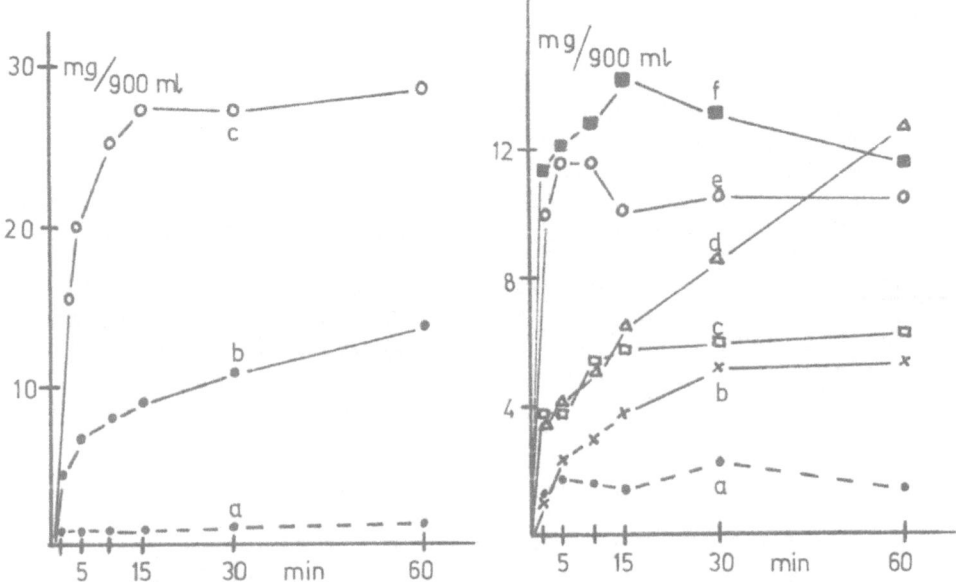

Fig. 3: Tofisopam   Fig. 4: Vinpocetine base

3. RESULTS AND DISCUSSION

Spironolactone is practically insoluble in water /3.86 mg/ /900 ml/h//Figure 1:a/. The increase of the dissolution depends on both the drug content and the method of making the products; for the 10.9% combinations, essentially more dissolves from the kneaded products and from tablets made with this kneaded product. The increase of the dissolution is 11 times, and 25 mg spironolactone is liberated within 5 minutes /1 tablet contains 25 mg drug/.

The internal diameter of γ-CD is larger than that of β-CD, and γ-CD yields complexes more easily with spironolactone. During 1 hour, the total pharmacon dissolves from the physical mix /Figure 1:b/; this quantity is 25 times more than that from pure spironolactone. The preparation of a physical mix is the simplest and cheapest drug technological operation.

DM β-CD also increases the solubility. In probability, the decrease of the surface tension plays a part in this. The dissolution starts more quickly, but after 15 minutes it slows down; overall, it is essentially better /Figure 1:c/.

The product made with a 1:1 mix of γ-CD and DMβ-CD combines the advantage of both CD derivatives: the total quantity of spironolactone dissolves, and quickly /Figure 1:d/.

Mebendazole is an antihelminthic agent used in veterinary medicine. Its solubility is less than 0.05% /Figure 2:a/. Mebendazole always dissolves in the highest quantity and most quickly from the 11.5% spray-embedded product /Figure 2:f/ [physical mix /b/, kneaded product with formic acid /c/ and with ethanol /d/, inclusion complex /e/]. The dissolution is 15 times better.

β-CD enhances the dissolution of all of the metronidazole products. The best are the 40.0% products, among them the kneaded products, spray-embeddings and physical mixes.

Tofisopam does not dissolve well: 1.22 mg/900 ml/h /Figure 3:a/. The commercially available 50 mg Grandaxin[R] tablets yield somewhat more dissolved product /b/, but from the tablets made with 25.2% kneaded β-CD product 23 times more Tofisopam is released than from the pure drug /c/.

The furosemide tablet contains 40 mg of active ingredient. During 1 hour, only one-tenth of this quantity dissolves. From the 11.3% kneaded β-CD and DMβ-CD products, the full 40 mg is liberated within 5 minutes.

Vinpocetine base is the pharmacon of Cavinton[R] tablets, the most successful Hungarian medicine in the past decade. It is a base that dissolves in water only very slightly /20 μg/ml/ /Figure 4:a/. One tablet contains 5 mg vinpocetine base. However, with 11.9% γ-CD it releases more than 7 times as much from kneaded products /f/, and 10 mg, twice as much as the usual dose of vinpocetine base, was liberated within 5 minutes [physical mix /b/, co-pulverizate /c/, ditto + Tween-20 /d/, and spray-embedding /e/].

In the case of hydrochlorothiazide, the rate of dissolution depends both on the drug content and on the method of making the products. The best liberation of hydrochlorothiazide was generally achieved from the kneaded products. The most drug was liberated and most quickly
- from the kneaded product for the 10.4% combinations,
- from the spray-embedding for the 20.8% combinations, and
- from the physical mix for the 31.2% combinations.

Besides these drugs, cinnarizine /10/, griseofulvin, hydrocortisone, iomeglamic acid, salicylic acid and triamcinolone were also investigated.

4. CONCLUSIONS

From about one thousand dissolution tests, it can established that the following factors influence the release of the pharmaca from the products containing CDs:
- the material characteristics of the pharmaca,
- the type of CD derivative / α-, β-, γ- or DM β-CD/,
- the combinations of two or more CDs and their ratio,
- the ratio of pharmacon and CD,
- other auxiliary substances /tensides or cellulose derivatives/,

- the content of the drug in the product, and
- the method of manufacturing the products.

Acknowledgement
The authors gratefully thank Prof. J. Szejtli for providing CD products and advice.

REFERENCES

1. M. L. Bender, and M. Komiyama: Cyclodextrin Chemistry, Springer Verlag, Berlin-Heidelberg-New York /1978/.
2. W. Saenger: Angewandte Chemie 92, 343 /1980/.
3. J. Szejtli: Cyclodextrins and their Inclusion Complexes, Akadémia Press, Budapest /1982/.
4. J. Szejtli: Proceedings of the 1st Intern. Symp. on Cyclodextrins, Reidel Publ. Co., Dordrecht /1982/.
5. J. L. Atwood et al.: Proceedings of the 3rd Intern. Symp. on Clathrate Compounds and 2nd Intern. Symp. on Cyclodextrins, Tokyo, Reidel Publ. Co., Dordrecht /1984/
6. Royal Society of Chemistry: Programme of the 4th Intern. Symp. on Inclusion Phenomena and 3rd Intern. Symp. on Cyclodextrins, Lancaster /1986/.
7. US Pharmacopoeias XX. /1980/ and XXI. /1985/.
8. O. I. Corrigan et al.: Pharmac. Acta Helv. 56, 204 /1980/.
9. K. Hódi, and M. Kata: Starch/Stärke 37, 205 /1985/.
10. M. Kata, and M. Wayer: Acta Chimica Hung. 118, 171 /1985/.

ABILITY OF THE ACETOTOLUIDES TO FORM CYCLODEXTRIN INCLUSION
COMPLEXES.

S P Jones[1] and G D Parr[2]
University of Nottingham, NOTTINGHAM NG7 2RD
Present addresses : 1. Beecham Pharmaceuticals
                       Research Division, WORTHING BN14 8QH
                    2. Reckitt & Colman
                       Pharmaceutical Division, HULL HU8 7DS

INTRODUCTION.

Cyclodextrins (CDs) are capable of conferring advantageous properties
on guest molecules that are able to penetrate and reside within the
CD cavity. Hence the increasing importance of CDs in the
pharmaceutical industry.
   When assessing the potential of a guest molecule to form a CD
inclusion complex, the ability of that guest molecule, or a portion
of the molecule, to penetrate the CD cavity is crucial. To help
elucidate the type of chemical structure best able to form
inclusion complexes with CDs, the behaviour of the acetotoluides
(o-, m- and p-ACT, see Figure 1) with α- and β-CD has been
investigated in aqueous solution and in the solid state.

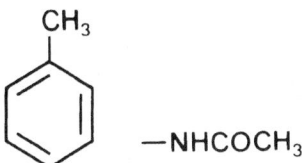

Figure 1   Structure of the Acetotoluides (ACTs)

METHODS.

Solubility Studies allow molecular interactions to be studied by
means of solubility measurements. Increasing amounts of CD are
added to a constant amount (in excess of its normal solubility)
of ACT in pH 5 acetate buffer. After equilibration, the solvent
may be analysed for ACT and the equilibrium constant of the reaction
may be calculated from the slope of the graph of concentration of
ACT against concentration of CD (presuming complex stoichiometry is

1:1, from Higuchi and Connors, 1965).

Filtration Cell Studies rely on a molecular weight-specific membrane retaining complexed molecules but allowing free passage of uncomplexed molecules. α-CD has a molecular weight of 972, β-CD of 1135. An ultrafiltration membrane with a molecular weight cut-off value of 1000 will therefore retain a large percentage, if not all, of the CD. The molecular weight of the ACTs is 149. The passage of ACT through such a membrane should therefore be unhindered. If, however, ACT forms a complex with CD, then it should be retained by the membrane. This is the basis of the method of detection of inclusion complexation by the filtration cell (see Jones and Parr 1983 and 1985).

Thermal Analysis (by differential scanning calorimetry) utilises the fact that if a crystalline inclusion complex is formed between CD and ACT, it will have different crystalline properties to the components of the complex. The difference in melting point between the complex components and the complex itself may therefore be detected.

RESULTS.

Solubility Studies. All the ACT isomers exhibited an increase in solubility with increasing concentrations of both α- and β-CD. Table 1 (below) contains the slope of each line (concentration of ACT against concentration of CD), as calculated by linear regression, and the equilibrium (stability) constant for each interaction.

ACT ISOMER	α-CYCLODEXTRIN		β-CYCLODEXTRIN	
	SLOPE	EQUILIBRIUM CONSTANT	SLOPE	EQUILIBRIUM CONSTANT
ORTHO	0.518	17.1	0.453	13.0
META	0.278	10.0	0.639	43.3
PARA	0.153	31.3	0.402	99.0

Table 1 Interaction of the ACT isomers with α- and β-CD in pH 5 buffer at 30°C

The interaction of o- and m-ACT with α-CD is just detectable, and although the equilibrium constant between p-ACT and α-CD is greater, the interaction is again weak. The ACT structure is therefore not compatible with the internal cavity of α-CD. However, interaction of β-CD with the ACTs is stronger and follows a more logical pattern : o- < m- < p-ACT. This seems to indicate that the benzene ring of the guest molecule will enter the cavity of the β-CD host molecule methyl group first. The acetamido group is too large to allow penetration of the entire ACT molecule into the CD cavity. If the acetamido group is in the para position however, this allows the benzene ring of some ACT guest molecules to reside within the β-CD cavity.

Filtration Cell Studies. By assaying the filtrate of the filtration cell, the percentage (by weight) of ACT isomer appearing in the filtrate when filtered alone and in the presence of α- and β-CD could be calculated. These results supported those of the solubility studies, i.e. there was little or no interaction of the ACTs with α-CD and a graded response of o- < m- < p-ACT with β-CD.

Thermal Analysis of the inclusion complex components, their physical mixtures and the freeze dried powders (used to prepare the inclusion complexes) confirmed that whilst all ACTs formed complexes with β-CD, none did with α-CD. It was difficult to conclude from the thermograms which, if any, ACT isomers were more compatible with the β-CD cavity than others.

CONCLUSION

These series of experiments have shown that the α-CD cavity was too small to allow stable inclusion complex formation. p-ACT is the isomer within this series that is best able to form inclusion complexes with β-CD, then m-ACT and finally o-ACT. This would seem to indicate that the benzene ring of the molecule is the part of the structure most likely to penetrate the cavity since (a) α-CD could not form stable complexes with any of the guest molecules and (b) less effective entry into the β-CD cavity is the result of the acetamido group moving from p- → m- → o- positions. Benzene ring penetration of the CD cavity is therefore required for stable inclusion complex formations in this group of compounds.

REFERENCES

Higuchi, T. and Connors, K.A., Phase Solubility Techniques. Adv. Anal. Chem. Instr., 4 (1965) 117 - 212.
Jones, S.P. and Parr, G.D., Detection of Beta-Cyclodextrin Complex Formation in Aqueous Solution. J. Pharm. Pharmac., 35 (1983) 5P.
Jones, S.P. and Parr, G.D., Study of the Inclusion Complex Formation of Barbitone and β-Cyclodextrin Using the Filtration Cell. Ibid, 37 (1985) 109P.

THE INCLUSION OF THE DRUG DIFLUNISAL BY ALPHA- AND BETA- CYCLODEXTRINS.
A NUCLEAR MAGNETIC RESONANCE AND ULTRAVIOLET SPECTROSCOPIC STUDY.

Stephen F. Lincoln, John H. Coates, Bruce G. Doddridge and
Andrea M. Hounslow
Department of Physical and Inorganic Chemistry,
University of Adelaide,
Adelaide, South Australia, 5001.

ABSTRACT. ^{19}F nmr and ultraviolet spectroscopic studies show that the inclusion of the anion of the drug diflunisal (DF) by alpha- and beta-cyclodextrins (αCD and βCD) in water produces the complexes: DF.αCD, DF.βCD and DF.(βCD)$_2$ characterized by stability constants of 17, 1.81×10^5 and 3.07×10^3 dm^{3}mol^{-1} respectively.

INTRODUCTION

The cyclodextrins are α-1,4-linked cyclic oligomers of D-glycopyranose which form inclusion complexes with a wide range of substrates.[1] Amongst such substrates are drug molecules whose cyclodextrin inclusion complexes are potentially valuable in drug delivery systems. As part of our studies[2] in this area we have used ultraviolet and ^{19}F nmr spectroscopy to investigate the inclusion, by alpha- and beta-cyclodextrin (αCD and βCD) of the anionic form of the anti-inflammatory drug diflunisal[3] (DF):

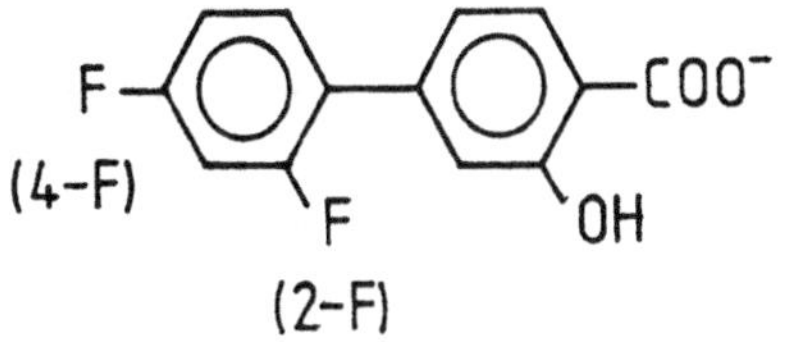

RESULTS

The spectroscopic studies were carried out in 10% D$_2$O KH$_2$PO$_4$/Na$_2$HPO$_4$ buffer solution of ionic strength 0.1 at pH 7.00 at 298.2 K. The DF (1.713×10^{-5} mol dm^{-3}) ultraviolet spectrum exhibited little variation at its 250 nm absorbance maximum in the [αCD] range 0-0.124 mol dm^{-3} consistent with the interaction between DF and αCD being weak. In contrast, DF exhibited a substantial variation in its spectrum as [βCD] varied in the range $(1.312-337.5) \times 10^{-5}$ mol dm^{-3}

consistent with the formation of 1:1 and 1:2 complexes shown in eqns (1) and (2):

$$DF + \beta CD \underset{}{\overset{K_1}{\rightleftharpoons}} DF \cdot \beta CD \qquad (1)$$

$$\beta CD + DF \cdot \beta CD \underset{}{\overset{K_2}{\rightleftharpoons}} DF \cdot (\beta CD)_2 \qquad (2)$$

and the K_1 and K_2 derived in the range 240-255 nm are given in Table 1.

In solutions containing either of the cyclodextrins, a ^{19}F broad band ^1H decoupled doublet resonance (J_{F-F} = 7.20 Hz) is observed for each of the 2-F and 4-F of DF consistent with exchange of DF between the free and included states being fast on the nmr timescale. The variation of the observed ^{19}F chemical shift (δ) of DF (4.81 × 10^{-3} mol dm^{-3}) with total [αCD] is shown in Figure 1, and is consistent with the formation of DF.αCD only in an equilibrium analogous to eqn (1). The K_1 derived from the simultaneous fit of δ for 2-F and 4-F to eqn (3), in which δ_0 and δ_1 are the ^{19}F chemical shift of DF and DF.αCD, is given in Table 1.

$$\delta = \frac{\delta_0 [DF] + \delta_1 [DF \cdot \alpha CD]}{[DF] + [DF \cdot \alpha CD]} \qquad (3)$$

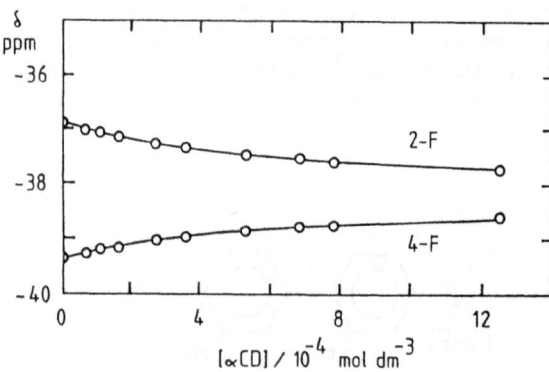

Figure 1. Variation of the ^{19}F chemical shift (δ) of DF (4.81 × 10^{-3} mol dm^{-3}) with total [αCD]. The negative shifts signify upfield shifts from a 2% sodium trifluoroacetate solution in D_2O external reference which is assigned a shift of zero. The solid curves represent the best fits of these data to eqn (3).

The variation of the observed ^{19}F chemical shift (δ) of DF (5.00 × 10^{-3} mol dm^{-3}) with total [βCD] is shown in Figure 2, and is consistent with the formation of DF.βCD and DF.(βCD)$_2$ as shown in eqns (1) and (2).

The variation of δ with [βCD] anticipated for equilibria (1) and (2) is given by eqn (4), in which δ_0, δ_1 and δ_2 are the ^{19}F chemical shifts of DF, DF.βCD and DF.(βCD)$_2$ respectively.

$$\delta = \frac{\delta_0[DF] + \delta_1[DF.\beta CD] + \delta_2[DF.(\beta CD)_2]}{[DF] + [DF.\beta CD] + [DF.(\beta CD)_2]} \quad (4)$$

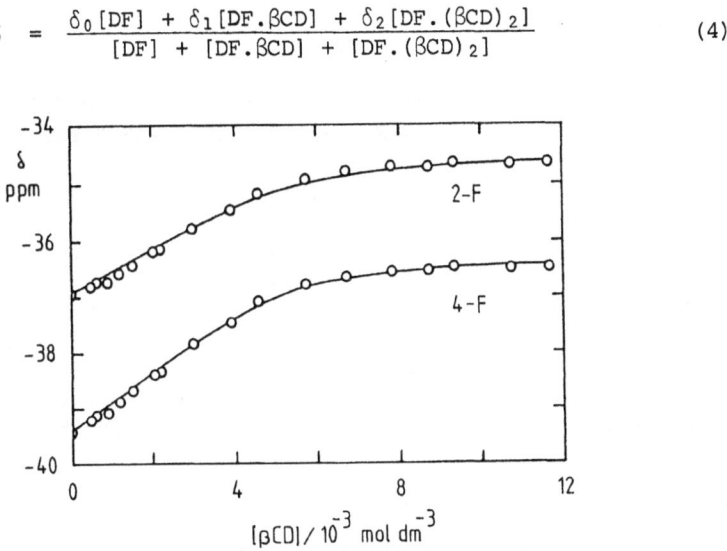

Figure 2. Variation of the ^{19}F chemical shift (δ) of DF (5.00 × 10^{-3} mol dm^{-3}) with total [βCD]. The negative shifts signify upfield shifts from a 2% sodium trifluoroacetate solution in D$_2$O external reference which is assigned a shift of zero. The solid curves represent the variation of δ predicted by eqn (4) using the K$_1$ and K$_2$ values determined from the ultraviolet spectrophotometric data.

A simultaneous non-linear least squares fit of the 2-F and 4-F data to eqn (4) yields: K$_1$ = (2.17 ± 9.16) × 10^5 dm^{3}mol^{-1} and K$_2$ = (5.1 ± 20.0) × 10^3 dm^{3}mol^{-1}, in which the large errors are a consequence of [DF] and [βCD] being very small compared to [DF.βCD] and [DF.(βCD)$_2$] at the total [DF] = 5.00 × 10^{-3} mol dm^{-3}. These K$_1$ and K$_2$ are in qualitative agreement with the more accurate values (Table 1) derived at much lower total [DF] using ultraviolet spectrophotometric methods. When K$_1$ and K$_2$ are set equal to the values derived from the ultraviolet spectrophotometric data, and the ^{19}F chemical shift data are again fitted to eqn (4), the best fit curves are seen to reproduce closely the experimental data (Figure 2). The corresponding δ_1 and δ_2 values are given in Table 1.

Table 1. Equilibrium constants and ^{19}F chemical shifts[a] for the diflunisal anion/cyclodextrin systems (298.2 K)

$K_1/10^3$ /dm^{3}mol^{-1}	$K_2/10^3$ /dm^{3}mol^{-1}		δ_0 ppm	δ_1 ppm	δ_2 ppm
		α-cyclodextrin			
0.0170±0.0009[b]	–	(2-F)	-36.89±0.01	-38.18±0.03	–
		(4-F)	-39.37±0.01	-38.27±0.03	–
		β-cyclodextrin			
181±20[c]	3.07±0.25[c]	(2-F)	-36.92±0.01	-34.91±0.05	-34.52±0.05
		(4-F)	-39.40±0.01	-36.71±0.05	-36.34±0.05

[a] A negative shift signifies an upfield shift from a 2% CF_3COONa solution in D_2O external reference which is assigned a shift of zero. The δ_0 values vary slightly with diflunisal concentration, and hence different values appear in the table for the αCD and βCD systems. The digital resolution was 0.007 ppm.

[b] Determined from ^{19}F shift data.

[c] Determined from ultraviolet spectrophotometric data.

DISCUSSION

The higher stability of DF.βCD by comparison with that of DF.αCD (Table 1) is consistent with DF fitting the βCD annulus (diameter 7-8 Å) better than the αCD annulus (diameter 5-6 Å). This probably also explains the high stability of DF.(βCD)$_2$, which contrasts with the absence of DF.(αCD)$_2$ at detectable concentrations. The differing variations of the DF ^{19}F chemical shifts characterizing DF.αCD and DF.βCD (Figures 1 and 2, and Table 1) indicate differing structural features in these inclusion complexes. It has been suggested that downfield ^{19}F shifts indicate transfer to a hydrophobic environment.[4] On this basis both 2-F and 4-F of DF encounter a hydrophobic environment in DF.βCD, whereas only 4-F experiences a hydrophobic environment in DF.αCD. Space-filling models indicate that if the fluorinated end of DF enters the cyclodextrin annulus first, both 2-F and 4-F in DF.βCD can interact with the hydrophobic regions of the annulus, as can 4-F in DF.αCD. However the steric hindrance of the smaller annulus of DF.αCD leaves 2-F in the hydrophilic region at the annulus entrance. Similar changes in environment may be invoked to explain the variation in ^{19}F chemical shifts accompanying the formation of DF.(βCD)$_2$.

This study demonstrates that annular size is a major determinant of the stoichiometry and stability of cyclodextrin inclusion complexes; and the inclusion of other drugs is now being studied.

REFERENCES

1. A general reference to cyclodextrin inclusion phenomena is:
   Inclusion Compounds, Eds., J.L. Atwood, J.E.D. Davis and D.D. MacNicol, (Academic Press, London, 1984) Vol. 2, W. Saenger, p. 231; Vol. 3, J. Szejtli, p. 331; R.J. Bergeron, p. 391; I. Tabushi, p. 445; R. Breslow, p. 473.

2. D.L. Pisaniello, S.F. Lincoln and J.H. Coates, *J.Chem.Soc., Faraday Trans.*, *1*, 1985, **81**, 1247.

3. 'Diflunisal in Clinical Practice', ed. K. Miehle, (Futura, New York, 1978).

4. T.M. Spotswood and B.C. Nicholson, *Aust.J.Chem.*, 1978, **31**, 2167.

Journal of Inclusion Phenomena 5 (1987), 55–58.
© 1987 by D. Reidel Publishing Company.

GEOMETRY OF α-CYCLODEXTRIN INCLUSION COMPLEX WITH m-NITROPHENOL DEDUCED FROM QUANTUM CHEMICAL ANALYSIS OF CARBON-13 CHEMICAL SHIFTS

Y. Inoue, M. Kitagawa, H. Hoshi, M. Sakurai, and R. Chûjô
Department of Polymer Chemistry
Tokyo Institute of Technology
O-okayama 2-chome, Meguro-ku, Tokyo 152
Japan

ABSTRACT. The host-guest orientation and the position of the guest m-nitrophenol(MNP) in the α-cyclodextrin(α-CD)-MNP inclusion-complex in aqueous solution has been determined by comparing the complexation induced carbon-13 NMR chemical shifts of MNP with those predicted by quantum chemical calculation. In the calculation, the non-polar environmental effect produced by the α-CD cavity on the carbon-13 shifts of included guest molecule has been formulated by the so-called NMR solvent effect theory. Here, carbon-13 shift displacements are assumed to be induced by transference of the guest from polar aqueous phase with higher dielectric constant to the non-polar α-CD cavity with lower dielectric constant. Among a variety of host-guest orientation investigated, only the geometry in which the nitrophenyl group is located in the α-CD cavity and the hydroxyl group is exposed to the aqueous phase can reproduce qualitatively the observed carbon-13 shift displacements of MNP. This geometry is consistent with that in the solid state determined by the X-ray method.

1. INTRODUCTION

High resolution carbon-13 NMR spectroscopy is one of the most useful methods in the analysis of the structure and molecular dynamics of cyclodextrin inclusion-complexes both in aqueous solution[1,2] and in the solid state[3]. Earlier carbon-13 NMR studies of α-CD inclusion complexes with benzoic acid, p-nitrophenol, and p-nitrophenolate in aqueous solution have shown that the included lead(head; see Fig.1A) carbons show high-field shifts compared to low-field shifts of corresponding para(tail; Fig.1A) carbons[4,5]. Similar distinctive patterns of carbon-13 displacements have been also observed for p-hydroxybenzoic acid and it has been concluded that the carboxyl group of p-hydroxybenzoic acid is directed into the α-CD cavity[4]. A variety of substituted benzenes are known to show quite similar carbon-13 high (head) and low(tail) field shifts, irrespective of the kinds of substituents if their size are matched to the α-CD cavity[4,5]. These characteristic carbon-13 displacements induced by complexation with α-CD

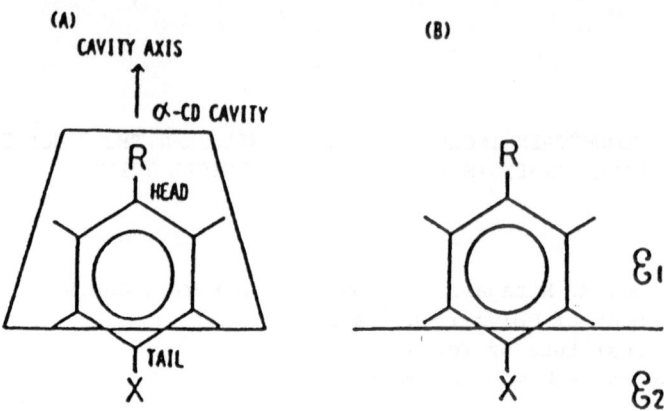

Figure 1. (A) Geometry of α-CD inclusion-complexes with benzene derivatives. (B) Double-layer model. The benzene derivative is situated on the borderline dividing the surroundings into two layers of dielectric constants, ε_1 and ε_2.

may be useful in inferring the extent of the host-guest complexation and the orientation of the guest in the α-CD cavity, if the origin of these carbon-13 displacements is accounted for.

In general, there are several kinds of non-bonded interactions, which may influence the carbon-13 chemical shifts. Among them, the electrical environmental effects are expected to be the major contribution to the α-CD complexation-induced carbon-13 shifts of head and tail carbons of the guest compounds, as these shifts are induced by moving the guest molecule from the free state surrounded by polar water molecules to the relatively non-polar α-CD cavity. In the case of NMR shielding, such environmental effects could be treated as solvent effects[6].

In a previous report[7], we have successfully applied the quantum chemical method to the determination of the geometry of α-CD inclusion-complexes with substituted benzenes such as p-nitrophenol, p-hydroxy-benzoic acid, and benzoic acid. This method was based on the so-called solvent-effect theory and it was assumed that the α-CD cavity has the environmental effect of lower dielectric constant(ε_1) on a included part of the guest and the other part of guest is exposed to the aqueous phase of higher dielectric constant(ε_2) as shown in Fig.1B.

In the case of α-CD—m-nitrophenol(MNP) inclusion-complex, there are two possible geometries as shown in Fig.2. Namely, the first is that the nitrophenyl group is included and the hydroxyl group is protruded from the α-CD cavity. This structure has been proposed based on the above-mentioned characteristic carbon-13 shifts of MNP observed in aqueous solution[4]. X-Ray crystallographic analysis of solid complex has also support this type of inclusion[8]. The second is the reverse of the first one, and has been proposed based on also carbon-13 chemical shifts in aqueous solution[9]. Here we tried to determine the host-guest orientation and position of the guest MNP in the α-CD inclusion-complex in aqueous solution by the quantum chemical analysis

GEOMETRY OF α-CYCLODEXTRIN INCLUSION COMPLEX

(A) (B)

Figure 2. Two possible geometries of α-CD—MNP complex.

of carbon-13 chemical shift displacements of
MNP induced by complexation with α-CD.

2. METHODS

To calculate carbon-13 chemical shifts, we used
the Karplus & Pople's average excitation energy
method using CNDO/2 parameters[10], since this
method was found to give good linearity between
the calculated and observed shifts. Molecular
orbital calculation were carried out on a HITAC-
M280 computer at the Information Processing
Center of Tokyo Institute of Technology.
As the model of solvent effect on carbon-13
chemical shifts, we used one developed by Ando

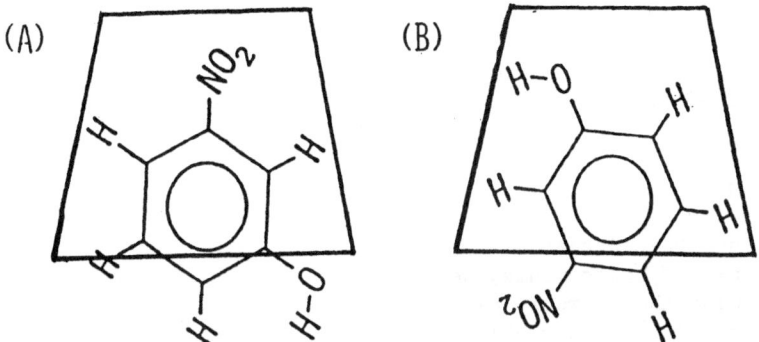

MNP

et al.[6] based on a Klopman's "solvaton" theory[11]. This model has
been successfully applied to interpret the dielectric solvent effect on
carbon-13 chemical shifts of many organic compounds. According to this
model, the interaction of solute with solvent molecules is incorporated
into semi-empirical MO calculations by an assumption of virtual particle
called a solvaton. The details of theory and calculation have been
reported[7]. Carbon-13 NMR spectra were recorded at 30 °C on a JEOL
PS-100 spectrometer operated at 25 MHz.

3. RESULTS and DISCUSSION

In order to take into account the heterogeneous nature of the surroundings around the guest molecule in the α-CD inclusion-complex, a realistic solvaton model was constructed. It was assumed that the α-CD cavity
has the environmental(solvent) effect of dielectric constant ε_1 on the
included part of the guest molecule, while the other part of the guest
is exposed to the aqueous layer of dielectric constant ε_2, as shown in
Fig. 1B. We called this "the double-layer model". It was assumed
that ε_1 and ε_2 are 2 and 80, respectively.

The carbon-13 shift displacements were calculated as the function of
the position of the borderline dividing the two layers for a number of
host-guest orientations. In the calculations, the accurate position

of the border line is unimportant, but what is important is which atom is included in the layer ε_1 and which remains in another layer of ε_2. A series of calculations were made by shifting the borderline at different positions and also by reversing the orientation of the guest relative to the α-CD cavity.

By comparing the calculated and observed carbon-13 displacements, we tried to determine the geometry of the inclusion complex. Among a lot of host-guest geometry investigated, the observed characteristic carbon-13 shifts, namely, high-field shift of C-3 resonance and low-field shift of C-6 resonance, are reproduced at the same time only when the borderline is on the shaded region as shown in Fig.3. Here, the host-guest orientation and the position of the MNP molecule in the α-CD cavity are

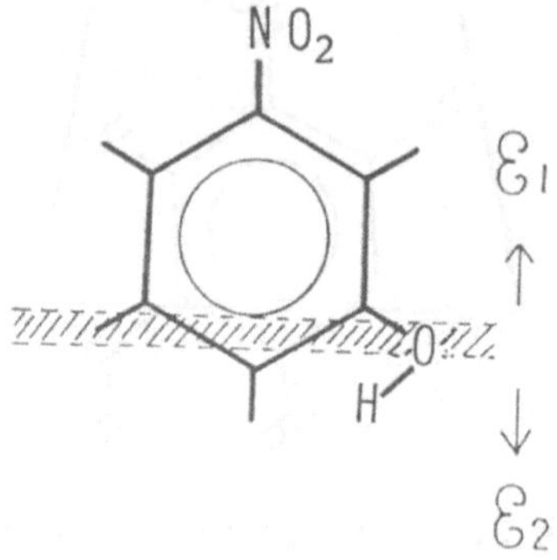

Figure 3. The position of the borderline (shaded region) dividing the surrounding into two layers, where the characteristic carbon-13 displacements are reproduced well by calculations.

the same as that found by X-ray method[8], if we assume that the border line corresponds to the α-CD's wider rim consisting of secondary hydroxyl groups.

References
1. Y. Inoue and Y. Miyata, Bull. Chem. Soc. Jpn., 54, 809(1981).
2. Y. Inoue, T. Okuda, and Y. Miyata, J. Am. Chem. Soc., 103, 7393(1981).
3. Y. Inoue, T. Okuda, and R. Chûjô, Carbohydr. Res., 141, 179(1985).
4. R. I. Gelb, L. M. Schwartz, B. Cardelino, H. S. Fuhrman, R. F. Johnson, and D. A. Laufer, J. Am. Chem. Soc., 103, 1750(1981).
5. Y. Inoue, T. Okuda, Y. Miyata, and R. Chûjô, Carbohydr. Res., 125, 65(1984).
6. I. Ando and G. A. Webb, Org. Magn. Reson., 15, 111(1981).
7. Y. Inoue, H. Hoshi, M. Sakurai, and R. Chûjô, J. Am. Chem. Soc., 107, 2319(1985).
8. K. Harata, H. Uedaira, and J. Tanaka, Bull. Chem. Soc. Jpn., 51, 1627 (1978).
9. M. Komiyama and H. Hirai, Bull. Chem. Soc. Jpn., 54, 828(1981).
10. M. Karplus and J. A. Pople, J. Chem. Phys., 38, 2803(1963).
11. G. Klopman, Chem. Phys. Lett., 1, 200(1967).

THE INCLUSION OF PYRONINE B AND PYRONINE Y BY BETA- AND GAMMA-CYCLODEXTRINS. A KINETIC AND EQUILIBRIUM STUDY.

Robert L. Schiller, Stephen F. Lincoln and John H. Coates
Department of Physical and Inorganic Chemistry,
University of Adelaide,
Adelaide, South Australia, 5001.

ABSTRACT. Equilibrium and temperature-jump spectrophotometric studies show that the mono-cation of pyronine B (PB) is included by beta- and gamma-cyclodextrins (βCD and γCD) to form the labile complexes PB.βCD, PB.γCD and (PB)$_2$.γCD in water. The equilibrium, kinetic and structural aspects of these complexes and those formed by pyronine Y are discussed in conjunction with data characterizing the inclusion of other dyes by cyclodextrins.

INTRODUCTION

The inclusion of organic dyes by cyclodextrins is a well known phenomenon, but the kinetics and mechanisms of such inclusion processes, particularly for the larger cyclodextrins, have not been much investigated.[1-4] In order to determine the effect of annular size on the dynamics of the inclusion process we have studied the inclusion of the mono-cations of pyronine B and pyronine Y (PB and PY):

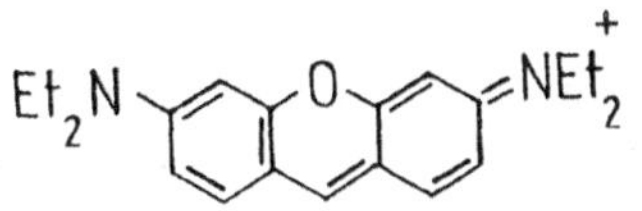

PB 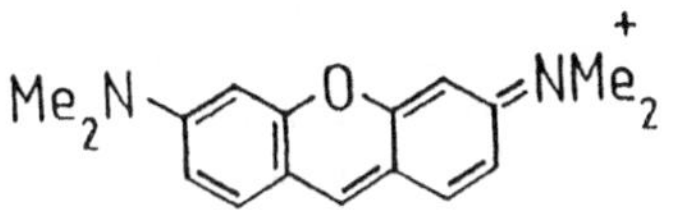   PY

by the beta- and gamma-cyclodextrins (βCD and γCD) which are α-1,4-linked cyclic heptamers and octamers of D-glucopyranose respectively characterized by internal annular diameters of 7-8 Å and 9-10 Å. The smaller αCD does not include PB and PY to a detectable extent.

RESULTS

All PB spectroscopic studies were carried out at pH 5.7 in aqueous 1.00 mol dm^{-3} NaCl at 298.2 K. The variation of the PB spectrum in the range 450-600 nm was consistent with the formation of a 1:1 inclusion complex only:

$$PB + \beta CD \underset{k_{-1}}{\overset{k_1}{\rightleftharpoons}} PB.\beta CD \qquad K_1 \qquad (1)$$

The variation of the relaxation time of this equilibrium, determined from temperature-jump studies at 555 nm, is given by:

$$1/\tau = k_1([PD] + [\beta CD]) + k_{-1} \qquad (2)$$

where all concentrations are equilibrium values existing prior to the temperature-jump. A linear regression of the variation of $1/\tau$ with total [βCD] in the range 10^{-4}-10^{-3} mol dm^{-3} yielded the parameters in Table 1.

A much larger variation of the PB spectrum was observed in the presence of γCD (Figure 1) consistent with the formation of both 1:1 and 2:1 inclusion complexes:

$$PB + \gamma CD \underset{k_{-1}}{\overset{k_1}{\rightleftharpoons}} PB.\gamma CD \qquad K_1 \qquad (3)$$

$$PB + PB.\gamma CD \underset{k_{-2}}{\overset{k_2}{\rightleftharpoons}} (PB)_2.\gamma CD \qquad K_2 \qquad (4)$$

Temperature-jump studies at 533 and 553 nm yielded the variation of the relaxation time (τ) with [γCD] shown in Figure 2. This variation is consistent with equilibrium (3) being particularly facile such that it is always at equilibrium while the less facile equilibrium (4), which incorporates the rate determining step for the processes producing the change in the spectrum, adjusts to the new temperature. (The temperature-jump was from 288.5 K to 298.2 K.) The variation of τ with [γCD] for this mechanism is given by eqn (5) in which all concentrations are the equilibrium values at 298.2 K.

$$1/\tau = \frac{k_2[PB]([PB.\gamma CD] + [PB] + 4[\gamma CD])}{([PB] + [\gamma CD] + 1/K_1)} + k_{-2} \qquad (5)$$

The inclusion of PY by γCD is characterized by spectral changes and τ variations consistent with equilibria analogous to (3) and (4), and a relaxation equation analogous to eqn (5). The derived parameters for both the PB/γCD and PY/γCD appear in Table 1 together with parameters from related systems.

THE INCLUSION OF PYRONINE B AND PYRONINE Y

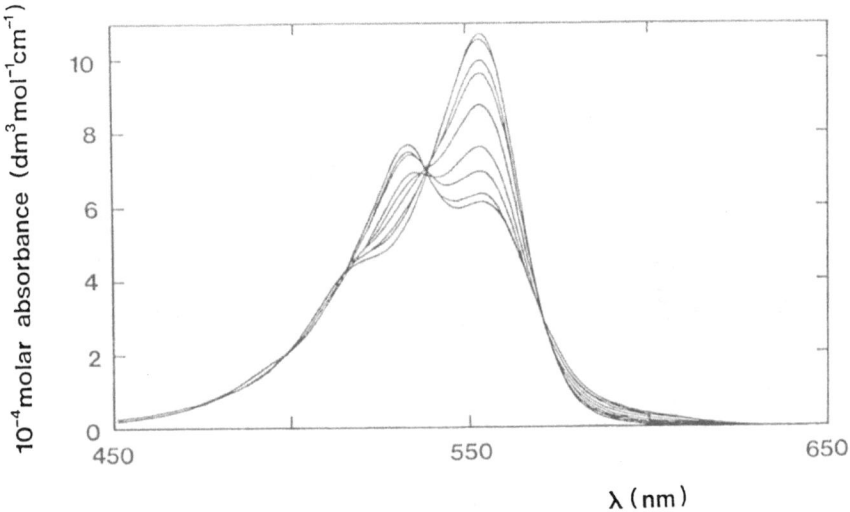

Figure 1. Variation of the PB spectrum in the presence of γCD at pH 5.70 in aqueous 1.00 mol dm^{-3} NaCl at 298.2 K. The molar absorbance at 550 nm decreases systematically as the total [γCD] increases sequentially from 0-4.093 × 10^{-3} mol dm^{-3}. Total [PB] = 9.7 × 10^{-6} mol dm^{-3}. These nine spectra exemplify the spectral variation observed for all thirty solutions studied.

Table 1. Dye-Cyclodextrin Inclusion Complex Equilibrium and Kinetic Parameters (298.2 K)

DYE	$K_1/10^2$ dm^{3}mol^{-1}	$K_2/10^5$ dm^{3}mol^{-1}	$k_2/10^9$ dm^{3}mol^{-1}s^{-1}	$k_{-2}/10^3$ s^{-1}
		γ-cyclodextrin		
pyronine B[a]	4.3±0.1	1.28±0.04	0.82±0.02	6.40±0.05
pyronine Y[a]	11.3±0.7	30±41	10±7	3.3±2.2
tropaeolin[b]	4.18±1.47	16.8±0.54	2.27±0.61	1.35±0.23
methyl orange[c]	0.45±0.07	20±11	9.4±5.1	4.8±0.8
crystal violet[d]	4.63±0.06	10.3±0.9	1.73±0.08	1.68±0.07
		β-cyclodextrin		
pyronine B[a] ($k_1/10^9$ = 0.11±0.01 dm^{3}mol^{-1}s^{-1}; $k_{-1}/10^3$ = 15±5 s^{-1})	73±31	-	-	-
tropaeolin[b]	7.1±0.7	40±70	5±6	1.3±1.5

[a] This work   [b] Ref. 4   [c] Ref. 2   [d] Ref. 3

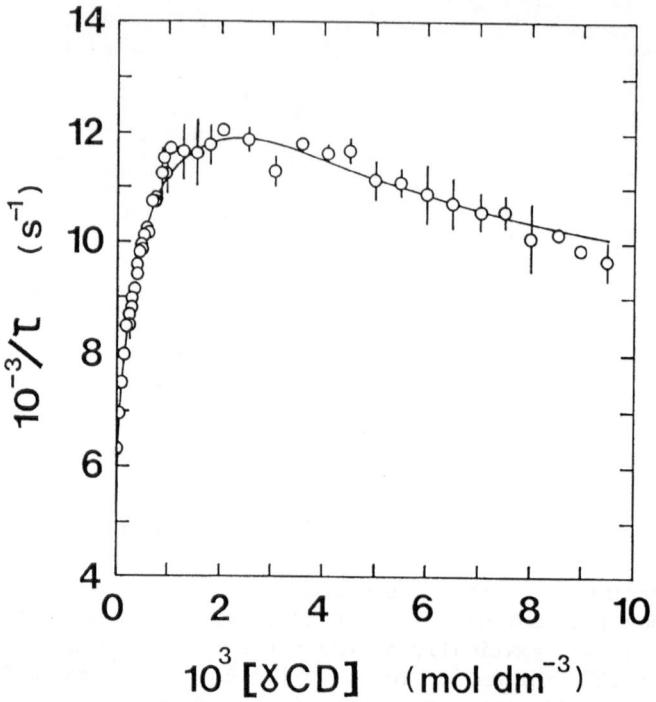

Figure 2. The variation of $1/\tau$ (298.2 K) for the PB/γCD system with the total [γCD]. Total [PB] varied in the range $(9.1-9.8) \times 10^{-6}$ mol dm^{-3}. The solid curve represents the best fit of the data to eqn (5).

DISCUSSION

The absence of detectable complex formation between PB and αCD, the formation of PB.βCD only, and the formation of PB.γCD and (PB)$_2$.γCD demonstrates the strong correlation between the size of the included dye and the annular diameters of αCD, βCD and γCD, which are 5-6 Å, 7-8 Å and 9-10 Å respectively. The replacement of the ethyl groups of PB by methyl groups in PY produces differences in K_1 and K_2 characterizing (dye)$_2\gamma$CD (Table 1) indicating the influence of changes in dye structure on stabilities. A more substantial demonstration of this is afforded by the formation of (tropaeolin)$_2\beta$CD, which indicates that the nature of the dye is also important in determining the stoichiometry of the inclusion complex (Table 1).

The dimerization constant (K_d) for PB:

$$2PB \underset{k_{-d}}{\overset{k_d}{\rightleftharpoons}} (PB)_2 \qquad K_d \qquad (6)$$

was determined to be $(1.3\pm0.5) \times 10^3$ dm^{3}mol^{-1}, and $K_d = (1.1\pm0.2) \times 10^3$

$dm^3 mol^{-1}$ for PY under the same conditions as the cyclodextrin studies. Thus inclusion by γCD increases the stability of $(PB)_2$ and $(PY)_2$ by *ca* 100 and 3000 fold. The dimerization relaxation for both dyes occurred within the heating time of our temperature-jump equipment (*ca* 2 μs), and in consequence k_d and k_{-d} were not measurable. Similarly increases in the stability of $(dye)_2$ are observed for the other dyes in Table 1.

The variation of K_1, K_2, k_2 and k_{-2} for the dye/γCD systems (Table 1) is surprisingly small when the diversity of the nature of the five dyes studied is considered. (PB, PY and crystal violet exist as monocations under the conditions of study, whereas tropaeolin and methyl orange exist as mono-anions.) There are probably several factors determining these parameters, and fortuitous combinations of these factors may produce the small range of stability and rate constants. However it is possible that the dispersion force interactions between the dye monomers included $(dye)_2$, and between this dimer and the interior of the cyclodextrin annulus may be a dominant factor determining and constraining the magnitude range of the parameters characterizing $(dye)_2$γCD.

REFERENCES

1. F. Cramer, W. Saenger and H-Ch. Spatz, *J.Am.Chem.Soc.*, 1967, **89**, 14.

2. R.J. Clarke, J.H. Coates and S.F. Lincoln, *Carbohydrate Res.*, 1984, **127**, 181.

3. R.L. Schiller, J.H. Coates and S.F. Lincoln, *J.Chem.Soc., Faraday Trans. 1*, 1984, **80**, 1257.

4. R.J. Clarke, J.H. Coates and S.F. Lincoln, *J.Chem.Soc., Faraday Trans. 1*, 1984, **80**, 3119.

^{2}H NMR STUDIES OF METALLOCENES IN HOST LATTICES

Nigel J. Clayden, Christopher M. Dobson,
Stephen J. Heyes and Philip J. Wiseman
University of Oxford
Inorganic Chemistry Laboratory
South Parks Road Oxford OX1 3QR

We are using ^{2}H NMR spectroscopy in conjunction with lineshape simulations to investigate the structure and dynamics of organometallic compounds incorporated into a variety of host lattices. Here, the potential of this approach is illustrated with results from studies of perdeuterated Fecp$_2$ and Cocp$_2^+$ in different crystalline environments.

Figure 1 shows spectra of Cocp$_2^+$PF$_6^-$. At low temperature (160K) the spectrum exhibits a characteristic Pake doublet pattern with a splitting of 65 kHz. This splitting, half that expected for a rigid lattice, arises from rapid motional averaging about the principal axes of Cocp$_2^+$, i.e. from rapid spinning of the cp rings, which is known to occur in metallocenes even at low temperatures [1]. At high temperature ($\geq$315K) a single narrow line is observed, indicating that rapid, effectively isotropic motion is experienced by the Cocp$_2^+$ ions. This can be understood using a model proposed for Febzcp$^+$PF$_6^-$ from Mössbauer studies [2]. Between 308 and 314K a phase change occurs from a distorted (low temperature) to a regular (high temperature) cubic structure. In the former the Cocp$_2^+$ molecules are restricted to a defined orientation within the distorted cube of PF$_6^-$ ions, whilst in the latter, fast hopping between different orientations, related by symmetry operations within the regular cubic structure, leads to effectively complete spherical averaging of the ^{2}H NMR spectrum. Another example where 'fast' motional averaging provides direct information about molecular orientation is the case of Cocp$_2^+$ in the intercalation compound TaS$_2$-(Cocp$_2$)$_{1/4}$, see figure 2. The spectrum at 296K shows two overlapping Pake doublets, one with a splitting of 65kHz, the other with a splitting of approximately half this value. This demonstrates that Cocp$_2^+$ is in two environments with different motional behaviour. A previous ^{1}H NMR study [3] concluded that the Cocp$_2^+$ ions were oriented with their principle axes parallel to the interlamellar plane, whereas an X-ray powder diffraction study of alkyl substituted metallocenes in TaS$_2$ and ZrS$_2$ [4] concluded that the axis was perpendicular to the plane. The ^{2}H NMR spectra are consistent with both orientations being present in the structure, providing that fast molecular motions are restricted in both cases to those about axes perpendicular to the layers. Then in the case of molecules with their

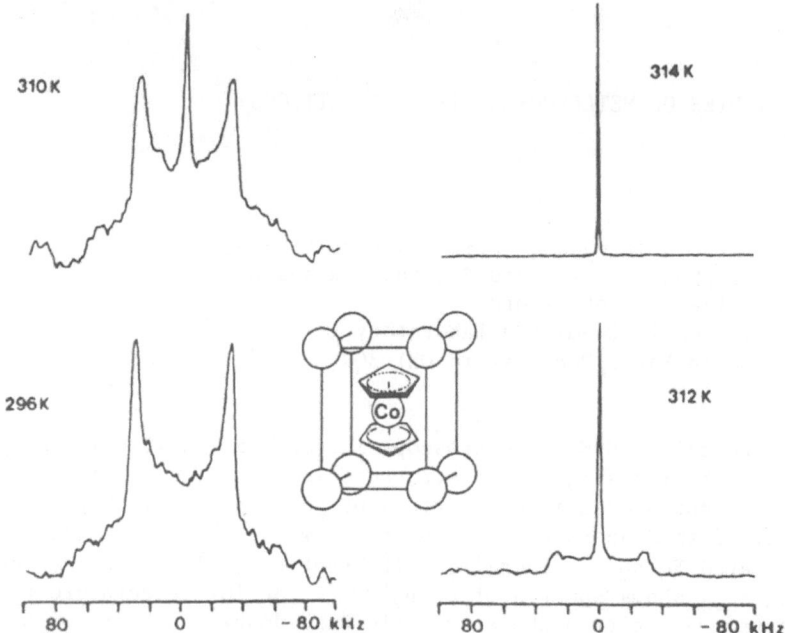

Figure 1: ^{2}H NMR spectra of $Cocp_2^+PF_6^-$. All spectra were recorded at 30.7MHz on a Bruker CXP200 spectrometer using a quadrupolar spin echo sequence with phase cycling.

principle axes parallel to the layers, calculations show that motional averaging results in a reduction by a factor of 2 in the splitting of the Pake doublet from the value of 65kHz anticipated for $Cocp_2^+$

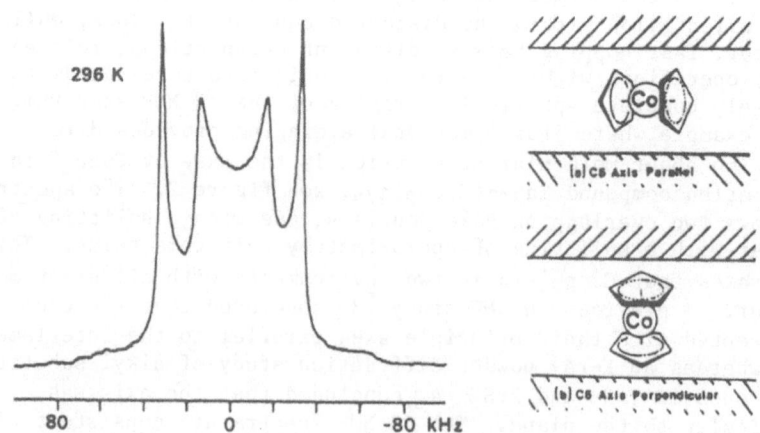

Figure 2: ^{2}H NMR spectrum of $2H-TaS_2-(d_{10}Cocp_2)_{1/4}$

experiencing motion only about the molecular axis [5].

Both these examples illustrate situations where the axes about which motion is possible are well defined by the lattice structures and the motion is always in the 'fast' limit compared to the ^{2}H nuclear quadrupole coupling constant (180kHz). ^{2}H NMR can, however, provide detailed evidence about the nature of more complex motions. Figure 3 shows an example of Fecp$_2$ in its thiourea clathrate. Here, as for Cocp$_2^+$PF$_6^-$, the ^{2}H NMR lineshape shows dramatic changes with temperature, but this time the changes are more gradual. At 160K the spectra show that the only fast motions involve the spinning of the cp rings about the molecular axis. At higher temperatures the changing lineshapes indicate the onset of more extensive motional averaging, although the motion is of limited amplitude until temperatures above 200K. The changes in the ^{2}H NMR spectra correlate well with the results of Mössbauer studies of this compound [6]. The existence of residual splitting in the ^{2}H NMR spectrum shows, however, that the motion is not entirely isotropic even at 300K, indicating that there are still preferred orientations of the rapidly reorienting ferrocene molecules.

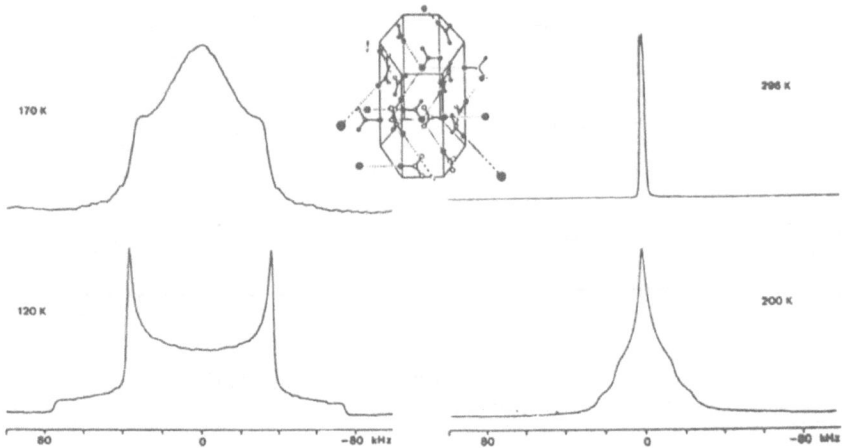

Figure 3: ^{2}H NMR spectra of thiourea:ferrocene (3:1) clathrate.

Similar spectral changes have been observed for Fecp$_2$ in its inclusion complexes with cyclodextrins. Of particular interest is a comparison of the motional behaviour of ferrocene included in α-, β- and γ-cyclodextrins, which have 6,7, and 8 rings respectively in the cyclic structures. The spectra of the Fecp$_2$ inclusion complexes of α- and β-cyclodextrins are closely similar, but marked differences are observed for the complex with γ-cyclodextrin. These differences are broadly consistent with proposals from CD studies of the inclusion complexes in solution which indicate that in both α- and β-cyclodextrins the Fecp$_2$ molecular axis is parallel to the cavity axis, but in γ-cyclodextrin it is perpendicular to it [7]. The different motional averaging experienced in the different complexes therefore relates closely to the different behaviour of $Cocp_2^+$ in its different orientations in TaS_2. In the case of the cyclodextrins, however, it is apparent from the NMR lineshapes that the included $Fecp_2$ molecules experience a considerably wider range of motions within the cavities.

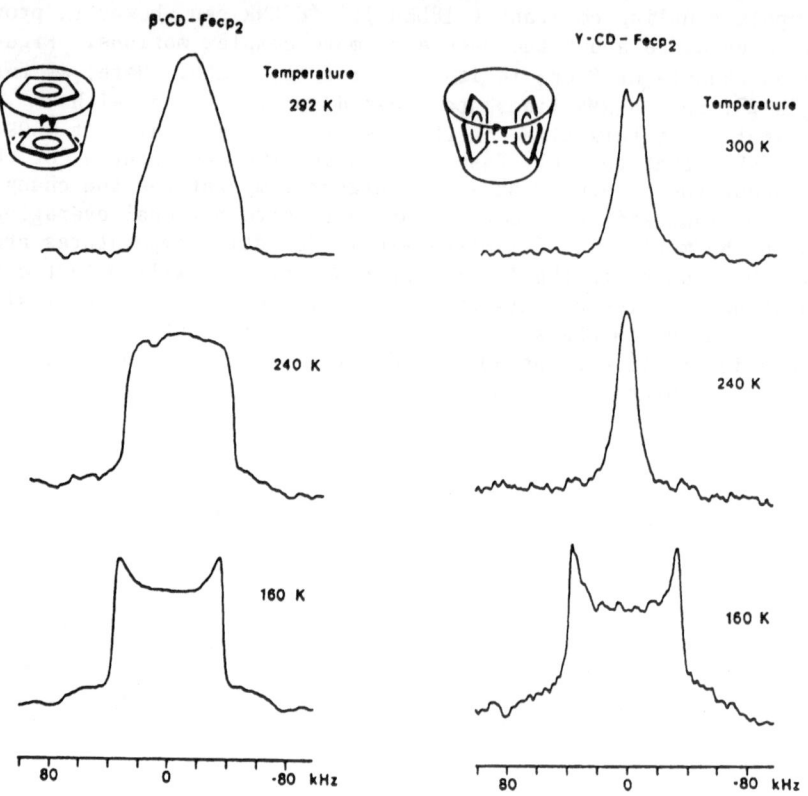

Figure 4: ^2H NMR spectra of the ferrocene inclusion complexes of β-cyclodextrin and γ-cyclodextrin

This work was supported by the SERC. We thank Dr. M.L.H. Green for valuable discussions.

References

1. S.E. Anderson, J. Organomet. Chem., 1974, **71**, 263
2. B.W. Fitzsimmons and A.R. Hume, J. Chem. Soc. Dalton, 1980, 180
3. B.G. Silbernagel, Chem. Phys. Lett., 1975, **34**, 298
4. R.P. Clement, W.B. Davies, K.A. Ford, M.L.H. Green and A.J. Jacobson, Inorg. Chem., 1978, **17**, 2754
5. N.J. Clayden, C.M. Dobson, M.L.H. Green, S.J. Heyes, P.J. Wiseman, J. Chem. Soc. Chem. Commun., (submitted for publication).
6. T.C. Gibb, J. Phys. C. Solid State Phys., 1976, **9**, 2627
   E. Hough and D.G. Nicholson, J. Chem. Soc. Dalton, 1978, 15.
7. A. Harada and S. Takahashi, J. Chem. Soc. Chem. Commun., 1984, 645

SEPARATION PROCESSES IN GAS-LIQUID CHROMATOGRAPHY BASED ON FORMATION OF α-CYCLODEXTRIN - CHIRAL HYDROCARBONS INCLUSION COMPLEXES

Tomasz Kościelski, Danuta Sybilska, and Janusz Jurczak
Institute of Physical Chemistry
Institute of Organic Chemistry
Polish Academy of Sciences
01-224 Warsaw, Kasprzaka 44/52
Poland

Chromatographic separation of enantiomeric hydrocarbons presents a number of difficulties caused by weak and not very selective intermolecular interactions in the common gas and liquid chromatographic systems. The use of optically active organic stationary phases cannot be realized on the basis of the "three-point-attachment" of Dalgliesh,[1] namely hydrocarbon molecules do not contain any functional groups suitable for forming the necessary hydrogen bonds for specific attachment of the enantiomer to the stationary phase. Chiral metal complexes were applied in RP-HPLC[2,3] and GLC[4] for resolution of chiral olefins into enantiomers; however in this case π donor - acceptor interactions being possible only for unsaturated hydrocarbons take place. We have recently discovered that under gas-partition chromatographic conditions, α-cyclodextrin (α-CD) complexation permits very efficient resolution of some terpenoic hydrocarbons (saturated and unsaturated) into enantiomers.[5] The present paper summarizes our further systematic studies on enantio- and diastereoselective separation of α- and β-pinene, and of four stereoisomers of pinane, whose formulae are shown below:

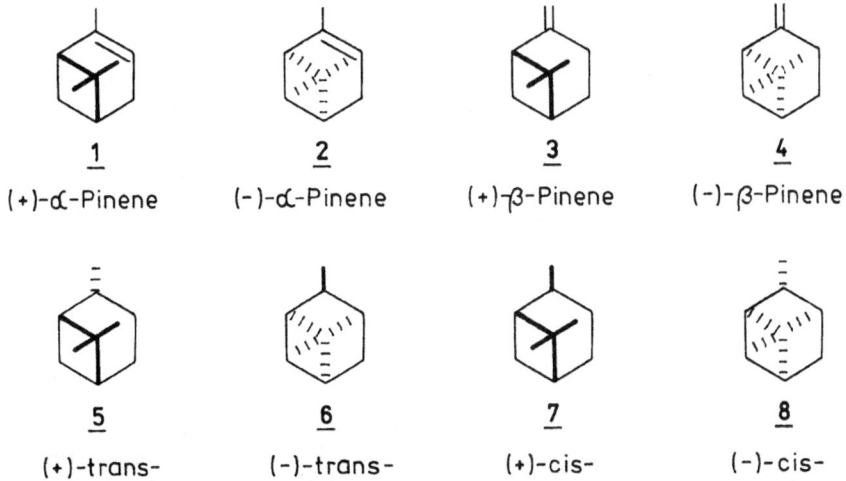

Pinanes

The ability of α- and β-CD complexation to modify the gas-partition chromatography system for separation of terpenoic hydrocarbons is exemplified in Fig. 1 for α-pinene.

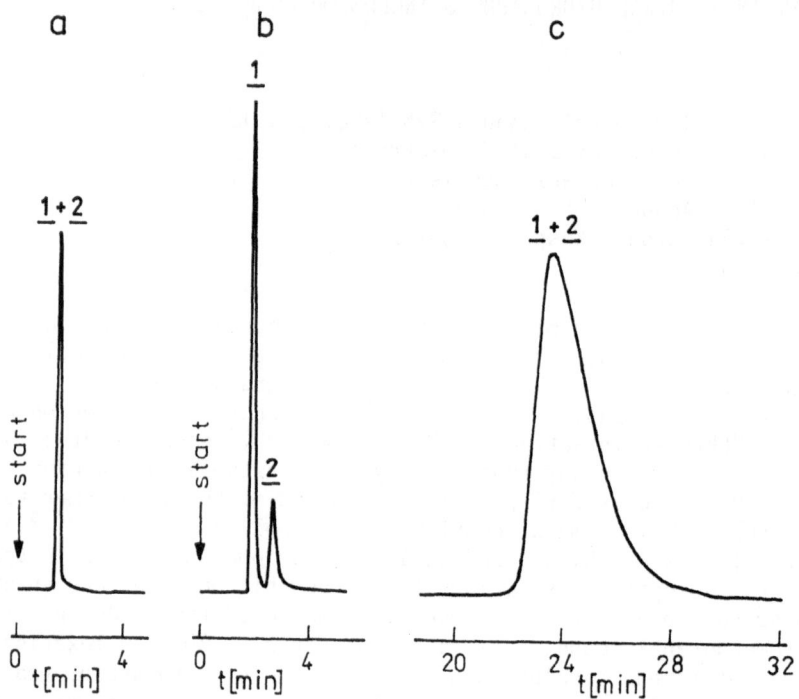

Fig. 1. Chromatograms of a nonracemic mixture of α-pinene, obtained at 50°C on a column packed with Celite covered by: (a) formamide, (b) 1.23 mol% of α-CD in formamide, (c) 1.18 mol% of β-CD in formamide.

As suggested by the chromatograms in Fig. 1, both α- and β-CD form inclusion complexes with α-pinene. Approximative estimation of the stability of CD complexes indicates that β-CD complex is much more stable than that of α-CD (the stability constant of the former complex is by one order greater than that of the latter). However, only α-CD complexation permits substantial chiral recognition of terpenoic hydrocarbons. This stereoselectivity originating from specific interactions between α-CD and guest molecules in inclusion complexes is markedly influenced by temperature and nature of the matrix medium.

So far, formamide is found to be the best matrix medium; its properties can sometimes be improved by addition of ethylene, propylene or tetramethylene glycol, or else of some inorganic salts as lithium or potassium nitrate. On the other hand, addition of diethylene or triethylene glycol or else silver nitrate destroys completely the selectivity of the column. On the basis of the above-presented results, optimal conditions were established for more complex mixtures containing enantiomers of β-pinene and stereoisomeric pinanes (Fig. 2).

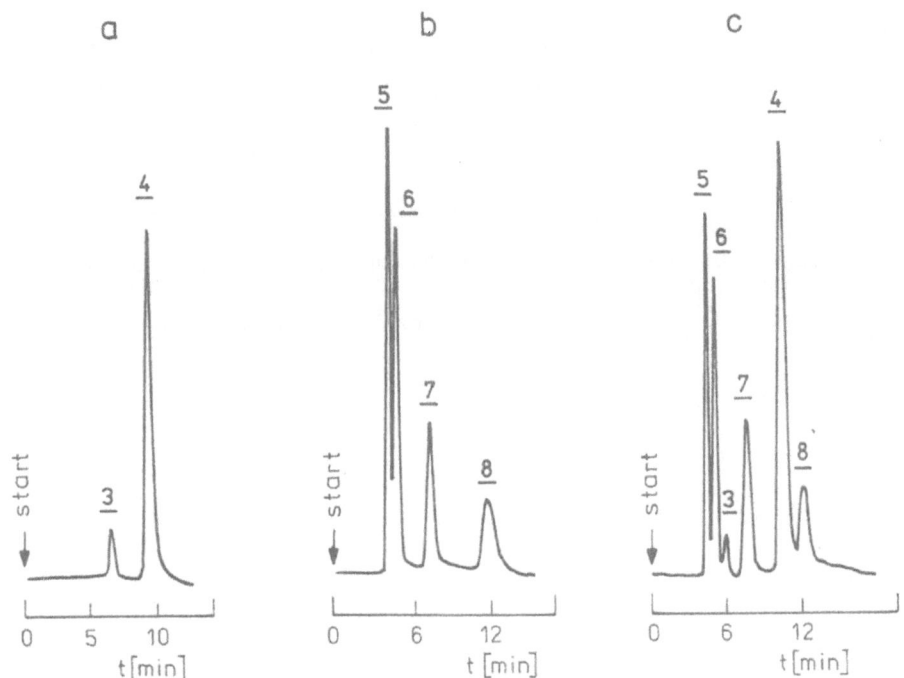

Fig. 2. Chromatograms (obtained at 35°C on a column packed with Celite covered by 0.6 mol% of α-CD in formamide) of mixtures containing: (a) enantiomeric β-pinenes (3, 4), (b) four stereoisomeric pinanes (5, 6, 7, 8), (c) all six above-mentioned compounds.

The results presented in Fig. 2 led to the conclusion that α-CD can readily be applied for gas chromatographic monitoring of the stereochemical course of β-pinene hydrogenation.[6] An analogous approach has earlier been successfully applied for monitoring of α-pinene hydrogenation.[7]

A simple and very convenient method for the determination of enantiomeric purity of chiral hydrocarbons, recently developed at this Laboratory,[8] is another interesting application of chromatographic separation of enantiomers via inclusion complexes with α-CD.

The present results fully confirm our claim that gas-liquid chromatography systems containing α-CD can be used as an analytical tool for monitoring stereocontrolled organic reactions. Moreover, the method seems to be promising for preparative purposes.

ACKNOWLEDGMENTS. The authors are indebted to Professor J. Szejtli (Chinoin, Budapest, Hungary) for kindly providing the sample of α-CD. Financial support from the Polish Academy of Sciences (Grants CPBR-3.20 and CPBP-01.13) is gratefully acknowledged.

REFERENCES

1. C. E. Dalgliesh, *J. Chem. Soc.*, 3940 (1952).
2. J. Köhler and G. Schomburg, *Chromatographia*, 14, 559 (1981).
3. J. Köhler, A. Deege, and G. Schomburg, *Chromatographia*, 18, 119 (1984).
4. V. Schurig, *Chromatographia*, 13, 263 (1980).
5. T. Kościelski, D. Sybilska, and J. Jurczak, *J. Chromatogr.*, 280, 131 (1983).
6. T. Kościelski, D. Sybilska, S. Belniak, and J. Jurczak, *Chromatographia*, 21, 413 (1986).
7. T. Kościelski, D. Sybilska, S. Belniak, and J. Jurczak, *Chromatographia*, 19, 292 (1984).
8. T. Kościelski, D. Sybilska, and J. Jurczak, *J. Chromatogr.*, 364, 297 (1986).

NOVEL HPLC ADSORBENTS BY IMMOBILIZATION OF MODIFIED CYCLODEXTRINS

Kenjiro Hattori and Keiko Takahashi
Department of Industrial Chemistry, Faculty of Engineering,
Tokyo Institute of Polytechnics,
Atsugi, Kanagawa 243-02
Japan

1. INTRODUCTION

Cyclodextrin(CD) has been successfully utilized in HPLC for the specific interaction between aromatic compounds and the hydrophobic cavity of the CD molecule.(Ref.1-3) In our present work(Ref.4-5), a novel packing material for HPLC was prepared with 6-deoxy-monoamino-β-CD immobilized through the epoxyl glyceride group on hydrophilic gel beads. The capacity factors of the guest molecules having an aromatic ring together with a carboxylic group were increased by the interaction of the hydrophobic effect and the ionic effect. The results suggest "host-guest chromatography" with multi-point molecular recognition.

2. EXPERIMENTAL
2.1 Preparation of bead gels

β-CD was immobilized on a matrix of gel beads. The bead gel(PW) was treated with epichlorohydrin in an alkaline solution at 45°C for 4 h to modify the active epoxyl group on the primary hydroxyl group of CD. Consequently, the product was treated with β-cyclodextrin in an alkaline solution of pH 12 at 37°C for 24 hr to give the β-CD immobilized bead gel(CD-PW).

Determination of the epoxyl content of the gel was determined by titration methods. The result indicated that there was the introduction of an active epoxyl group of 620 μmole/(1 g dry gel). Also, the amount of immobilized β-CD on CD-PW was determined by HPLC analysis of the glucose produced by the hydrolysis with the acid. It showed that 540μ mole/(1 g dry gel) of β-CD was immmobilized in this preparation.

The immobilization of amino-β-CD and amino-methylated-β-CD on PW gel was carried out. Amino-CD gel (ACD-PW) was obtained by the procedure with the epoxyl bead gel and 6-monoazide- β-CD which was prepared by regiospecific tosylation and substitution by sodium azide. The obtained monoazide-β-CD gel was treated with triphenylphosphin in DMF, then with ammonium hydroxide which gave amino-β-CD gel(ACD-PW). Amino-methyl-β -CD gel(A-Me-CD-PW) was prepared starting with the 6-monoazide-β-CD gel. The gel was methylated with dimethylsulfate in the mixed solvent of chloroform and an aqueous sodium hydroxide solution with shaking at room temperature for 72 hrs. Monoazide-methyl-β-CD gel was treated with triphenylphosphine in DMF, then with ammonium hydroxide which gave A-Me-CD-PW. The amino group content of both resulting gels was determined by neutral titration with a pH stat apparatus. There were amino groups of 530 and 230 μmole per 1 g dried ACD-PW and A-Me-CD-PW, respectively.

2.2 Column Chromatography

The above obtained bead gels were packed by the slurry packing method in a stainless column of 4.0 mm inner diameter and 300 mm length. In the experiment of pH dependency, the eluent was usually a mixture of 10 v/v% acetonitrile and 90 v/v% 1/40 M-phosphate buffer between pH 4-7 or 90 v/v% 1/40 M carbonate buffer between pH 9-11.5. Flow rate was approximately 0.5-0.7 ml/min. In the experiment for the effect of organic solvent, the eluent was a mixture of x v/v% acetonitrile and (100-x) v/v% 1/40 M-phosphate buffer of pH 7.0 at the flow rate of 0.6-0.8 ml/min.

3. RESULT AND DISCUSSION

3.1 Introduction of amino-CD on the gel beads

As the PW gel polymer has hydroxyl groups on the side chain, CD units were able to be introduced by the treatment of the gel with epichlorohydrin using active epoxyl groups on the gel surface. It contained 620 μmole/(1 g dry gel). More than 87% of these groups were reacted with CD to prepare CD-PW gel beads. This report is the first for a HPLC adsorbent which contains a CD cavity on the synthetic polymer. In order to enforce the capacity factor, two kinds of modification of CD molecules were carried out. To attach the amino group on CD, the usual preparation methods by way of tosylation and azidification were adopted. This "pre-procedure method" gave a sufficient amount of amino groups on the CD-PW gel of 530 μmole/(1 g dry gel). Another improvement of this HPLC adsorbent was methylation of the hydroxyl group on CD molecules. With the addition of $CHCl_3$, dimethylsulfate and an aqueous alkaline solution, the methylation, at least on the C-6 hydroxyl group, was easily performed, though there was a 57% decrease of amino groups through this procedure. Both ACD-PW and A-Me-CD-PW gel beads were new types of HPLC adsorbents. These gel beads can be used at high speed flow in a wide range of pH for the separation of various aromatic acids such as mandelic acid and N-protected phenylalanine.

3.2 pH Dependence of Capacity Factors

These four types of adsorbents, i.e., PW, CD-PW, A-CD-PW and A-Me-CD-PW were compared with capacity factors k' ($=t_R/t_0-1$) towards the various guest molecules by changing the pH of the eluent in order to elucidate the interaction between the amino group on CD and the carboxylic group on the substrates. There was no retention for Asp, and only by A-Me-CD-PW was there a slight retention in the eluent of the alkaline region. In the case of methyl mandelate(Me-M), the PW gel showed no retention. For the gels of ACD-PW and A-Me-CD-PW, there was a tendency of increased retention in the alkaline region. In the case of the Phe substrate, PW and CD-PW guests did not show any retention, but ACD-PW gel showed retention factors up to 1.0 in alkaline solution.

Mandelic acid(MA) was not thoroughly retained by PW gel, but a slight retention by the CD-PW gel was indicated. In the case of ACD-PW gel, the capacity factor decreased at pHs of 3.6 and 10.1, near the pKa values of MA and ACD-PW. This result indicated an ion-exchange type of retention behavior and a strong electrostatic interaction between the amino group and the carboxylic group which was not the case for the Asp substrate. A-Me-CD-PW gel then showed an extroadinary high capacity factor especially at acidic regions. The enhanced hydrophobic interaction by the methylation at the hem of CD ring was mainly at the C-6 positions of the glucose residue. Another substrate, benzyloxycarbonyl(Z)-Ala, is similar in behavior to MA. The k' values for ACD-PW and A-Me-CD-PW was much higher up to 150 at acidic regions.

3.3 Dependence of Capacity Factor on Organic Eluents

In order to clarify the interaction between the hydrophobic CD ring and each guest molecule, the concentration of organic solvent was changed and the dependency of the retention behavior was examined. In these experimental runs, an aqueous buffer of pH 7 was mixed with the organic solvents. The concentration of acetonitrile was changed in the range of 0 to 30 v/v%, and the capacity factors for the guest molecules of Me-M, Phe and Asp were examined.

For the guest molecules Me-M, three CD-immobilized gels, CD-PW, ACD-PW, and A-Me-CD-PW, showed a slight retention change from 0 to 3.5 depending 30 to 0 v/v% of acetonitrile content, although the matrix gel PW showed no change. This change of k' seemed to reflect the change of merely hydrophobic interaction caused by the CD ring. Asp, which has no aromatic moiety and is not expected to interact with the CD cavity, showed no retention change for all the gel adsorbents. This suggests that the presence of a phenyl ring was the key point for the interaction with CD's hydrophobic cavity.

In the case of the MA substrate, the organic solvent effect on the capacity factor was observed with a CD-PW gel compared with a PW gel which showed no effects. This indicated that MA effectively interacted with CD's hydrophobic cavity. With ACD-PW gel, larger changes of capacity factor were observed. Also, the result suggested an interaction between the CD cavity and the guest. The extroadinary large effect was observed for a A-Me-CD-PW gel which was obtained by the methylation of the β-CD ring, and CD cavity was improved to the more hydrophobic circumstance. These results supported the idea that the hydrophobic cavity involved the retention behavior and also the amino group increased the retention behavior.

In the case of Z-Ala, for a PW gel, there was no change in retention. For CD-PW and ACD gels, a small change was observed. For the ACD-PW gel, a larger change in retention than for the CD-PW gel was observed. The A-Me-CD-PW gel had the highest capacity factor and was the most sensitive to the change of organic content.

From the above solvent effect on the retention capacity, it is presumed that the hydrophobic cavity of CD evidently interacted with guest molecules.

3.4 N-Substitution Effect of the Guest Molecules

Aromatic amino acids showed low capacity factors because of ionic repulsive force between the amino moiety of amino acid such as Phe, Tyr and dihydroxyphenylalanine, the inclusion effect was not observed. But

N-substituted amino acids showed high capacity factors; N-acetylated Phe and N-formylated Phe gave 6 and 10 times higher capacity factors than Phe in ACD-PW respectively, and 30 and 50 times higher in A-Me-CD-PW. These N-substitution effects on retention behavior were also observed for a nonaromatic amino acid such as Ala. The capacity factors of N-substituted amino acids depended on the ionic strength of eluent. The dependency was observed clearly in A-Me-CD-PW. Methylation of hydroxyl groups on the CD ring enhanced the interaction with amino acids.

The inclusion effect was observed clearly in A-Me-CD-PW gel. The aromatic guests such as Z-Phe and Z-Ala showed high capacity factors than those in ACD-PW.

4. Conclusion

According to the CPK scale molecular model, an amino group is embedded inside the CD ring. So it may be presumed that the ionic interaction between carboxylic group and amino group shoud be difficult from the outside of the CD ring. Thus, it is necessary that the guest molecule is included into the cavity by hydrophobic interaction, then the included molecule can be held tightly by both ionic and hydrophobic interaction.

All these observations supported the operational interaction in the present packings which were prepared by immobilization of CD derivatives on the hydrophilic polymer gel. Especially in the case of Z-Ala and MA, as they have both phenyl and carboxylic groups, a stable complex can be formed by the double forces through the hydrophobic and ionic interactions. This caused the high retention power. It may be concluded that the present amino-CD gel and amino-methyl-CD gel indicated both hydrophobic and ionic interactions, and it showed novel and unique adsorbent properties for practical HPLC use.

REFERENCES
1. W.L.Hinze, Sepn.Purifn.Methods., 10(1981)159.
2. E.Smolkova-Keulemansova, J.Chromatogr., 251(1982)17.
3. a)D.W.Armstrong, W.DeMond, B.P.Czeck, Anal.Chem., 57(1985)481.
 b)W.L.Hinze, T.E.Riehl,D.W.Armstrong, W.DeMond, A.Alak, T.Wand, Anal.Chem., 57(1985)237. c)D.W.Armstrong, W.DeMand, A.Alak, W.L.Hinze, T.E.Riehl, K.H.Bui, Anal.Chem., 57(1985)234. d)D.W.Armstrong, W.DeMand, J.Chromatogr.Sci., 22(1984)411.
4. K.Hattori, K.Takahashi, M.Mikami, H.Watanabe, J.Chromatogr., 355 (1986)383.
5. K.Takahashi, S.Nakada, M.Mikami, K.Hattori, Nippon Kagaku Kaishi, (1986) 1032.

CONFORMATIONAL AND ENANTIOMERIC DISCRIMINATION IN CYCLODEXTRIN INCLUSION COMPOUNDS

G. Tsoucaris, G. Le Bas, N. Rysanek and F. Villain
Laboratoire de Physique
centre Pharmaceutique
Universite Paris-Sud
92290 - Chatenay-Malabry

ABSTRACT:Cyclodextrins open great possibilities for studying both the crystal structure and the spectroscopic properties in solution of the same compound. In the present paper new inclusion compounds are studied. Two conformationaly labile molecules (cyclopentanone, bilirubin), the pheromone of the olive fly with chiral stable conformation and p-nitroaniline with non linear optic properties are investigated.

Cyclodextrins (cyd) open great possibilities for studying both the crystal structure and the spectroscopic properties in solution of the same compound.
The interaction between the host cavity and the guest causes a structural perturbation of the guest and consequently a modification of its physical, chemical and biological properties. Circular dichroism is one of the properties which are particularly sensitive to conformational changes induced by the cyd environment (1).
It was shown that bilirubin, biliverdin and 4-helicene in solution acquire a preferential chiral conformation exhibiting a very strong circular dichroism spectrum upon interaction with cyd ; crystallographic study has provided experimental evidence that the CD spectrum of benzil molecule in β cyd complex is to be ascribed mostly if not entirely to a conformational isomerism (2,3). This provides a novel method of studying the chiroptical properties of conformationaly labile molecules.
In the present paper we report new inclusion compounds:

1.Cyclopentanone $\underline{1}$ in α cyd

$\underline{1}$ has two enantiomeric conformations in solution. The CD spectrum of $\underline{1}$ complexed with α cyd shows a minimum at λ = 290 nm. Such an extremum for the n-π* transition was attributed to the R isomer (4).

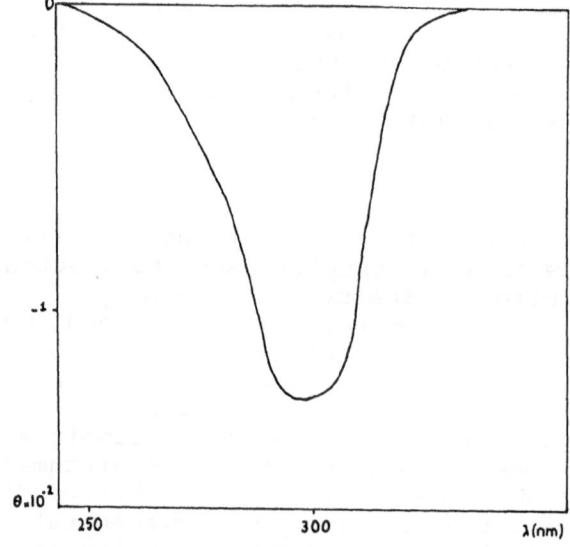

Figure 1. The two enantiomeric conformations of 1 , looking behind the axis of C=O bond.

Figure 2. CD spectrum of the complex α-cyd-1.

In crystal structure , conformational analysis is often impaired by disorder exhibited in the guest and the host molecule as well, however significant phenomena can be observed: host/guest interaction and the nature of the formed bonds.
A typical example is the complex α-cyd - 1.

α-cyd - 1, 6(H_2O) complex crystallizes as a head to tail channel (figs.3,4).

CRYSTALLOGRAPHIC DATA

space group	P6
a = b (Å)	23.64
c (Å)	8.00
V (Å^3)	3941
Z	3
Nref	1940
Nobs	1713
R	0.06

It is noteworthy to observe the high symmetry of this structural arrangement. There are two symmetrically independant molecules in the cell, one hexagonal, the other trigonal. To our knowledge, these ideal symmetries for α-cyd molecules have not been found, up to date.

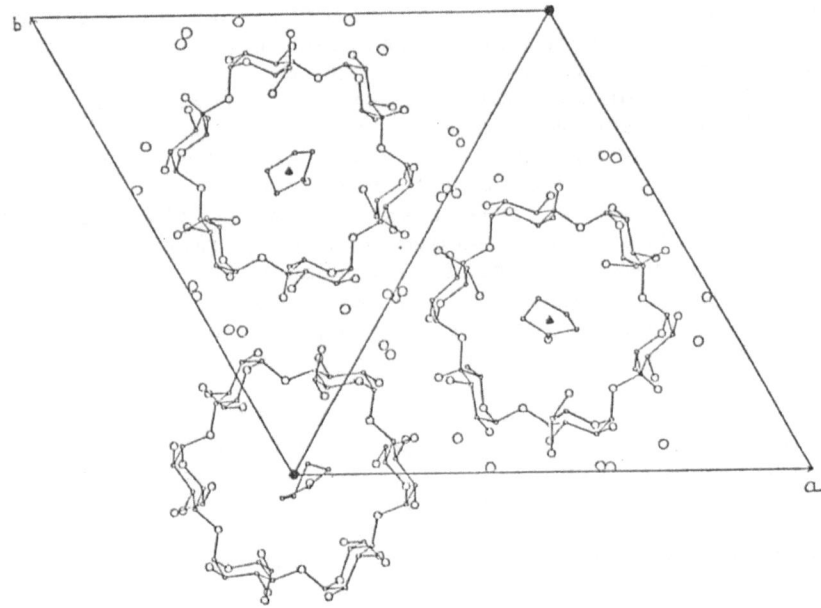

Figure 3. Projection of the structure on the (a,b) plane.

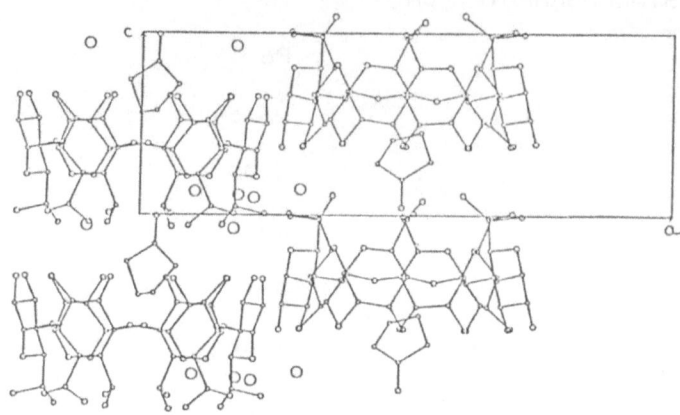

Figure 4. Projection of the structure on the plane (a,c).

The two guest molecules in the unit cell are disordered, one along the six fold axis, the other along the three fold axis.

The primary hydroxyl groups are disordered between the gauche-gauche and the gauche trans conformation. In the model suggested by X-ray analysis, the occupancy factor of the gauche-trans conformation is about:
- 1/6 for the cyd molecule at the 6 fold axis,
- 1/3 for the molecule at the 3 fold axis, but for only three primary hydroxyl groups, the other three being entirely in the gauche-gauche conformation.

The most remarkable fact exhibited by this crystal structure is that the disorder of the primary hydroxyl groups of the host depends directly upon the orientations of the guest molecule. In all cases H-bonds are established between the primary hydroxyl groups of the host and the carbonyl groups of the guests.

The positions of most of the hydrogen atoms were determined. Secondary hydroxyl groups of the hexagonal cyd molecules O (2) H and O (3) H are respectively acceptor and donor. In the trigonal molecule, the two secondary hydroxyl groups are donors in one glucose residue and acceptors in the other residue of the asymmetric unit.

In order to specify the hydrogen positions and the dynamic disorder, neutron diffraction and X-Ray diffraction at low temperature will be undertaken.

2. Bilirubin in permethyl ß cyd

Let us describe the biological implication of this study. Bilirubin is the end product of heme catabolism in man and most animals. This lipophilic pigment is normaly carried in blood by serum albumin until it is excreted by liver. All forms of jaundice are a manifestation of an excess of bilirubin over the binding capacity of serum albumin. Then the aqueous insoluble pigment leaves the intravascular system, diffuses into the lipophilic tissues and, in the case of newborn babies, it may reach the brain and cause irreversible damage. In order to urgently remove the bilirubin excess, massive U.V. irradiation is applied. The generally admitted mechanism involves transformation into "photobilirubins" which are more soluble in water and readily excretable(5).

We have shown that a solubilization effect is also obtained by association with cyd. The detection of this complex becomes effective through the remarkable chiroptical properties of bilirubin. The two enantiomeric conformations stabilized by six intramolecular hydrogen bonds, as found in crystalline bilirubin (6) interconvert rapidly in solution at room temperature by breaking and remaking all six H-bonds (fig.5).

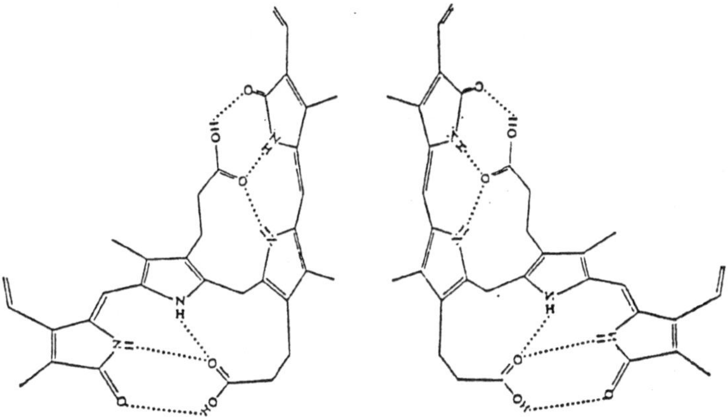

Figure 5. The two enantiomeric conformations of cristalline bilirubin.

Selective complexation of one enantiomer is achieved by cyd in aqueous alkaline solution which leads to optical activity (fig.6) (2).

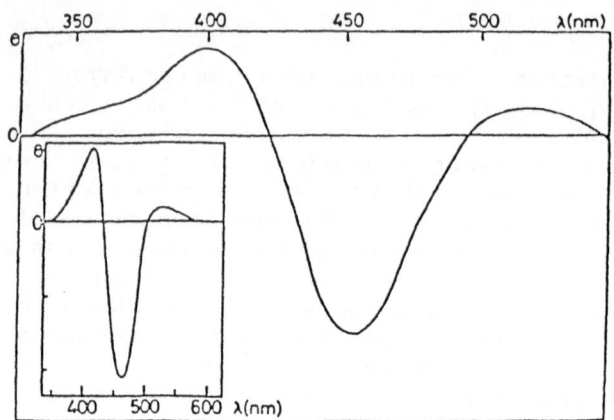

Figure 6. CD spectrum of bilirubin-β cyd, pH=10.2, (λ=455nm, θ=-2.7 10^4 deg.cm^2.dmole^{-1}). Inset shows the CD spectrum of the complex bilirubin-ligandin.

We report now the chiral recognition of bilirubin by permethylated β cyd. A basic requirement for the complexation of bilirubin with cyd derivatives is a common solvent. The CD spectrum shown below has been obtained in CH Cl$_3$. Besides, it seems that CH Cl$_3$ may be enclathrated and therefore, there is a competition with bilirubin for cyclodextrin cavity. This fact can explain the poor quality of the spectrum compared to that of β cyd (fig.7). We are presently pursuing the experimental study.

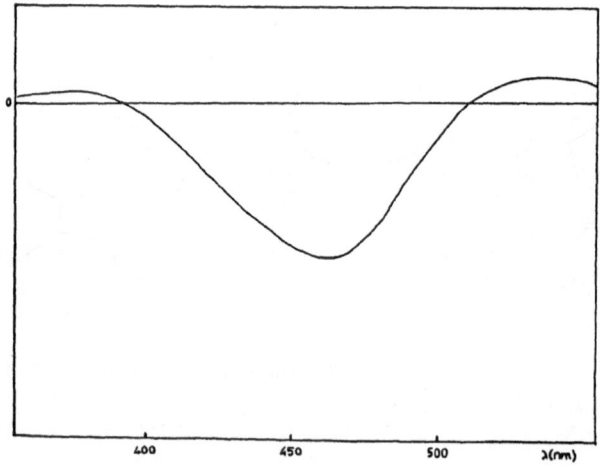

Figure 7. CD spectrum of permethyl β cyd - bilirubin at a molar ratio of permethyl β cyd to bilirubin of 50 : 1 in chloroform, (λ = 455nm, θ = -5.0 10^3 deg.cm^2.dmole^{-1}).

3. The pheromone of the olive fly: 1,7 dioxaspiro (5,5) undecane 2.

The first biological experiments have shown that the pheromones are considerably stabilized and therefore slowly released after enclathratation in cyd.

The cell parameters of 2 complexed with β-cyd are given:

space group	P1
a(Å)	15.60
b	15.72
c	15.93
α(°)	101.4
β	101.7
γ	103.2

2 is chiral and cyd can be used in order to isolate or to enrich the more biologically active enantiomer R-(-)-2. (fig.8)

S   R

Figure 8. Two enantiomeric isomers of 2.

4. Non linear materials.

A prerequisite for non linear optic material is the absence of inversion center. Cyd can be used as a tool in order to achieve a non centrosymmetric crystal. Cyd offers a new procedure for the evaluation of the order of a molecular non linearity.

p-nitroaniline 3 is one of the most interesting molecules. The CD spectrum demonstrates a complexation with α cyd.

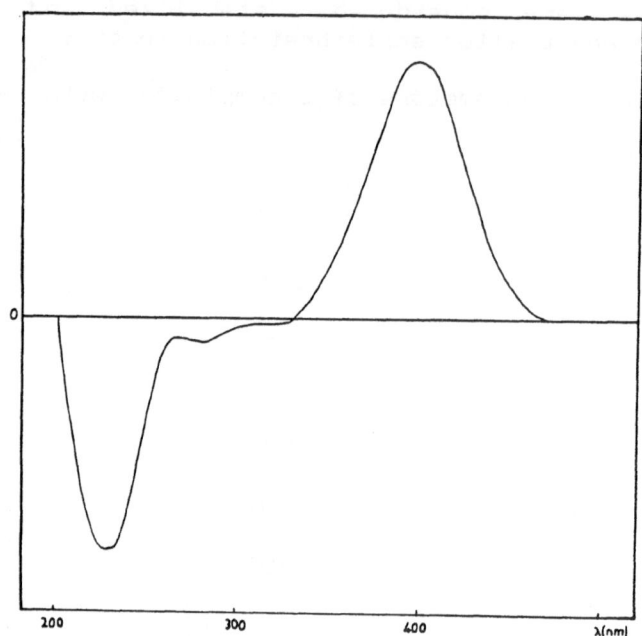

Figure 9. CD spectrum of α-cyd-3 with the molar concentrations: α-cyd = 10.10^{-3} M/l, 3 = 1.5.10^{-3} M/l. (λ = 400nm, θ = 7.6 10^3 deg.cm^2.dmole^{-1}) .

(1). K. Sensse und F. Cramer; Chem. Ber.,102, 509-521 (1969).
S. Takenaka, N. Matsuura and N. Tokura; Tetrahedron Letters, 26, 2325-2328 (1974).
J. Szejtli; Cyclodextrins and their inclusion complexes, Akademiai Kiado, Budapest (1982).
(2). G. Le Bas, C. de Rango and G. Tsoucaris; 1st Int. Symposium on Cyclodextrins, Budapest (1981).
(3). G. Le Bas, C. de Rango, N. Rysanek and G. Tsoucaris; J. of Inclusion Phenomena,2,861(1984)
(4). W. Klyne .Tetrahedron 13, 29, (1961)
C. Djerassi; P.N.A.S USA, 48, 1093 (1962).
(5) A.F. McDonagh, L. Palma and D.A. Lightner; Sciences 208,145 (1980)
(6) G. Le Bas, A. Allegret, Y. Mauguen, C. de Rango and M. Bailly; Acta Cryst., B36, 3007-3011 (1980).

REACTION OF CYCLODEXTRIN-NICOTINAMIDE AS A NADH COENZYME MODEL

Chul-Joong Yoon, Hiroshi Ikeda, Ryoichi Kojin, Tsukasa Ikeda, and Fujio Toda
Department of Bioengineering and Science, Faculty of Engineering, Tokyo Institute of Technology
O-okayama, Meguro-ku, Tokyo 152, Japan.

ABSTRACT. β-cyclodextrin-1,4-dihydronicotinamide can reduce cytochrome c in aqueous solution, by adding redox dyes as mediators. In the reduction of cytochrome c mediated by redox dyes, the speeds of reduction differ depending on the activities of the dyes. Inhibition using cyclohexanol occured in the presence of nile blue or methylene blue, and showed competitive inhibition. In the case of using neutral red as a mediator, the reduction was accelerated by adding cyclohexanol. Reaction rate constants of this reaction were independent of the redox potential of the redox dyes.

1. INTRODUCTION

Cyclodextrins(CDs), forming inclusion complexes with a wide variety of substrates in aqueous media[1], affect the rate of various chemical reactions and exhibit substrate selectivity, so that they are often conveniently and successfully utilized as an enzyme model[2]. Although cyclodextrins exhibit rate enhancement, stereospecificity, enantiomeric specificity, etc. in organic reactions, their catalytic activities are not high enough for them to act as a true enzyme.

To improve their catalytic activity, many modified CD derivatives were prepared, especially as models of chymotripsin-catalyzed hydrolysis. We have already reported an effective modification: the regioselective monosubstitution of α- or β-CD by histamine. α-CD-histamine accelerated the hydrolysis of p-nitrophenyl acetate 15 times more than CD itself at pH 8.0 [3]. The catalytic rate constant of α-CD-histamine is close to an actual enzyme, chymotripsin, in the hydrolysis of p-nitrophenyl acetate.

Also we have already reported syntheses of α- and β-CD-nicotinamide as models for an NADH dependent enzyme. The corresponding reduced forms were more stable in aqueous solution, and showed a large rate enhancement (15-50 times greater) in the reduction of ninhydrin, compared with monomeric NADH[4].

This paper describes the nonenzymatic reduction of cytochrome c with β-CD-1,4-dihydronicotinamide (β-CD-NAH), mediated by redox dyes as a mediator.

2. MATERIALS AND METHODS

Mono-tosylated-β-CD was prepared as described previously and recrystallized from water[5]. This compound is identical in all respects with mono-tosylated-β-CD substituted at C-6 position of glucose ring[6].

Iodination of mono-tosylated-β-CD(20g) with sodium iodide(24g) was carried out in methanol(300ml) at 70°C for 50 hours. After reprecipitation with acetone, β-CD-iodide was purified by a column of highly porous polystyrene gel(DIAION HP-20). The purity was confirmed by HPLC. A solution of β-CD-iodide(5g) and nicotinamide(5g) in DMF(50ml) was stirred for 2 days at 100°C and then diluted with acetone to precipitate the product. The precipitated crude product was dissolved in water and evaporated to dryness until no odour of DMF could be detected. The product was dissolved in water and applied to a column of HP-20. The column was eluted with water, followed by 5% aqueous methanol and 20% aqueous methanol. The 20% methanolic eluate was evaporated to dryness and the dried material was dissolved in water and applied to a column of CHP-20P. The column was eluted with water, followed by 5% aqueous methanol, and 10% aqueous methanol. The final eluate was evaporated to dryness, and the chromatographic procedure repeated until HPLC indicated that the material was pure. The yield of β-CD-NA was ca.10% based on the starting mono-tosylated-β-CD. Anal.calc.for $C_{48}H_{75}O_{35}N_2I$: C,42.17; H,5.53; N,2.05. Found: C,42.96; H,5.55; N,1.98.

Reduction at the C-4 position of the nicotinamide moiety in β-CD-NA was carried out according to the general procedure of Hynes and Todd[7], using hydrosulphite as the reducing reagent, and giving β-CD-NAH. The reacting solution was applied to a column of CHP-20P, and eluted with water followed by 10% aqueous methanol to remove unchanged β-CD-NA, and then 40% aqueous methanol. The final eluate was evaporated to remove methanol, and the concentrated solution was freeze-dried, to give purified β-CD-NAH in ca.40% yield. β-CD-NAH was checked spectrometrically at 358nm, and the purity was confirmed by HPLC(Toyo Soda, Toyosoda TSK Ods gel, φ5 X 500mm, solvent; acetonitrile-water system).

Rates of reduction were followed spectrometrically using a Hitachi model 220A spectrophotometer. The increase of reduced cytochrome c was followed at 550nm. All rates were determined using 3ml quartz cells with a 1cm light path, under anaerobic conditions. The reaction were carried out in pH 7.0(±0.01) phosphate buffer solutions at 25°C, and initiated by addition of β-CD-NAH dissolved in DMF. All reagent were purchased from commercial suppliers and were used without further purification.

3. RESULTS AND DISCUSSION

To simulate dehydrogenase catalysis, system comprising β-CD-NAH as the hydrogen donor and cytochrome c as the hydrogen acceptor was studied in aqueous media. Since no direct reduction of cytochrome c occured by β-CD-NAH, the use of redox as a mediator promoted the reduction as shown in Scheme 1. Four kinds of dyes[nile blue(NB), methylene blue(MB), neutral red(NR), and phenosafranine(PS)] having good solubility were used in this experiment.

Scheme 1.

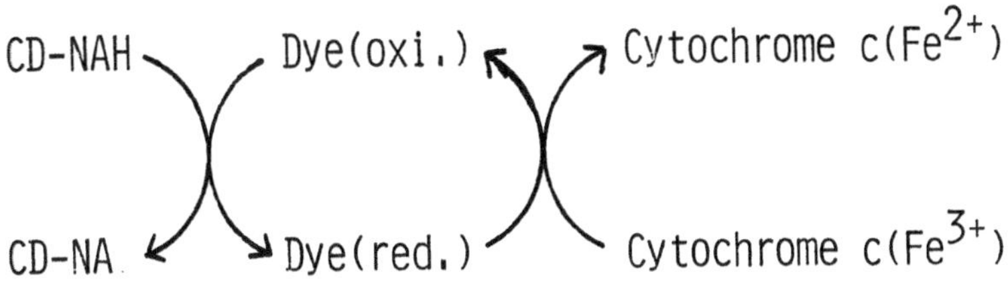

$$CD\text{-}NAH \searrow \quad \nearrow Dye(oxi.) \searrow \quad \nearrow Cytochrome\ c(Fe^{2+})$$
$$CD\text{-}NA \nearrow \quad \searrow Dye(red.) \nearrow \quad \searrow Cytochrome\ c(Fe^{3+})$$

A typical reduction of cytochrome c with β-CD-NAH, as mediated by various redox dyes, is given in Figure 1. The reduction rate was obtained spectrometrically by measuring the intensity of the absorption band at 550nm, under anaerobic condition. This result shows that NB as a mediator accelerates the reduction of cytochrome c more significantly than other dyes, whereas there was scarcely alteration of reduction rate in the case of using PS as a mediator.

The inhibition experiments were carried out in order to investigate the relationships between rate enhancements and properties of dyes. The change of reactivity by adding cyclohexanol as inhibitor in the presence of NB, MB, or PS is shown in Figure 2. The reduction mediated by NB or MB was largely depressed in the presence of cyclohexanol. In the case of PS, there was no inhibition effect. On the other hand, Figure 3 shows that reduction was greatly accelerated by adding cyclohexanol in the presence of NR. Cyclohexanol acts as a catalyst in this case, not inhibitor.

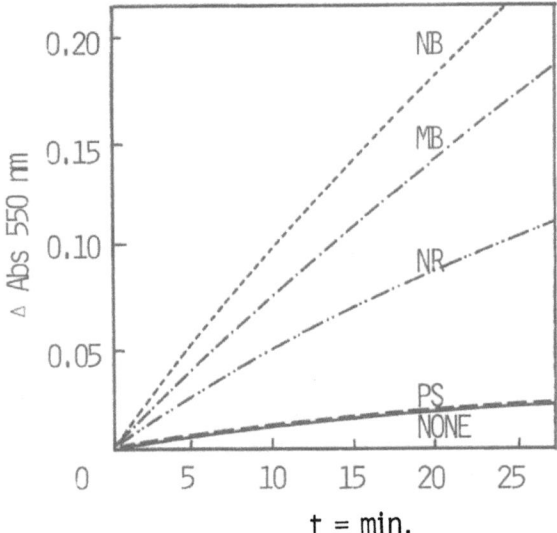

Figure 1. Effects of various dyes on the reduction of cytochrome c with β-CD-dihydronicotinamide.
[β-CD-NAH] = 2.0 X 10^{-4}M   [cytochrome c] = 3.6 X 10^{-5}M
[dye] = 5.0 X 10^{-6}M

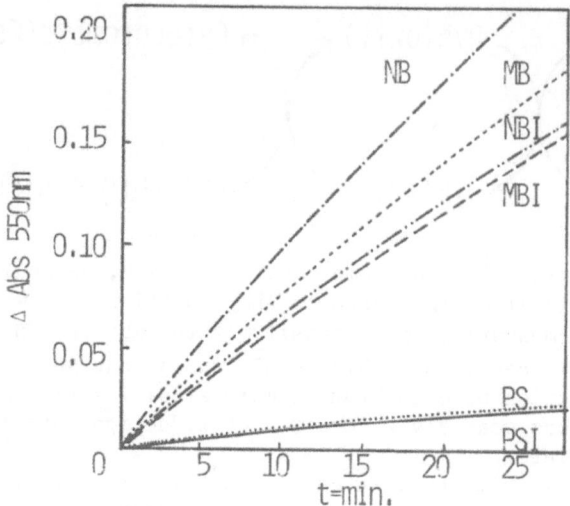

Figure 2. Effects of added cyclohexanol on the reduction of cytochrome c with β-CD-dihydronicotinamide.

[β-CD-NAH] = 2.0 X 10^{-4}M [cytochrome c] = 3.6 X 10^{-5}M
[dyes] = 5.0 X 10^{-6}M [cyclohexanol] = 1.0 X 10^{-3}M

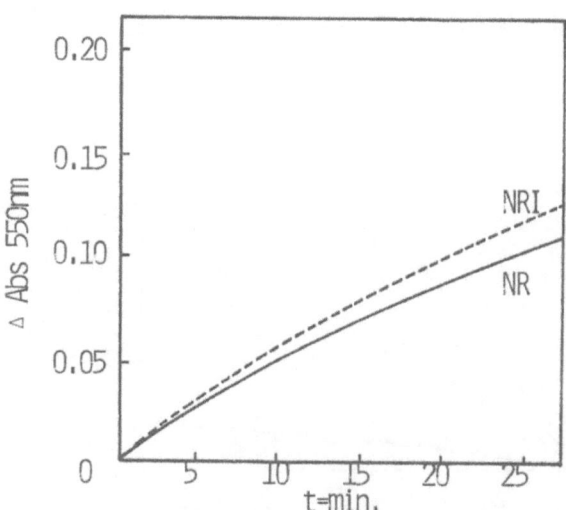

Figure 3. Effects of added cyclohexanol on the reduction of cytochrome c with β-CD-dihydronicotinamide.

[β-CD-NAH] = 2.0 X 10^{-4}M [cytochrome c] = 3.6 X 10^{-5}M
[dyes] = 5.0 X 10^{-6}M [cyclohexanol] = 1.0 X 10^{-3}M

Since these reductions are of the Michaelis-Menten type(coenzyme-like), catalytic rate constants according to Lineweaver-Burk plots which are alteration of first order rate constants depending on the concentration changes of dyes, were calculated and summarized in Table 1. Where, k-cat, Km, and k-cat/Km mean reduction rate constant, dissociation constant, and second order rate constant. NB as a mediator has a larger value of k-cat/Km and a smaller value of Km than the other dyes. These results indicate that NB is a better mediator in the reduction of cytochrome c with β-CD-NAH and is included into the cavity of β-CD-NAH easily. In the case of NR, Table 1 clearly shows an acceleration effect by adding cyclohexanol. This result can be explained by assuming Scheme 2.

Table 1. Rate constants for the reduction of cytochrome c by β-CD-dihydronicotinamide.

Mediator	Cyclohexanol	$Km(10^{-5}M)$	$k_{cat}(10^{-4}s^{-1})$	$k_{cat}/Km(M^{-1}s^{-1})$
NB	None	1.4	1.6	12
	Added	2.2	1.4	6.0
MB	None	8.2	4.6	5.7
	Added	6.3	3.6	5.6
NR	None	6.3	2.4	3.9
	Added	2.9	2.4	8.4

Scheme 2.

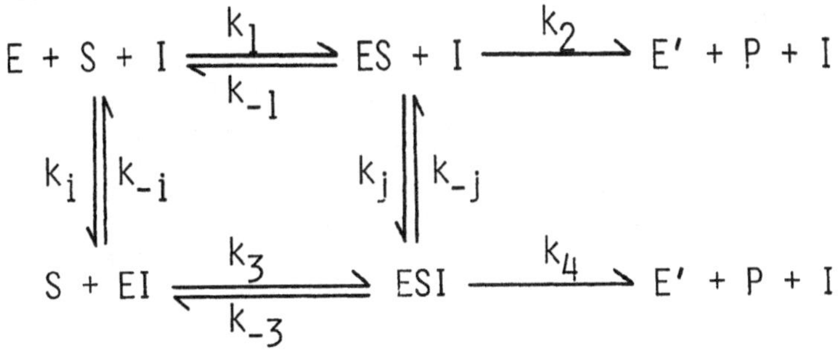

$$E + S + I \underset{k_{-1}}{\overset{k_1}{\rightleftharpoons}} ES + I \overset{k_2}{\longrightarrow} E' + P + I$$

$$k_i \updownarrow k_{-i} \qquad\qquad k_j \updownarrow k_{-j}$$

$$S + EI \underset{k_{-3}}{\overset{k_3}{\rightleftharpoons}} ESI \overset{k_4}{\longrightarrow} E' + P + I$$

E(β-CD-NAH) and S(dyes) reversibly form an aggregate of ES(complex) to produce E'(β-CD-NA) and P(product). When I(cyclohexanol) is concerned in the reaction, it is considered to form complex EI and ESI with ES.
Overall reaction rate : $k_{cat} = (1-X)k_2 + Xk_4$
Where, X means partial ratio of reaction rate occuring by adding cyclohexanol, and depends on the concentration of

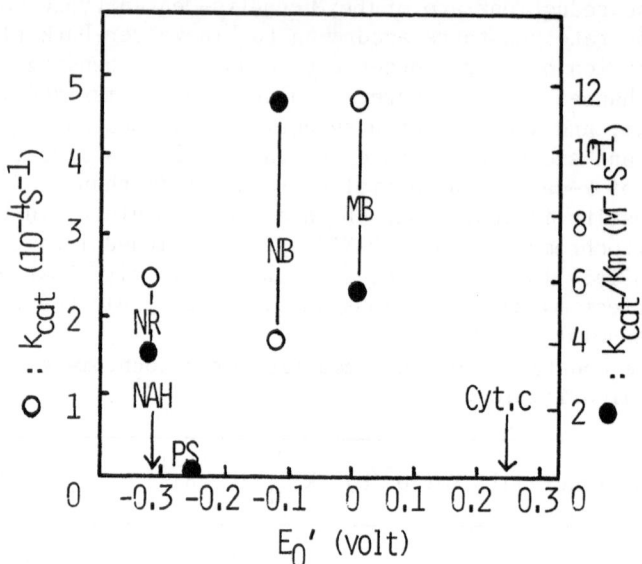

Figure 4. Relations of catalytic rate constants and redox potential in the reduction of cytochrome c mediated by redox dyes with β-CD-dihydronicotinamide.

cyclohexanol. Therefore, $X = [I] / (k_{-j}/k_j + [I])$
without cyclohexanol, $X = 0$ $k_{cat} = k_2$.
When cyclohexanol was added, $X = 0$ $k_{cat}(obs) = k_2$
Therefore, $k_2 = k_4$.
This assumption indicates that β-CD-NAH forms complex of identical structure with dyes in the absence or presence of cyclohexanol.

The relations of catalytic rate constants with redox potential of dyes are given in Figure 4. The reduction of cytochrome c was accelerated even in the presence of NR which has higher redox potential than β-CD-NAH, while the reduction mediated by PS has extraordinarily lower activity than that expected from its redox potential. These results make clear that the reduction mediated by redox dyes in the presence of β-CD-NAH depends upon the other elements, not redox potential.

4. CONCLUSION

The most important step in the reduction of cytochrome c mediated by various redox dyes was ascertained to be the complex formation between β-CD-NAH and each dye. When there was sufficient complex formation between β-CD-NAH and a mediator then the reduction of cytochrome c could be enhanced greatly.

That is, these nonenzymatic cytochrome c reduction systems can be

regarded as an artificial respiratory electron transport chain *in vitro*.
We are grateful to Nihon Shokuhin Kako Co., LTD. for providing us with sample of β-cyclodextrin.

5. ACKNOWLEDGMENT

This work was supported by the Grant-In-Aid for the Special Project Research on the Properties of Molecular Assemblies from the Ministry of Education, Science and Culture.

6. REFERENCES

[1] M.L.Bender and M.Komiyama, *Cyclodextrin Chemistry*, Springer-Verlag (1978); F.Cramer and F.M.Henglein, *Quart.Rev.(London)*, 68, 649, (1956)

[2] R.Breslow, P.Bovy, and C.L.Hersh, *J.Am.Chem.Soc.*, 102, 2115, (1980); I.Tabushi, *Tetrahedron*, 40, 269, (1984); C.Sirlin, *Bull.Soc.Chim.Fr.*, II-5, (1984); T.Ikeda, R.Kojin, H.Ikeda, C-J.Yoon, M.Iijima, K.Hattori, and F.Toda, *J.Incl.Phenom.*, 2, 669, (1984).

[3] Y.Iwakura, K.Uno, F.Toda, S.Onozuka, K.Hattori, and M.L.Bender, *J.Am.Chem.Soc.*, 97, 4432, (1975).

[4] C-J.Yoon, H.Ikeda, R.Kojin, T.Ikeda, and F.Toda, *J.Chem.Soc.,Chem.Commun.*, 1080, (1986).

[5] S.Onozuka, M.Kojima, K.Hattori, and F.Toda, *Bull.Chem.Soc.Jpn.*, 53, 3221, (1980).

[6] A.Ueno and R.Breslow, *Tetrahedron Lett.*, 23, 3451, (1982); K.Takahashi, K.Hattori, and F.Toda, *Tetrahedron Lett.*, 25, 3331, (1984).

[7] L.J.Hynes and A.R.Todd, *J.Chem.Soc.*, 303, (1950).

CATALYTIC ACTIVITY OF β-CYCLODEXTRIN-HISTAMINE

Tsukasa Ikeda, Ryoichi Kojin, Chul-joong Yoon, Hiroshi Ikeda, Masao Iijima and Fujio Toda
Department of Bioengineering and Science,
Faculty of Engineering, Tokyo Institute of Technology
O-okayama, Meguro-ku, Tokyo 152, Japan

ABSTRACT. We modified cyclodextrin (CD) by a histamine group to make a model of α-chymotrypsin. Enzymatic turnover reaction was realized with CD-histamine at around neutral pH value. Compared with amino-CD, it is ascertained that this catalytic activity of CD-histamine is caused by an imidazole group. Using several substrates in the hydrolytic reactions, it shows that CD-histamine has a structural selectivity for substrates which are structurally different to each other.

1. INTRODUCTION

CD has a hydrophobic cavity which acts like a binding site of an actual enzyme. To attach a reactive functional group, we modified β-CD by a histamine group and we realized enzymatic turnover reaction in the hydrolysis of p-nitrophenyl acetate with β-CD-histamine at around neutral pH value. Catalytic rate constant of this reaction was close to an actual enzyme, α-chymotrypsin [1].

We thought that this catalytic activity was caused by an imidazole group. But β-CD-histamine has also an amino group which has a possibility to act as an active site.

This paper reports that we synthesized amino-β-CD to compare catalytic activity with β-CD-histamine and it was clear that an imidazole group acted as an active site in the hydrolysis of p-nitrophenyl acetate. And it also shows that in the hydrolytic reactions, β-CD-histamine has a structural selectivity for substrates which are structurally different to each other.

2. MATERIALS AND METHODS

C-6 mono-tosylated-β-CD and β-CD-histamine used in this work were prepared as reported previously [1]. C-6 mono-amino-β-CD was prepared as follows. A solution of C-6 mono-tosylated-β-CD (10g) and sodium azide (1g) in DMF was treated for 70 min. at 80°C. The reaction mixture was evaporated to dryness. The dried material (crude C-6 mono-azide-β-CD) was dissolved in water and reduced by sodium borohydride (0.5g). After the reaction, the reaction mixture was evaporated to dryness to give crude mono-amino-β-CD. Mono-amino-β-CD was purified by cation exchange chromatography with a CM-Sephadex C-25 column. The purity of the product was confirmed by TLC (Rf:0.11, solvent:butanone-1M acetic acid-methanol, 2:5:3, detecting reagent:ninhydrin). Yield; 8%. Anal.Calc.for $C_{42}H_{71}O_{34}N_1$ $2H_2O$: C,43.1; H,6.5; N,1.2. Found: C,42.9; H,6.3; N,1.3.

All reagents for synthesis and hydrolytic reactions were purchased from commercial suppliers and were used without further purification.

The hydrolytic reactions were followed by the appearance of p-nitrophenol spectrometrically at 400nm using a HITACHI model 220A. The reaction medium was pH 7.2 phosphate buffer and kept at 25°C.

3. RESULTS AND DISCUSSION

CD-histamine has an imidazole group and a secondary amino group (Figure 1). It is thought that the active site in the hydrolysis of p-nitrophenyl acetate is an imidazole group which has a high activity at around neutral pH value. But a secondary amino group has a possibility to act as a catalytic site.

Figure 1.   β-CD-histamine   amino-β-CD

[Scheme 1]

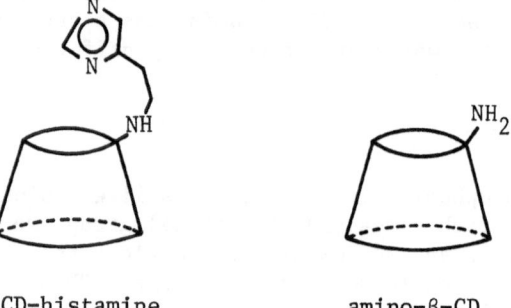

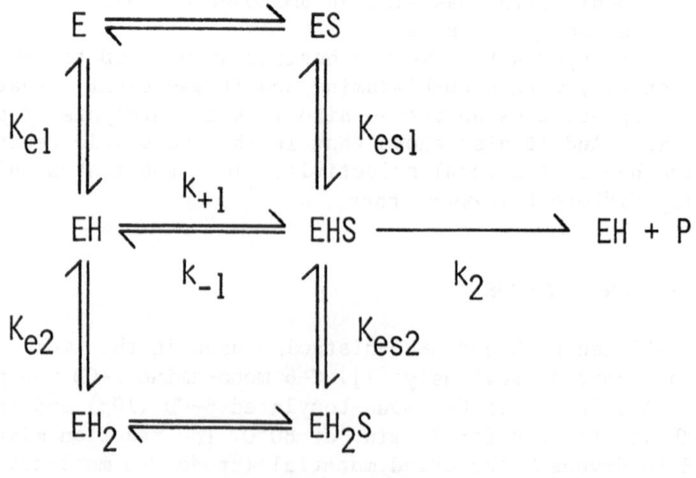

Scheme 1 shows a reaction scheme in the hydrolysis of p-nitrophenyl acetate by β-CD-histamine considering that an imidazole group acts as the active site. Ke1 and Ke2 are disociation constants for hydrogen ion of an imidazole group of β-CD-histamine. Kes1 and Kes2 are dissociation constants for hydrogen ion of complex of β-CD-histamine and substrate (p-nitrophenyl acetate).

The rate equation of this reaction scheme is shown in scheme 2. Km is the apparent Michaelis constant including k1 and k2. This scheme shows that the reaction rate vary with pH and that we can obtain dissociation constants (pKe2,pKes2) by measurement of the rate constants.[2]

[Scheme 2]

$$V = \frac{k_2[E]_0[S]}{K_m(1+[H]/K_{e1}+K_{e2}/[H])+[S](1+[H]/K_{es1}+K_{es2}/[H])}$$

Figure 2 shows the pH dependence of kcat / Km in the hydrolysis of p-nitrophenyl acetate with β-CD-histamine. In the previous report [1], pH dependence of kcat was already described. Both curved lines are in good agreement with the rate equation. From these two graphs pKe2 and pKes2 are obtained and shown in table 1.

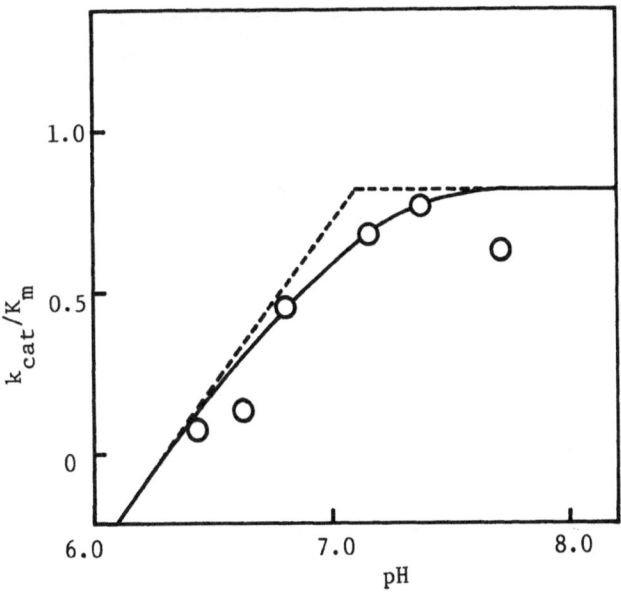

Figure 2. pH dependence of k_{cat}/K_m

Table 1 shows that pKe2, pKes2 and pKa (obtained by titration with dilute hydrochloric acid) are all at around 7. These values correspond to the dissociation constant of an imidazole group. The reason why pKa is lower than pKe2 or pKes2 is thought as follows. By forming a complex with p-nitrophenyl acetate and β-CD-histamine, intramolecular interaction with an imidazole group and CD is weakened. And then pK becomes a little bit large.

Table 1. The negative logarithms of the dissociation constant

β-CD-hiatamine	pK_{e2}	7.1
	pK_{es2}	7.2
	pK_a	6.8
Histamine	pK_a	6.14*

* Lange's handbook of chemistry

Catalytic activity of amino-β-CD was obtained by hydrolysis of p-nitrophenyl acetate in the same manner as β-CD-histamine. Figure 3 shows a Linewewver-Burk plot of this result. The data give a straight line, so this reaction is also a type of Michaelis-Menten mechanism.

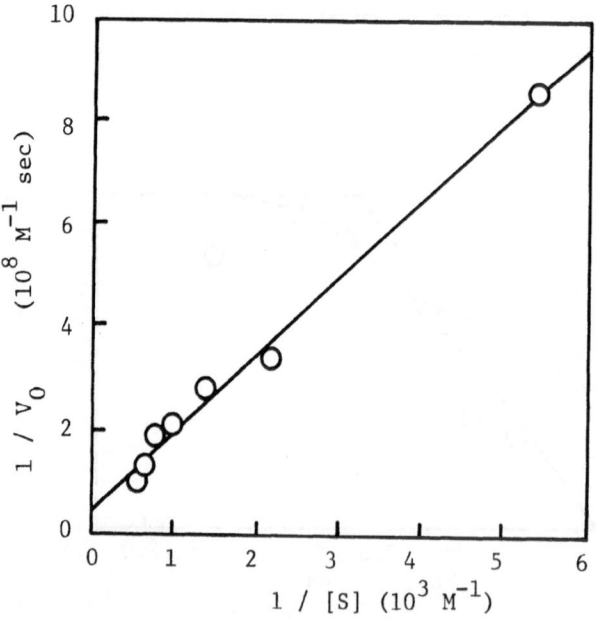

Figure 3. Lineweaver-Burk plot of hydrolysis of p-nitrophenyl acetate with amino-β-CD

The pKa value of amino-β-CD obtained by titration with dilute hydrochloric acid and back titration with dilute sodium hydroxide solution was 6.3. It is thought that this lower pKa value for the amino compound is due to the effect of the hydrophobic cavity and hydroxyl groups of β-CD. Table 2 shows catalytic rate constants in the hydrolysis of p-nitrophenyl acetate. Compared with β-CD-histamine, the value of k_{cat}/K_m of amino-β-CD is about 1/20. The catalytic activity of amino-β-CD is too low to explain that the catalytic activity of β-CD-histamine depends on its amino group.

These results suggest that it is an imidazole group, not an amino group, acts as the active site of β-CD-histamine in the hydrolysis of p-nitrophenyl acetate.

Table 2. Catalytic rate constants in the hydrolysis of p-nitrophenyl acetate.

Catalyst	k_{cat} (10^{-4} sec^{-1})	K_m (10^{-3} M)	k_{cat}/K_m (M^{-1} sec^{-1})
amino-β-CD	1.29	3.60	0.04
β-CD	≈ 0	-	-
β-CD-histamine	21.9	2.84	0.77

To confirm a selectivity of β-CD-histamine for substrates, experiments of the hydrolysis with β-CD-histamine were made using several kinds of substrates as follows; N-tert-butoxycarbonyl-L-glutamine-p-nitrophenyl ester (Boc-Gln-ONP), Boc-L-asparagine-p-nitrophenyl ester (Boc-Asn-ONP), Boc-Glycine-p-nitrophenyl ester (Boc-Gly-ONP), Boc-L-Leucine-p-nitrophenyl ester (Boc-Leu-ONP). In these cases, all hydrolytic reactions with β-CD-histamine were detected as a reaction type of Michaelis-Menten mechanism. Catalytic rate constants in the hydrolysis with β-CD-histamine are shown in Table 3. Compared with Boc-Gln-ONP and Boc-Asn-ONP, the structural difference is only one carbon in the side chain of amino acid. But reactivity (k_{cat}/K_m) is different about three times. In these four substrates, in the case of Boc-Asn-ONP, the re-

Table 3. Catalytic rate constants in the hydrolysis with β-CD-histamine

Substrates	k_{cat} (10^{-2} sec^{-1})	K_m (10^{-3} M)	k_{cat}/K_m (M^{-1} sec^{-1})
Boc-Gln-ONP	0.323	0.47	6.96
Boc-Asn-ONP	1.607	0.82	19.54
Boc-Gly-ONP	0.227	0.24	8.94
Boc-Leu-ONP	0.037	1.14	0.33

activity is the highest and in the case of Boc-Lue-ONP, it is the lowest. It is thought that these differences of activities depend on the fittness to form a complex and also the structural configuration between the substrate and the active site of β-CD-histamine when the complex is formed. At present, only the data by spectroscopic analysis are taken, but more information about the structural selectivity of β-CD-histamine for substrates will be available from X-ray analysis and NMR analysis. Such experiments are now in progress. These studies will be applied to design new artificial enzymes.

4. ACKNOWLEDGMENT

We are grateful to Nihon Shokuhin Kako Co., LTD. for providing us with a sample of β-CD.
This work was supported by the Grant-In-Aid for the special Project Research on the Properties of Molecular Assemblies from the Ministry of Education, Science and Culture.

5. REFERENCES

[1] T.Ikeda, R.Kojin, C-j.Yoon, H.Ikeda, M.Iijima, K.Hattori and F.Toda, *J.Incl.Phenom.*, $\underline{2}$, 669, (1984)
[2] E.Zeffren and P.L.Hall, *The Study of Enzyme Mechanisms*, John Wiley & Sons,inc., New York, (1973)

MODIFICATIONS OF BENZOXAZOLE RING SUBSTITUENTS IN A.23187 (CALCIMYCIN). EFFECT ON CATION CARRIER PROPERTIES.

M. Prudhomme, G. Dauphin, J. Guyot and G. Jeminet
Université de Clermont II, Laboratoire de Chimie
Organique Biologique, U.A. 485 du CNRS, BP 45
63170 AUBIERE, FRANCE
N. Gresh
Institut de Biologie Physico-Chimique, Laboratoire de
Biochimie Théorique associé au CNRS, 13 rue Pierre et
Marie Curie, 75005 PARIS, FRANCE

EXTENDED ABSTRACT

Introduction. Calcimycin (or A.23187) belongs to the growing family of bacterial carboxylic polyether ionophores(1). It was isolated from a strain of Streptomyces Chartreusis NRRL 3882 (2).

This ionophore is able to carry selectively alkaline-earth cations through membrane phases by an antiport mechanism ($M^{++} \Leftrightarrow 2H^+$); with a transport efficiency in favour of Ca^{++} versus Mg^{++} in biological membranes(3). Owing to its specificity, calcimycin is universally used as a tool to investigate the role of the second messenger calcium in physiological processes.

However, the different steps of the transport are not well understood, not what make this molecule so efficient. Our purpose was to get information, especially as regards to the formation of the $(A.23187)_2 \cdot M$ lipophilic complex which diffuses through the membrane and its dissociation at the interface, via a chemico-structural approach.

Solid state studies for 2:1 complexes revealed that magnesium (see below)(6) and calcium (7) coordination spheres are different. Nevertheless, in both cases the benzoxazole moiety appears to play a key role in the specific complexation of the cationic guest. In a first set of experiments we carried out chemical modifications of this part of the molecule obtained by fermentation.

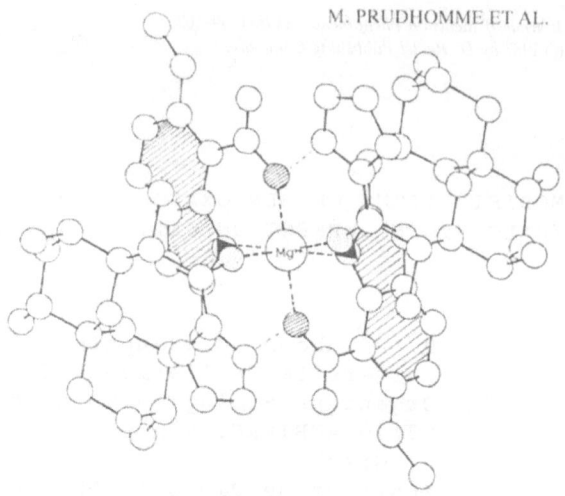

Calcimycin X = NHCH$_3$ Y = Z = CH$_3$
microbial analogues recently isolated:
X 14885A X = OH Y = CH$_3$ Z = H (4)
CP 61405 X = OH Y = Z = H (5)

Representation of the
(A 23187)$_2^{--}$. Mg++ complex
from X-ray data (6).
(▶, ⊘, ⊙, coordination sites)

<u>Semi-synthesis</u>. From a selective cleavage of the oxazole ring we have worked out a semi synthetic approach in several steps which provided suitably designed analogues, closely related with respect to their overall stereochemistry, with only the benzoxazole ring substituents modified (8).

<u>Cation carrier properties</u>. Calcimycin has been studied in detail using the biphasic model system water/toluene-butanol, 70:30 (9); accordingly we chose this system for the sake of consistency, to get an overall characterization of the analogues with regard to their cation extraction and liberation abilities.

From the physico-chemical results, the prominent role of the 2-carboxyl benzoxazole sequence in the formation and stability of complexes was emerging.

Further, electron-donating substituents H-bonded with the -COOH group brought an additional stabilizing factor for the 2:1 associations.

"ab initio" computations carried out independently on models confirmed that binding energies were very sensitive to the nature of the substituents(10).

Bulky substituents, hindering the carboxylic group, strongly disturbed the coordination sphere and destabilized the complexes.

A difference was observed between calcium and magnesium essentially for the initial rates of decomplexation giving in all cases $Ca^{++} \gg Mg^{++}$ (8).

Conformation. The preferential conformations of the analogues, in the acid form, were investigated by ^{1}H NMR(1D and 2D, 300MHz); a common behaviour was noted. In $CDCl_3$, a compact globular form was postulated with a head-to-tail intramolecular chelation, similar to that one depicted for calcimycin by X-ray analysis (2). In CD_3OD, the C_8-C_9-C_{10} part was affected, corresponding to a facilited rotation of the benzoxazole ring, this conformational change could be also induced at the water/membrane interface. In both solvents, H18-H19 stayed strictly antiperiplanar suggesting a more rigid C18-C19-C20 arm. In order to examine the possible role of the 1-3 interaction between Me17 and Me19 for the preferential conformation observed, we have undertaken the synthesis of a biomimetic model with free rotating arms. Work is in progress.

In conclusion, calcimycin which is a small molecule compared to proteinic systems for instance, appears to be precisely biosynthetized for the recognition of calcium and magnesium by a specific molecular design and constitutes a model which is different from the multidentate macrocycles under investigation at the moment.

REFERENCES

1- J.W. Westly: Notation and classification.(Polyether antibiotics, naturally occuring ionophores V-1, Ed. J.W. Westly), pp. 1-20, Marcel Dekker(1982)
2- M.O. Chaney, P.V. Demarco, N.D. Jones and J.L. Occolowitz: J. Am. Chem. Soc. 96, 1932(1974).
3- M. Prudhomme, J. Guyot and G. Jeminet: J. antibiotics. 39, 934(1986)
4- J.W. Westley, C.M. Liu, J.F. Blount, L.H. Sello, N. Troupe and P.A. Miller: J. antibiotics. 36, 1275(1983)
5- W.D. Celmer, W.P. Cullen, H. Madea and J. Tone: U.S. Patent 4,547,523(1985)
6- M. Alléaume and Y. Barrans: Can. J. Chem 63, 3482(1985)
7- G.D. Smith and W.L. Duax: J. Am. Chem. Soc 98, 1578(1976)
8- M. Prudhomme, G. Dauphin and G. Jeminet: J. antibiotics. 39 922(1986)
9- D.R. Pfeiffer and H.A. Lardy: Biochem. 15, 935(1976)
10- N. Gresh: Nouv. J. Chim. 10, 201(1986)

Journal of Inclusion Phenomena 5 (1987), 103–108.
© 1987 *by D. Reidel Publishing Company.*

ION TRANSPORT, ION EXTRACTION, AND ION BINDING BY SYNTHETIC CYCLIC OCTAPEPTIDE

Toshimi Shimizu, Yoshio Tanaka, and Keishiro Tsuda
Research Institute for Polymers and Textiles, 1-1-4 Yatabe-Higashi, Tsukuba, Ibaraki 305, Japan

ABSTRACT. Cyclic octapeptide, cyclo[Gly-L-Lys(Z)-Sar-L-Pro]$_2$ (CGLSP2) was synthesized as an ionophore model. Its ion-transport ability through a chloroform membrane was investigated in connection with ion extractability (K_{ex}) and conformational properties. CGLSP2 transported the picrate salts of Ba^{2+} and Ca^{2+} efficiently. The K_{ex} sequences were $Ba^{2+} > Ca^{2+} \gg Mg^{2+}$ and $K^+ > Rb^+ > Na^+$, showing good agreement with the selectivity in ion transport. In addition, cation-binding properties of CGLSP2 to alkali and alkaline earth metal ion were investigated in acetonitrile by CD and NMR spectroscopy. Titration curves obtained from CD data revealed three kinds of CGLSP2/cation complexes. The values of 1:1 complex-formation constants (K_1) decreased in the order $Ba^{2+} > Ca^{2+} > Mg^{2+} > Li^+ \gg Na^+ \sim K^+$. ^{1}H- and ^{13}C-NMR data showed that free CGLSP2 exists in at least five different conformational states in acetonitrile. After the addition of equimolar amounts of Ba(ClO$_4$)$_2$, these conformations converged into a single C$_2$-symmetric conformation with all-trans peptide bonds.

1. INTRODUCTION

Transmembrane ion transport by cyclic depsipeptide, macrotetrolide actins, and polyether antibiotics is based on "a mobile carrier" [1]. Only few studies, however, have been carried out with regard to carrier-mediated ion-transport by linear or cyclic peptides [2,3]. In comparison, cation-binding properties by synthetic cyclic peptides have been studied extensively [4-6]. We have also synthesized cation-binding cyclic peptides with many N-substituted amino acid residues [7-10]. A new cyclic octapeptide, cyclo[Gly-L-Lys(Z)-Sar-L-Pro]$_2$ (CGLSP2) was synthesized as an ionphore model [11]. It includes two sets of two successive N-substituted peptide bonds, i.e. Lys-Sar and Sar-Pro bonds. This arrangement of peptide bonds is different from those of synthetic cyclic octapeptides studied so far [7,12,13]. Therefore, CGLSP2 is expected to have a ring skeleton that shows a fine balance between flexibility and rigidity. This paper describes an ion transport across chloroform liquid membrane by this cyclic octapeptide. In addition, ion-transport capacity of CGLSP2 is discussed in connection with ion extractability and conformational change at the ion binding.

2. EXPERIMENTAL

2.1. Synthesis of cyclic octapeptide and related peptides

The preparation and physical characteristic of the cyclic octapeptide, CGLSP2 and related peptides are reported in our preceding paper [11].

2.2. Method of ion transport

Transport of metal picrate across a chloroform liquid membrane was examined in a U-shaped tube at 25°. The source aqueous phase (pH 7.2) contains a metal chloride and picric acid. The alkali metal cations used were Na^+, K^+, and Rb^+, and the alkaline earth metal cations Mg^{2+}, Ca^{2+}, and Ba^{2+}. The chloroform phase contains an ion carrier. The second aqueous phase was adjusted to pH 7.2. Linear octapeptides Boc-[Gly-L-Lys(Z)-Sar-L-Pro]$_2$-OH (LGLSP2-OH) and Boc-[Gly-L-Lys(Z)-Sar-L-Pro]$_2$-OCH$_3$ (LGLSP2-OM), cyclic tetrapeptide c-[Gly-L-Lys(Z)-Sar-L-Pro] (CGLSP1), and linear tetrapeptide Boc-Gly-L-Lys(Z)-Sar-L-Pro-OH (LGLSP1-OH) were used as related petidic carriers. In addition, valinomycin and dibenzo-18-crown-6 were employed as a K^+ ionophore. The amount of cation transported to the second aqueous phase through membrane was determined spectrophotometrically at 355 nm.

2.3. Procedure of ion extraction

Equal volumes of a chloroform solution containing CGLSP2 and an aqueous solution (pH 7.2) containing a metal picrate were stirred vigorously for 30 min. The mixture was allowed to stand at 25° for 1 h. The absorption of picrate anion in chloroform phase was then determined at 410 nm at 25°. Although a similar extraction was performed with chloroform free from ion carrier, no change of absorption was observed in the chloroform phase.

2.4. Measurement

^{1}H-NMR spectra were obtained at 360 MHz on a NICOLET NT-360 spectrometer with a NIC 1180 Computer Data System. ^{13}C-NMR spectra were recorded on a Varian CFT-20 spectrometer at 20 MHz. CD spectra were recorded on a JASCO J-40A automatic recording spectropolarimeter.

3. RESULTS AND DISCUSSION

3.1. Cation transport across a chloroform liquid membrane

The cumulative amount of transported cation after 10 h by various ion carriers is summarized in Table I. CGLSP2 transported 10.6 μmol Ba^{2+} or 5.5 μmol Ca^{2+} for 10 h. Between cations having similar ionic

Table I Micromoles of cation transported after 10 h

Carrier (diameter, Å)	Na^+ (1.90)	K^+ (2.66)	Rb^+ (2.96)	Mg^{2+} (1.30)	Ca^{2+} (1.98)	Ba^{2+} (2.70)
CGLSP2	0.30	0.54	0.47	0.28	5.5	10.6
LGLSP2-OH	--	0.15	--	0.27	2.6	5.5
LGLSP2-OM	--	0.17	--	--	0.26	1.25
CGLSP1	--	0.14	--	--	--	0.80
LGLSP1-OH	--	0.13	--	--	--	0.10
Valinomycin	--	3.9	--	--	--	0.41
Dibenzo-18-crown-6	9.8	--	--	--	--	0.18

Phase I : [metal chloride] = 10 mM, [picric acid] = 25 mM, [HEPES] = 10 mM (pH 7.40).
Chloroform Phase : [carrier] = 2.00 X 10^{-4} M.
Phase II : [HEPES] = 10 mM (pH 7.40).

diameter, Ba^{2+} and K^+, the amount of Ba^{2+} transported is about 20 times larger than that of K^+. A similar trend can be seen between Ca^{2+} and Na^+. In addition, CGLSP2 can transport Ba^{2+} at least 2-fold more than LGLSP2-OH, LGLSP2-OM, and CGLSP1. This finding suggests that the number of residues and the cyclic structure is an important factor when designing a peptidic ion carrier. In this way, CGLSP2 was found to be most selective for Ba^{2+}. However, among the physiological cations, it was most selective for Ca^{2+}.

3.2. Cation extraction

Extraction equilibrium constants of CGLSP2 with metal picrates were obtained and shown in Table II. It was confirmed before that CGLSP2 forms a 1:1 complex with cation studied. Consequently, it was found that CGLSP2 can extract Ba^{2+} and Ca^{2+} more efficiently than Mg^{2+}, K^+, Rb^+, and Na^+. The order of the extractability showed good agreement with the selectivity in ion-transport ability.

Table II Extraction equilibrium constants (K_{ex}) of CGLSP2

Cation	K_{ex}	
Na^+	2.28 X 10^2	
K^+	4.73 X 10^2	(M^{-2})
Rb^+	3.65 X 10^2	
Mg^{2+}	2.65 X 10^3	
Ca^{2+}	4.04 X 10^5	(M^{-3})
Ba^{2+}	8.88 X 10^5	

Aqueous Phase : [metal chloride] = 10 mM, [picric acid] = 25 mM, [HEPES] = 10 mM (pH 7.40).
Chloroform Phase : [CGLSP2] = 1.40~7.00 X 10^{-4} M.

This implies that ion transport by CGLSP2 is controlled by ion extraction equilibrium between the source aqueous phase and the chloroform liquid membrane phase.

3.3. Cation binding in acetonitrile

In confirmation of the complexation properties of CGLSP2 in solution, cation-binding studies were accomplished in acetonitrile solution. By analyzing the titration curves obtained from CD spectra [14], 1:1 complex-formation constants (K_1) were evaluated and shown in Table III. In addition, the rate constant of 1:1 complex-formation constants (k_f) was evaluated and also shown in Table III. The K_1 values decrease in the order $Ba^{2+} > Ca^{2+} > Mg^{2+} > Li^+ >> Na^+ \sim K^+$. Doubly charged cations are bound preferentially over monovalent cations. The affinity of Ba^{2+} for CGLSP2 is about 10-fold smaller than that of K^+ for valinomycin. In addition, the Ca^{2+} binding of CGLSP2 is about 10-fold smaller than that of antamanide which is naturally occurring cyclic decapeptide. Between cations having similar ion diameter, the ratio of the binding constants $K_1(Ba^{2+})/K_1(K^+)$ is approximate 10^4. Similar trend can be seen between Ca^{2+} and Na^+. The rate constant (k_f) of 1:1 complex-formation between CGLSP2 and Ba^{2+} is at least 300-fold larger than that reported for complex formation between c-[L-Lys(Z)-L-Pro]$_4$ and Ba^{2+} [6]. However, it is lower than that reported for valinomycin/K^+ complex [15].

Conformational change of the cyclic backbone of CGLSP2 upon complexation was investigated by NMR spectroscopy [14]. ^{1}H- and ^{13}C-NMR data showed that free CGLSP2 in acetonitrile exists in at least five different conformational states. This feature is ascribed to a cis-trans isomerism around

Table III 1:1 complex-formation constants (K_1) and rate constants (k_f) of various compounds at 25° in acetonitrile

Compound/Cation	K_1 (M^{-1})	k_f (M^{-1}min^{-1})
CGLSP2/Ba^{2+}	2.7 X 10^4	>4 X 10^3
CGLSP2/Ca^{2+}	7.6 X 10^3	--
CGLSP2/Mg^{2+}	5.6 X 10^3	--
CGLSP2/Li$^+$	2.4 X 10^3	--
CGLSP2/Na$^+$	very low	--
CGLSP2/K$^+$	very low	--
c-[Lys(Z)-Sar]$_4$/Ba^{2+}	1.3 X 10^3(*)	12(*)
c-(Gly-Pro)$_4$/Ba^{2+}	1.6 X 10^6	--
Antamanide/Ca^{2+}	1 X 10^5	--
Valinomycin/K$^+$	3 X 10^5	2.1 X 10^9(**)
Nonactin/Na$^+$	3.8 X 10^2(**)	~1.2 X 10^{10}(**)

(*) in 95% C_2H_5OH (**) in CH_3OH

CYCLIC OCTAPEPTIDE AS AN IONOPHORE MODEL

two sets of two successive N-substituted peptide bonds. A predominant conformer has a C_2-symmetric structure containing two cis Lys-Sar peptide bonds. After the addition of equimolar amounts of $Ba(ClO_4)_2$, those conformations converged into a single C_2-symmetric one with all-trans peptide bonds. This corresponds to a 1:1 complex species. On further addition of $Ba(ClO_4)_2$, CGLSP2 changed the conformation into an asymmetric structure with one cis Lys-Sar peptide bond. This corresponds to a 1:2 complex species. Similar results were obtained for the Ca^{2+}-binding by CGLSP2 in acetonitrile. From the above, the following binding scheme may be proposed, as shown in Fig. 1.

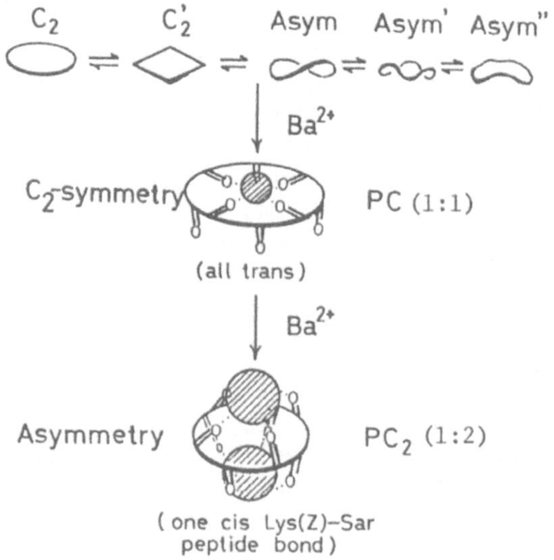

Fig. 1. Scheme for the complexation of CGLSP2 with Ba^{2+}. C_2 and C_2' represent distinct C_2-symmetric conformers. Asym, Asym', and Asym'' represent asymmetric conformers.

Therefore, the slow process of the complex formation between CGLSP2 and Ba^{2+} may be related to a cis-trans isomerization around peptide bonds [16].

Figure 2 shows a proposed conformation of 1:1 complex between CGLSP2 and Ba^{2+}. The Ba^{2+} or Ca^{2+} cation may be bound favorably by the carbonyl groups of the two Sar and the two Gly residues. Here, the

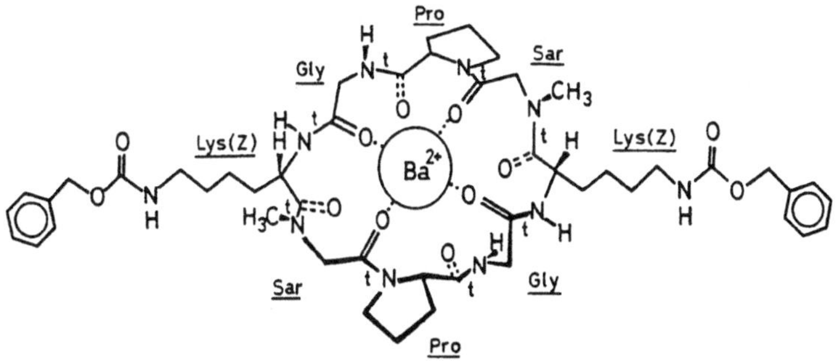

Fig. 2. Schematic representation of a proposed conformation of the 1:1 complex between CGLSP2 and Ba^{2+}.

doubly charged cations enhances the stability of 1:1 complex more effectively than the fitness of the cation size to cavity. However, the Ca^{2+}/Mg^{2+} selectivity of CGLSP2 in the extraction equilibrium disagrees with that of complex formation. It would be attributed to the different solubility of the CGLSP2/cation/picrate-anion into the chloroform phase. Therefore, cation selectivity in transport by CGLSP2 should be closely related to the net solubility [17] of the various cation picrate/CGLSP2 complexes in chloroform, as well as the size and charge number of cation and the cavity size of peptide.

4. REFERENCES

[1] Y.A. Ovchinnikov, V.T. Ivanov, and A.M. Shkrob, 'Membrane Active Complexones', Vol. 12, Elsevier, New York, 1974.
[2] G. Spach, Y. Trudelle, and F. Heitz, *Biopolymers*, 22, 403 (1983).
[3] C.M. Deber, *Can. J. Biochem.*, 58, 865 (1980).
[4] L.V. Sumskaya, T.A. Balashova, I.I. Mikhaleva, T.S. Chumburidze, E.I. Melnik, V.T. Ivanov, and Y.A. Ovchinnikov, *Bioorg. Khim.*, 3, 5 (1977).
[5] E.R. Blout, *Biopolymers*, 20, 1901 (1981).
[6] S. Kimura and Y. Imanishi, *Biopolymers*, 22, 2383 (1983).
[7] T. Shimizu and S. Fujishige, *Biopolymers*, 19, 2247 (1980).
[8] T. Shimizu, Y. Tanaka, and K. Tsuda, *Bull. Chem. Soc. Jpn.*, 55, 3817 (1982).
[9] T. Shimizu, Y. Tanaka, and K. Tsuda, *Int. J. Biol. Macromol.*, 5, 179 (1983).
[10] T. Shimizu, Y. Tanaka, and K. Tsuda, *Int. J. Peptide Protein Res.*, 22, 194 (1983).
[11] T. Shimizu, Y. Tanaka, and K. Tsuda, *Int. J. Peptide Protein Res.*, 27, 344 (1986).
[12] C.M. Deber and P.D. Adawadkar, *Biopolymers*, 18, 2375 (1979).
[13] S. Kimura and Y. Imanishi, *Biopolymers*, 23, 563 (1984).
[14] T. Shimizu, Y. Tanaka, and K. Tsuda, *Bull. Chem. Soc. Jpn.*, 58, 3436 (1985).
[15] Th. Funk, F. Eggers, and E. Grell, *Chimia*, 26, 637 (1972).
[16] L.N. Lin and J.F. Brandts, *Biochemistry*, 19, 3055 (1980).
[17] F. Vögtle and E. Weber, 'Host Guest Complex Chemistry III', p 1, Springer-Verlag, Berlin, 1984.

SYNTHETIC ANALOGS OF PEPTIDE-BINDING ANTIBIOTICS

Nalin Pant, Michael Mann, and Andrew D. Hamilton[*]

Department of Chemistry, Princeton University

Princeton N.J.   08544   USA

The vancomycin family of antibiotics offers an attractive target in synthetic host-guest chemistry. Vancomycin (1) has been shown by ^{1}H nmr to form discreet complexes with the -D-Ala-D-Ala fragment found in bacterial cell walls. The structure of the active complex (2) involves six hydrogen bonds between the antibiotic and dipeptide substrate. The complexation is strongly substrate- and stereoselective due to the concave nature of the cavity formed by the biphenyl and triphenyl diether components. Our interest lies in designing synthetic analogs of vancomycin both as novel receptors for peptide carboxylate substrates and as potentially interesting antibiotics.

In simplifying the complex structure of (1) we are eliminating all features that do not play a direct role in binding (e.g. carbohydrate and benzylic hydroxyl groups). We are also, in our early models, focusing on the right hand portion of (1) which forms, in (2), five of the six hydrogen bonds to the substrate.

Our first synthetic analog (3) contains the three amide bonds involved in H-bonding to the carboxylate terminus as well as the key diphenyl ether functionality. The synthesis of (3) shown in scheme 1, involves a multistep sequence from 3-cyanophenol and D-(or L-) dinitrotyrosine. The ^{1}H nmr of (3) shows the 2-H proton resonance of the benzylamine ring to be shifted upfield to 5.85 ppm. This is very similar to the resonance of the equivalent proton in vancomycin which occurs at 5.65 ppm and is due to the cyclic peptide constraining the 2-H proton to lie under the adjacent benzene ring. Despite the spectroscopic, and thus probable conformational, similarity between (3) and the right hand ring in (1) no complexation between (3) and carboxylate substrates was detected. This is most probably due to the absence, in (3), of the key N-terminal ammonium group of vancomycin which has been shown to play, via electrostatic interactions, an important role in carboxylate binding.

Our second model compound (4) incorporates the N-terminal ammonium group as a dimethylglycine residue. Inspection of CPK models suggested that the addition of a methyl group to the N-terminus would not hinder the approach of a substrate or the formation of the complex. However the use of a tertiary amine considerably simplifies the synthetic strategy to (4). This is shown is scheme 2 and involves preliminary protection of dinitrotyrosine as its t-butoxycarbonyl derivative (5) followed by diphenyl ether formation and deprotection to amine (6). Addition of dimethylglycine acid chloride to (6) formed the tripeptide (7) which could be deprotected and cyclized, as with (3), to form analog (4). Preliminary studies of the interaction of the hydrochloride salt of (4) with carboxylate substrates indicate ^{1}H nmr behavior very similar to that of vancomycin, particularly involving loss of a single amide resonance and formation of a new resonance at 11.9 ppm. Further details of this study as well as the synthesis of (3) and (4) were presented in Lancaster.

References

1. M.P. Williamson and D.H. Williams, J. Am. Chem. Soc. 1981, 103, 6580.
   M.P. Williamson, D.H. Williams, and S.J. Hammond, Tetrahedron, 1984, 40,

MOLECULAR GRAPHICS IN THE STUDY OF THE CALCIUM-BINDING SITES OF CARP PARVALBUMIN AND OTHER PROTEINS

J.C. Lockhart and H. Grey,
The University of Newcastle upon Tyne,
Department of Inorganic Chemistry,
The University,
Newcastle upon Tyne, NE1 7RU,
U.K.

INTRODUCTION

Calcium ions play a very significant role in biological control mechanisms e.g. in the contractile apparatus of smooth muscle, in regulatory action of calmodulin etc.[1] The mechanism by which Ca^{2+} ions enter or leave the Ca^{2+} specific binding sites and channels in biological systems is of the utmost importance. Having an interest in the production of Ca^{2+} selective ionophores for ion-selective electrodes (ISE) which would be at the same time selective and give the ISE a fast response time, we looked for inspiration to molecular modelling studies of the Ca^{2+} binding sites of proteins, for which selectivity coupled with fast release is also a prerequisite.

We obtained atomic coordinates for modelling studies on carp parvalbumin, bovine-intestinal calcium binding protein, and also troponin-C (for which only the α-carbons are available at present) from the Brookhaven Protein Data Bank[2] (October, 1985 release). The molecular graphics package used was Chemgraf[3] (January, 1985 release), running on a VAX 11/780 at NUMAC[4], and the display was on a Sigma 5688 terminal.

RESULTS AND DISCUSSION

Carp parvalbumin contains two calcium-binding sequences called the CD and EF sites, established from the original X-ray structural determination of Kretsinger et al.[5] The amino-acid sequences of the loops are shown in Table I, together with the Ca^{2+} binding site sequences for troponin C. The donor groups are underlined for the CD and EF sites.

The difference in metal release rates for the CD and EF sites[6,7] is of great significance in relation to control processes. We used the techniques of molecular graphics to investigate differences in site geometry which might be contributory to such rate differences.

Coordinates for the CD and EF loops (with added hydrogens) were used for the displays in Figures 1 and 2. These were manipulated (e.g.

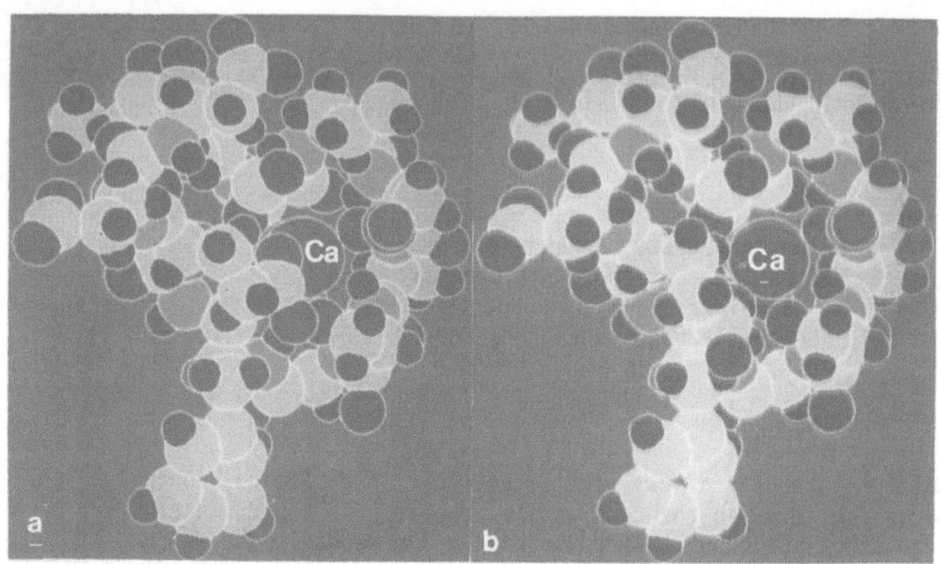

Figure 1. View of CD site down Ca-ASP51 a) as in crystal structure b) GLU59 rotated around $C_\beta - C_\alpha$ by 40°

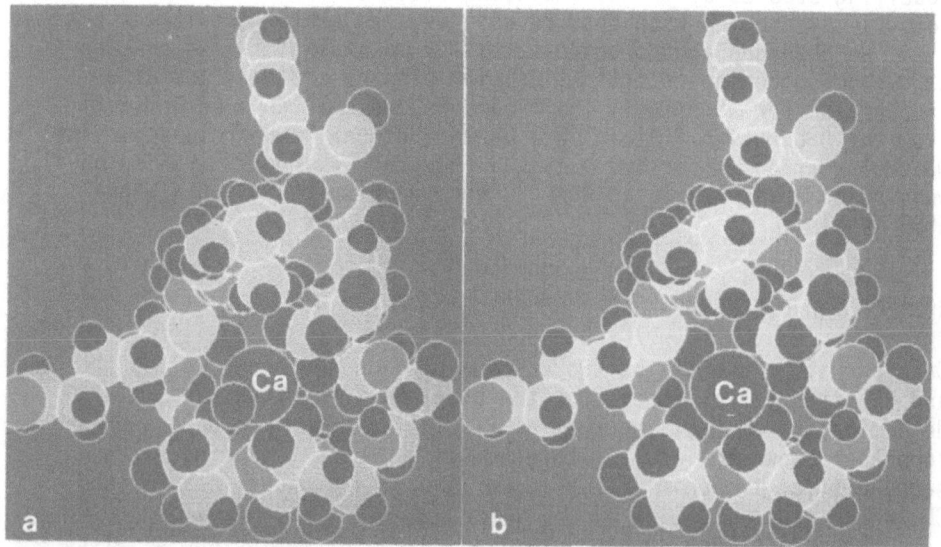

Figure 2. View of EF site down Ca-ASP90 a) as in crystal structure[5] b) ligating water removed.

to simulate rotation around the C_β-C_α bond of each side-chain). One ligand in each site seemed to be specially significant. Figure 1a shows the CD site, viewed down the Ca^{2+} to ASP51 OD direction (observer's view being from the solvent to the protein surface): the Ca ion is partly obscured by the uppermost donor oxygen, GLU59 OE1. Figure 1b shows (as a result of rotation around the C_β-C_α bond of GLU59) a clear view of the Ca^{2+}. The CD site may thus be considered a capped site, with Ca^{2+} exposed to solvent after twisting of the capping ligand. The corresponding situation for the EF site is shown in Figure 2 (the observer's view being again from solvent to surface, but from Ca^{2+} to ASP90 OD2). The corresponding uppermost ligand on the EF site (Figure 2a) is a water molecule. In Figure 2b this has been removed, affording a clear view of the Ca^{2+}. Simple VDW and Coulombic calculations show that it is easier in energy terms to remove Ca^{2+} with the donor water from the EF site, than to remove Ca^{2+} from the CD site after twisting round the GLU59. Moving any of the other donor residues leads to considerable VDW repulsions elsewhere in the site.

Consideration of the location of donor groups within the binding loops was also revealing. The CD and EF sites are ostensibly similar (Table I), but apart from the GLU/H_2O mismatch, there are interactions of Ca^{2+} in the EF site with both oxygens of ASP92 and GLU101 carboxyls. It is noteworthy that the outer (capping) region of the CD site is more constricted, having shorter Ca-O distances than the anterior; this site has a smaller VDW volume. The opposite arrangement is found in the EF loop where the anterior is more constricted with shorter Ca-O distances; the outer region is looser. All of these features are believed to be contributory to the rate differences for cation release between the CD and EF sites. Computing facilities available to us did not permit the question of cooperativity between these sites to be addressed.

The hypotheses rest on the accuracy of the model for parvalbumin and the reliability of the VDW parameters used; as these are refined the hypothesis will have to be reviewed. The lessons for the design of Ca^{2+} selective ligands are obvious, but the synthetic problems are considerable and may be more amenable to genetic engineering[8] on CD and EF type loops.

REFERENCES

1. 'Calcium Binding Proteins' (E. Carafoli et al. eds) Elsevier, Amsterdam, 1974.
2. Brookhaven Protein Data Bank, F.C. Bernstein, T.F. Koetzle, G.J.B. Williams, E.F. Meyer, Jr., M.D. Brice, J.R. Rodgers, O. Kennard, T. Shimanouchi, and M. Tasumi, J.Mol.Biol., **112**, 535-542, 1977.
3. Chemgraf (now Chem-X) created by E.K. Davies, Chemical Crystallography, Oxford University, developed and distributed by Chemical Design Ltd., Oxford.

TABLE I

AMINO ACID SEQUENCES OF Ca^{2+} BINDING SITES

CD[a]		EF[a]		TNCIII[b]		TNCIV[b]		TNCI[b]		TNCII[b]	
ASP	51	ASP	90	ASP	106	ASP	142	ASP	30	ASP	66
GLN	52	SER	91	LYS	107	LYS	143	ALA	31	GLU	67
ASP	53	ASP	92	ASN	108	ASN	144	ASP	32	ASP	68
LYS	54	GLY	93	ALA	109	ASN	145	GLY	33	GLY	69
SER	55	ASP	94	ASP	110	ASP	146	GLY	34	SER	70
GLY	56	GLY	95	GLY	111	GLY	147	GLY	35	GLY	71
PHE	57	LYS	96	PHE	112	ARG	148	ASP	36	THR	72
ILE	58	ILE	97	ILE	113	ILE	149	ILE	37	ILE	73
GLU	59	GLY	98	ASP	114	ASP	150	SER	38	ASP	74
GLU	60	VAL	99	ILE	115	PHE	151	THR	39	PHE	75
ASP	61	ASP	100	GLU	116	ASP	152	LYS	40	GLU	76
GLU	62	GLU	101	GLU	117	GLU	153	GLU	41	GLU	77

[a] Resolution 1.9Å, ref. 5
[b] Resolution 2.8Å, D. Herzberg and M.N.G. James, Nature, **313**, 653, 1985.

4. NUMAC Northern Universities Multiple Access Computer, Newcastle upon Tyne.
5. R.H. Kretsinger and C.E. Nockolds, J.Mol.Biol., **91**, 201, 1975.
6. J. Haiech, J. Derancourt, J.F. Pechere, J.G. Demaille, Biochem., **18**, 2752, 1979.
7. T.C. Williams, D.C. Corson, B.D. Sykes, J.Am.Chem.Soc., **106**, 5698, 1984.
8. We thank participants at the Lancaster Symposium for their suggestions.

SYNTHESIS AND CONFORMATIONAL STUDIES OF A NEW HOST SYSTEM BASED ON CHOLIC ACID*

Cynthia J. Burrows** and Richard A. Sauter
Department of Chemistry
State University of New York at Stony Brook
Stony Brook, New York 11794-3400
U.S.A.

ABSTRACT. New synthetic hosts have been designed incorporating two molecules of cholic acid linked by a rigid diamine. Proton NMR studies indicate that the compounds exist in a rigid conformation with the steroid hydroxyl groups intramolecularly hydrogen-bonded. Heat or addition of methanol leads to conformational isomerism due to insertion of methanol into the cavity.

Synthetic molecular receptors command widespread interest as mimics of membrane transport agents and enzyme active sites.[1,2] The development of new host systems for the selective complexation of organic and inorganic cations and anions has mushroomed in recent years; however, examples of solution phase coordination of neutral organic molecules are few. Most of these examples involve inclusion of an aromatic hydrocarbon into the hydrophobic pocket of a water soluble cyclodextrin or cyclophane receptor.[3-6] A smaller subset of synthetic receptors possess lipophilic exteriors and hydrophilic cavities for the association of polar substrates in nonpolar media. [7-9] We report here the synthesis and characterization of a new system of the latter type designed to bind polar guests in chloroform solution.

Cholic acid forms the architectural unit for the construction of amphiphilic receptors that are chloroform soluble while possessing a cavity lined with hydroxyl groups. Cholic acid is an ideal building block for artificial receptors because of the following features: (i) rigidity of the steroid framework insures formation of a cavity, (ii) the two faces of the steroid differ dramatically in their properties-- the alpha face displays three hydrogen bonding groups while the beta face is entirely hydrophobic, (iii) the cis A-B ring junction imparts a curvature to the steroid ring system, (iv) the hydroxyl group are directed convergently toward the center of the concave face, (v) the

---

*A preliminary report of this work was presented at the 4th International Symposium on Inclusion Phenomena, July 20-25, 1986, Lancaster, U.K.
**Author to whom correspondance should be addressed.

side chain carboxylate is readily derivatized, (vi) cholic acid is chiral, and (vii) it is nearly as inexpensive as cholesterol. Thus, cholic acid displays a great deal of complexity and molecular information in a readily available material.

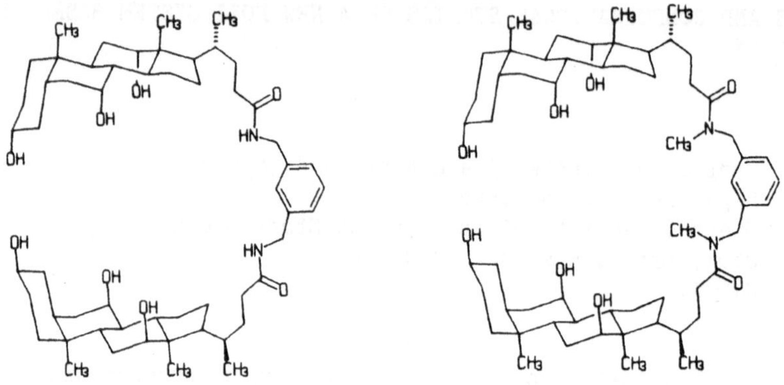

(1)   (2)

The first molecules investigated were compounds **1** and **2**, bis-cholamides joined by an aromatic link. Although a number of more rigid linking groups were prepared, we became intrigued with the unusual conformational properties of the m-xylylene derivatives. Compounds **1** and **2** were synthesized as shown in Scheme I by a method analogous to that of Kunitake.[10] Both were isolated in high yield (60-80%) as white solids after flash silica chromatography (10% MeOH, $CHCl_3$). The new compounds were fully characterized by spectroscopic and analytical methods. Both compounds were chloroform soluble (0.05 M) but fairly insoluble in non-hydrogen bonding solvents such as methylene chloride.

Scheme I

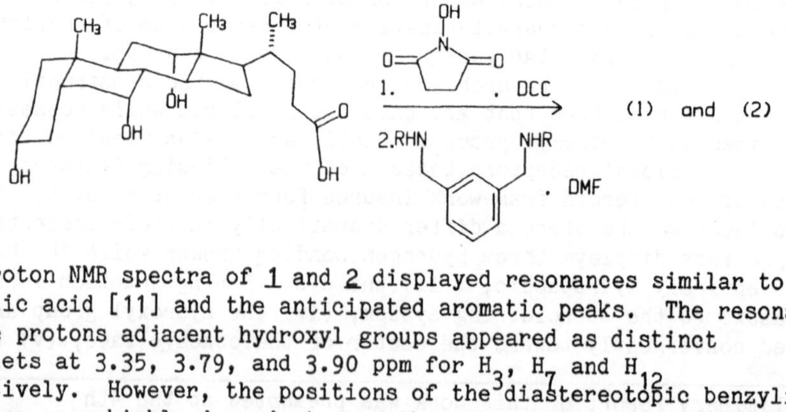

Proton NMR spectra of **1** and **2** displayed resonances similar to that of cholic acid [11] and the anticipated aromatic peaks. The resonances for the protons adjacent hydroxyl groups appeared as distinct multiplets at 3.35, 3.79, and 3.90 ppm for H_3, H_7 and H_{12} respectively. However, the positions of the diastereotopic benzylic hydrogens were highly dependent upon the solvent and the presence of small quantities of polar substances such as caffeine. In pure $CDCl_3$

A NEW HOST SYSTEM BASED ON CHOLIC ACID

119

two doublets of doublets appear at 3.95 and 4.77 ppm. These resonances collapse to a single narrow multiplet upon addition of small quantities of CD_3OD. Figure 1 shows the 3 to 5 ppm region of the spectrum for compound $\underline{1}$ as a function of added methanol. Concentrations of $\underline{1}$ were varied between 0.02 M and 10^{-4} M with similar results. The same phenomenon was observed in the variable temperature spectrum of $\underline{1}$ upon warming the solution. Figure 2 indicates a few of the temperatures studied. Coalescence occurs near 29°C (300 MHz) suggesting a dynamic process with a 14 kcal/mole energy barrier.

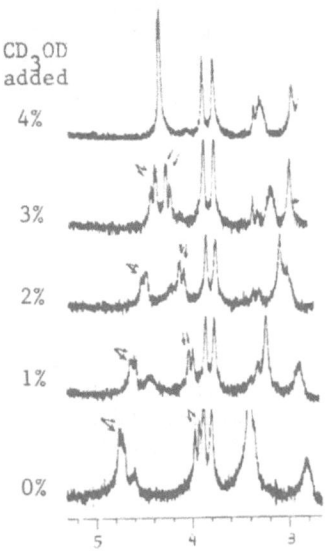

Figure 1. 300MHz spectrum of $\underline{1}$ in $CDCl_3$ at 294° as a function of added CD_3OD. Arrows indicate benzylic H's.

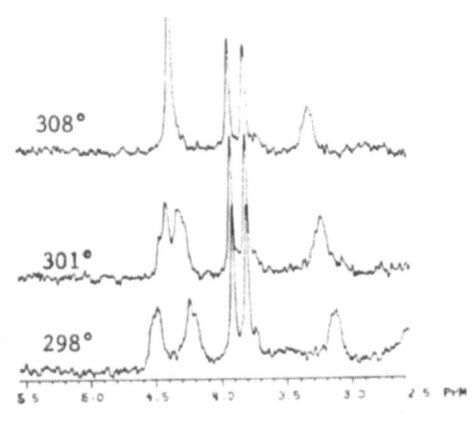

Figure 2. 300MHz variable temp. NMR study of $\underline{1}$ in $CDCl_3$.

An explanation consistent with the data involves solvent or temperature-induced isomerization of two conformational isomers, a closed form in which the two steroid halves of $\underline{1}$ are face-to-face and held together by direct intramolecular hydrogen bonding, and an open form in which rotations about bonds in the linking group are fast on the NMR time scale. Rigidity in the closed form of $\underline{1}$ would explain the large difference in chemical shift between the benzylic hydrogens. In a freely rotating form, these protons are likely to be in similar chemical environments. For comparison, compound $\underline{3}$ was prepared and investigated in the same fashion. Under all experimental conditions, the benzylic protons of $\underline{3}$ appeared as a narrow multiplet at 4.42 ppm. Thus, the dynamic process is a result of two cholate groups being present in the molecule.

(3)

The CDCl$_3$ solution spectrum of 2, an N-methylated analog of 1, suggests that there are at least three major conformers present at room temperature (21°C.) Incremental addition of CD$_3$OD produced a similar convergence of the benzylic proton signals, but now nearly an equal volume of methanol was necessary to bring about free rotation. (See Figure 3.) Variable temperature studies failed to show convergence up to 65°C for 2.

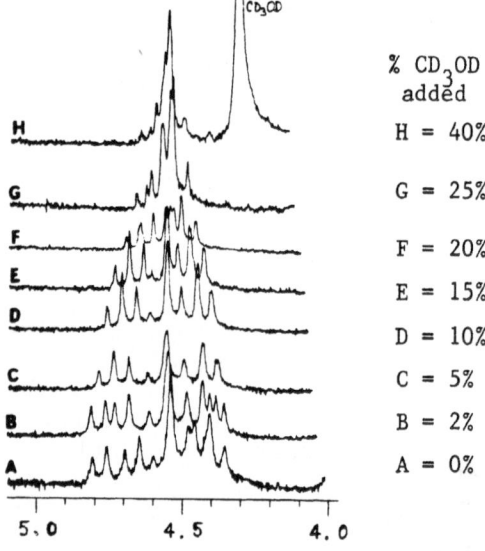

Figure 3. 300MHz spectrum of 2 in CDCl$_3$ at 294° with increments of CD$_3$OD added.

% CD$_3$OD added
H = 40%
G = 25%
F = 20%
E = 15%
D = 10%
C = 5%
B = 2%
A = 0%

closed 1 $\xrightleftharpoons{\text{MeOH or } \Delta}$ open 1

While conformational isomerism about the amide C(O)-N bond may account for some of the complexity of the spectrum of 2, the interpretation most consistent with compound **1**'s sensitivity to the presence of a hydrogen-bonding solvent is that of intramolecular bonding of a closed form interconverting with a freely rotating form. These results suggest that this new class of synthetic hosts displays the appropriate solubility and hydrogen-bonding characteristics required for association of polar molecules such as methanol. Further studies are being directed toward the characterization of molecular complexes and the elaboration of **1** and **2** to macrocyclic analogs.

Finally, it is interesting to note that the biological function of cholic acid and other bile acids is to form inclusion complexes with hydrocarbons in aqueous solution and the excretion of cholesterol.[12] Indeed, others have mimicked this phenomenon in synthetic systems where cholic acid forms a hydrophobic pocket.[13,14] We have attempted to reverse the role of cholic acid and to take advantage of its singular features as a convergently functionalized binding group.

Acknowledgements. We thank the Donors of the Petroleum Research Fund administered by the American Chemical Society for partial support of this work. Funds for the purchase of a 300 MHz NMR spectrometer from the National Science Foundation are gratefully acknowledged.

References.

1. J. M. Lehn, *Science* 1985, *227*, 849-56.
2. D. J. Cram, *Science*, 1983, *219*, 1177-83.
3. M. L. Bender and M. Komiyama, 'Cyclodextrin Chemistry'; Springer: Berlin, 1978.
4. I. Tabushi and K. Yamamura, *Topics Curr. Chem.*, 83, *113*, 145-82.
5. K. Odashima and K. Koga, in 'Cyclophanes'; P. M. Keehn, St. M. Rosenfeld, Ed.; Academic Press: New York, 1983, Vol. 2, pp 629-78.
6. J. Franke and F. Vogtle, *Angew. Chem., Int. Ed. Engl.* 85, *24*, 219-21.
7. J. Rebek, Jr., B. Askew, N. Islam, M. Killoram, D. Nemeth, and R. Wolak, *J. Am. Chem. Soc.* 1985, *107*, 6736-8.
8. B. J. Whitlock and H. J. Whitlock, *J. Am. Chem. Soc.* 1985, *107*, 1325-9, and references therein.
9. D. O'Krongly, S. R. Denmeade, M. Y. Chiang, and R. Breslow, *J. Am. Chem. Soc.* 1985, *107*, 5544-5.
10. Y. Okahata, R. Ando, and T. Kunitake, *Bull. Soc. Chem. (Japan)* 1979, *52*, 3647-53.
11. D. V. Waterhaus, S. Barnes, and D. D. Muccio, *J. Lipid Res.* 1985, *26*, 1068-78.
12. G. A. D. Haslewood, 'The Biological Importance of Bile Salts', North-Holland: Amsterdam, 1978.
13. J. McKenna, J. M. McKenna, and D. W. Thornthwaite *J. Chem. Soc., Chem. Commun.* 1977, 809-11.
14. J. P. Guthrie, P. A. Cullimore, R. S. McDonald, and S. O'Leary, *Can. J. Chem.* 1982, *60*, 747-71.

SOLID STATE STUDIES ON p.t-BUTYL-CALIX[6]ARENE DERIVATIVES

G.D. Andreetti, G. Calestani*, F. Ugozzoli
Institute of Structural Chemistry, University of Parma
A. Arduini, E. Ghidini, A. Pochini, R. Ungaro*
Institute of Organic Chemistry, University of Parma
Viale delle Scienze
I-43100 Parma, Italy

Calix[n]arenes[1] (1) are a class of phenol-formaldehyde macrocyclic oligomers which exhibit different sizes and shapes depending on the number of phenolic units in the cyclic array and on the nature of substituents on the aromatic nucleus (R^1) and on the phenolic oxygen (R^2).

(1)

(2) R^3 = H
(3) R^3 = $CH_2CON(C_2H_5)_2$

They are particularly attractive in Host-Guest Complex Chemistry because of their ability to form inclusion complexes with ions or/and neutral molecules[2], so they can function as ion and neutral molecule receptors in which the shape and the dimension of the cavity determine the selectivity toward the guest.

In calix[4] and calix[5]arenes[3,4] the macrocycle is blocked in a "cone" conformation, both in solution and in solid state, through the formation of strong intramolecular hydrogen bonds.

Calix[6] and calix[8]arenes are more mobile in solution[5] and these macrocyclic hosts have shown little ability to form inclusion complexes.

Recently an X-ray crystal structure of the cyclic octamer derived from p.t-butyl-calix[8]arene has been reported[6] but no data are available on the correspondent calix[6]arene (2) which has been used as starting materials for the synthesis of various host compounds[7].

The crystal structure of compound (2) has been obtained by X-ray diffraction, following the same experimental conditions reported previously[8] on a colourless transparent single crystal obtained from chloroform.

The following crystal data have been obtained: a=18.344; b=19.945; c=17.079 Å; V=6248 Å^3; Z=4. Laue Group D_{2h}, Space Group Pna2$_1$ or Pnma. Although the structure is characterized by high disorder mainly in the t-butyl groups, which makes it difficult to obtain a completely satisfactory R value, the resolution is sufficient to establish the conformation of the molecule (Fig. 1).

No guest molecule has been evidenced inside the macrocycle which shows a distorted cone structure with the t.butyl groups pointing outside the macroring and the six phenolic oxygens pointing toward the interior.

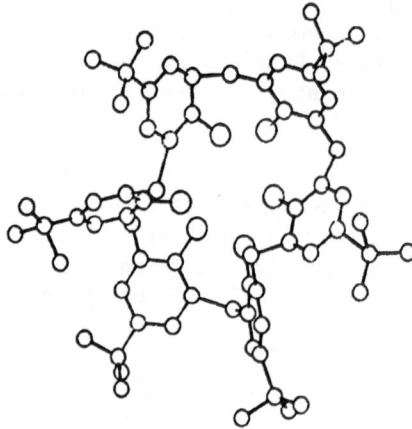

Fig. 1.
Perspective view of
p.t-butylcalix[6]arene

By treating p.t-butylcalix[6]arene (2) with α-chloro-N,N-diaethylacetamide (NaH, THF/DMF=5 v/v) the ligand (3), m.p. 238-240°C was obtained in 90% yield.

The ^{1}H and ^{13}C NMR spectra show a complex pattern indicating a mixture of conformational isomers to be present in solution and a reduced mobility of the macrocycle at room temperature.

The X-ray crystal structure of compound (3) has been solved following improved tecniques[8].

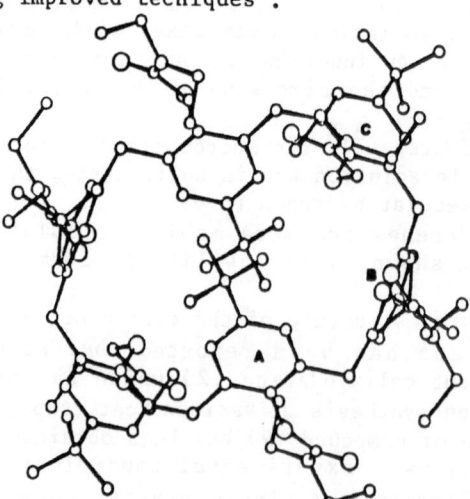

Fig. 2.
The X-Ray crystal
structure of p.t-butyl-
calix[6]arene amide:
view perpendicular to the
mean molecular plane.

Crystal data are as follows: a = 24.564(4); b = 14.509(3); c = 14.380(5) Å; β = 91.3(1)° Monoclinic, Space Group P 2₁/n, V = 5111.9 Å^3; Z = 4; D_{calc} = 1.09 g.cm^{-3}.

The structure has been refined up to R = 0.10 using 1987 independent observed reflections from 8740 collected reflections.

The molecule, shown in Fig. 2, possesses a center of symmetry which coincides with a crystallographic center of symmetry.

The conformation can be referred to the mean molecular plane through the CH_2 bridging groups, that can be taken as a reference plane.

The dihedral angles formed by the three adjacent phenolic units A, B, C, with respect to the reference plane are = 35.09, 82.03, 40.36° respectively.

The amide chains and the t-butyl groups of the units B and C do not point toward the interior of the macroring. The phenolic unit A points its amide chains outside the macroring below the reference plane, while its t-butyl group points inside the macroring above the reference plane (and its centrosymmetrically related t-butyl group lies below) partially filling the intramolecular cavity.

Two methanol molecules for each macrocycle are guested in the intermolecular holes of the crystal lattice.

The situation is quite different if compared to that observed in the p.t-butylcalix[6]arene hexapodand[9], where the intramolecular cavity was partially filled by two centrosymmetrically related ethereal chains and no guest molecules were observed.

With the aim of blocking the conformational mobility of the parent p.t-butylcalix[6]arene (2) creating a species able to perform molecular inclusion and to be active in catalysis, the p.t-butylcalix[6]arene-titanium(IV) complex has been synthesized, by treating (2) with $Ti(OiPr)_4$ in boiling toluene. A compound of molecular formula $C_{66}H_{78}O_7Ti_2$ · $3C_7H_8$ crystallizes from toluene in the form of orange prisms suitable for X-ray analysis[8].

Crystal data are: a = 26.334(7); b = 17.052(7); c = 21.462(7) Å, β = 94.89°. Monoclinic, Space Group P 2/n. V = 9600(1) Å^3, Z = 2.

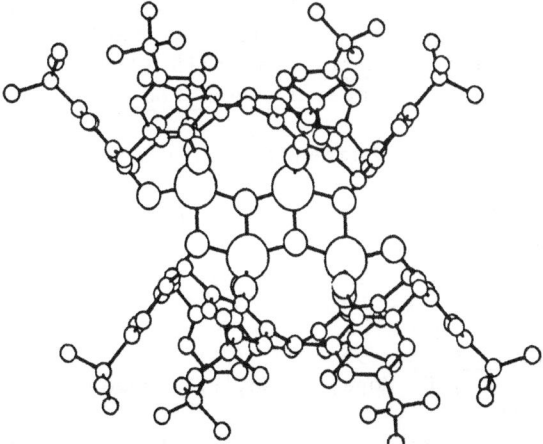

Fig. 3. X-Ray crystal structure of p.t-butylcalix[6]arene Ti(IV) complex.

The refinement is now performed up to R = 0.199. The molecule (Fig. 3) is dimeric and consists of two p.t-butylcalix[6]arene molecules in a "cone" structure, created by two Ti(IV) atoms which are bonded to six phenolic oxygen atoms and share a Ti-O-Ti bond.

The Ti atom is penta-coordinated in the form of a slightly distorted trigonal bi-pyramid as already observed in some Titanium(IV) oxygenated complexes[10], although their preferred geometry is generally octahedral[11]. The Ti-O bond distances in the rather unusual Ti-O-Ti bridge (it has not been observed in the other two calixarene - Ti(IV) complexes previously reported[12,13]) range from 1.922(5) to 1.944(7) Å and the bond angle is 110.8°. A toluene molecule is guested in the intramolecular apolar cavity of each macrocycle, whereas two other molecules occupy intermolecular voids of the crystal lattice. These two molecules are rather disordered and this causes probably problems in the final refinement of the crystal structure.

Interestingly the rigid structure shown in the solid state is also retained in solution as inferred from the ^{1}H NMR spectrum which shows three doublets for the equatorial protons of the bridging methylenes centered at 3.25, 3.35, 3.53δ and three doublets 4.66, 4.75, 5.14δ for the axial ones.

References

1. For a review article see G.D. Gutsche, Top. Curr. Chem., **123**, 1 (1984).
2. A. Arduini, A. Pochini, S. Reverberi, R. Ungaro, G.D. Andreetti, F. Ugozzoli, Tetrahedron, **42**, 2089 (1986) and references therein.
3. G.D. Andreetti, R. Ungaro, A. Pochini, J. Chem. Soc. Chem. Commun., 1006 (1979).
4. M. Coruzzi, G.D. Andreetti, V. Bocchi, A. Pochini, J. Chem. Soc. Perkin Trans 2, 1777 (1983).
5. C.D. Gutsche and L.J. Bauer, J. Am. Chem. Soc., **107**, 6052 (1985).
6. C.D. Gutsche, A.E. Gutsche, A.I. Karaulov, J. Inclusion Phenom., **3**, 447 (1985).
7. V. Bocchi, D. Foina, A. Pochini, R. Ungaro, G.D. Andreetti, Tetrahedron, **38**, 373 (1982); S. Shinkai, S. Mori, T. Tsubaki, T. Sone, O. Manabe, Tetrahedron Lett., **25**, 5315 (1984); S. Shinkai, H. Koreishi, K. Veda, O. Manabe, J. Chem. Soc. Chem. Commun., 233 (1986).
8. R. Ungaro, A. Pochini, G.D. Andreetti, F. Ugozzoli, J. Inclusion Phenom., **3**, 409 (1985).
9. R. Ungaro, A. Pochini, G.D. Andreetti, P. Domiano, J. Inclusion Phenom., **1**, 35 (1985).
10. K. Watenpaugh, C.N. Coughlan, Inorg. Chem., **5**, 1782 (1966).
11. G.W. Svetich, A.A. Voge, Acta Cryst., B,**28**, 1970 (1972); J. Chem. Soc., Chem. Commun., 676 (1971); K.B. Sharpless, S.S. Woodvard, N.G. Finn., Pure Appl. Chem., **55**, 1823 (1983).
12. M.M. Olmstead, G. Sigel, H. Hope, X. Xu, P.P. Power, J. Am. Chem. Soc., **107**, 8087 (1985).
13. S.G. Bott, A.W. Coleman, J.L. Atwood, J. Chem. Soc., Chem. Commun., 610 (1986).

TOPOLOGY OF N-ETHYLMALEIMIDE IN NORMAL HUMAN ERYTHROCYTE MEMBRANE

F. Severcan
Department of Electrical Engineering
Dokuz Eylül University
Bornova, Izmir
Turkey

ABSTRACT. Normal human, hemoglobin free erythrocyte ghosts were labelled with ^{14}C-N-Ethylmaleimide and incubated at two different temperatures 4°C and 37°C. Gel electrophoresis, autoradiography and nuclear scintillation counting techniques were used in analysing which proteins were labelled. The results of the experiments with incubation at 4°C and 37°C indicated that 35 % of the NEM was associated with spectrin and 50 % was bound to spectrin- actin complex.

1. INTRODUCTION

Electrophoresis of ghost membrane proteins in polyacrylamide gels containing 1 % SDS yields a pattern in which six well resolved bands are observed [1]. These bands are separated according to their molecular weight. Of the total membrane protein, about one-third are contained in two major bands, bands I and bands II, weighing 240,000 and 220,000 respectively. These bands are called spectrin. Bands 5 weighing 43,000-39,000 are called actin. Spectrin, together with actin play a major role in stabilising the membrane and maintaining its discoid shape [2]. To study the protein components of erythrocyte membranes, the intact membranes were labelled N-ethylmaleimide (NEM) [3,4] and maleimide anologue sulfhydryl spin labels [5-9]. However very few studies are available in the literature in mapping the binding sites of NEM and malemide spin labels to erythrocyte membranes [4,10].

In this research, NEM binding sites to normal human erytrocyte membranes have been investigated by labelling the membranes with ^{14}C-NEM. Gel electrophoresis, autoradiography and nuclear scintillation counting tecniques have been used to show which protein components were labelled.

2. MATERIALS AND METHODS

Hemoglobin-free white membrne ghosts were prepared from normal human erythrocytes according to the methods in [11,12]. The red blood cells were washed three times with 150 mM NaCl saline solution buffered with 5 mM sodium phosphate at pH 8 (1 volume of cell + 10 volume of buffer solution).

1 ml of packed cell was lysed by resuspension in 40 ml of cold 5 mM sodium phosphate (pH 8) and was centrifuged at 24,000 g for twenty minutes to form pellets which were washed 3 more times by suspension in the same buffer at the same speed. The resulting pellets were hemoglobin free white unsealed erythrocyte ghosts.

Spectrin-actin was purified according to the method of Bennett and Branton [13]. White membrane ghosts were washed in 0,3 mM Na_2HPO_4, pH 7.6 at $0^{\circ}C$. The pellets are resuspended in a final volume of 5 ml in 0,3 mM Na_2HPO_4, pH 7.6, and incubated for 25 min. at $37^{\circ}C$ and the suspension centrifuged for 15 min. at 225,000 g. The resulting supernant contains approximately 80 % of the erytrocyte spectrin and almost all bands 5 or erythrocyte actin. The spectrin-actin solution was concentrated to 3 mg/ml by using aquacide (Calbiochem, CA.). Protein concentration was determined by using Lowry method [14]. Hemoglobin free erythrocyte ghosts were labelled with ^{14}C-N-ethylmaleimde with specific activity 23.7 mCi/mmol (New England Nuclear) in hypotonic buffer. Some of the samples were incubated at $37^{\circ}C$ and the others were incubated at $4^{\circ}C$ for one hour. They then were washed three times by suspension in hypotonic buffer, followed by centrifugation (10,000 g-30,000 g). Radioactive labelled spectrin-actin was prepared from labelled erythrocyte ghosts or spectrin-actin complex was purified first then spectrin-actin was labelled with ^{14}C-NEM.

All samples were subjected to 7.5 % polyacrylamide slab gel electrophoresis containing 1 % sodium dodecyl sulfate with similar apparatus in [15] The gels were run at 27 mA until bromophenol blue tracking dye migrated to 1 cm of the bottom of the gel. The gel was nudged into a big crystallization dish and covered wth Coomasie blue stain. It was stained overnight with gentle stirring, and then was destained with destaining solution. The gel of the labelled samples was dried under the vacuum and autoradiography was performed. The bands of the gel were then sliced and were put in different tubes. H_3PO_4 was added and incubated in an oven at $60^{\circ}C$ for 24 hours. Then they were counted by nuclear scintillation counter (Beckman LS-3150 T Model Scintillation Counter).

3. RESULTS AND DISCUSSION

Lenard labelled intact membranes from different animal species with ^{14}C-NEM (see Figure 1. a) and showed that 50 % of the total NEM was associated with the spectrin-actin complex [4]. In this work membranes were incubated with ^{14}C-NEM at $37^{\circ}C$.

Early EPR studies of normal human erythrocyte membranes labelled with N-(1 oxyl-2,2,6,6- tetramethyl 4-piperidinyl) maleimide (see Figure 1. b) which is a derivative of NEM indicated that 75 % of the total labelled sites belong to the spectrin-actin complex [10]. Membranes were incubated with N-Maleimide at $4^{\circ}C$. It was also indicated that the differences between the results of [4] and [10] may be due to the difference of the molecules and different incubation temperatures used.

In order to investigate whether different temperature used in early studies may cause these discrepancies, in this research, the human white erythrocyte ghosts were labelled with ^{14}C-NEM and were incubated at two different temperatures, $4^{\circ}C$ and $37^{\circ}C$.

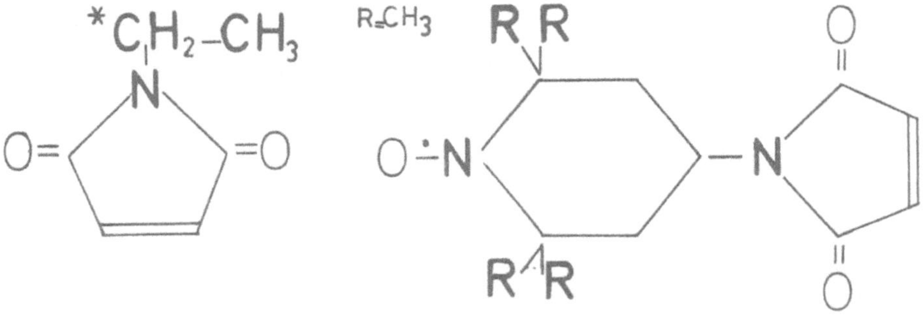

Figure 1 a) Schematic representation of Ethylmaleimide, N- Ethyl-1-^{14}C and b) N-Maleimido Tempo.

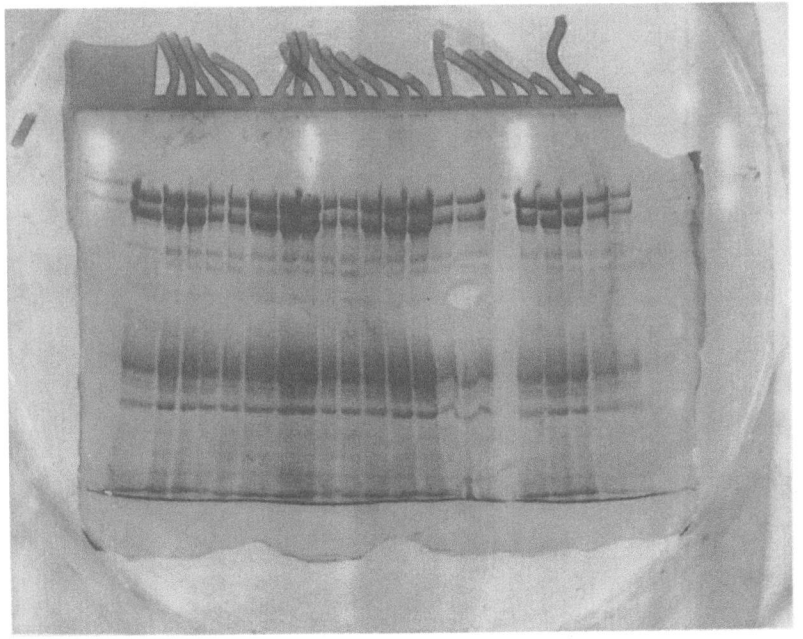

A B C

Figure 2. SDS-Polyacrylamide slab gel electrophoresis of samples A) hemoglobin free erythrocyte ghost. B) and C) hemoglobin free erythrocyte ghost labelled with ^{14}C-NEM and incubated at 4°C and 37°C respectively.

Figure 2 shows the SDS-polyacrylamide slab gel electrophoresis of radioactive labelled hemoglobin free normal human erythrocyte ghosts incubated at two different temperatures, $4^{\circ}C$ and $37^{\circ}C$. More information about bands observed, can be obtained from [1]. No variation in the characteristic pattern of the radioactive labelled ghost membrane have been detected at the two different incubation temperatures.

Figure 3 shows the SDS-polyacrylamide slab gel electrophoresis of labelled spectrin-actin. Spectrin-actin can be extracted from radioactive labelled erythrocyte ghosts as easily as from unlabelled ghosts.

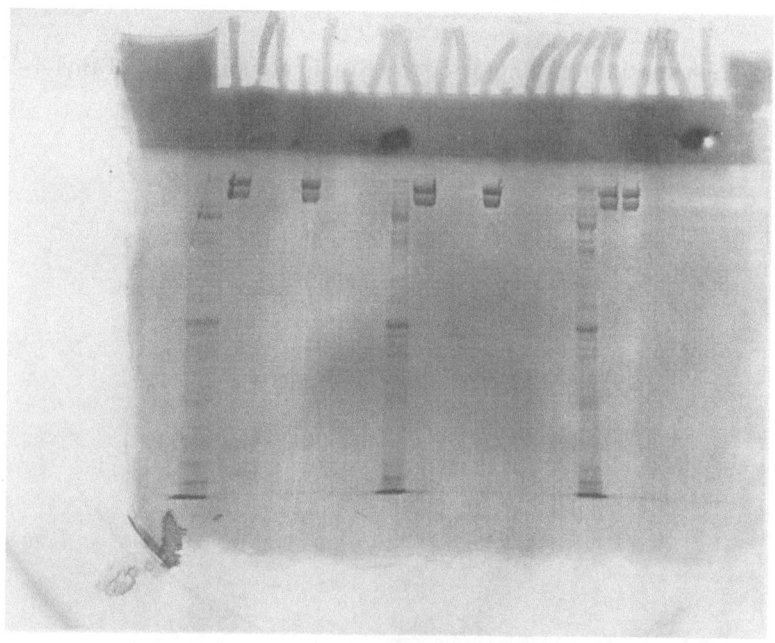

B A   A   B

Figure 3. SDS-polyacrylamide slab gel electrophoresis of radioactive labelled A) spectrin-actin B) spectrin depleted white erythrocyte ghost.

Figure 4 shows the autoradiogram of the gel of erythrocyte ghosts, labelled with ^{14}C-NEM, incubated at two different temperatures, $4^{\circ}C$ and $37^{\circ}C$ respectively. Empty columns belong to unlabelled ghosts. This picture represents the radioactivity distribution in the gels.

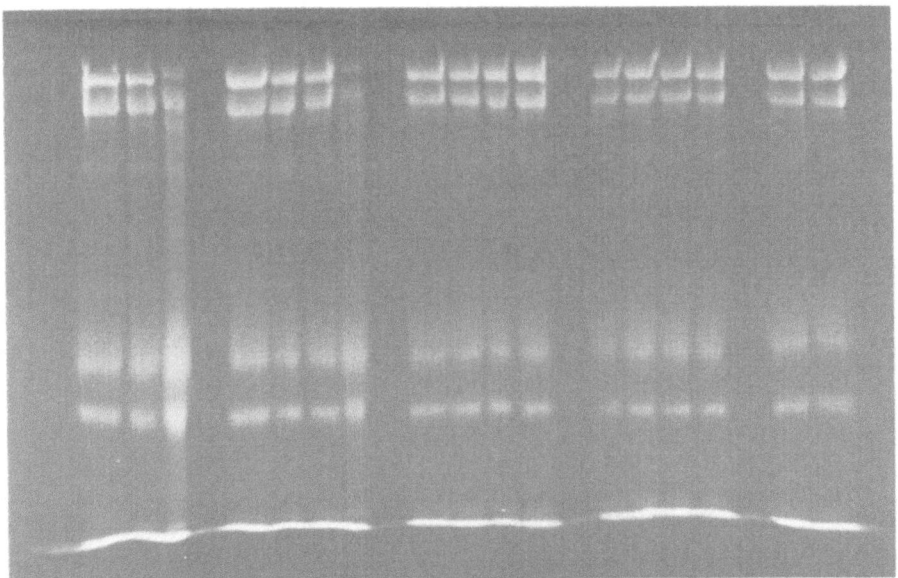

Figure 4. Autoradiogram of erythrocyte ghosts. Empty columns belong to unlabelled ghosts.

Results of nuclear scintillation counting for samples incubated at $4^{\circ}C$ are shown in Figure 5. It was observed that 35 % of the total NEM was bound to spectrin and 50 % of the NEM was bound to spectrin-actin complex.

Results of radioactivity counting for samples incubated at $37^{\circ}C$ (see Figure 6) also indicated that 35 % of the NEM was associated with spectrin and 50 % of the NEM was bound to spectrin-actin. The results are in agreement with [4] and we can say that different incubation temperatures may not cause any variation .

ACKNOWLEDGEMENTS

I am grateful to Professor W. Huestis for providing the opportunity to work in her laboratory, and to James E. Ferrel. Without their help this work would not have been done.

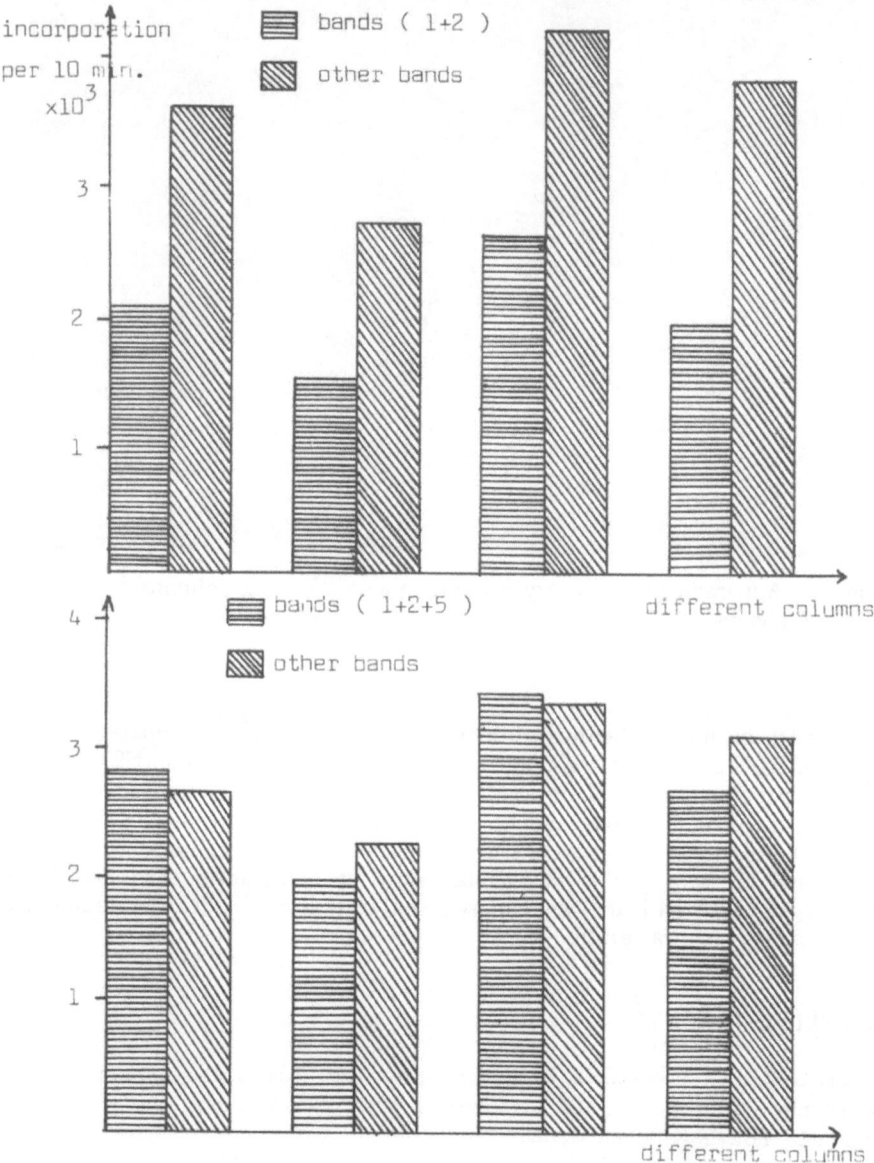

Figure 5. Results of incorporation per 10 min. for samples prepared at 4°C. 35 % of NEM is bound to spectrin, 50 % of the NEM is bound to spectrin-actin.

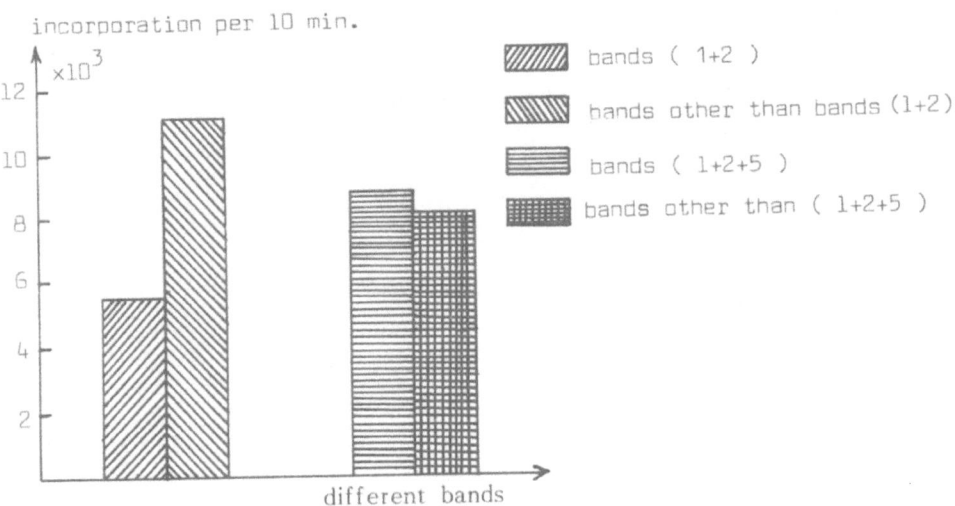

Figure 6. Results of incorporation per 10 min. for samples prepared at 37°C. 35 % of the NEM is associated with spectrin. 50 % of the NEM is bound to spectrin-actin.

REFERENCES

1. G. Fairbanks, T. L. Steck, and D. F. H. Wallach : Biochemistry, **13**, 2606 (1971).
2. W. Birchmeier and S. J. Singer : J. Cell Biol., **73**, 647 (1977).
3. J. Lenard : Biochemistry, **9**, 1129 (1970).
4. J. Lenard : Biochemistry, **9**, 5037 (1970).
5. H. E. Sandberg, R.G. Bryant, and L. H. Piette :Arch. Biochem. Biophys. **133**, 144 (1969).
6. K. W. Berger, M.D. Barratt, and V. B. Kamat : Chem. Phys. Lipids, **6**, 351 (1971).
7. D. Adams, M. E. Markes, W. J. Lewis, and K. L. Carraway : Biochim. Biophys. Acta, **426**, 38 (1976).
8. L. W, - M. Fung, M. J. Soo Hoo, and W. A. Meena : FEBS Lett., **105** 379 (1979).
9. R. Cassoly, D. Daveloose, and F. Leterrier : Biochim. Biophys. Acta, **601**, 478 (1980).
10. L. W,-M. Fung and M.J. Simpson : FEBS Lett., **108**, 269 (1979).

11. J. T. Dodge, C. Mitchell, and D. J. Hanahan : Arch. Biochem. Biophys., **100**, 119 (1963).
12. T. L. Steck and J. A. Kant : 'Methods in Enzymology. Biomembranes', (eds. S. Fleischer and L. Packer), pp. 172 - 180, Academic Press, New York (1974).
13. V. Bennett and D. Branton : J. Biol. Chem., **252**, 2753 (1977).
14. O. H. Lowry, N. J. Rosebrough, A. L. Farr, and R. J. Randall : J. Biol. Chem. **193**, 265 (1951).
15. A. Lemay : Anal. Biochem., **92**, 130 (1979).

ANNOUNCEMENT

Announcing a major new book series . . .

INCLUSION SCIENCE

The D. Reidel Publishing Company is proud to announce the launch of a major series of monographs to accompany its *Journal of Inclusion Phenomena*. The monographs, to be published in the series Inclusion Science, will deal in depth with the same topics as are covered in the journal, namely:

- Ion Transport through Inclusion Compound Membranes;
- Cyclodextrins and their Uses in Industry;
- Enzyme Modelling and Artificial Enzyme Systems;
- Inclusion Polymerisation;
- Intercalates in Biological and Nonbiological Systems;
- Chromatography;
- Zeolites.

Editorial Board:
J. L. ATWOOD, University of Alabama, U.S.A.
J. E. D. DAVIES, University of Lancaster, U.K.
T. IWAMOTO, University of Tokyo, Japan
N. N. LI, UOP Inc., Illinois, U.S.A.
J. LIPKOWSKI, Academy of Sciences, Poland
D. D. MacNICOL, University of Glasgow, Scotland
W. SAENGER, Free University, Berlin

The first book in the series, to be published in early 1987, will be the definitive work by J. SZEJTLI:

CYCLODEXTRIN TECHNOLOGY

In this masterly survey, the acknowledged expert on cyclodextrins, Professor J. Szejtli of Chinoin Pharmaceutical Company, Budapest, gives a comprehensive and detailed account of the most up-to-date aspects of the multitude of uses to which cyclodextrins can be put in all facets of research, development and production.

Brief Table of Contents:
Chapter 1: Chemistry, Production, Analysis, Metabolism, Biological Effects of Cyclodextrins, Derivatisation, and Polymerisation;
Chapter 2: Inclusion Effects in Solution, Preparation and Characterisation of Inclusion Complexes in Solid and Solution Phase;
Chapter 3: Application of Cyclodextrins in the Pharmaceutical Industry;

Chapter 4: Cyclodextrins in Foods and Toiletries;
Chapter 5: Cyclodextrins in Cosmetics and Pesticides;
Chapter 6: Cyclodextrins in Organic Synthesis, Biotechnology, and in Separation Processes;
Chapter 7: Cyclodextrins in Analytical Chemistry and in Diagnostic Systems

Price to be annnounced.   ISBN 90-277-2314-1

AN EQUILIBRIUM AND KINETIC STUDY OF THE COMPLEXATION OF LITHIUM AND SODIUM IONS BY THE CRYPTAND 4,7,13-TRIOXA-1,10-DIAZABICYCLO-[8.5.5]-EICOSANE ($C21C_5$).

Amira Abou-Hamdan,[a] Ian M. Brereton,[b] Andrea M. Hounslow,[a] Stephen F. Lincoln[a] and Thomas M. Spotswood[b]
Departments of Physical and Inorganic Chemistry,[a] and Organic Chemistry,[b] University of Adelaide, South Australia 5001.

ABSTRACT. In the solid and solution state Li^+ and Na^+ form inclusive and exclusive cryptates respectively with $C21C_5$, in which Li^+ resides inside and Na^+ resides outside the $C21C_5$ cavity. Similar inclusive and exclusive structures are observed for $[Li.C211]^+$ and $[Na.C211]^+$. The logarithms of the stability constants in dimethylformamide for $[Li.C21C_5]^+$, $[Li.C211]^+$, $[Na.C21C_5]^+$ and $[Na.C211]^+$ are: 2.80, 6.99, 2.87 and 5.20; and the corresponding decomplexation rate constants are: 107, 0.013, 28800 and 12 s^{-1} at 298.2 K. The relationships between cryptate structure, stability and lability are considered, as are solvent influences.

INTRODUCTION

In the solid state the cryptates $[Na.C21C_5.NCS]$ and $[Na.C211.NCS]$ exist in the exclusive form in which Na^+ resides outside the cryptand cavity adjacent to the fifteen membered 4,7,13-trioxa-1,10-diaza ring;[1] and $[Li.C21C_5]NCS^2$ and $[Li.C211]I^3$ exist in the inclusive form in which Li^+ resides inside the cryptand cavity. ^{13}C nmr studies indicate that $[Na.C21C_5]^+$ and $[Na.C211]^+$ (NCS^- within bonding distance of Na^+ in the solid state is substantially displaced by solvent in solution) exist in the exclusive form; and $[Li.C21C_5]^+$ and $[Li.C211]^+$ exist in the inclusive form in solution.[1,2] (The existence of the exclusive and inclusive cryptates is a consequence of the ionic diameters of Na^+ and Li^+, 2.04 Å and 1.52 Å respectively, being larger than and similar to the $C21C_5$ and C211 cavities of diameter 1.6 Å.) Cryptand $C21C_5$ has one less oxygen than C211:

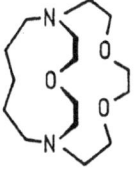 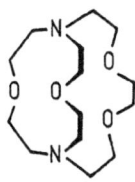

           C211

but is otherwise structurally similar, and thus affords an opportunity to study the effect of this difference on the stability and lability of the corresponding alkali metal cryptates.

RESULTS

The stability constant ($K_S = k_f/k_d$), and formation (k_f) and decomplexation (k_d) rate constants characterizing the equilibrium:

$$M^+ + C21C_5 \underset{k_d}{\overset{k_f}{\rightleftharpoons}} [M.C21C_5]^+ \qquad (1)$$

where $M^+ = Li^+$ or Na^+, are related to the lifetimes of M^+ in the cryptate (τ_c) and the solvated state (τ_s) through eqns (2) and (3), where X_c and X_s are the mole fractions of M^+ in the cryptate and solvated state respectively, and all other symbols have their usual meanings.

$$\tau_c/X_c = \tau_s/X_c \qquad (2)$$

$$1/\tau_c = k_d = (k_BT/h)\exp(-\Delta H^\#/RT + \Delta S^\#/R) \qquad (3)$$

The K_S values for $[Li.C21C_5]^+$ and $[Na.C21C_5]^+$ were determined by ^7Li nmr and potentiometric methods respectively; and the rates of Li^+ and Na^+ exchange were determined by nmr complete lineshape methods. An example of the coalescence of the ^{23}Na resonances arising from solvated Na^+ and $[Na.C21C_5]^+$ as the temperature increases is shown in Figure 1. For both $[Li.C21C_5]^+$ and $[Na.C21C_5]^+$, and their C211 analogues,[4,5] k_d is independent of concentration in a range of solvents and hence the dominant cryptate decomplexation process does not involve solvated M^+ in the transition state.

Cryptate stability and lability vary substantially with the nature of M^+ and cryptand as is exemplified by the data obtained in dimethylformamide shown in Table 1. It is also found that cryptate stability and lability are considerably affected by the nature of the solvent as is exemplified by the $[Na.C21C_5]^+$ data in Table 2.

TABLE 1. Rate and apparent stability constants for cryptate systems in dimethylformamide solution at 298.2 K.

cryptate	$10^{-3}k_f/$ $dm^3mol^{-1}s^{-1}$	$k_d/$ s^{-1}	log K_S/dm^3mol^{-1}
[a] $[Li.C21C_5]^+$ (inclusive)	67	107	2.80
[b] $[Li.C211]^+$ (inclusive)	127	0.013	6.99
[a] $[Na.C21C_5]^+$ (exclusive)	21400	28800	2.87
[c] $[Na.C211]^+$ (exclusive)	1920	12	5.20

[a] This work;   [b] Ref. 4;   [c] Ref. 5.

THE COMPLEXATION OF LITHIUM AND SODIUM IONS

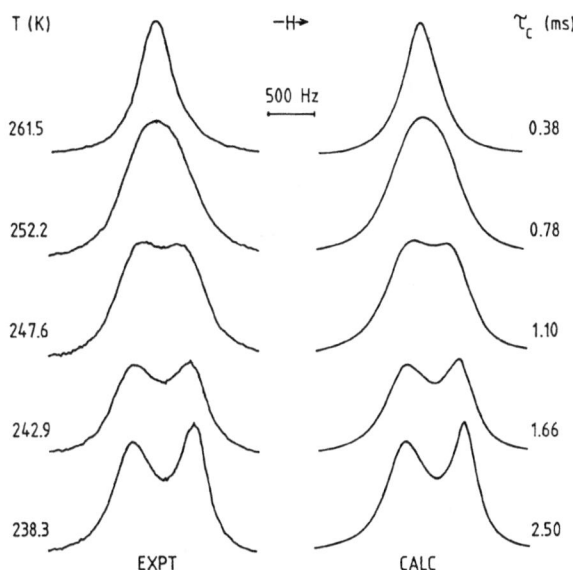

Figure 1. Typical exchange modified 79.39 MHz ^{23}Na nmr spectra of a dimethylformamide solution of NaClO$_4$ (0.106 mol dm^{-3}) and C21C$_5$ (0.053 mol dm^{-3}). Experimental temperatures and spectra appear at the left of the Figure and best fit calculated lineshapes and the corresponding τ_c values appear at the right. (A total of ten spectra at different temperatures were analysed for this solution.) The resonance of [Na.C21C$_5$]$^+$ appears downfield of the resonance of solvated Na$^+$.

DISCUSSION

The much higher stabilities of [Li.C211]$^+$ and [Na.C211]$^+$, by comparison with those of [Li.C21C$_5$]$^+$ and [Na.C21C$_5$]$^+$ (Table 1), demonstrate the considerable importance of the fourth oxygen of C211 in rendering its cryptates substantially more stable than the C21C$_5$ cryptates irrespective of their exclusive or inclusive structure. The greater stability of [Li.C211]$^+$ by comparison with that of [Li.C21C$_5$]$^+$ is largely a consequence of the greater k_d of the latter cryptate; and similarly the greater stability of [Na.C211]$^+$ by comparison with that of [Na.C21C$_5$]$^+$ is largely a consequence of the greater k_d of the latter cryptate (Table 1). This suggests that the rate determining step controlling the magnitude of k_d for [M.C211]$^+$ involves the disruption of the interaction between the fourth oxygen of C211 and M$^+$, while the absence of this interaction in [M.C21C$_5$]$^+$ results in greater k_d values. (A structural manifestation of these interaction differences is observed in the solid state where Na$^+$ is 0.14 Å and 0.37 Å above the face of the O$_3$ plane of the 4,7,10-trioxa-1,10-diaza ring to which it is bound in

Table 2. Parameters[a] for $[Na.C21C_5]^+$ in various solvents

Solvent	D_N^b	$\log K_s/\text{dm}^3\text{mol}^{-1}$ (298.2 K)	$10^{-5} k_f$ $\text{dm}^3\text{mol}^{-1}\text{s}^{-1}$ (298.2 K)	k_d s^{-1} (298.2 K)	$\Delta H_d^\#$ kJ mol^{-1}	$\Delta S_d^\#$ $\text{J K}^{-1}\text{mol}^{-1}$
acetonitrile	14.1	5.08	100	84.8±1.6	57.9±0.7	-13.8±2.1
propylene carbonate	15.1	5.12	25.5	19.4±0.5	70.3±0.5	15.3±1.4
acetone	17.0	3.98	84	878±6	54.4±0.4	-6.1±1.2
methanol	19.0	3.76	104	1800±50	44.9±0.1	-31.9±0.4
dimethylformamide	26.6	2.87	214	28800±300	40.0±0.1	-25.3±0.5
pyridine	33.1	3.72	4.9	93.5±0.5	62.8±0.2	3.3±0.5

[a] Quoted errors represent one standard deviation obtained from a linear regression analysis of the temperature dependence of experimental τ_c data through eqn 3.
[b] Gutmann donor number from ref. 6. The dielectric constants from the same reference are: acetonitrile 38.0, propylene carbonate 69.0, acetone 20.7, methanol 32.6, dimethylformamide 36.1, dimethyl sulfoxide 45.0, and pyridine 12.3. It should be noted that other authors have used D_N values for some solvents which differ from those originally derived by Gutmann - e.g. B.O. Strasser and A.I. Popov, *J.Am.Chem.Soc.*, 1985 <u>107</u>, 7921.

$[Na.C211.NCS]$ and $[Na.C21C_5.NCS]$ respectively.[1])

Exclusive $[Na.C21C_5]^+$ is characterized by larger k_f and k_d values than inclusive $[Li.C21C_5]^+$, which probably indicates that for the latter cryptate these rate constants are largely determined by entry to, and exit from the $C21C_5$ cavity. Thus Li^+ exchange between the solvated and inclusive cryptate environments is envisaged to proceed through a reactive exclusive intermediate as shown in the two major steps:

$$Li^+ + C21C_5 \underset{}{\overset{\text{fast}}{\rightleftharpoons}} [Li.C21C_5]^+ \underset{}{\overset{\text{slow}}{\rightleftharpoons}} [Li.C21C_5]^+ \qquad (4)$$
$$\qquad\qquad\qquad\quad \text{(exclusive)} \qquad\quad \text{(inclusive)}$$

In contrast Na^+ exchange between the solvated and exclusive cryptate environments proceeds through a single major step:

$$Na^+ + C21C_5 \underset{k_d}{\overset{k_f}{\rightleftharpoons}} [Na.C21C_5]^+ \qquad (5)$$
$$\qquad\qquad\qquad\quad \text{(exclusive)}$$

Corresponding mechanisms are envisaged for M^+ exchange on $[M.C211]^+$.

The data in Table 2 show that K_s characterizing equilibrium (5) is solvent dependent largely due to the variation of k_d. This variation appears to depend on the electron donating power of the solvent, as expressed through the Gutmann donor number (D_N),[6] and the steric characteristics of the solvent. Thus as D_N increases in the sequence: acetonitrile < acetone < methanol < dimethylformamide, k_d increases, which suggests that the ability of solvent to compete with C21C$_5$ for bonding sites on Na^+ in $[Na.C21C_5]^+$ is important in the rate determining decomplexation step. On this basis it appears that the lower than expected k_d values observed for propylene carbonate and pyridine may be a consequence of their effective electron donating abilities being decreased, below that expected from their D_N values, due to their bulkiness preventing a close approach to the sterically crowded Na^+ in $[Na.C21C_5]^+$. There is no apparent correlation between k_d and solvent dielectric constant (Table 2).

REFERENCES

1. S.F. Lincoln, E. Horn, M.R. Snow, T.W. Hambley, I.M. Brereton and T.M. Spotswood, *J.Chem.Soc., Dalton Trans.*, 1986, 1075.

2. A. Abou-Hamdan, T.W. Hambley, A.M. Hounslow and S.F. Lincoln, *J.Chem.Soc., Dalton Trans.*, (in press).

3. D. Moras and R. Weiss, *Acta Crystallogr., Sect. B*, 1973, 29, 400.

4. S.F. Lincoln, I.M. Brereton and T.M. Spotswood, *J.Chem.Soc., Faraday Trans. 1*, 1985, 1623; 1986, 1999.

5. Y.M. Cahen, J.L. Dye and A.I. Popov, *J.Phys.Chem.*, 1975, 79, 1292.

6. V. Gutmann, *Coordination Chemistry in Nonaqueous Solutions*, Springer-Verlag, Wien, 1968.

METAL-FREE MACROCYCLES VIA TEMPLATE METHOD: A STARTING POINT FOR SELECTIVE COMPLEXATION STUDIES.

D.E.Fenton,[a,*] B.P.Murphy,[a,b] R.Price,[c] P.A.Tasker[c] and D.J.Winter[a].
a Department of Chemistry, The University, Sheffield, U.K.
b School of Chemistry, Polytechnic of North London, London, U.K.
c ICI plc, Organics Division, Blackley, Manchester, U.K.

ABSTRACT. A new range of structurally-related macrocyclic ligands have been prepared and their potential as metal-ion discriminating agents examined.

INTRODUCTION. The recognition and binding of metal ions by organic ligands has obvious implications for many aspects of chemistry and biochemistry. Valinomycin and the actin series of cyclic ionophores [1] are examples of how Nature has made use of the special properties of macrocycles to achieve discriminatory behaviour. The chemist, on the other hand, has through the years developed many 'classical' analytical reagents [2] which show reasonable specificity for particular ions, but their introduction has often resulted from chance observation rather than a systematically designed program.
  While much elegant work has been produced on the selective behaviour of the polyether macrocycles towards alkali and alkaline earth metal ions [3], studies involving the transition and base metal ions have been relatively scant. We report here a versatile route to a new range of structurally similar macrocycles as the first part of a longer term project aimed at the elucidation of the principles governing metal ion selectivity.

DISCUSSION. There are basically two routes to the synthesis of macrocyclic ligands. These are:
(1) The <u>in situ</u> or template method, where the intervention of a metal ion directs the course of reaction towards cyclic rather than polymeric products. The product is usually a macrocyclic complex of the template metal used, and the macrocyclic cavity is normally of a size commensurate with the metal's ionic radius. While this technique has resulted in the isolation of many macrocycles which are not available by other means its main drawback is that demetallation to produce free ligand is not always possible; often the macrocycles prove unstable in the unco-ordinated state.

* To whom correspondence should be addressed.

Scheme 1

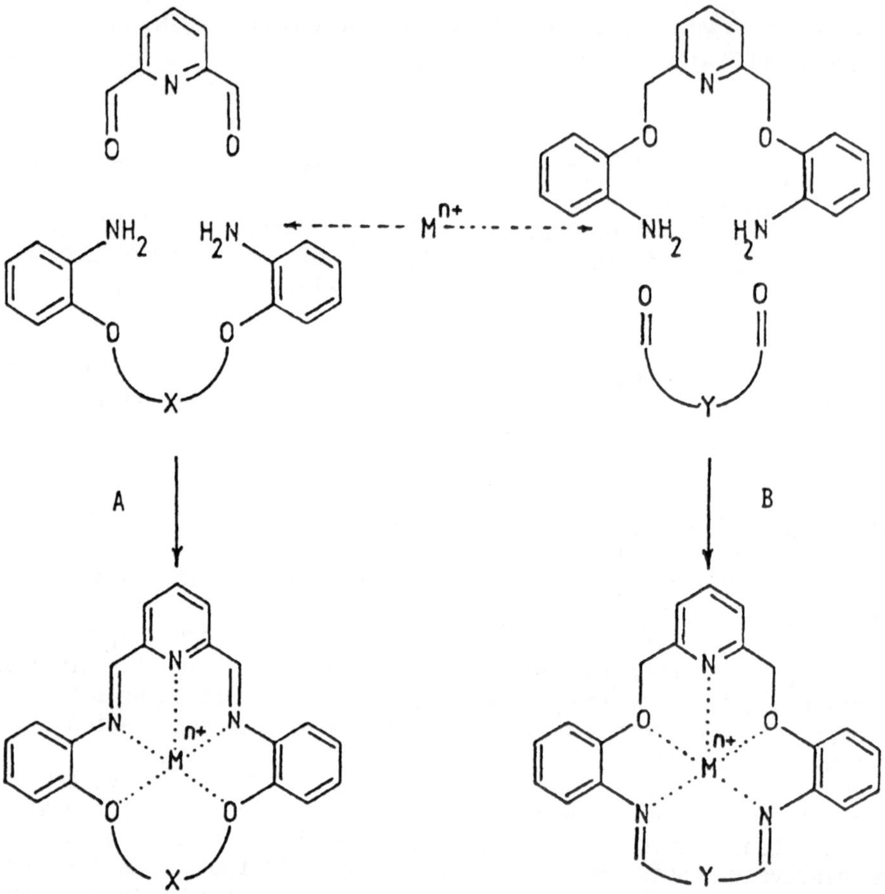

TABLE 1

F.A.B. m.s. of The Template Products

m/e	Assignment
496	$[Mn(\underline{1})(ClO_4)]^+$
511	$[Mn(\underline{2})(ClO_4)]^+$
525	$[Mn(\underline{3})(ClO_4)]^+$
-	$[Mn(\underline{4})(ClO_4)]^+$
527	$[Mn(\underline{5})(ClO_4)]^+$
-	$[Pb(\underline{6})(ClO_4)]^+$
497	$[Mn(\underline{7})(ClO_4)]^+$
727	$[Pb(\underline{8})(ClO_4)]^+$
614	$[Pb(\underline{9})]^+$
716	$Pb(\underline{10})(ClO_4)]^+$
655	$[Pb(\underline{11})]^+$
1104	$[Pb(\underline{11})_2]^+$
1205	$[Pb(\underline{11})_2(ClO_4)]^+$
684	$[Pb(\underline{12})(ClO_4)]^+$

F.A.B. Mass Spectroscopy

M/M'	$M(\underline{1R})^+$	%
Cu^{63}/Ni^{58}	409/404	100/67
Cu^{63}/Co^{59}	409/405	100/9
Cu^{63}/Mn^{55}	409/401	100/-
$Cu^{63}/Ag^{107}_{109}/Pb^{208}$	409/$^{453}_{455}$/554	100/$^-_-$/-

$(\underline{1R}) = 347$

Table 2

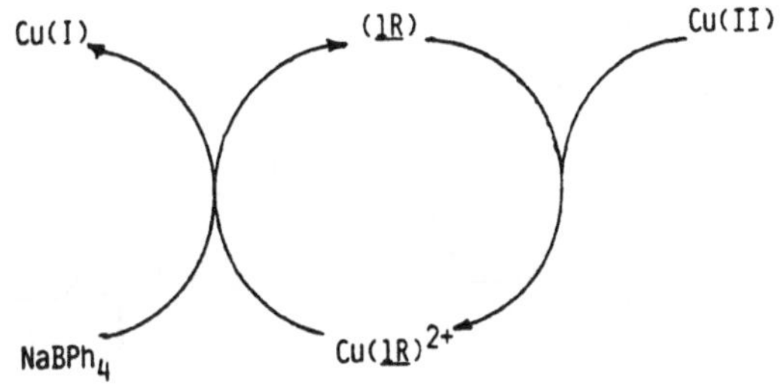

Scheme 2

(2) Direct synthesis, usually employing high dilution conditions, can be used. This approach produces the macrocycle in a form which not only makes spectroscopic characterisation easier, but which is also more amenable to organic modification.

The methods reported herein make use of the more favourable aspects of both routes.

Two approaches have been adopted (Scheme (1)) A - condensation of 2,6-diformylpyridine with various diamines and B - condensation of various dicarbonyls with a pyridyl diamine. Mn(II) or Pb(II) were used to template the reactions to produce the di-imine macrocyclic complexes.

The F.A.B. mass spectra results (Table 1) demonstrate the mononuclear di-imine nature of all but one of the products. The exception

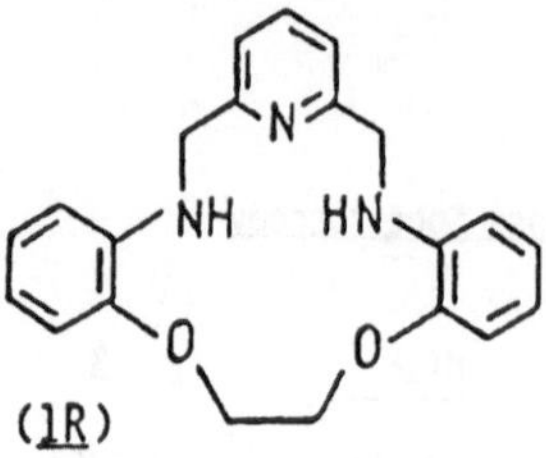

(1R) 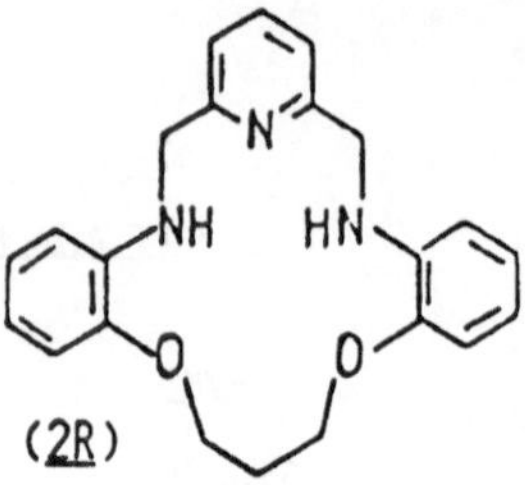 (2R) 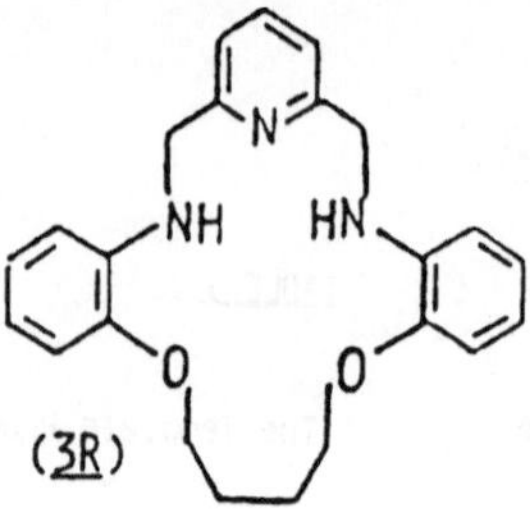 (3R)
(4R) 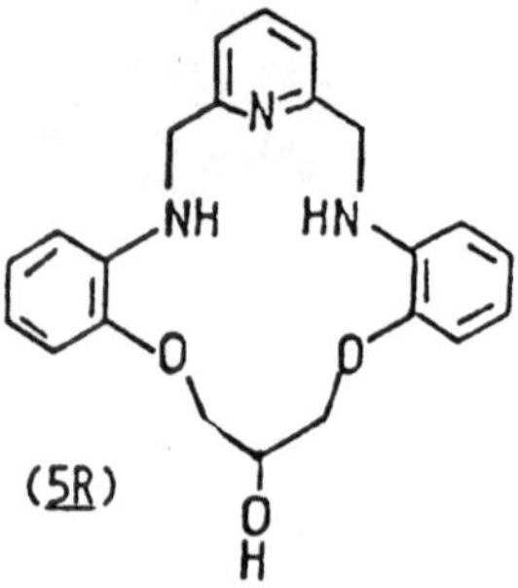 (5R) 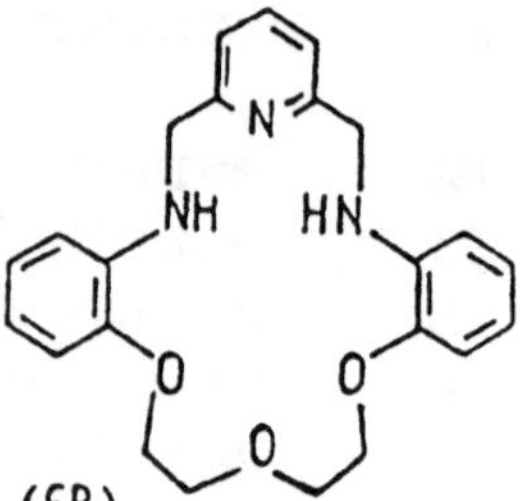 (6R) 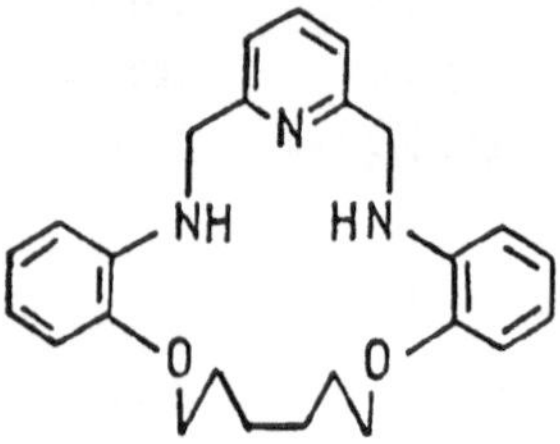
(7R) 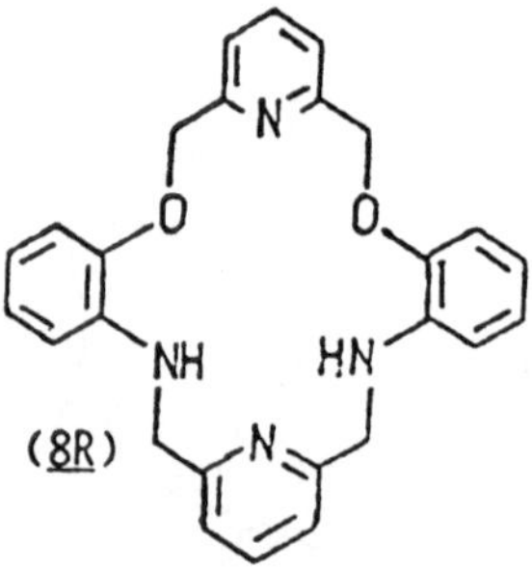 (8R) 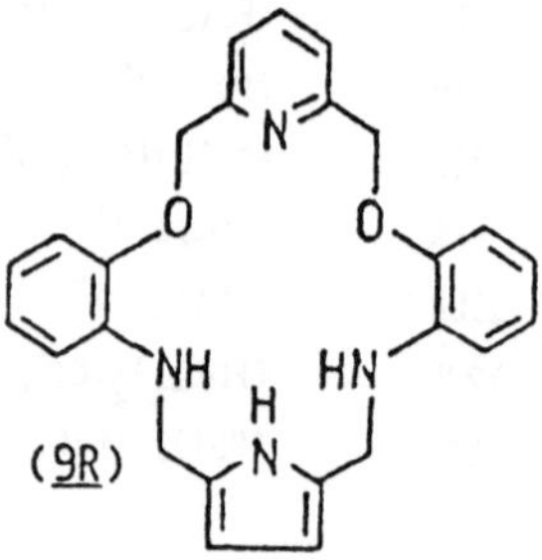 (9R) 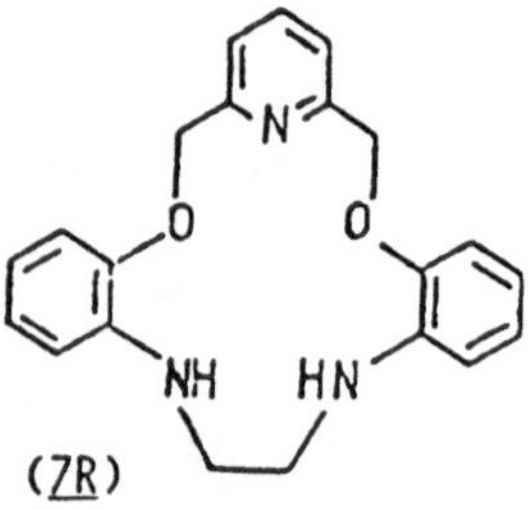
(10R) 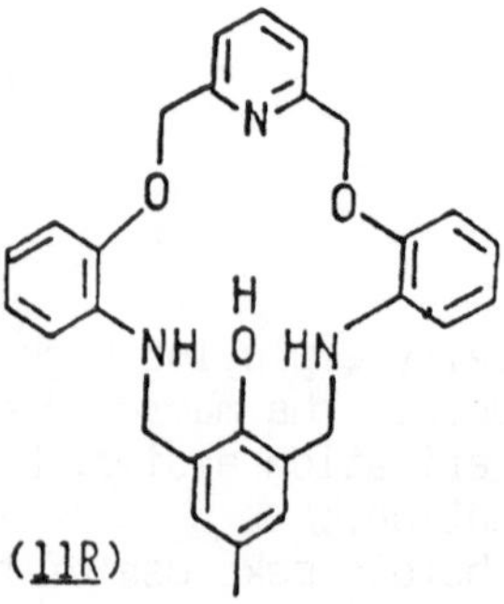 (11R) 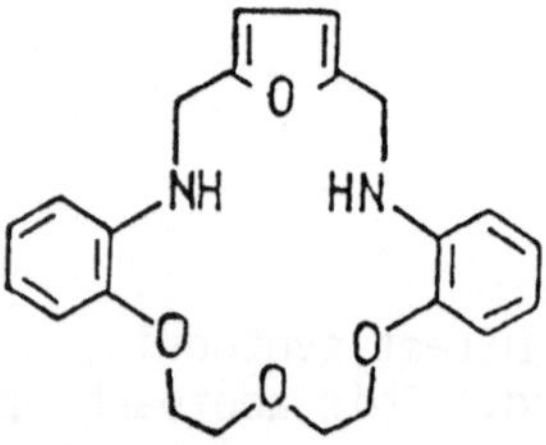 (12R) 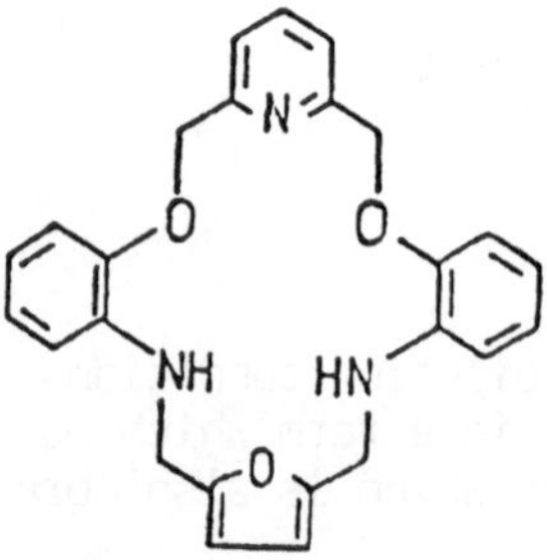

Matrix of Structural Types

Figure 1

Fig. 2

The structure of [Cu(<u>7R</u>)(NO$_3$)]NO$_3$ 0.5H$_2$O (Log K 6.84)

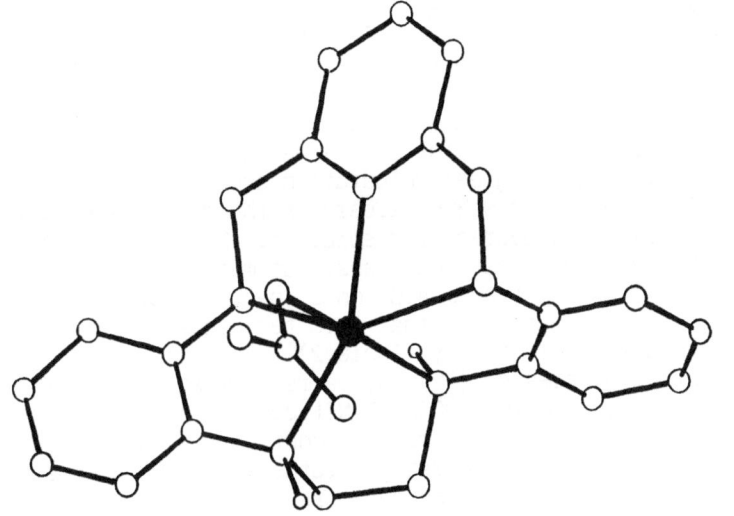

The structure of [Ni(<u>7R</u>)(NO$_3$)$_2$] 0.5H$_2$O (Log K < 3.0)

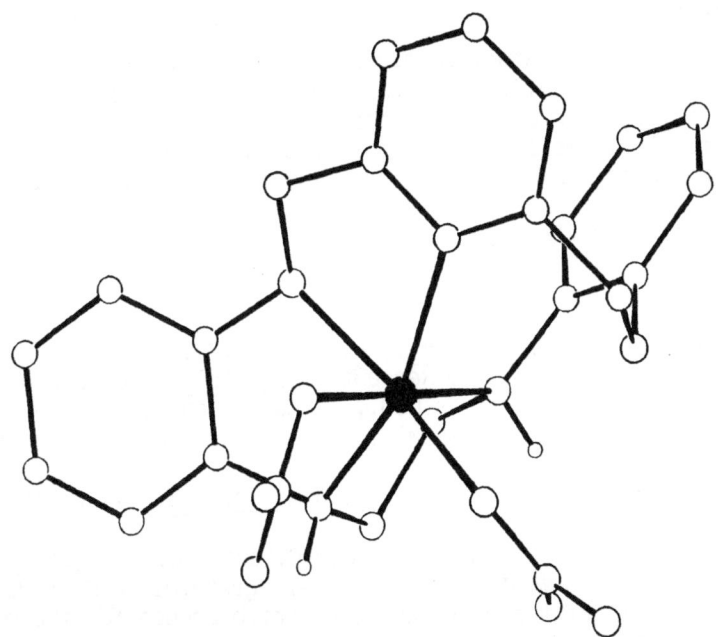

is thought to exist as a 'triple decker' - macrocyclic sandwich. The 2:1 and 1:1 [(11):Pb] fragments of $[Pb_2(11)_3][ClO_4]_2 \cdot 3H_2O$ can be seen.

The complexes of (1) to (12)* can be demetallated and reduced (using sodium borohydride) to give, in economic yields, the free macrocycles (1R) to (12R) which constitute a matrix of ligand structural types (Fig. 1) systematic variation in ring size, the number type and disposition of donors has been achieved.

Empirical discrimination for Cu(II) has already been observed with (1R) (Scheme 2). Intense green colouration occurs on mixing MeCN solutions. Treatment with $NaBPh_4$ liberates the metal as soluble Cu(I); the macrocycle being recovered from solution. No colour changes are observed with the other first row transition metals.

Preliminary results from F.A.B. mass spectroscopy experiments have demonstrated the potential that this technique may have as a screening method for selectivity (Table 2).

Equimolar quantities of (1R) and pairs or mixtures of metal ion chlorides were dissolved in 3-NOBA (3-Nitrobenzyl alcohol), the matrix for the F.A.B. experiment. The results confirm the above empirical observation in that Cu(II) consistly returns with 100% abundance for the species $[Cu(1R)]^+$ over all competitors.

Complementary X-ray diffraction and stability constant data (Fig.2) for the macrocycle (7R) indicate an enhanced ligand propensity for Cu(II) relative to Ni(II). The origins of this discriminatory behaviour can be seen from the crystal structures to lie in the ability of Cu(II) and the inability of Ni(II) to accommodate the complete array of macrocyclic donor atoms. A complete study of this matrix of ligands is currently underway.

CONCLUSION. Facile and economic routes to a new series of macrocyclic ligands have been achieved. Within this series the deliberate and systematic variation of many of the features of the macrocycles lends them particularly well to a designed program aimed at rationalizing the factors which govern discriminatory behaviour towards metal ions.

* (12R) obtained via route A (Scheme 1) using 2,5-diformylfuran as dicarbonyl.

REFERENCES.
(1) D.E.Fenton: Chem. Soc. Rev., 6, 325 (1977).
(2) See for example Organic Reagents for Metals, W.C.Johnson, Ed., Hopkin and Willimas Ltd., Chadwell Heath, 1964 and Solvent Extraction of Metals, R.A.Chalmers, Ed., Van Nostrand Reinhold, London, 1970.
(3) See for example B.Dietrich, J. Chem. Ed., 62, 954 (1985).

ACKNOWLEDGEMENTS. We thank the S.E.R.C. and I.C.I. Organics Division for a co-operative award to B.P.M. We are grateful to Dr M.McPartlin and Mr A.Bashall of the Polytechnic of North London for the preliminary crystal structures.

TORANDS: PLANAR POLYAZAMACROCYCLIC LIGANDS FOR METAL IONS

Thomas W. Bell, Albert Firestone, Frieda Guzzo and Lain-Yen Hu
Department of Chemistry, State University of New York,
Stony Brook, New York 11794-3400 USA

ABSTRACT. A series of macrocyclic ligands related to hexaaza[18]-annulene form stable complexes with alkali metal and alkaline earth ions. A planar, substituent-solubilized "torand", consisting of multiply fused pyridine rings, has been synthesized and has been found to sequester calcium from a dilute source.

Since the discovery of crown ethers by Pedersen[1] there have been many attempts to design and synthesize hosts whose ion affinities and selectivities surpass those of the original cyclic polyethers. Particularly important examples are the cryptands[2] and spherands[3]. These polycyclic receptors form stronger complexes and are generally more selective than crown ethers, however their complexes equilibrate more slowly as a consequence of their more rigid, encapsulating structures. This report is focused on the development of a new class of macrocyclic hosts, the torands, whose rigid toroidal structures should permit rapid equilibration of complexes.

Replacement of the six oxygen atoms of 18-crown-6 (1) with six sp^2 hybridized nitrogen atoms affords the hypothetical hexaaza[18]-annulene (2), an impractical host from the perspective of configurational and chemical stability. Fusion of six-membered rings to the periphery of 2 generates more feasible structures, such as 3, 4 and 5. Of these potential hosts, only 5 would have the shape of a rigid torus although we have also investigated 3 and 4 as more accessible model systems for 5. These model studies are important because saturated nitrogen analogues of crown ethers form considerably weaker alkali metal complexes.[4] This series of hosts also makes it possible to systematically examine the effects of rigidity on complexation properties.

Torand model 3 was previously known only as alkaline earth, lead and cadmium perchlorate complexes, which were prepared by templated condensation of o-phenylenediamine with 2,6-pyridinedicarboxaldehyde.[5] We have successfully converted strontium perchlorate and trifluoromethanesulfonate (triflate) complexes of 3 to the corresponding potassium complexes (3·K$^+$) using potassium fluoride.[6] Decomplexation of 3·K$^+$ with [2.2.2]-cryptand or excess 18-crown-6 then affords the free ligand (3), which in the solid state adopts the roughly elliptical conformation shown.[7] NMR data indicate that the solution conformation of 3 also deviates from the circular conformation observed in complexes[5]. Thus, macrocycle 3 does not exhibit the rigidity of a torand, nevertheless it may demonstrate how complex stabilities change when sp^2 hybridized nitrogen replaces oxygen in a flexible system. We proceeded to synthesize 3·Na$^+$ from 3 and probed the stability constants of 3·Na$^+$ and 3·K$^+$ first by NMR[6] and more recently by means of ion selective electrodes. The latter, more accurate method affords stability constants in DMSO as follows: logKs(3·K$^+$) = 3.7; logKs(3·Na$^+$) = 3.0. Torand model 3 is accordingly the first nitrogen analogue of 18-crown-6 to form stronger complexes with alkali metal ions (18-crown-6: logKs(K$^+$) = 3.2; logKs(Na$^+$) = 1.5).[8]

The flexibility observed in macrocycle 3 led us to search for more rigid, yet easily synthesized, torand models, such as diphenanthroline-diimines (4). Diamine 6 was prepared from the dihydrochloride[9] using alkaline Dowex 1-X8 50 anion exchange resin, then was condensed with dialdehyde 7[9] in refluxing methanol in the presence of strontium triflate. Concentration of the reaction mixture, addition of benzene and filtration of the grey precipitate afforded 4·Sr^{2+} in 39% yield. Microanalytical data for Sr, H and N all agreed with the expected structure, whereas the carbon analysis was slightly high because of a trace impurity of benzene. The ^{1}HNMR spectrum of 4·Sr^{2+} in DMSO-d$_6$ revealed a complex pattern that was temperature independent (20-80°C). Homonuclear decoupling experiments permitted the assignment of two structures, as shown. Symmetry arguments allowed identification of the minor isomer as the expected product, since it has two types of symmetrical phenanthroline

TORANDS: PLANAR POLYAZAMACROCYCLIC LIGANDS FOR METAL IONS

rings. The isomer ratio always remained 1:2, even when the mixture was reprecipitated, suggesting that interconversion may occur by a prototropic equilibrium. Our attempts to directly remove strontium from $4 \cdot Sr^{2+}$ have not succeeded, so we reduced the unsaturated bridges using sodium borohydride in methanol, then isolated diphenanthrolinediamine 8 as it picrate salt.

These torand model studies demonstrate that effective complexing agents for alkali metal and alkaline earth ions may be constructed using sp^2 hybridized nitrogen binding sites. In addition, the low solubilities of the model compounds led us to incorporate flexible, solubilizing substituents in our synthetic approaches to fully fused torands (5). We have already reported the synthesis of a soluble heptacyclic terpyridyl bearing n-butyl groups[10], Summarized here is an extension of this methodology to the successful synthesis of a torand.[11] The n-butyloctahydroacridine 9 was converted to benzylideneketone 10 (46%, three steps), which was dimerized by Newkome-Fischel pyrolysis of the trimethylhydrazonium salt[10], yielding dibenzylideneheptacycle 11 (31%). Ozonolytic cleavage of the benzylidene groups afforded diketone 12 in 60% yield. The complementary segment, bis(β-dimethylaminoenone) 14 was also prepared from 9 via dibenzylidene derivative 13. Macrocyclization was effected by treating a 1:1 mixture of 12 and 14 with trifluoromethanesulfonic acid in hot acetic acid followed by ammonium acetate. Neutralization of the reaction mixture with lithium hydroxide and chromatography of the chloroform extract gave torand 15 as the calcium triflate complex in 12% yield. Calcium is introduced as a 0.3% impurity in the triflic acid used in this experiment. The spectroscopic properties of 15 are similar to those reported for the unsubstituted parent system.[12]

Efforts are now in progress to isolate the new torand as a free ligand and to survey the stabilities of alkali metal and alkaline earth complexes. At this point it is clear that this new host has unusual and potentially useful properties. Calcium is apparently sequestered from a dilute source yielding a lipophilic complex. Strong metal-ligand interaction is also implied by the FAB mass spectrum of 15, which displays m/z 975 (15-CF_3SO_3) and m/z 826 (15-(CF_3SO_3)$_2$) as the two most abundant ions above m/z 400. An investigation of the solution stability constant of 15 is now in progress. In conclusion, the model studies and the pilot synthesis of a torand complex described here indicate that the torands represent a promising new class of ionophores.

Acknowledgment is made to the National Institutes of Health and to the Petroleum Research Fund, administered by the American Chemical Society, for support of this research.

1. Pedersen, C.J. J. Am. Chem. Soc. **1967**, _89_, 2495-2496.
2. Dietrich, B.; Lehn, J.M.; Sauvage, J.P. Tetrahedron Lett. **1969**, 2885-2888.
3. Cram, D.J.; Kaneda, T.; Helgeson, R.C.; Lein, G.M. J. Am. Chem. Soc. **1979**, _101_, 6752-6754.
4. Frensdorff, H.K. J. Am. Chem. Soc. **1971**, _93_, 600-606.
5. Drew, M.G.B.; Cabral, J. de O.; Cabral, M.F.; Esho, F.S.; Nelson, S.M. J. Chem. Soc. Chem. Commun. **1979**, 1033-1035.
6. Bell, T.W.; Guzzo, F. J. Am. Chem. Soc. **1984**, _106_, 6111-6112.
7. Bell, T.W.; Guzzo, F. J. Chem. Soc., Chem. Commun. **1986**, 769-771.
8. Kolthoff, I.M.; Chantooni, M.K., Jr. Anal. Chem. **1980**, _52_, 1039-1044.
9. Chandler, C.J.; Deady, L.W.; Reiss, J.A. J. Heterocycl. Chem. **1981**, _18_, 599-601.
10. Bell, T.W.; Firestone, A. J. Org. Chem. **1986**, _51_, 764-765.
11. Bell, T.W.; Firestone, A. J. Am. Chem. Soc., in press.
12. Ransohoff, J.E.B.; Staab, H.A. Tetrahedron Lett. **1985**, _26_, 6179-6182.

Journal of Inclusion Phenomena **5** (1987), 153–156.

SYNTHESIS OF SYMMETRICAL N-TOSYLDIAZAMACROCYCLES AND COMPLEXATION PROPERTIES OF THEIR DERIVATIVES

Sebastiano Pappalardo, Francesco Bottino
Dipartimento di Scienze Chimiche, Università di Catania
Viale A. Doria 8, 95125 Catania, Italy

Paolo Finocchiaro*, Antonino Mamo
Istituto Chimico, Facoltà di Ingegneria, Università di Catania
Viale A. Doria 6, 95125 Catania, Italy

Frank R. Fronczek
Department of Chemistry, Louisiana State University
Baton Rouge, LA 70183-1804

1. INTRODUCTION

Synthetic azamacrocycles have attracted considerable attention in recent years because of their ability to selectively bind both metal [1] and organic ammonium cations [2].

We have recently developed a general synthetic procedure which offers an easy access to a variety of multifunctional N-tosylazamacrocycles (azacrown ethers, azacyclophanes, and azaheterophanes) [3]. Detosylation of these materials has afforded multifunctional (poly)aminomacrocycles, which have proved to be a suitable matrix for structurally defined polycondensates. As a result, two novel polyamide ligands 7 and 8, containing cyclic or acyclic bipyridinediyl diamino units 5 and 6 in the polymer backbone, have been prepared and their complexing properties towards bivalent ions of the first transition series explored [4].

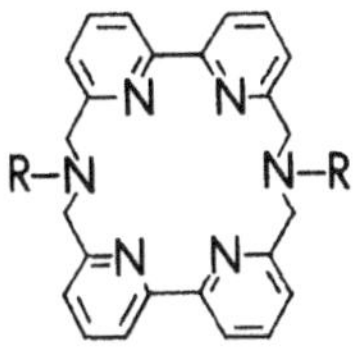

a: R=Ts
b: R=H
c; R=C_6H_5CO

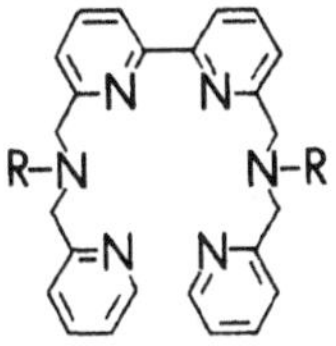

5 6

2. RESULTS

The synthesis of symmetrical multifunctional N-tosylazamacrocycles 4 has been accomplished by coupling readily available bis(halomethyl) or bis(tosylate ester) compounds 1 with two equiv of tosylamide monosodium salt (TsNHNa) in anhydrous dimethylformamide (DMF) under moderate dilution. The reaction is best envisioned as proceeding through the monoalkylated key intermediate 2, which quickly undergoes self-condensation (path A) in the presence of an excess of TsNHNa to give the desired 1:1 (n=0) and/or 2:2 (n=1) macrocycles. Alternatively, 2:2 macrocycles can arise from the N,N'-ditosyl intermediate 3 through path B (Scheme 1).

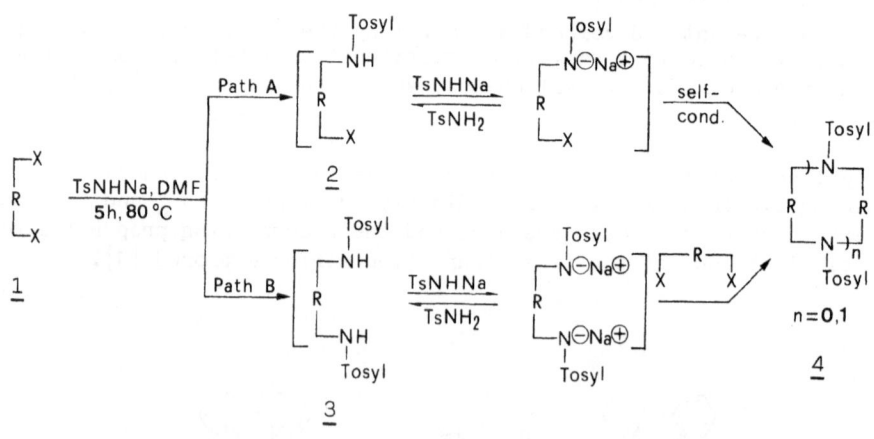

Scheme 1

Detosylation of 4 to the corresponding (poly)aminomacrocycles was effected by conc. H_2SO_4 or Na/liquid NH_3.

Polymers 7 and 8 were obtained by polycondensation of the appropriate diamino-precursor with one equiv of terephtaloyl chloride in anhydrous chloroform. Model compounds 5c and 6c were synthesized by similar routes.

Complexation of models 5c and 6c as well as of polyamides 7 and 8 with Co(II), Ni(II), and Cu(II) has been achieved. The ligands and their complexes have been characterized by conventional techniques. The [Co(6c)](NO$_3$)$_2$·2H$_2$O complex has been further characterized by a single crystal X-ray analysis.

3. CONCLUSIONS

^{1}H-VTNMR analysis on macrocyclic diamide 5c indicates a fixed syn conformation in solution, with a 'face to face' arrangement of the dipyridinyl moieties, as shown in Fig. 1.

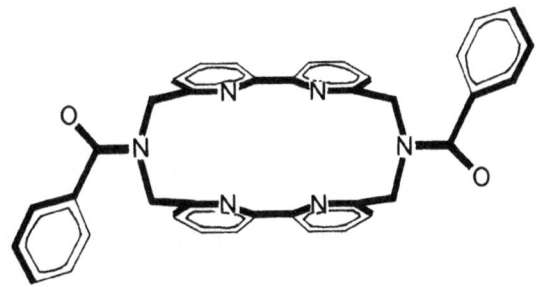

Fig. 1. The syn conformation of compound 5c.

Model compound 6c and polyamide 8 form stable 1:1 complexes with transition metals Co(II), Ni(II), and Cu(II). Conversely, macrocyclic model 5c and polyamide 7 exhibit the unique property to specifically complex Cu(II) even in the presence of sizeable amounts of Co(II) and Ni(II).

The IR data show that Co(II) and Ni(II) complexes exhibit a large shift (55-60 cm^{-1}) of the carbonyl absorption to lower frequencies, strongly suggesting that the carbonyl groups are involved in the complexation; in sharp contrast, the C=O frequency band is not affected in the Cu(II) complexes, indicating for the latter different binding sites.

According to the IR data, the single crystal X-ray structure determination of [Co(6c)](NO$_3$)$_2$·2H$_2$O indicates an octahedral environment around the complex cation, as shown in Fig. 2. The square planar coordination positions are occupied by the bipyridine nitrogens and by the two carbonyl groups, while two water molecules reside at the axial positions.

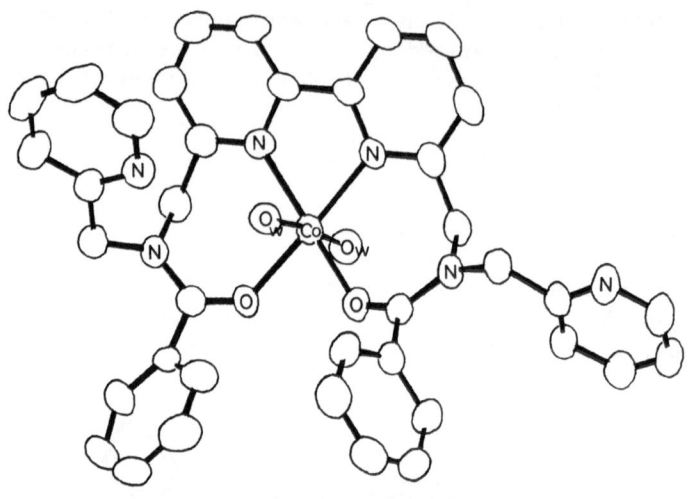

Fig. 2. ORTEP drawing of $[Co(6c)](NO_3)_2 \cdot 2H_2O$ complex

Acknowledgement. The authors gratefully acknowledge the Ministero della Pubblica Istruzione (Fondi 40%) for partial support of this work.

References

1. J. J. Christensen, D. J. Eatough, and R. M. Izatt: Chem. Rev. 74, 351 (1974); J. D. Lamb, R. M. Izatt, and J. J. Christensen: 'Chemistry of Macrocyclic Compounds', ed. by G. Melson, Plenum Press, New York 1979.
2. J. M. Lehn and P. Vierling: Tetrahedron Lett. 1323 (1980).
3. S. Pappalardo, F. Bottino, M. Di Grazia, P. Finocchiaro, and A. Mamo: Heterocycles 23, 1881 (1985).
4. S. Pappalardo, F. Bottino, P. Finocchiaro, and A. Mamo: J. Polym. Sci., Polym. Chem. Edn., in press.

SIZE AND CHARGE DEPENDENCE OF BINDING BY AZACYCLOPHANES

Guy Lepropre and Jacques Fastrez*
Laboratoire de Biochimie Physique et des Biopolymères,
Place L. Pasteur, 1/1B,
1348 Louvain-la-Neuve, Belgium

ABSTRACT. Tetrameric and hexameric azacyclophanes, water soluble in the acidic or in the full pH range, and bearing functional side chains are shown to bind aromatic molecules or transition metal ions. The effect of the macrocycle size and charge on binding is investigated.

For some time, the molecular inclusion of an organic guest molecule into an organic host was mainly confined to the cyclodextrin chemistry [1]. In the last decade, however, quite a number of different systems have been proposed among which cyclophanes appear to form a promising family [2,3,4]. To understand what controls the binding and predict the selectivity, more work needs to be done on the effect of the structure of the host on its properties. It is the purpose of this communication to report on the effect of the size and the charge of the host on the binding.

Several tetrameric or hexameric aza-paracyclophanes bearing functional side chains have been prepared by coupling under high dilution conditions (continuous addition of 150ml of .04 molar solutions of N,N'-dimethoxycarbonylmethyl-p-xylilenediamine and αα'-dibromoxylene to 600 ml of boiling acetonitrile in the presence of potassium carbonate in the course of 5 hours). The isolated yield of Ia (white crystals, mp: 211-2°C) and IIa (oil) is 18% and 13% respectively. The alanine derivative (mp: 204-5°C) is obtained in 12% yield. The valine derivative could not be prepared.

Hydrolysis of Ia by .25 molar sodium hydroxide in methanol/water (80/20% vol), neutralization and desalting on G25 Sephadex yields Ic (80%).

Reduction with borane in THF or lithium aluminium hydride in dioxane yields after the usual work up [5] Id or IId in 65% yield.

Hydroxylaminolysis of Ia in methanol/THF (75/25% vol), .5 molar in hydroxylamine hydrochloride and 1 molar in sodium methoxide yields after neutralization and desalting 70% of Ie.

All compounds have been characterized by IR, ^{1}H and ^{13}C-NMR spectroscopy and Ia and IIa further by analysis, mass specrometry and/or gel permeation chromatography on a 100Å-5μ-Styragel (500/7.7 mm column in

THF calibrated with cyclic oligoesters of known molecular weight.

$$\text{I} \qquad\qquad \text{II}$$

a: $R = CH_2COOCH_3$, b: $R = CH(CH_3)COOCH_3$, c: $R = CH_2COOH$, d: $R = CH_2CH_2OH$, e: $R = CH_2CONHOH$.

Ic is soluble in the entire pH range between pH 1 and 13. The pKas have been determined by potentiometric titration and the protonation scheme checked by proton NMR. It is zwitterionic at pH 5, the carboxylates become protonated below this pH and the amino functions above it. The following pKas corresponding to the equation:

$$H_4A_4 \underset{}{\overset{Ka_1}{\rightleftharpoons}} H_3A_4^- \underset{}{\overset{Ka_2}{\rightleftharpoons}} H_2A_4^{2-} \underset{}{\overset{Ka_3}{\rightleftharpoons}} HA_4^{3-} \underset{}{\overset{Ka_4}{\rightleftharpoons}} A_4^{4-}$$

have been obtained in water (0.3M KCl): $pKa_1 = 6.34$, $pKa_2 = 7.43$, $pKa_3 = 8.28$, $pKa_4 = 9.30$.

The pKas of the carboxylic functions cannot be determined accurately. The first pKa lies around 2.5. The macrocycle is not fully protonated at pH 1.

Ie is water soluble below pH 2.5 and above pH 9.5. It is easily acylated by p-nitrophenylacetate at pH 10. There is no indication of a turnover, a prerequisite for catalysis. This may be related to the fact that, underbasic conditions, O-acyl-hydroxamic acids decompose with Lossen rearrangment [6]. Comparison of the acylation rate (k = 300/M.s) with that of a simple hydroxamic acid, acethydroxamic acid (k = 93/M.s) [7], suggests that the macrocyclic structure does not contribute to facilitate the acylation significantly, either because the ester is very poorly bound to the macrocycle or because the bound substrate is not correctly oriented for reaction. From the observations on the binding ability of the macrocycles reported below, the first hypothesis appears to be more likely.

The binding of several organic molecules has been measured by fluorimetry either by measuring the influence of the macrocycles on the fluorescence of the guest directly (for 1-anilino-8-naphthalene sulfonate (ANS) or 2-toluidino-6-naphthalene sulfonate (TNS) anions) or or by measuring the inhibition of the binding of the formers. The technique has been used previously [2]. The data are collected in Table I.

From the data presented, it is clear that the tetrameric macrocycles bind alpha substituted naphthalene derivatives, simple aromatic compounds or 1-amino-adamantane very poorly, a beta substituted naphthalenic compound (TNS) is bound more efficiently.

Table I.
Dissociation constants of complexes of organic molecules
to azacyclophanes.

Host	Guest	pH	Kd (M)	Note
Ic	1,8-ANS	1.0	$3.4 \; 10^{-2}$	a
"	2,6-TNS	2.0	$1.4 \; 10^{-2}$	a
"	2,6-TNS	5.5	$1.7 \; 10^{-2}$	a
Ie	1,8-ANS	2.0	$2.0 \; 10^{-2}$	a
Id	1,8-ANS	1.0	$1.5 \; 10^{-2}$	-
"	2,6-TNS	2.0	$1.0 \; 10^{-3}$	-
"	1-Naphthalene sulfonate	2.0	$1.4 \; 10^{-2}$	b
"	Tosylate	2.0	$2.6 \; 10^{-2}$	b
"	1-aminoadamantane	2.0	$4.5 \; 10^{-2}$	b
IId	1.8-ANS	1.5	$\leqslant 3 \; 10^{-4}$	-
"	2,6-Naphthalene disulfonate	1.5	$\leqslant 1 \; 10^{-4}$	b

a. Saturation not reached, dissociation constant estimated on the assumption that the effect of the macrocycle on the fluorescence is the same for all cyclophanes.
b. measured by inhibition of the binding of ANS or TNS.

Compared to Id, the hexameric compound IId is more efficient in binding ANS or a beta sustituted naphthalene derivative. It is not clear why the tetrameric macrocycles investigated in this work have a lower affinity for ANS than the corresponding cyclophanes reported by Tabushi et al [2] . (I with R = CH_3. Kd = 1.8 mM). Either the side chains are filling the cavity to some extent, or they are forcing a conformation of the macrocycle where the cavity is essentially closed. This question is being addressed by X-Ray crystallography. The cyclophanes introduced by Koga [3], form more stable complexes, this is likely to be due to the fact that the cavity is larger and that the diphenylmethane unit used to construct them confers some rigidity.
There is some indication that the charge of the macrocycle influences the binding as shown by the fact that the tetraprotonated Id binds TNS better than Ic at pH 2 (where the host is essentially monoprotonated) and at pH 5.5 where it is neutral. Binding by Ic at pH 12 is very weak,if present. The electrostatic effect is nevertheless relatively weak. This observation is consistent with the conclusions to be drawn from the the titration data, where the pKa splitting corrected for the statistical effect (a factor of 16) covers less than two orders of

magnitude.

Ic is also able to form a 1-1 complex with a cupric ion at pH 5.5, with a dissociation constant smaller than 10 µM. Comparison of the spectrum of this complex (λmax = 520 nm) with that of other cupric complexes [8] suggests a structure in which two nitrogens from the skeleton of the macrocycle and two oxygens from its side chains are ligated to the metal ion. In this complex, the macrocycle would be in a "closed form". In agreement with this proposal, the cupric complex is unable to bind ANS.

Acknowledgements.
G.L. is grateful to the Belgian Institute for Scientific Research in Industry and Agriculture for a Fellowship. J.F. is a Research Associate of the Belgian National Fund for Scientific Research.

References.

1. M.L. Bender and M. Komiyama: Cyclodextrin Chemistry (Reactivity and Structure Concepts in Organic Chemistry, vol. 6), Springer Verlag, West-Berlin (1978).
2. I. Tabushi, Y. Kuroda, I. Kimura: Tetrahedron Lett. 37, 3327 (1976); I. Tabushi: Acc. Chem. Res. 15, 66 (1982) and references therein.
3. I. Takahashi, K.Odashima, K. Koga: Tetrahedron Lett. 25, 973 (1984) and references therein.
4. Y. Murakami, J.-I. Kikuchi and H. Tenma: Chem. Letters 1036 (1985).
5. J.M. Lehn, J. Simon and J. Wagner: Nouv. J. Chim. 1, 77 (1977).
6. D.G. Doare, A. Olson and D.E. Koshland: J. Am. Chem. Soc. 90, 1638 (1968).
7. M. Dessolin, M. Laloi-Diard, M. Vilkas: Bull. Soc. Chim. France 2573 (1970).
8. B.J. Hathaway, D.E. Billing: Coord. Chem. Rev. 5, 143 (1970).

MULTIPLE RECOGNITION IN POLYTOPIC ANION HOSTS

Franz P. Schmidtchen
Lehrstuhl für Organische Chemie und Biochemie der TU München
D-8046 Garching, Lichtenbergstr. 4, FRG

Selectivity in guest binding by biogenic or abiotic receptors depends on the type, number, spatial orientations and overall flexibility of the ensemble of recognition sites interacting with the substrates. The putatively optimal approach to construct a selective host compound is to arrange anchor groups on a rigid concave molecular framework capable of inclusion of the guest species. This route, however, may soon reach the limits of synthesizability if polyfunctional biologically relevant guests (nucleotides, hormones, coenzymes) are to be selectively bound.

An alternative is to connect modular receptor sites in a linear (branched or unbranched) open chain fashion to constitute a receptor which must be folded by the guest template for binding. Though the selectivity features may be inferior to a rigid host compound possessing the same types and numbers of anchor groups, substrate specificity in the former approach may easily be altered or enforced by simple attachment of supplementary recogniton sites.

In order to probe this concept the substrate specificities of two artificial ditopic receptors $\underline{1}$ and $\underline{2}$ [1] have been studied, each of which is composed of two different and independent anchor groups interlinked by a freely rotatable p-xylene spacer unit. The monotopic receptor $\underline{3}$, a common building block of either ditopic receptor $\underline{1}$ or $\underline{2}$ is known [2] to bind hydrophobic preferentially anionic guest species whereas the other subsites of $\underline{1}$ and $\underline{2}$ have been shown to complex hydrophic anions [2a] and prim. ammonium cations [3a], respectively.

The selectivity advantage of the ditopic design of $\underline{1}$ in relation to its monotopic parent compound $\underline{3}$ was investigated using the dimensional probes $\underline{4}$ - $\underline{8}$ each containing two anionic functions at a fixed distance. Analysis of the UV-absorption band shifts experienced by these solvatochromic probes on inclusion into the molecular cavity of the larger tetrahedral receptor site $\underline{3}$ yielded the 1 : 1 association constants in water.

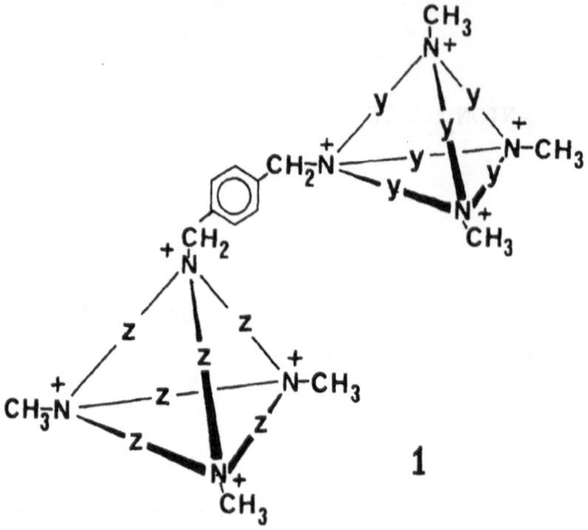

$y = -(CH_2)_6-$

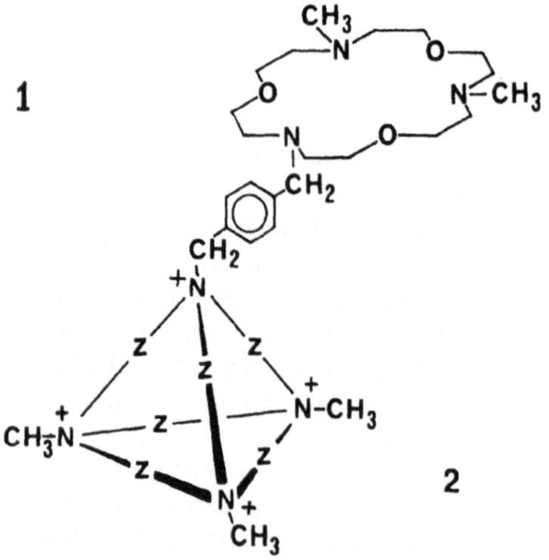

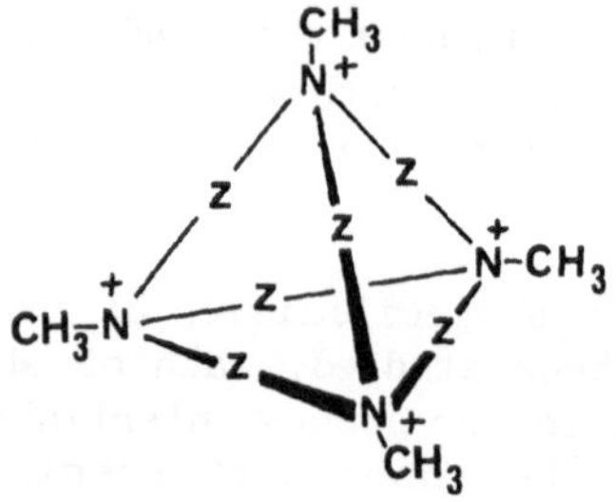

$z = -(CH_2)_8-$

The complexes with <u>1</u> are generally more stable than those with <u>3</u> owing to increased electrostatic attractions. In addition to this unspecific binding enhancement the specific interaction of suitably large substrates with <u>both</u> receptor subsites becomes apparent from inspection of the selectivity factor Q. Whereas the smaller substrates <u>4</u>, <u>5</u>, <u>6</u> are confined to the interaction with the big subsite in <u>1</u> due to insufficient spacing of the anionic moieties. The more extended probes <u>7</u> and <u>8</u> can span the gap between the receptor subsites in <u>1</u>. This additional interaction mode translates as a threefold increase in Q, although rotation of the subunit with respect to each other could paralyze simultaneous recognition of both anionic moieties of the probe.

MULTIPLE RECOGNITION IN POLYTOPIC ANION HOSTS

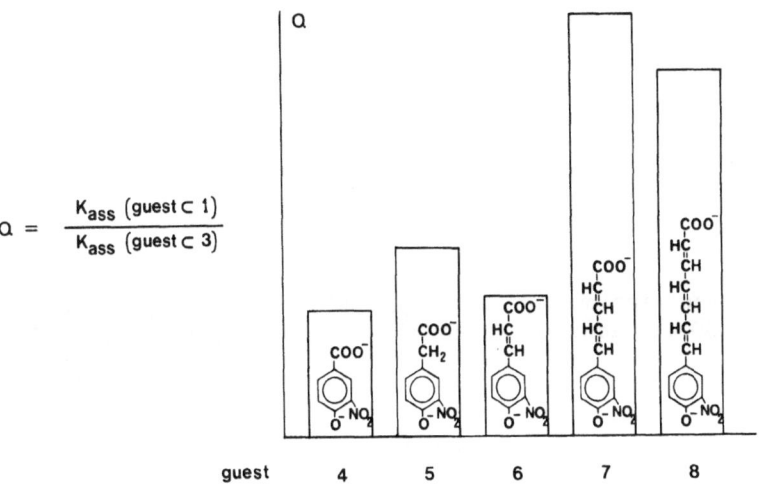

$$Q = \frac{K_{ass}\,(guest \subset 1)}{K_{ass}\,(guest \subset 3)}$$

Association constants K_{ass} $[M^{-1}]$ and selectivity factors Q of the dimensional probes 4 - 8 with the monotopic (3) and ditopic (1) receptor in water, pH 8.8, 27°C.

substrate		4	5	6	7	8
K_{ass}	1	714	322	2041	5265	10 000
	3	208	62	556	476	1042
	Q	3.4	5.1	3.7	11.1	9.6

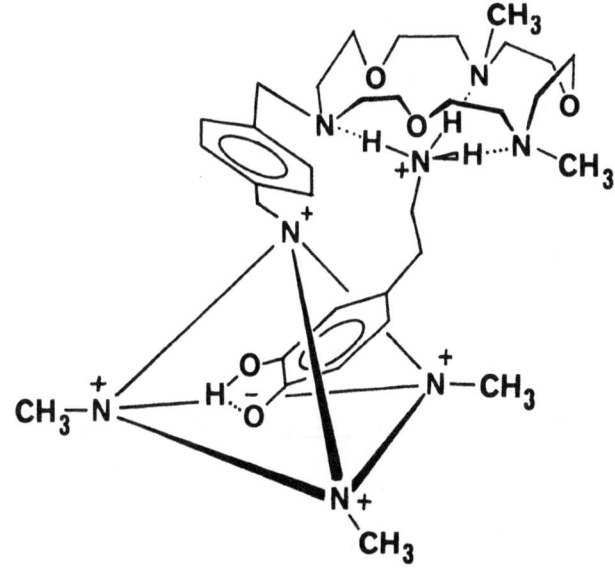

Fig. 1:

The combination of a receptor unit for hydrophobic (anionic) species (3) with a prim. ammonium anchor group (the azacrown ether [3]) is expected to yield a ditopic receptor for biogenic amines. A possible complex structure is depicted in Fig. 1.

Preliminary measurements of the association constants by a competition method with K^+ ion using K^+ selective electrodes reveal that the ditopic receptor 2 preferentially binds hydrophobic amines.

Association constants K_{ass} [M^{-1}] of amine guest compounds in 80 wt% aqueous methanol; pH_{obs} = 10.00; μ = 1.0, 25°C.

substrate	K_{ass}
K^+	246
5-aminopentanol	83
tyramine	184
dopamine	134
tyrptamine	183

Conclusion:

Even the first and simplest member of the family of linear modular receptors, the ditopic host 1, exhibits a measurable selectivity advantage solely attributable to the ditopic design. It amounts to a factor of 3 with respect to binding homologous substrates.

References:

1) F.P. Schmidtchen, Tetrahedron Lett. 27, 1987 (1986); see also: F.P. Schmidtchen, ibid. 25, 4361 (1984)

2) F.P. Schmidtchen, Chem.Ber. 114, 597 (1981); F.P. Schmidtchen J.Chem.Soc.Perkin Trans. II, 1986, 135.

3) J.-M. Lehn, P. Vierling, Tetrahedron Lett. 21, 1323 (1980)

Journal of Inclusion Phenomena 5 (1987), 165–168.
© 1987 by D. Reidel Publishing Company.

CRYSTAL STRUCTURES OF 1:1 COMPLEXES BETWEEN UREA AND TWO CROWN ETHER DERIVATIVES OF PHTHALIC ACID

Franco Benetollo,[a] Gabriella Bombieri,[b] and Mary R. Truter.[c]

[a]Istituto di Chimica e Tecnologia dei C.N.R., 35100 Padua, Italy

[b]Istituto di Chimica Farmaceutica e Tossicologica, Universita di Milano, 20131 Milan, Italy

[c]Department of Chemistry, University College London, 20 Gordon Street, London WC1H 0AJ, U.K.

INTRODUCTION

Interest in urea inclusion compounds has changed in three decades from its properties as a host to its potential as a guest, more or less directly in the hope that improvement in artificial kidneys can be achieved. There is a fine balance between the tendency of urea molecules to hydrogen bond to each other and to form external bonds, as required for encapsulation. Further, it may form uronium salts with strong acids.

We have determined the crystal structures of 1:1 adducts of urea with the host molecules A and B based on phthalic acid as a functionalised crown ether.

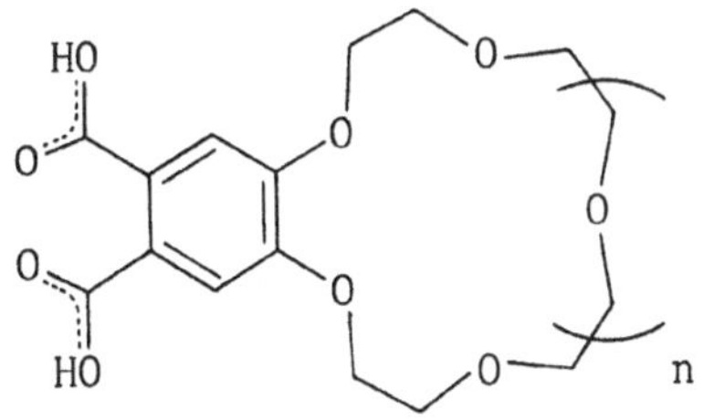

n = 1 A
n = 2 B

CRYSTAL DATA

Urea-A, $C_{17}H_{24}N_2O_{10}$, M=416.4. Orthorhombic, Space group $P2_12_12_1$, a=8.750(2), b=10.844(3), c=21.215(3) Å. U=2013(1) Å^3. λ=0.71069 Å. Z=4, D_C=1.37 g cm^{-3}, R=0.083 for 599 observations (I⩾3σ[I]).

Urea-B, $C_{19}H_{28}N_2O_{11}$, M=460.4. Triclinic, Space group $P\bar{1}$, a=8.336(2), b=11.009(2), c=13.313(2) Å, α=105.55(3), β=103.62(3), γ=104.63(3) °. U=1076.9(5) Å. Z=2, D_c=1.420 g cm^{-3}. λ=0.71069 Å. R=0.072 for 2105 observations (I⩾3σ[I]).

RESULTS AND DISCUSSION

The location of all the hydrogen atoms in urea-B shows unequivocally that it is not a uronium salt. A similar conclusion can be drawn from the hydrogen-bonding pattern of urea-A. In both structures N-H···O bonds lie in the range 2.9 to 3.2 Å and O-H···O bonds in the range 2.4 to 2.6 Å.

The environment of the urea molecules.

In each structure the urea molecules have three double hydrogen-bonding interactions, donor bonds by the two 'endo' hydrogen atoms to ether oxygen atoms and two sets of N-H··· donor and ···O receptor giving eight-membered rings.

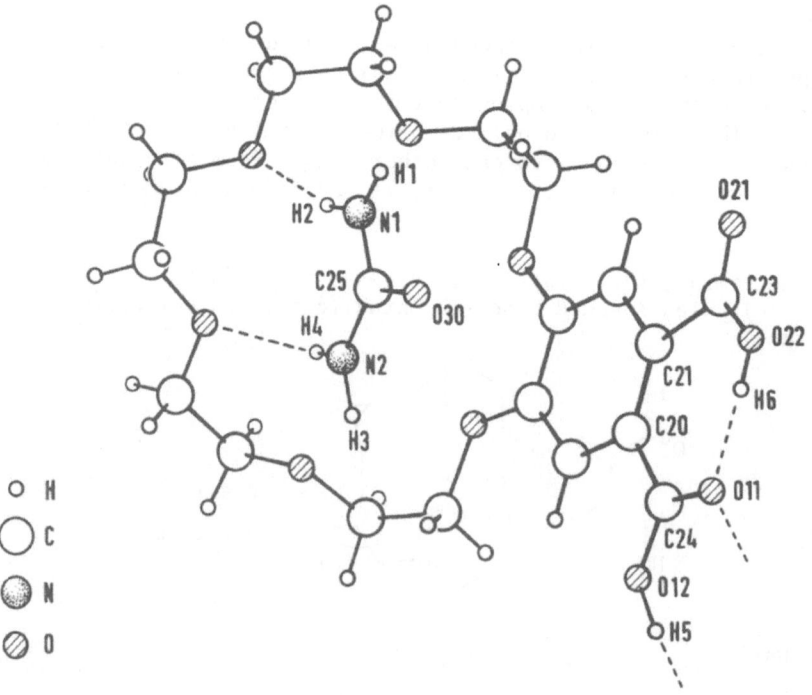

Fig. 1. One molecule of B and of urea with numbering of some atoms. Broken lines indicate hydrogen bonds.

In Fig. 1 the view is down the N-H···O hydrogen bonds from the 'endo' hydrogen atoms in urea to the ether oxygen atoms of one host molecule. Urea molecules are hydrogen bonded about centres of symmetry, N1-H1···O30' and O30···H1'-N1' to give dimers: each dimer is sandwiched between two host molecules. The third hydrogen bonding interaction holds the (urea-B)$_2$ dimers in a chain and is from N2-H3 to the carbonyl oxygen O11" with an acceptor at O30 from O12"-H5". Broken lines in Fig. 1 indicate the direction of the urea molecule in the next dimer in the chain.

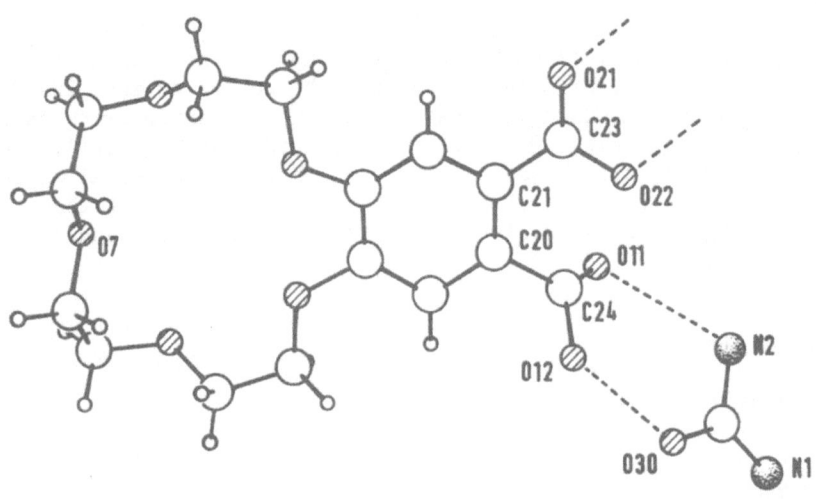

Fig. 2. One molecule of host A and of urea with the numbering of some atoms. Broken lines indicate hydrogen bonds.

In Fig. 2 one set of carboxylic acid-urea hydrogen bonding is shown. In this structure there is no urea-urea interaction, each molecule of urea is surrounded by three host molecules, one as shown, a second hydrogen bonded to the other O21,O22 carboxylic acid and the third by two 'endo' N-H bonds to ether oxygen atoms. These hydrogen bonding interactions hold the crystal in three dimensions but the crystal diffracts poorly.

Dimensions of the molecules.

In neither complex do the bond lengths and angles of the urea molecule differ significantly from those in urea itself.[1] The benzo-15-crown-5 entity of A is essentially the same as in the uncomplexed molecule or one of the very weakly bound ligands[2] in Na(benzo-15-crown-5)$_2$BPh$_4^-$ while the benzo-18-crown-6 entity of B has the same bond lengths and angles but a subtly different conformation from those in comparable complexes.[3]

It is in the ortho-dicarboxylic acid entities that A and B differ most significantly. B shows a strong internal hydrogen bond, O22-H6···O11 2.462(8) Å such as is usually found in hydrogen phthalates, leading to near coplanarity of the carboxylic acids and the benzene ring. In A there is no internal hydrogen bond, the distance O22···O30' 2.55 Å corresponds to a hydrogen bond while O11 to O22 is 2.88(2) Å. The separation results from a twisting of the carboxylic acids from the plane of the benzene ring, the angle between them being 64 °. This twisting allows the planes of the carboxylic acid groups to be at about 20 ° to those of the urea molecules to which they are hydrogen bonded.

Chirality?

Although the formulae of both urea and A suggest the possibility of planes of symmetry (for A through O7 and the centre of the C20-C21 bond) the inclusion compound crystallises in an enantiomeric space group. It is the twist of the carboxylic acids which gives 'handedness' to the molecules, all those in our crystal have the dihedral angle C23-C21-C20-C24 -10(3) ° (or all may be +10 °). In urea-B the corresponding angle is -0.3(0.9) ° in half the molecules and +0.3 ° in the other half.

ACKNOWLEDGEMENTS

We thank Dr D.G. Parsons for the crystals.

REFERENCES

1. S. Swaminathan, B.M. Craven, and R.K. McMullan, *Acta Cryst.*, B40, 1984, 300.
2. J.D. Owen, *J. Chem. Soc. (Dalton)*, 1980, 1066.
3. J.A. Bandy and M.R. Truter, *Acta Cryst.*, B38, 1982, 2639.

TRANSITION METAL COMPLEXES OF HOMOLEPTIC POLYTHIA CROWNS

M.N. Bell, A.J. Blake, R.O. Gould, A.J. Holder, T.I. Hyde,
A.J. Lavery, G. Reid and M. Schröder*
Department of Chemistry,
University of Edinburgh,
West Mains Road,
Edinburgh EH9 3JJ,
Scotland.

The binding of transition metal ions to polydentate macrocyclic ligands to give mono-, bi- and poly-nuclear complexes is well known. We have been investigating the complexation of transition metal ions, particularly those of the platinum group metals, by the homoleptic polythia crown ligands 1,4,7,10,13,16-hexathiacyclooctadecane (L^1), 1,4,8,11-tetrathiacyclotetradecane (L^2) and 1,4,7-trithiacyclononane (L^3). These ligands were attractive since they would be expected to bind effectively to the relatively soft second and third row metal ions and lead to the formation of complexes exhibiting unusual stereochemical, electronic and redox properties.

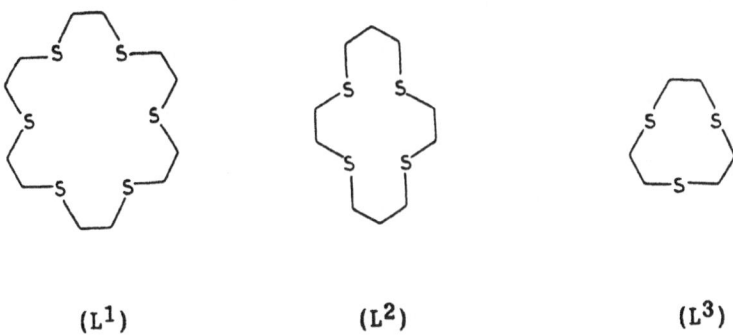

(L^1)　　　　　　　(L^2)　　　　　　　(L^3)

It was shown originally by Black and coworkers (*Tet. Lett.* 1969, 3961) that the potentially hexadentate ligand (L^1) (the thia analogue of 18-crown[6]) could readily encapsulate first row transition metal ions such as Ni^{2+} and Co^{2+} in an octahedral manner. The ability of (L^1) to act as a binucleating ligand had not however been demonstrated. We therefore initiated a study on the reactivity of (L^1) with a variety of metal substrates with a view to investigating its coordination to polymetallic centres.

Reaction of (L^1) with two molar equivalents of $[Cu(NCCH_3)_4]^+$ gave the di-copper(I) complex $[Cu_2(L^1)(NCCH_3)_2]^{2+}$ (1); the single crystal X-ray structure of (1) shows each tetrahedral copper(I) ion bound to *three* thia donors of (L^1), Cu-S = 2.32-2.34Å, and one molecule of CH_3CN, Cu-N = 1.94Å, with a Cu..Cu distance of 4.25Å.

Treatment of the isoelectronic, carbocyclic dimers $[M(cp^*)Cl_2]_2$ (M = Rh, Ir) and $[MCl_2(arene)]_2$ (M = Ru, Os; arene = p-cymene, hexamethylbenzene, benzene) with (L^1) affords the binuclear species $[M_2(cp^*)_2Cl_2(L^1)]^{2+}$ and $[M_2Cl_2(arene)_2(L^1)]^{2+}$ respectively. The single crystal X-ray structure of the di-rhodium(III) product $[Rh_2(cp^*)_2Cl_2(L^1)]^{2+}$ (2) shows the metal ions bound to only *two* of the thia donors of (L^1) with Rh-S = 2.377, 2.365, Rh-Cl = 2.387, Rh-C = 2.161-2.188Å.

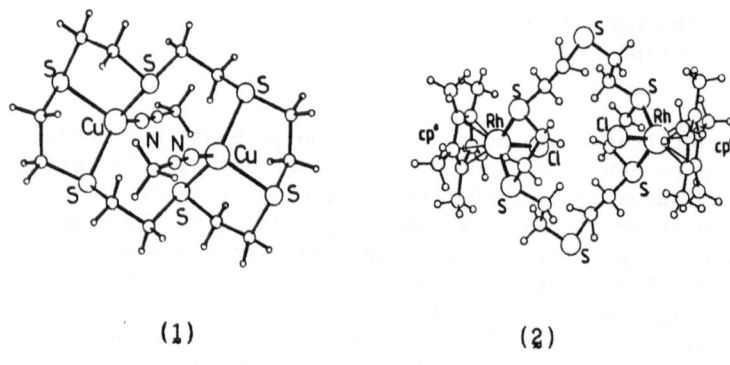

(1)　　　　　　　　(2)

A series of mononuclear platinum metal complexes have also been prepared.

Reaction of $PdCl_2$ or $PtCl_2$ with (L^1) gave 1:1 complexes $[M(L^1)]^{2+}$ (M = Pd, Pt) (3). The crystal structure analyses of these complexes confirm square planar coordination of the metal ions to *four* thia donors of (L^1) (Pd-S = 2.309, Pt-S = 2.296Å) with the two remaining sulphur donors of (L^1) being essentially non-bonded (Pd-S' = 3.273, Pt-S' = 3.380Å; <S'PdS = 75.1°, 104.9°, <S'PtS = 74.2°, 104.8°). The dangling thia donors S' are therefore unable to complete octahedral coordination around Pd(II) and Pt(II) due to the relatively large radii of these metal ions.

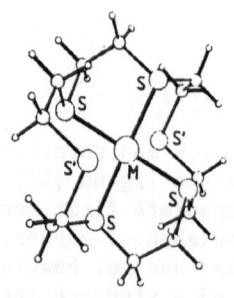

(3)

The small ring trithia macrocycle (L^3) has been shown to bind to first row transition metal ions in a facial manner. Thus bis-macrocyclic species of type $[M(L^3)_2]^{2+}$ (M = Co(II), Ni(II), Cu(II)) have been generated previously and shown to have octahedral MS_6 stereochemistries (Glass and coworkers, Inorg. Chem., 1983, 22, 266). Reaction of (L^3) with $PdCl_2$ and $PtCl_2$ in a 2:1 molar ratio yielded complex cations of stoichiometry $[M(L^3)_2]^{2+}$; their structural and redox properties were assessed.

The single crystal X-ray structure of $[Pt(L^3)_2]^{2+}$ (4) shows the complex to have an unusual square-based pyramidal stereochemistry. The Pt(II) ion is coordinated by four thia donors in a square plane, Pt-S = 2.25-2.30Å, with one of the remaining sulphur ligands bound apically Pt-S' = 2.88Å, <SPtS' = 84.0-97.2°. The sixth thia donor is not coordinated to the metal centre, Pt...S'' = 4.04Å. In contrast to the yellow Pt(II) complex (4), the isoelectronic Pd(II) species $[Pd(L^3)_2]^{2+}$ (5) is green and not isostructural. The crystal structure of (5) shows the centrosymmetric cation to have an unexpected distorted octahedral stereochemistry around Pd(II) with Pd-S_{equ} = 2.332, 2.311 and Pd-S'_{ax} = 2.952Å.

(4)   (5)

The complexes (4) and (5) each show, by cyclic voltammetry, one, one electron oxidation at $E_{\frac{1}{2}}$ = +0.39V. ΔE_p = 145mV. and $E_{\frac{1}{2}}$ = +0.605V. ΔE_p = 84mV. vs. Fc/Fc$^+$ respectively in CH_3CN at platinum electrodes. Controlled potential electrolysis of the complexes at +0.5V. and +0.7V. respectively at a platinum gauze affords the corresponding oxidation products $[M(L^3)_2]^{3+}$ which have been identified by esr spectroscopy as formally metal(III) species; $g_\parallel$ = 1.987, $g_\perp$ = 2.044, $A_\parallel$ = 85G, $A_\perp$ = 30G (^{195}Pt, I=½, 33.8%) for $[Pt(L^3)_2]^{3+}$, $g_\parallel$ = 2.009, $g_\perp$ = 2.049, $A_\parallel$ = 5G, $A_\perp$ = 20G (^{105}Pd, I=$\frac{5}{2}$, 22.2%) for $[Pd(L^3)_2]^{3+}$. Interestingly, the Pd(II) and Pt(II) complexes of (L^1) and (L^2), ((3) and (6)), show no oxidative redox processes by cyclic voltammetry in CH_3CN. The electrochemical inactivity of these latter species may be rationalised by the inability of the macrocycles (L^1) and (L^2) to form octahedral complexes with Pd and Pt centres. The ligand (L^1) may be regarded as being too small to fully encapsulate octahedrally the relatively large Pd(II) and Pt(II) ions. (L^2) would be expected to bind equatorially to give square planar complexes; this has been confirmed by the single crystal X-ray structure of $[Pd(L^2)]^{2+}$ (6); Pd-S = 2.23-2.33Å. By contrast, coordination of two molecules of (L^3) to Pd(II) and Pt(II) enables a preferred (distorted) octahedral stereochemistry to be achieved on oxidation to the metal(III) species. The coordinative flexibility of (L^3) in this system appears therefore to be crucial in stabilising the d^7 metal centre. In addition, the positive charge in $[M(L^3)_2]^{3+}$ would be expected to be stabilised further by delocalisation onto the thia ligands. The extent of positive charge on the thia donors is currently being assessed.

(6)   (7)

The homoleptic hexathia complexes $[Rh(L^3)_2]^{3+}$ and $[Ru(L^3)_2]^{2+}$ have also been synthesised. The single crystal X-ray structures of these products confirm their octahedral stereochemistries with Rh-S = 2.330, 2.332; Ru-S = 2.327-2.336Å. An unexpected feature of the structure of $[Ru(L^3)_2](BPh_4)_2 \cdot 2dmso$ (7) is the approach of the dmso solvate molecules towards the outer face of the coordinated trithia ligands. This occurs via H-bonding of the O-donor of the dmso solvates with the protons of the methylene groups of (L^3), O...H = 2.201, 2.419, 2.790, 3.291Å. This secondary interaction between the dmso molecules with the rear cone/cavity of the coordinated trithia ligand may be regarded as a weak inclusion of solvent; this is supported by the observation that dmso may be replaced by two molecules of other donor solvents such as CH_3CN and CH_3NO_2. The development of related systems incorporating larger and deeper cavities is under investigation.

This is an extended abstract of a poster presented at the 4th. *International Symposium on Inclusion Phenomena*, University of Lancaster, 20-25 July 1986.

We thank BP Chemicals and SERC for a CASE Award to (TIH), SERC for support, and Johnson Matthey Plc for generous loans of platinum metals.

ISOLATION, PROPERTIES AND ASSOCIATION PHENOMENA OF ALKALINE SALTS AND THEIR CROWN COMPLEXES OF A RADICAL ANION

J. Veciana* and A. Durán.
Departamento de Materiales Orgánicos Halogenados. Centro de Investigación y Desarrollo (C.S.I.C.) C./ Jorge Girona Salgado, 18-26, 08034 Barcelona. Spain.

ABSTRACT. The obtention of (tetradecachloro-4-oxidotriphenylmethyl)$^{-}$ M^{+} (M=Li, Na, K, n-Bu$_4$N) salts in ethereal solution and the isolation of some alkaline complexed salts (M= Li-12C4, Na-18C6, K-18C6, K(THF)$_1$-(H$_2$O)$_{3-4}$, n-Bu$_4$N) are described and discussed. The association phenomena of these salts has been studied by electronic spectroscopy, osmometry and electron spin resonance. Linear correlations between radii counterions and the position maxima of the electronic spectra bands permit the study of the species present in solution (free ions, ion pairs and quadrupolar aggregates).

1. INTRODUCTION

Most of the existing work concerning the structure of organic anions in solution has been performed with carbanions[1], enolates and fenoxides[2], ketyls[3] and semiquinone derivatives[3]. Such studies were based principally on ESR, NMR and UV-visible techniques, and in particular the latter have been limited to very dilute solutions in a narrow range of concentrations, due to the very high molar extinction coefficient of the anions.

We now report a study of the structure of ionic species of a radical anion of the quinone methide type, a vinylog of a ketyl anion radical, using UV-visible and ESR spectroscopies in a wide range of concentrations[3].

2. OBTENTION AND ISOLATION OF SALTS

Several salts of tetradecachloro-4-oxidetriphenylmethyl radical (<u>1</u>) (M= Li, Na, K, n-Bu$_4$N) have been obtained quantitatively in ethereal solution (diethyl ether, Et$_2$O; tetrahydrofuran, THF; and dimetoxiethane, DME) by reaction of tetradecachloro-4-hydroxitriphenylmethyl radical[4] (<u>2</u>) with alkaline metals or hydroxides or by reduction of perchlorofuchsone (<u>3</u>) with metals.

$$\begin{array}{c} \text{[Scheme showing reactions]} \end{array}$$

M= Li, Na, K, n-Bu$_4$N, Li-12C4, Na-18C6, K-18C6

Solutions of salts **1** have been found to be persistent in Et$_2$O, THF, DME and DMSO/Et$_2$O and very stable toward oxygen. However, isolation of such salts in solid state is only possible when the alkaline counterion is completely solvated; i.e., K(THF)$_1$-(H$_2$O)$_{3-4}$, K-18C6, Na-18C6, Li-12C4 or when the counterion is a poor electron acceptor like n-Bu$_4$N. The solvation sphere of alkaline cations is a major factor on the stabilities and even on the ionicities of such salts, as evidenced in the thermal decomposition.

$$(\underline{1})^{\overline{\cdot}}\ K^+(THF)_1(H_2O)_{3-4} \xrightarrow[\text{Ar}]{200°C} \underline{3} + (K°) + THF + H_2O$$

Such decompositions formally imply an electron transfer from the radical-anion to potassium and is similar to those observed in perchlorotriphenylmethide salts[5].

3. ASSOCIATION PHENOMENA

The association phenomena of salts **1** in solution - free ions (M$^+$+A$^{\overline{\cdot}}$), quadrupolar ions (M$^+$A$^{\overline{\cdot}}$)$_2$ - have been studied by electronic spectroscopy, electron spin resonance and osmometry.

The electronic spectra of salts **1** show four bands, whose wavelength maxima and absortivities depend on : a) Nature of counterion (Li, Na, K, n-Bu$_4$N), b) Solvent (Et$_2$O, THF, DME, DMSO/Et$_2$O),

c) Concentration (10^{-6}–10^{-2} M) and d) Complexing agents (12C4, 18C6); the highest wavelength band being the most sensitive. Figure 1 shows the curves of the wavenumber for the latter band vs the concentration in Et_2O of salts 1 with different counterions. Similar curves are obtained in THF and DME.

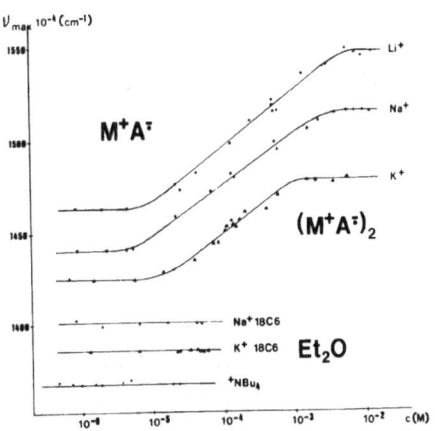

Figure 1. Dependence of wavenumber absortion maxima, $\bar{\nu}$ max, of salts 1 on concentration, C.

Such results can be interpreted on the basis of formation of ion pairs at lower concentrations and quadrupolar aggregates at higher ones. For both species a batochromic shift is observed when the solvation capability of solvents is increased (Et_2O < THF < DME), suggesting an externally solvated species like $A^{\bar{\cdot}} M^+$, S_n and $(A^{\bar{\cdot}} M^+)_2 S_n$. The quadrupolar nature of these species at higher concentrations has been independently confirmed by osmometry and by the presence of electron-electron dipolar interaction in e.s.r. spectra. The absortion maximum of the highest wavelength band in $DMSO/Et_2O$ is not affected neither by the counterions nor concentration indicating the presence of totally solvated ion pairs, $M^+//A^{\bar{\cdot}}$ (or free ions, $M^+ + A^{\bar{\cdot}}$). The addition of 18-crown-6 to Et_2O solutions of Na and K salts 1 gives rise to a batochromic shift of the maxima whose positions indicate a replacement by the crown ether of the ethereal external solvation sphere on $A^{\bar{\cdot}} M^+$, S_n giving species like $A^{\bar{\cdot}} M^+$-18C6. By contrast 12-crown-4 is not able to perform such replacement on Li salt 1 in Et_2O, due to the higher hardness of Li.

Linear correlations between the counterion radii, $(2+r_m)^{-1}$, and the wavenumbers, $\bar{\nu}$ max, of the species $A^{\bar{\cdot}} M^+$, S_n and $(A^{\bar{\cdot}} M^+)_2 S_n$, in several solvents and also for the species $M^+//A^{\bar{\cdot}}$ (or $M^+ + A^{\bar{\cdot}}$) in $DMSO/Et_2O$ have been observed (Fig. 2).

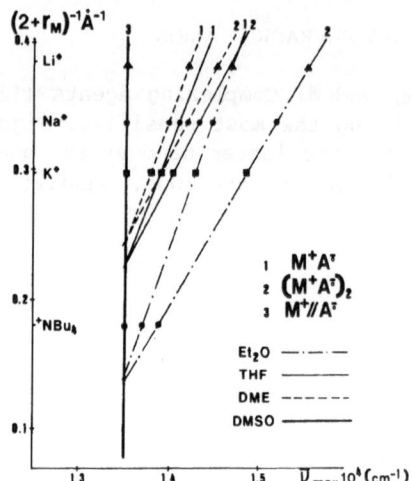

Figure 2. Linear correlations between wavenumber, $\bar{\nu}$ max, and cationic radii, $(2+r_m)^{-1}$ for $A^{\bar{\cdot}} M^+$, S_n, $(A^{\bar{\cdot}} M^+)_2 S_n$ and $M^+//A^{\bar{\cdot}}$ species.

It is remarkable that such linearity is not limited only to ion pairs and free ions as has been described elsewere[2], but is possible to expand it to the quadrupolar species. Even more noticiable is that the slopes of the straight lines are similar for the ion pairs in all the studied solvents (slope $\simeq 1.9 \cdot 10^4$) as occurs also for the quadrupolar ions (slope $\simeq 1.3 \cdot 10^4$), being both values different from that observed for the free ions (slope = ∞). Therefore, such magnitude (slope) appears to be independent from the nature of the solvent and seems to be only dependent on the state of aggregation, reflecting the "anion susceptibility to the counterions perturbations". Thus electronic spectroscopy appears to be a reliable tecnique even for the identification of the state of aggregation of radical anions.

REFERENCES

1) E. Buncel, T. Dorst (Eds.), Comprehensive Carbanion Chemistry, Part A; Elsevier, New York, (1980).

2.a) H.E. Zaugg; A.D. Schaefer. J. Am. Chem. Soc., 87 (1965) 1857 b) T. Miyashita; T. Aoki; M. Matsuda. Bull. Chem. Soc. Jpn. 49 (1976) 231 c) J.F. Garst; R.A. Klein; D. Walmsley; E.R. Zabolotny. J. Am. Chem. Soc. 87 (1965) 4080.

3.a) Szwarc, M. Ed., Ions and Ion Pairs in Organic Reactions Wiley-Interscience, New York (1973). Vols. 1,2. b) Screttas, C.G.; Micha-Screttas, M. J. Org. Chem. 46 (1981) 993.

4) M. Ballester, J. Riera, J. Castañer, A. Rodriguez, C. Rovira and J. Veciana. J. Org. Chem., 47 (1982) 4498.

5) J. Veciana, J. Riera, J. Castañer, N. Ferrer. J. Organomet. Chem., 297 (1985) 131.

SYNTHESIS AND COMPLEXING PROPERTIES OF A CHIRAL MACROCYCLIC MOLECULAR RECEPTOR WITH CONVERGENT BINDING SITES

Marek Pietraszkiewicz
Institute of Physical Chemistry, Polish Academy
of Sciences, 01-224 Warszawa, Kasprzaka 44/52
Poland

ABSTRACT. Two novel hosts: 15-crown-5 and N-benzyl-aza-15-crown-5 incorporating a boron-containing D-mannopyranosidic unit form more stable cascade complexes with (S)-amino acid sodium or potassium salts than with the respective (R)-enantiomers. Complexes with sodium salts are more stable than the corresponding complexes with potassium salts as revealed by variable-temperature NMR measurements. Strong non-bonded interaction between the sugar unit and the α-substituents of the amino acids results in enantiomeric differentiation and destabilization of the complex. Complex formation is interpreted in terms of ion-pair inclusion by macrocyclic ring and nitrogen-boron interaction.

1. INTRODUCTION

Considerable work has been done in the field of chiral hosts capable of discriminating between enantiomeric guest molecules. A variety of chiral residues either synthetic or derived from natural products |1| (sugars, tartaric acid, amino acids etc.) have been fused to macrocyclic rings. In most cases, chiral primary ammonium cations have been the guest species of choice, although the complexation of chiral carboxylates has been examined to some extent |2|.

Amino acids possess two functional groups able to participate in complexation. Appropriate design of the host molecule can lead to a particular host containing two binding sites specific for amino and carboxylic groups.

It is well known that crown ethers form inclusion complexes with ion pairs in aprotic solvents. Phenylboronic cyclic esters form adducts with amines |3|, making them good candidates as convergent binding sites for amino groups. Linking these specific binding sites with a chiral residue was the aim of this work as well as an examination of the complexing properties of the new molecules as enantioselective complexers for amino acid carboxylates by means of VT proton NMR spectroscopy.

Two new chiral macrocyclic compounds containing both a crown ether and a boronic ester were prepared from methyl 4,6-O-isopropylidene- α -D-mannopyranoside |4| as shown in Figure.

1': $(ClCH_2CH_2)_2O$, $Bu_4^nNHSO_4$, NaOH 50%, PTC, 25°, 20h, 75%, |5|
2': $PhCH_2NH_2$, MeCN, refl., Na_2CO_3, 30%, |6|
3': HCl/MeOH, 25°, 2h
4': $PhB(OH)_2$, 75%

1: $TsOCH_2(CH_2OCH_2)_2CH_2OTs$, NaH, DMSO,
2: Dowex-50, MeOH, 2h
3: $PhB(OH)_2$, 90%

2. MATERIALS AND METHODS

The reagents were purchased from Fluka and used without purification except dimethyl sulphoxide and acetonitrile, which were dried and distilled over calcium hydride.
Both new compounds A and B gave satisfactory microanalyses and NMR spectra. Selected amino acids: (R)- and (S)-alanine (Ala), (R)- and (S)-phenylglycine (PhGly) and (R)- and (S)- β -phenylalanine (β -PhAla) as their sodium and potassium salts were dried in vacuo over phosphorus pentoxide prior to use and served as guest species. The host/guest ratio was 2:1 in deuteriochloroform solutions. Dissolution of solid salt was accomplished in an ultrasonic bath within 20 to 100 min. The host concentration was ca. 0.1 M. All the variable-temperature spectra were recorded with JEOL-JNM-4H-100 spectrometer at 100 MHz with TMS as internal standard and lock. The anomeric proton served as an NMR probe. The free energy of activation at the coalescence temperature was calculated from the expression |7|:

$$\Delta G_c^{\ddagger} = 4.575 \times 10^{-3} T_c (9.972 + \log T_c/\Delta\nu).$$

in kcal mol^{-1}.

3. RESULTS

The ^{1}H NMR spectra of solutions containing the free ligand and the complex in a 1:1 ratio showed one signal for the anomeric proton at ambient temperature, indicating fast exchange between nonequivalent sites. Cooling the samples resulted in line broadening and finally splitting of the anomeric proton signal into two components attributed to the 1:1 complex and the free ligand. Generally, higher coalescence temperatures were observed for the samples containing (S)-enantiomers. The complexes with the potassium were less stable than those with the sodium carboxylates. Thermodynamic data are collected in the Table.

Table I. Kinetic stabilities of inclusion complexes involving the hosts A and B and alkali metal carboxylates of the amino acids

Amino acid salt	HOST A			HOST B		
	Tc^a	$\Delta\nu^b$	$\Delta G_c^{\ddagger c}$	Tc	$\Delta\nu$	$\Delta G_c^{\ddagger}$
(R)-Ala$^-$Na$^+$ ^{d}	228	11	11.8	225	10	11.6
(S)-Ala$^-$Na$^+$	245	15	12.5	240	15	12.3
(R)-Ala$^-$K$^+$	218	10	11.3	212	9	11.0
(S)-Ala$^-$K$^+$	226	16	11.5	220	13	11.3
(R)-PhGly$^-$Na$^+$ ^{e}	214	9	11.1	210	10	10.9
(S)-PhGly$^-$Na$^+$	225	13	11.5	222	17	11.3
(R)-PhGly$^-$K$^+$	-	-	-	207	9	10.7
(S)-PhGly$^-$K$^+$	231	12	11.9	220	14	11.2
(R)-β-PhAla$^-$Na$^+$ ^{f}	225	14	11.5	222	16	11.3
(S)-β-PhAla$^-$Na$^+$	240	21	12.1	232	22	11.6
(R)-β-PhAla$^-$K$^+$	215	12	11.0	209	11	10.8
(S)-β-PhAla$^-$K$^+$	227	18	11.5	223	16	11.3

^{a} Coalescence temperature (°K). ^{b} $\Delta\nu$ is the frequency separation (in Hz) of the two signals in the slow exchange limit. ^{c} Free energy of activation in kcal mol^{-1} (see Ref. |7|). ^{d} Ala$^-$Na$^+$; CH$_3$CH(NH$_2$)CO$_2^-$Na$^+$. ^{e} PhGly$^-$Na$^+$; PhCH(NH$_2$)CO$_2^-$Na$^+$. ^{f} β-PhAla$^-$Na$^+$; PhCH$_2$CH(NH$_2$)CO$_2^-$Na$^+$.

4. DISCUSSION

It was assumed that the amino acid carboxylates form complexes with A and B in a cascade mode, i.e. the alkali cation is held in the cavity of the macrocycle and ion-paired with carboxylate group, and the amino group is linked to the boron site. This model of complexation would lead to highly defined complexes in which remarkable differentiation between enantiomers is achieved through non-bonded interactions between substituents at the α-carbon atom of the guest and the chiral unit. On the other hand, if these steric interactions are too strong, they will lead to destabilization of the complex. In fact, the observed coalescence temperatures are not very much higher than those reported for the inclusion complexes with primary ammonium cations |8|. Probably, steric interactions contribute significantly to destabilization of the complexes.

This interpretation can be supported by an observed slow dissolution of the solid salts in the guest solutions.

Potassium carboxylates form weaker complexes, probably due to weaker complexation of the potassium cation which is too large to fit into the cavity of the 15-membered macrocyclic ring. There is also a very small difference between hosts A and B in their strength of complexation.

The (S)-enantiomers form stronger complexes with hosts A and B. Although some enantioselectivity was achieved, it is difficult to draw a final conclusion about the stereochemical nature of these complexes. An attempts are made to obtain a good crystals for an X-ray analysis.

5. ACKNOWLEDGEMENT

This work was supported by grant CPBR 3.20 from the Polish Academy of Sciences.

6. REFERENCES

1. S. T. Jolley, J. S. Bradshaw and R. M. Izatt, J. Heterocycl. Chem., 19, 3 (1982).
2. J.-M. Lehn, J. Simon, and A. Modrapour, Helv. Chim. Acta, 61, 2407 (1978).
3. M. Yalpani, R. Köster, G. Wilke, Chem. Ber., 116, 1336 (1983).
   M. Yalpani and R. Boese, ibid, 116, 3347 (1983).
4. C. Copeland and R. V. Stick, Aust. J. Chem., 31, 1371 (1978).
5. P. Di Cesare and B. Gross, Synthesis, 458 (1979).
6. S. Kulstad and L. A. Malmsten, Acta Chem. Scand., B33, 469(1979).
7. J. Sandström: Dynamic NMR Spectroscopy, p. 96, Academic Press, London 1982.
8. S. L. Baxter and J. S. Bradshaw, J. Heterocycl. Chem., 18, 223 (1981).
   M. Pietraszkiewicz and J. F. Stoddart, J. Chem. Soc., Perkin Trans. 2, 1559 (1985).

ORGANOMETALLIC IONOPHORE FOR ALKALI METAL CATIONS

I. GOLDBERG[*], H. SHINAR AND G. NAVON[*]
School of Chemistry, Tel Aviv University, 69978 Ramat Aviv, Israel

W. KLAUI
Institut fur Anorganische Chemie, Technischen Hochschule Aachen, 5100 Aachen, West Germany

Abstract. Complexes of a novel synthetic organometallic ionophore with lithium and sodium cations have been characterized by single-crystal X-ray diffraction. The crystal structure of the lithium complex consists of cation-ligand <u>dimers</u> with a tetrahedral coordination around Li. The sodium complex reveals a different structure type consisting of cation-ligand <u>trimers</u>, with water molecules being included between the trimeric entities. The coordination sphere around the Na ions has a distorted octahedral symmetry. It is anticipated that the observed structures of dinuclear Li and trinuclear Na complexes represent possible modes of aggregation of the cation-ligand entities in lipophilic media.

1. INTRODUCTION

Synthetic ionophore ligands, mostly based on cyclic and branched polyethers, have been widely used in the last decade to study carrier facilitated cation transport across membranes. It has recently been observed that the organometallic ligand $(C_5H_5)Co[PO(OC_2H_5)_2]_3^-$ also has

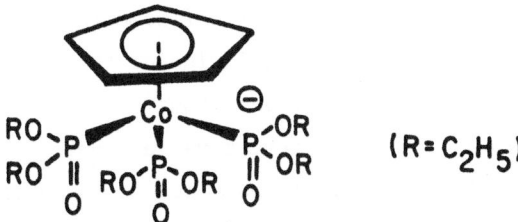

$(R = C_2H_5)$

efficient ionophoric properties for alkali metal cations, with a specificity toward Li^+ [1]. The association constants in aqueous [3] as well as methanol [2] solutions follow the trend $H^+>Li^+>Na^+>K^+$. The rate of transport of the ions across a lipid membrane was monitored by ^{23}Na and ^7Li nuclear magnetic resonance, revealing a significant preference for Li^+ over Na^+ [1]. In order to elucidate the structural basis of the ionophoric behaviour we analysed crystallographically the lithium and sodium complexes of this ligand.

2. EXPERIMENTAL DATA

Single crystals of both compounds suitable for the crystallographic study were obtained from water as well as from wet n-heptane and iso-butylacetate solutions. Diffraction data were measured at ca. 18°C on a CAD4 diffractometer, using MoKα radiation. The crystal data are:

<u>Lithium complex</u>: triclinic, space group $P\bar{1}$, $\underline{a}$=11.944(3), $\underline{b}$=12.154(6), $\underline{c}$=20.468(3) Å, $\underline{\alpha}$=92.79(3), $\underline{\beta}$=91.05(1), $\underline{\gamma}$=118.07(4)°, $\underline{R}$=0.094 for 3694 observations above 3σ(I) out to $2\theta_{max}$=46°.

<u>Sodium complex</u>: monoclinic, space group $P2_1/c$, $\underline{a}$=10.828(5), $\underline{b}$=25.008(3), $\underline{c}$=31.307(6) Å, $\underline{\beta}$=98.76(2)°, $\underline{R}$=0.091 for 3005 observations above 3σ(I) out to $2\theta_{max}$=40°.

The asymmetric unit of the lithium compound contains two units of the 1:1 complex, while that of the sodium compound contains three units of the 1:1 complex and four molecules of water. The sodium compound was found also to crystallize from water in a different polymorphic form [monoclinic, $\underline{a}$=10.969(5), $\underline{b}$=53.226(4), $\underline{c}$=14.731(3) Å, $\underline{\beta}$=103.38(3)°], but good crystal of this phase could not be grown.

The two structures were solved by a combination of direct methods and Fourier techniques (MULTAN80). Their refinements were carried out by large block least-squares (SHELX76), including the positional and thermal parameters of all the nonhydrogen atoms. The refinement calculations indicated clearly that both structures are partially disordered. In the sodium compound the disorder is confined to the peripheral OEt groups of the ligand. In the lithium compound an additional twofold orientational disorder of an entire phosphonate group is present.

3. DISCUSSION OF RESULTS

As stated at the outset, the organometallic ligand can transport alkali metal cations across phospholipid membranes. Since the molecular surface of the 1:1 complex is polar at one end and apolar at the other, it will be reasonable to assume that transfer of the cations through lipophilic media is more likely to occur in the form of polymeric cation-ligand aggregates than in the form of monomeric entities. The observed structures of the dinuclear Li^+ and the trinuclear Na^+ complexes may represent in fact two possible modes of such an aggregation during the process of transport.

The smaller Li^+ ions are sufficiently well enclosed within a cavity formed between two ligands, leading to the formation of dimers.

The Li^+ ions lie near the inversion centers of the crystal and bridge between the centrosymmetrically related ligands. They are tetracoordinated to four P=O ligating sites, three O-nucleophiles from one ligand and an additional P=O from the other part of the dimer. The Li-O distances vary from 1.88 to 1.95 Å in the ordered dimer and from 1.82 to 2.01 Å in the orientationally disordered entity. Within the dimers the distances between the two cations are 2.61 and 2.63 Å. The exterior surface of the dinuclear assembly consists of lipophilic OEt and cyclopentadienyl hydrocarbon groups, providing a lipophilic environment for the adjacent units in the crystal.

The sodium compound reveals a new structure type consisting of <u>trimers</u>. The trimeric entity is formed by triangularly arranged species of the ligand. It has a polar interior lined with P=O groups which provide a series of ion binding sites and an apolar exterior consisting of hydrocarbon fragments. The metal cations are contained within the central cavity of such unit. Each one of them is bound to four P=O groups, providing an effective bridge between adjacent molecules of the ligand. All Na-O bonds are within the 2.23-2.44 Å range; the distances between adjacent Na$^+$ ions within the trimeric unit vary from 3.26 to 3.42 Å.

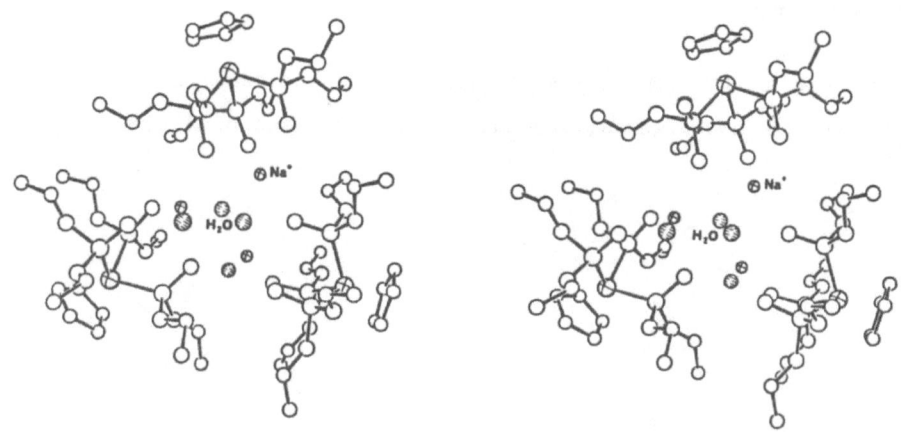

In the crystal the trimeric cation-ligand assemblies are stacked one on top of the other, forming a channel-type structure. Molecules of water are included along the channels, linking between adjacent entities of the complex and supplementing the (pseudo octahedral) coordination sphere of the metal cations. Two of the water molecules lie in close proximity to the sodium ions, the corresponding Na$^+$-O(water) distances varying from 2.33 to 2.85 Å. The exterior of the stack consists of hydrocarbon fragments and is lipophilic; side packing of the stacked units is thus stabilized by van der Waals forces. In this respect the structure of the sodium complex (shown below) resembles a solid state model of a molecular channel similar to that of a functionalized macrocyclic polyether previously described in the literature [4]. The two crystal types contain channels with a hydrophilic interior which are suitable for the accommodation of polar species as metal cations and water.

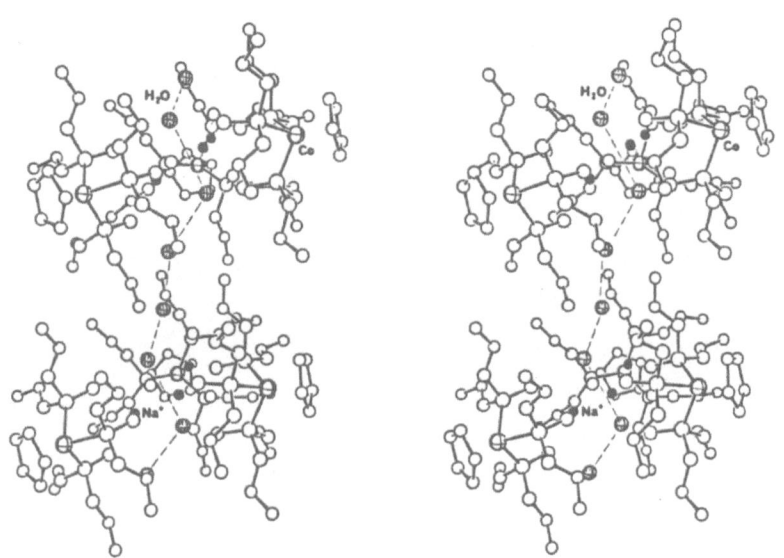

It is important to note that molecular weight determinations for the sodium compound indicate the presence of trimeric aggregates in a benzene solution as well, which is consistent with the above discussion. Further structural investigations of related complexes and chemical modifications of the organometallic ligand are under way in order to improve the ionophoric properties of this type of molecular hosts.

Acknowledgements. The authors are grateful to Mrs. Z. Stein and Mr. E. Neria for their invaluable assistance.

References

1. H. Shinar, G. Navon and W. Klaui: J. Am. Chem. Soc. **108**, 5005 (1986).
2. G. Anderegg and W. Klaui: Z. Naturforsch. **B36**, 949 (1981).
3. G. Navon, H. Shinar and W. Klaui: Rev. Port. Quim. **27**, 27 (1985).
4. J.P. Behr, J.M. Lehn, A.C. Dock and D. Moras: Nature **295**, 526 (1982).
5. A short account of this work has been presented at the Tenth European Crystallographic Meeting, Wrocław (Poland), 7 August, 1986.
6. **Supplementary data:** The atomic coordinates of the two compounds are deposited with the British Library as Supplementary Publication No. SUP 82047.

SOLID CLATHRATE SOLUTIONS

Yu.A.Dyadin, G.N.Chekhova, N.P.Sokolova
Institute of Inorganic Chemistry, Academy of Sciences of the
USSR, Siberian Branch, Novosibirsk 630090,
U.S.S.R.

ABSTRACT. The classification of solid clathrate solutions may be subdivided into three types: interstitial solutions, those with the substitution of one guest by another and those with the substitution of the particles in a host framework is given. All these types of solutions are illustrated by experimental (or computed) state diagrams of binary and ternary systems of guest-host and host-guest1-guest2 kinds, where host components are water, urea, thiourea and hydroquinone.

1. INTRODUCTION

Three types of solid solutions: 1) iskhoric solutions, i.e. interstitial ones, formed by filling the host framework cavities with any appropriate particles to some degree, $y(0<y<1)$ ("iskhoric" is derived from Greek εἰσχωρῶ - to intrude), 2) alloxenic solutions, i.e. those with the substitution of one guest by another ("alloxenic" is derived from the greek words αλλος - another, ξεNOS - guest), 3) allokiric solutions, i.e. those with the substitution of particles in a host framework ("allokiric" is also derived from the greek words αλλος - another, cυρios - host) may be observed in the clathrate systems.

The interstitial solutions are typical of clathrate compounds and they have been considered [1]. Studying water [2,3], urea [4,5], thiourea [6] and hydroquinone [7-9] clathrates we have found that only the latter (in the presence of the limited guest set [10-12]) forms the solutions of this type.

Allokiric solutions are formed in H_2O - Bu_4NA - Bu_4NB systems (where Bu_4N^+ is tetra-n-butylammonium cation, A and B are single charged anions). These clathrate frameworks consist of water molecules and anions, the guest being organic cation.

Alloxenic solutions are wide-spread in clathrate chemistry. These are the natural gas hydrates, mixed hydroquinone clathrates and mixed channel inclusion compounds. The conditions of their formation seem to be sufficiently simple and are in fact the same as those of the formation of the individual clathrate compound.

2. EXPERIMENTAL

2.1. Reagents

Urea and thiourea were of especially pure grade. Tetra-n-butylammonium (TBA) hydroxide was prepared and purified as in [13]. All the operations with it were carried out in the free from carbon dioxide atmosphere and by using boiled water.

TBA fluoride was synthesized by neutralizing diluted TBA hydroxide solution with hydrofluoric acid, followed by recrystallization of TBA fluoride as the clathrate hydrate (all the operations were carried out by using polyethylene or teflon vessels).

TBA bromide was purified by means of thrice-repeated recrystallization of pure grade reagent out of ethyl acetate.

Hydroquinone was purified by thrice-repeated recrystallization out of aqueous solutions which were slightly acidified by sulfuric acid.

Benzene, cyclohexane and acetonitrile were of chromatography grade, n-paraffins were of chemical pure one.

Carbon tetrachloride and chloroform were purified by means of conventional methods.

The formic and acetic acids were purified from water and other contaminants in different ways: the former was recrystallized many times out of aqueous solution at 8,5°C, the latter was kept together with acetic anhydride for 24 hours followed by the distillation.

2.2. Analysis

The concentration of hydroxyl ions was determined by means of titration by 0,1N nitric acid solution with methylorange as an indicator, that of Br^- anions and thiourea - by 0,05N $Hg(NO_3)_2$ titration (in the presence of ethyl alcohol) with diphenylcarbazone as an indicator, that of F^- anions - by direct potentiometric method [14], that of hydroquinone - by 0,1N $Ce(SO_4)_2$ solution titration in the presence of sulfur acid with phenylanthranilic acid as an indicator, that of formic and acetic acids - by 0,07N NaOH solution titration with thymol blue as an indicator.

2.3. Methods

The DTA method has been used for binary systems study. Two versions of Schreinemakers's method have been utilized: the classical one [15]("wet residue" method has been applied to hydroquinone and water systems) and the modified one ("initial weighed portion" method [16] has been used for thiourea systems). In the latter the liquid phase composition has been determined by means of $CCl_4-C_6H_6$, $CCl_4-C_6H_{12}$, $CHCl_3-C_6H_{12}$, $CCl_4-n-C_{10}H_{22}$ solution density measurements (the thiourea concentration in these equilibrium solutions has been shown by means of analyses to be less than 0,1%). The Schreinemakers's ray has been drawn through the figurative points of the liquid and "initial weighed portion" compositions. The solid solution composition was obtained by combining the usage of Schreinemakers's and Cameron's [17] methods (acetic acid was used as an inert and easily analyzed admixture).

Particular attention was given to equilibrium data obtaining. So, for example, the samples for DTA placed into the sealed ampoules were being kept for 1-1,5 months, "initial weighed portions" samples in three-component thiourea systems being kept for 7-8 months (by agitation), in the water systems the phase equilibrium was set at the period ranging from 4 hours to some days, in hydroquinone systems - from 5 to 20 days.

3. RESULTS AND DISCUSSION

3.1. Interstitial solutions

In the foregoing it has been mentioned that the hydroquinone clathrates in the presence of the limited set of guests (such as inert gases or their analogs) form the solutions of this type. It should be noted that hydroquinone possesses the unique property: in its initial stable α-modification there are cavities of molecular dimensions, which may contain such kind of guests as the clathrate β-modification. Since α- and β-modification structures [18,19], $\Delta\mu^{\alpha\to\beta}$ [20] and the necessary thermodynamic data [21,22] are known for hydroquinone one can simulate these systems easily enough and estimate them quantitatively up to computing the total phase diagram of the binary hydroquinone - guest system [11,12]. The agreement of the computed and experimental data is illustrated in Fig. 1 for the hydroquinone-krypton and hydroquinone-xenon systems.

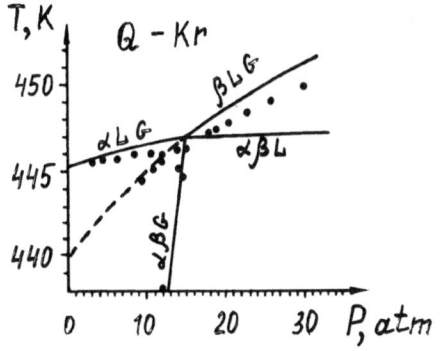 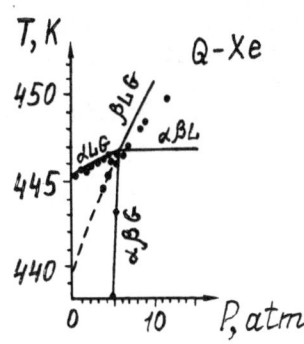

Figure 1. The phase P,T-diagrams of the hydroquinone - inert gas systems. The computed (—— stable state and - - - unstable one) and experimental (•) data [23].

The comparison of the computed data with the experimental ones indicates that the former not only reflects qualitatively the phenomenon nature (namely the existence of clathrates in inert gas (except He and Ne) - hydroquinone systems and the dissolving of guests in the host α-modification) but describes the phase diagram quantitatively. Computed P,x- and T,x-sections of the hydroquinone - xenon system diagram (see Fig.2) confirm the berthollide nature of hydroquinone clathrates which helps to clarify the essence of Kurnakov's imaginary compounds [17].

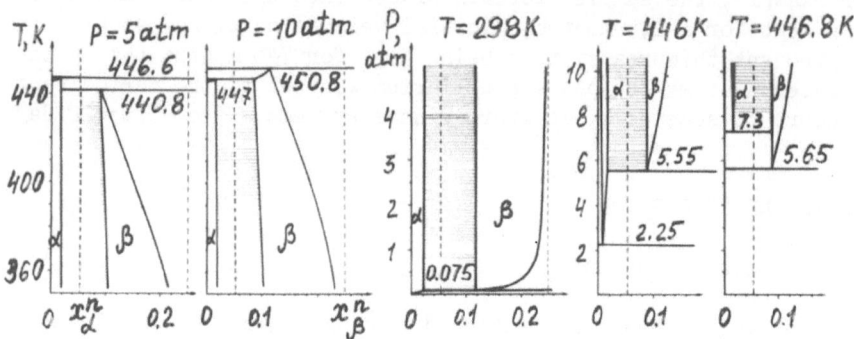

Figure 2. The isobaral and isothermal sections of P,T-diagrams of the hydroquinone - xenon system.

3.2. Allokiric solutions

The solutions with the substitution of particles in a host framework are rather rare in clathrate chemistry. The present data are not yet enough to draw the final conclusions concerning the regularities of the formation of the solutions of this kind. Nevertheless some features attract our attention.

At 20°C in TBA hydroxide - TBA fluoride - H_2O system (see Fig.3) the continuous series of the solid solutions of the typeI (according to Roozeboom) on the basis of the cubic 1:28,6(salt:water) hydrate is being crystallized out of the liquid solutions impoverished by water. When crystallized out of the liquid solutions enriched by water, the solid solutions with the break in continuity of the typeV on the basis of 1:28,6 TBA hydroxide hydrate and that of 1:32,8 TBA fluoride hydrate are formed.

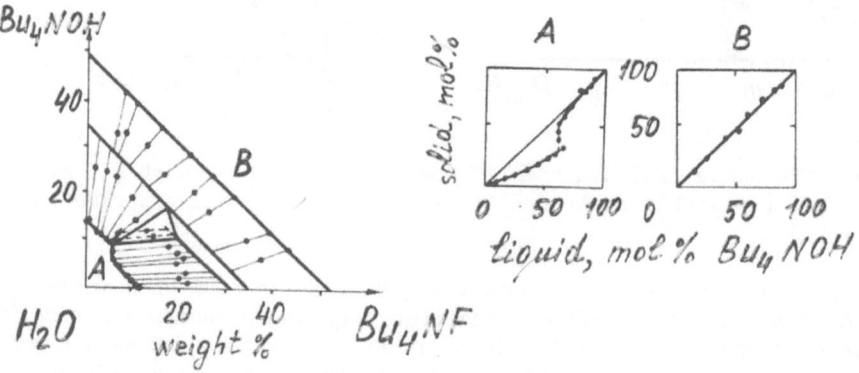

Figure 3. The solubility isotherm (20°C) in the $(C_4H_9)_4NF - (C_4H_9)_4NOH - H_2O$ (a) and the diagram of the component distribution among the solid and liquid phases (without taking account of water) (b).

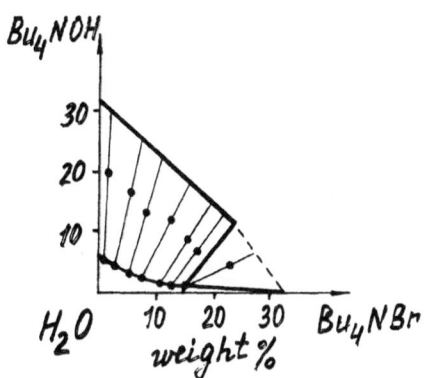 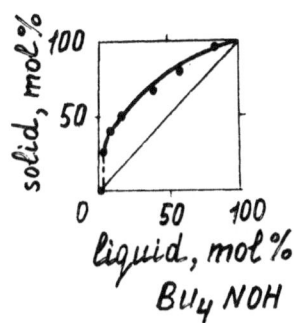

Figure 4. The solubility isotherm (8°C) in the $(C_4H_9)_4NOH$ - $(C_4H_9)_4NBr$ - H_2O (a) and the diagram of the component distribution among the solid and liquid phases (without taking account of water) (b).

Such a behaviour may result from the stability of the individual hydrates [13,24]: the fluoride and hydroxyl anions stabilize the 28,6 hydrate framework to the same extent (which is verified by the equality of the congruent melting points (as high as 27,4°C) of the corresponding hydrates) and therefore hydrates form with each other the practically ideal solid solutions whatever crystallization region. The hydrate solid solutions having tetragonal cell are enriched with the fluoride component because of its greater stability. So, TBA fluoride hydrate has the congruent melting point as high as 27,2°C, whereas TBA hydroxide hydrate has the incongruent melting point as high as 19°C. For the same reason in the TBA hydroxide - TBA bromide - H_2O system the limited solid solutions of type IV (according to Roozeboom) are realized (see Fig.4).

3.3. Alloxenic solutions

The conditions of the formation of the solution of such kind for channel inclusion compounds are simplified substantially and seem to be reduced to those of the formation of the individual clathrates, i.e. to the conformity of the guest molecules dimensions to the cross-section of the clathrate framework channel (for urea 5,25Å, for thiourea 6,9-7,4Å). n-Paraffins ranging from pentane to eicosane form with one another the continuous series of the solid solutions of the typeI in the urea framework channels whatever the number of the carbon atoms, while the same paraffins without the urea framework form the solutions with the break of the continuity. As noted in [25] for the latter case the problem of the molecule packing may be subdivided into two (namely that of the molecule packing in a layer and that of the layer packing - the difference of the latter resulting in the well-known variety of the n-paraffin structures). In the case of the clathrates these restrictions are eliminated substantially and the solid solutions are formed without any difficulties. In the same way CCl_4 with cyclohexane (C_6H_{12}) forms the solutions of the typeI

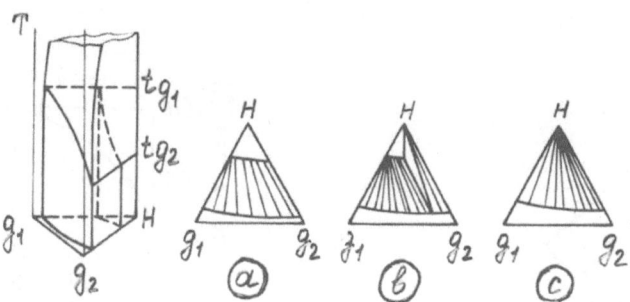

Figure 5. The scheme of the solubility polytherm of the host(H) - guest1 (g1) - guest2(g2) system: t_{g1} - the peritectic point of the clathrate with guest1, t_{g2} - the peritectic point of the clathrate with guest2. The isothermal sections at the temperatures: a) lower than t_{g2}; b) lower than t_{g1} and higher than t_{g2}; c) higher than t_{g1}.

in thiourea channels, whereas the CCl_4-C_6H_{12} system is of eutonic type [28]. In the thiourea - guest1 - guest2 systems (where guests are: CCl_4, C_6H_{12}, $CHCl_3$, C_6H_6) the solutions of the typeI enriched with the guest component making the more stable clathrate (C_6H_{12}>CCl_4>$CHCl_3$>C_6H_6 [6]) are formed. At 20°C n-decane does not form the clathrate with thiourea [27], but together with CCl_4 it is cocrystallized being up to 50 mol% in solid (without taking account of the host). In all the studied systems the interstitial solutions are not formed: the guests in the channel form an extremely dense one-dimensional packing and the framework is stable only under this condition. Thereby as well as in the case of the urea clathrates a number of the rather "severe" restrictions (according to A.I.Kitaigorodsky for the molecular crystals) is eliminated because by the joint inclusion of the above mentioned guest pairs into the rhomboedric thiourea the one-dimensional packing takes place.

On the ground of the studied isotherms the generalized polythermal phase diagram of the thiourea - guest1 - guest2 system has been made in which one can observe "the transformation" of the spontaneous adduct (according to the earlier adopted terminology [28]) into the unspontaneous one (and vice versa) at the temperature change.

In Fig.5 the situation when the peritectic temperature of the clathrate with the guest1 is higher than that of the clathrate with the guest2 is given. When t<t_{g2} in the whole concentration region the continuous series of the solid solutions enriched with the guest1 seems to be realized (see Fig.5, section a). This case was observed when the isotherms (20°C) of the thiourea systems (where the guest pairs were: CCl_4-C_6H_{12}, CCl_4 - C_6H_6, $CHCl_3$ - C_6H_{12}) were being studied. When t_{g2}<t<t_{g1} there appears the region of the host crystallization (see Fig.5, section b); it is realized in the thiourea - CCl_4 -n-$C_{10}H_{22}$ system; when the temperature rises the region of the solid solutions is getting narrow and when t>

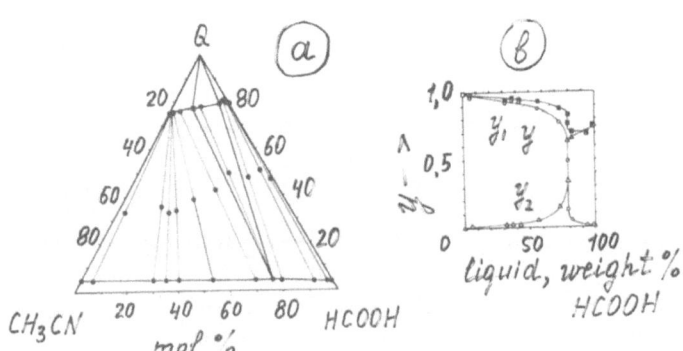

Figure 6. The solubility isotherm in the hydroquinone - HCOOH - CH₃CN system (a); b - the variation of the total (y) and the constituent degree of the clathrate framework filling (y_1 - for acetonitrile, y_2 - for formic acid).

>t_{gl} the pure thiourea is crystallized (see Fig.5, section c). All this allowed us to establish, for instance, the fact of existing individual thiourea clathrate with benzene as a guest at the usual temperatures[29]. At 20°C in the hydroquinone - HCOOH - CH₃CN system the alloxenic solid solutions (see Fig.6,a) with the continuity break (the typeIV) are formed which is also confirmed by comparison of the behaviour of the external vibrations in Raman spectra of the individual and mixed clathrate samples [30].

The stoichiometry of the mixed clathrates is represented in Fig.6,b, from which one can deduce that the degree of filling the clathrate phase varies depending on the formation conditions of the latter, when HCOOH: :CH₃CN=3,4 in the liquid it undergoes the leap. It is seen especially clearly when examining the character of the change of the filling degree with each of the guests separately (see Fig.6,b). The break in the continuity of the solid solutions which seems to be due to the different symmetry of the unit cells of hydroquinone clathrates with acetonitrile and the formic acid as guests is a quite unexpected phenomenon. It may be explained by different distortions of β-framework with each of the guests and the greater interaction among guest molecules of the same kind.

Thus, due to the nature of the clathrate compounds one can observe the rather distinct division into two subsystems: a guest and a host one. The former is more "mobile" as its properties are similar to those of liquid at least at usual temperatures. Therefore in this subsystem the solutions are formed with the same ease as in the liquid phase. The latter posesses all the properties of solid with the corresponding "strict" restrictions of the substitution of the particles.

REFERENCES

1. J.H.Van-der-Waals, J.C.Platteeuw: Advan.Chem.Phys., 2, 1 (1959).
2. Yu.A.Dyadin, L.S.Aladko, I.I.Yakovlev: Izv.Sib.Otd.Akad.Nauk SSSR, 14, 49 (1971).
3. Yu.A.Dyadin, L.S.Aladko: Zh.Strukt.Khim., 18, 51 (1977).
4. Yu.A.Dyadin, G.N.Chekhova: Izv.Sib.Otd.Akad.Nauk SSSR, 14, 18 (1975).
5. Yu.A.Dyadin, G.N.Chekhova, T.Ya.Arapova: Izv.Sib.Otd.Akad.Nauk SSSR, 9, 68 (1979).
6. G.N.Chekhova, Yu.A.Dyadin, T.V.Rodionova: Izv.Sib.Otd.Akad.Nauk SSSR, 12, 78 (1979).
7. Yu.A.Dyadin, G.N.Chekhova, I.I.Yakovlev: Izv.Sib.Otd.Akad.Nauk SSSR, 14, 49 (1973).
8. G.N.Chekhova, Yu.A.Dyadin: Izv.Sib.Otd.Akad.Nauk SSSR, 12, 75 (1978).
9. G.N.Chekhova, Yu.A.Dyadin: Polish J.Chem., 2, 407 (1982).
10. G.N.Chekhova, T.M.Polyanskaya et al.: Zh.Strukt.Khim., 16, 1054 (1975).
11. V.R.Belosludov, Yu.A.Dyadin et al.: Izv.Sib.Otd.Akad.Nauk SSSR, 5, 49 (1984).
12. V.R.Belosludov, Yu.A.Dyadin et al.: J.I.Ph., 1, 251 (1984).
13. L.S.Aladko, Yu.A.Dyadin et al.: Izv.Sib.Otd.Akad.Nauk SSSR, 2, 41, (1977).
14. B.S.Smolyakov, T.Ya.Arapova, V.V.Kokovkin: Izv.Sib.Otd.Akad.Nauk SSSR, 5, 93 (1984).
15. F.A.H.Schreinemakers: Z.Phys.Chem., 11, 75 (1893); 25, 71 (1906).
16. B.Angla: Ann.chim., 4 (12), 639 (1949).
17. V.Ya.Anosov, M.I.Ozerova, Ju.Ya.Phialkov. Osnovi Phyzikokhimitsheskogo Analiza. - M.: Nauka, 1976, 503p.
18. H.M.Powell: J.Chem.Soc., 61 (1948).
19. S.C.Wallwork, H.M.Powell: J.Chem.Soc., Perkin II, 641 (1980).
20. J.H.Helle, D.Kok et al.: Rec.trav.chim., 81, 1068 (1962).
21. R.L.Deming, T.L.Carlisle et al.: J.Phys.Chem., 73, 1762 (1969).
22. D.E.Evans, R.E.Richards: J.Chem.Soc., 3932 (1952).
23. Yu.N.Kazankin, A.A.Palladiyev, A.M.Trophimov: Zh.Ob.Khim., 42, 2607 (1972).
24. Yu.A.Dyadin, I.S.Terekhova et al.: Zh.Strukt.Khim., 17, 655 (1976).
25. A.I.Kitaigorodsky. Molecular Crystals. - M.: Nauka, 1971, 424p.
26. R.C.Makitra, Ja.M.Tsikantshuk: Zh.Ob.Khim., 46, 2189 (1976).
27. R.L.McLaughlin, W.S.McClenahan: J.Amer.Chem.Soc., 74, 5804 (1952).
28. L.S.Fetterly. In: Non-Stoichiometrie Compounds. Ed. by L.Mandelcorn. Acad.Press. New York-London, 1964.
29. Yu.A.Dyadin, G.N.Chekhova, T.Ya.Arapova: Izv.Sib.Otd.Akad.Nauk SSSR, 2, 45 (1977).
30. V.R.Belosludov, Yu.A.Dyadin et al.: J.I.Ph., 3, 243 (1985).

CLATHRATE THERMODYNAMICS FOR THE UNSTABLE HOST FRAMEWORK

Yu.A. Dyadin, V.R. Belosludov, G.N. Chekhova,
M.Yu. Lavrentiev
Institute of Inorganic Chemistry of the USSR Academy of
Sciences, Siberian Branch, 630090 Novosibirsk, USSR.

ABSTRACT. We have introduced the concept of clathrates whose empty host framework is unstable. In contrast to the Van der Waals theory, according to which the empty host framework is metastable, we believe it to become metastable when, in the cellular clathrates, certain type of cavities are fully occupied or, in the channel clathrates, the guest molecules are closely packed. The free energy of the channel and cellular types of clathrates has been determined using statistical thermodynamics methods. The obtained chemical potentials allowed us to describe the equilibrium of the clathrate with the stable host α-phase and the gaseous guest phase. For the cellular clathrates the equations have been obtained determining the dependence of the degree of filling of small cavities upon temperature and the gaseous phase pressure. In the case of the channel clathrates the set of equations on the composition and parameter of the orientational ordering is found. These equations enable us to describe quantitatively compressed state of the guest molecules in the channel and to find temperatures of orientational ordering.

1. INTRODUCTION

At present, clathrate compounds are considered as both ideal [1] and nonideal solid solutions [2-3] of the guest component in a crystalline framework of the host component. In the absence of the guest component, clathrate β-modification is considered to be metastable with respect to the stable α-modification under ordinary conditions, i.e. $\Delta\mu=\mu_\alpha-\mu_\beta<0$, where μ_i is the chemical potential of the corresponding modification. A number of experimental data on water, urea and thiourea clathrates, hardly consistent with the classical theory, made us study the formation of clathrates with unstable (labile) host frameworks. A comprehensive motivation of such possibility is given in [4].

2. THERMODYNAMICS OF CLATHRATES WITH UNSTABLE HOST FRAMEWORKS

Empty host framework instability may be due to the disturbance of the mechanical equilibrium conditions and the framework dynamic instability.

If the mechanical equilibrium conditions are disturbed, some of (or all) the framework atoms (molecules) will be affected by noncompensated forces causing the destruction of the cavities. When the dynamic equilibrium conditions are upset vibrations with complex frequencies resulting in an irreversible framework destruction exist among eigenfrequencies of the empty framework or appear with the temperature increase at the expense of anharmonicity. The guest molecules getting into the framework cavity restore the mechanical equilibrium and cause the disappearance of the "vibrations" with complex frequencies.

2.1. Cryptato-Clathrates

In our model we shall assume the framework to become at least metastable when the cavities of a labile type are completely filled by the guest molecules. For simplicity, clathrates with two types of cavities (labile and stable in the absence of the guest) and one type of non-dipolar guest molecules will be considered.

Let us assume that when the guest molecules fill the labile cavities a metastable framework[*]) forms, whose free energy F can be determined using a number of suppositions accepted for classical clathrates.

  a) A cavity cannot hold more than one guest molecule.
  b) Classical statistics is valid.
  c) We will assume that the forces influencing atoms (molecules) compensated at the expense of the forces created by the guest molecules are small. (The case of mechanically unstable framework.) Weak forces imply a small change of a portion of the free energy of the metastable framework concerned, when guest molecules are included into the labile cavities.

In the event of dynamic instability let us assume that the real frequencies of the empty host framework change negligibly with the inclusion of the guest molecules into the cavities and they provide a major contribution to that part of the free energy of the metastable framework which is connected with the host molecules.

  d) The cell model is true, i.e., the guest molecules are in the cavity statical potential (an adiabatic approximation for the description of the guest molecules relative to the host framework).

On the basis of the suppositions a)-d), the expression for the metastable framework free energy $\bar{F}$, both for mechanical and dynamic instability of the empty host framework can be written as follows:

$$\bar{F} = F_Q - kTN_1 \ln h_1 + \frac{1}{2} N_1 U_1, \qquad (1)$$

where on the basis of the c-th supposition the part of the free energy F_Q dependent upon the host molecule is separated, N_1 is the number of

─────────────
[*]) In this case the framework is a clathrate compound, such as hydrates $THF \cdot 17H_2O$, $CHCl_3 \cdot 17H_2O$, in which only one type of cavity is filled.

the labile cavities. The second term describes the behaviour of the guest molecules in the cavity, whose expression will look as follows if one uses the designation of [1-3]:

$$h_1 = 2\pi a_1^3 g_1 \exp[-W_1(0)/kT] \cdot \Phi_1(T), \qquad (2)$$

$$\Phi_1(T) = \sum_{\{n\}} \int \frac{d^3p}{(2\pi\hbar)^3} \exp[-(\varepsilon_n + p^2/2m)/kT], \qquad (3)$$

$$g_1 = (2\pi a_1^3)^{-1} \int d^3r \cdot \exp\{-[W_1(\vec{r}) - W_1(0)]/kT\}, \qquad (4)$$

where ε_n is the n-th inner level of the guest molecule energy in the cavity; $W_1(\vec{r})$ is the potential energy of the guest-host interaction; a_1 is the size of the cavity.

The third term in (1) describes the interaction among the guest molecules, where U_1 is the energy of the interaction between a guest molecule and other molecules.

If one knows the expression for the free energy of a metastable structure consisting of a framework whose labile cavities are filled with guest molecules, the further theory is created just as the traditional clathrate solution theory [1-3]. The expression for the clathrate free energy will look as follows:

$$F = \bar{F} + kTN_2[-y\ln h_2 + \tfrac{1}{2} y^2 U_2/kT + (1-y)\ln(1-y) + y\ln y], \qquad (5)$$

where N_2 is the number of stable cavities, $y = \bar{N}_g/N_2$ is the filling degree of these cavities; $\bar{N}_g$ is the total number of the guest molecules, situated in the stable cavities; U_2 is the energy of the interaction of a guest molecule, which completely fill the stable cages.

The expression for h_2 is derived from (2)-(4), where the expression for the potential energy of the interaction between the guest molecules, which is in the stable cavity, with the host molecules and other guest molecules in the labile cavities, should be inserted instead of $W_1(\vec{r})$, a_2 should be inserted instead of a_1, where a_2 is the size of the stable cavity; and ε'_n should be inserted instead of ε_n (ε'_n is the inner energy level of the stable cavity). The chemical potentials of the guest molecules are determined by the appropriate differentiation over the number of the host molecules N_Q and the number of the guest molecules $N_g = N_2 + \bar{N}_g$:

$$\mu_Q = \left(\frac{\partial F}{\partial N_Q}\right)_{N_g,T,V} = \left(\frac{\partial F_Q}{\partial N_Q}\right)_{T,V} -kT(\nu_1 \ln h_1 - \nu_2 \ln h_2) + \tfrac{1}{2}\nu_1[U_1 - 2yU_2] + \nu_2 kT[\ln(1-y) + \frac{\nu_1}{\nu_2}\ln\frac{1-y}{y}] - \frac{\nu_2}{2} y^2 U_2, \qquad (6)$$

$$\mu_g = \left(\frac{\partial F}{\partial N_g}\right)_{N_Q,T,V} = -kT \cdot \ln\frac{y}{(1-y)h_2} + yU_2, \qquad (7)$$

where ν_1, ν_2 are determined in the usual way: $N_i = \nu_i N_Q$, $i = 1,2$. Expressions (1)-(7) solve this task.

2.2. Tubulato-Clathrates

For the canal type of clathrates we will assume that to stabilize the lattice the guest molecules arrange in densely packed chains and that in addition to the suppositions b)-d) (see section 2.1), the following suppositions are realized:

e) The canal is considered to be smooth, i.e. the energy of a guest molecule does not depend on its coordinate along the canal axis. Furthermore, we can divide the energy levels connected with transversal and longitudinal motion of a guest in the canal.

f) The guest molecules are considered to be incompressible, i.e. their length does not change when the inner degree of freedom is excited. This shows that we discuss harmonical molecule vibrations with small amplitudes.

The above suppositions allowed us to consider the contributions to the free energy, related with the framework and the guest molecules, and in the last item we can single out the part responsible for the longitudinal motion of the guest molecules in a clathrate. Thus, the clathrate free energy is the sum of three parts:

$$F = F_Q + F_{g-h} + F_2. \tag{8}$$

The expression for F_{g-h} is found as in reference [1]:

$$F_{g-h} = -N_g kT \cdot \ln h, \tag{9}$$

where

$$h = g(T)\phi(T) \cdot \exp\{-W(0)/kT\}, \tag{10}$$

$$g = \int d^2 r \cdot \exp\{-(W(\vec{r}) - W(0))/kT\}. \tag{11}$$

In this expression N_g is the number of the guest molecules, $W(\vec{r})$ is the guest molecule energy depending on its mass centre remoteness from the canal axis, $\vec{r}$ is a two-dimensional vector in the plane perpendicular to the canal axis. The expression for the free energy portion connected with the longitudinal motion of the guest molecules in a canal looks as follows:

$$F_2 = -kT \cdot \ln \sum_{\{l\}} \exp(-E_l/kT). \tag{12}$$

Here l numbers the totality of the quantities, describing the configuration of the guest molecules in the canal, i.e. the orientation of molecules and the distance between them. E_l can be divided into two parts, one of them depending on the orientation of the interacting molecules. Thus for the interaction energy of the chain of the guest molecules located in the same canal one has:

$$E_l = \sum_i \varphi(a_i) - \sum_i J(a_i)\sigma_i \sigma_{i+1} - 6\bar{J} \sum_i \langle \sigma_i \rangle \sigma_i, \tag{13}$$

where a_i is the distance between the ends of the i-th and (i+1)-th guest molecules; $\varphi(a_i)$ is the potential energy of the interaction between them; $\sigma_i = \pm 1$ depending on the i-th molecule orientation; $J(a_i)$ is the quantity determining the strength of the interaction between the adjacent molecules in the canal; $\bar{J}$ is the same as $J(a_i)$ but for the interaction between the molecules situated in the neighbouring canals; $\langle\sigma_i\rangle = \sigma$ is the mean value of the order parameter. The substitution of the sum over the molecules from the adjacent chains by σ corresponds to the mean field approximation. The last two items in the expression for E_1 determine the orientation contribution to the guest chain interaction energy. Unlike in [5] we neglect the dependence of the guest-guest interaction upon the guest orientation (in [5] it was shown that this item changes the transition temperature negligibly) and consider J to depend on a. For F_2 we have

$$F_2 = F_2^a + F_2^\sigma, \quad F_2^{a,\sigma} = -kT \cdot \ln Z_2^{a,\sigma}, \qquad (14)$$

$$Z_2^a = \sum_{\{a_i\}} \exp[-\frac{1}{kT} \sum_i \varphi(a_i)], \qquad (15)$$

$$Z_2 = \sum_{\{\sigma_i\}} \exp\{-[\sum_i J(a_i)\sigma_i\sigma_{i+1} + 6\bar{J}\sigma\sum_i \sigma_i]/kT\}, \qquad (16)$$

where $\{a_i\}$ and $\{\sigma_i\}$ are arbitrary chosen sets of quantities $a_1,\ldots, a_{N_g}$ and $\sigma_1,\ldots, \sigma_{N_g}$, respectively; the summation in (15),(16) is carried out over all possible sets. Having substituted $J(a_i)$ by $J(a)$ where $a = \langle a_i \rangle$ is the average distance between the ends of the adjacent guest molecules, one can single out both F_2^a and F_2^σ. Thus, F_2^a is in a quasiharmonic approximation [6] which is true in the event of a weak anharmonicity, i.e. when $|a-a_0|/a_0 \ll 1$, where a_0 is the distance between the last carbon atom of one molecule and the first one of the adjacent molecule, corresponding to the minimum of the potential energy of the interaction between them. The orientational contribution to the free energy F_2^σ in the mean field approximation accepted by us is determined accurately as the free energy of one-dimensional Ising model in the external field. The final expression for the free energy of the clathrate type in question will have the following appearance:

$$F = F_Q + N_g\{-\frac{1}{2} kT \ln [2\pi kT h^2 f^{-1}(a)] - J(a) + \varphi(a) - \qquad (17)$$
$$- kT \cdot (-\frac{\bar{h}kT}{8\bar{f}^2} + \frac{\bar{g}^2 kT}{12\bar{f}^3}) - kT \cdot \ln[\text{ch} x + A(\sigma)]\},$$

$$x = \frac{6\bar{J}\sigma}{kT}, \quad y = \frac{2J(a)}{kT}, \quad f(a) = \frac{d^2\varphi}{da^2}, \quad \bar{f}=f(a_0), \quad \bar{g}=\varphi'''(a_0), \text{*)}$$

$$\bar{h} = \varphi''''(a_0), \quad A(\sigma) = \text{ch}^2 x - 2\text{sh } y \cdot \exp(-y),$$

where σ is determined according to the equation

---

*) The prime over $\varphi(a)$ here and below means the differentiation over the argument.

$$\sigma = \mathrm{sh}\, x\, [A(\sigma) + \frac{kT}{6\bar{J}}\, \mathrm{ch}\, x]/A(\sigma)\cdot[\mathrm{ch}\, x + A(\sigma)]. \tag{18}$$

To determine the chemical potentials of the host and the guest components in a clathrate it is necessary to differentiate the free energy accordingly over the number of the host and guest molecules. Doing so one should take the following expression into account:

$$N_Q = N_g[a + L(n-1)]/d, \tag{19}$$

where L is the distance between the adjacent carbon atoms in the same molecule (along the canal axis), n is the number of carbon atoms in a molecule, d is the distance between the framework layers, perpendicular to the 6-th order axis.

Equation (19) is obtained by calculating the number of the guest molecules per one host molecule. Thus, assuming $\partial F_2^\sigma/\partial a = 0$ [*)] in the first approximation we obtain

$$\mu_Q^\beta = \mu_Q + d\varphi'(a) + \frac{1}{2}\, dkT\, \frac{\varphi'''(a)}{\varphi''(a)} \tag{20}$$

where $\mu_Q = \partial F_Q/\partial N_Q$ is the contribution to the chemical potential, connected with the host structure;

$$\mu_g^\beta = \varphi(a) - [a + L(n-1)]\varphi'(a) - \frac{1}{2}\, kT\cdot\ln(2\pi kTh^2 f^{-1}(a)) - \frac{1}{2}[a+L(n-1)]kT\, \frac{\varphi'''(a)}{\varphi''(a)} - kT[-\frac{\bar{h}kT}{8\bar{f}^2} + \frac{\bar{g}^2 kT}{12\bar{f}^3}], \tag{21}$$

(17),(20),(21) comprise the distance between molecules, whose expression is deduced from the equilibrium conditions of the phases concerned. As an example we shall discuss the host α and β phases equilibrium and the gaseous and clathrate phases of the guest molecules. Equating the chemical potentials of different host and guest phases, we obtain the equation, describing this equilibrium:

$$\Delta\mu = d\varphi'(a) + \frac{1}{2}\, dkT\, \frac{\varphi'''(a)}{\varphi''(a)}, \tag{22}$$

$$P = \frac{kT}{g}\, \frac{\Phi_g}{\Phi}\, (\frac{f(a)}{2\pi kT})^{1/2}\, \exp\{\frac{W(0) + \varphi(a) - [a+L(n-1)]\varphi'(a)}{kT} - \frac{1}{2}(a+L(n-1))\, \frac{\varphi'''(a)}{\varphi''(a)} - (-\frac{\bar{h}kT}{8\bar{f}^2} + \frac{\bar{g}^2 kT}{12\bar{f}^3})\}, \tag{23}$$

where P is the guest component pressure in the gaseous phase, $\Phi_g(T)$ is the guest molecular partition function in the gaseous phase, $\Delta\mu = \mu_Q^\alpha - \mu_Q$, μ_Q^α is the host chemical potential in the α-phase.

Notice, that the distance between the guest molecules a and, con-

---
[*)] As will be shown further σ differs from 0 at low temperatures, when the thermal expansion of the chain is negligible.

sequently, the clathrate composition, too, are determined by the guest-guest interaction and depend only on temperature, while the guest-host interaction determines the equilibrium vapour pressure of the guest.

To determine the thermodynamic clathrate functions it is necessary to find $a(T)$ and $\sigma(T)$. The order parameter σ is determined according to eq.(18). The solution for $\sigma \neq 0$ is well known to become possible for this equation at $T<T_c$, where T_c is the phase transition temperature, determined by the interaction constants J and $\bar{J}$. In the event of urea clathrates with n-paraffins, as it follows from the calculations and the experiments [5], the transition temperature $T_c \sim 100K$, i.e., it is rather low. This allows us to single out two temperature regions and find $a(T)$ and $\sigma(T)$ in these regions.

a) High temperatures ($T>T_c$).

For several values of $\Delta\mu$ the $a(T)$ dependence has been calculated according to equation (22) (in the approximation of incompressible molecules). Lennard-Jones potential ($\varphi(a)=4\varepsilon[(\frac{\sigma}{a})^{12}-(\frac{\sigma}{a})^{6}]$) has been chosen for the function $\varphi(a)$. The parameter ε had the value of 148.2K - the same as for the CH_4-CH_4 interaction, and the parameter σ has been chosen so that $\sigma \cdot 2^{1/6}$ should give 4.06 Å - the equilibrium distance between the carbon atoms in the crystalline lattice of n-paraffins. The distance between the urea molecules along the sixth order axis $d=1.834$ Å.

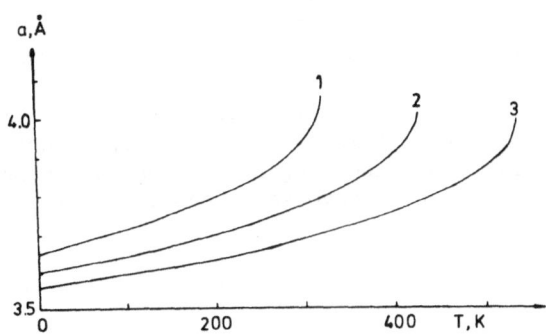

Figure 1. $a(T)$ dependence for $\Delta\mu=-2.96$ kcal/mole (1); $\Delta\mu=-3.94$ kcal/mole (2); $\Delta\mu=-4.93$ kcal/mole (3).

The diagrams $a(T)$ for three values of $\Delta\mu$ are shown in figure 1. Each $\Delta\mu$ has its own value of $T=T_0$, above which the clathrate cannot exist any longer, and the reason for this is not the structure instability but the fact that the equilibrium condition $\mu_0^\alpha = \mu_0^\beta$ is not realized at any a. The value of $\Delta\mu$ at which the maximum clathrate existence temperature T_0 is equal to the destruction temperature of the urea clathrate with polyethylene ($T_0 = 148°C$, $\Delta\mu = -3.94$ kcal/mole [7]) has been accepted accurate. The equilibrium distance between the neighbouring molecules at room temperature appears to be equal to 3.76 Å, which is very similar to the experimental value of 3.74 Å [8,9].

b) Low temperatures ($T < T_c$).
The equation for T_c follows from equation (18) and has the form:

$$\left(\frac{kT_c}{6\bar{J}}\right) = \exp\left(\frac{2J(a)}{kT_c}\right). \tag{24}$$

To solve equation (24) we neglect the dependence $a(T)$, which is possible in the temperature region concerned. The values of T_c obtained are in good agreement with the calculations and experiment [5].

3. CONCLUSIONS

The expression (17) obtained for the free energy of the clathrate of canal type makes possible to restore all the thermodynamics functions. Thus, the chemical potentials obtained allowed us to write equations (22), (23) for the equilibrium concerned, determining a and the equilibrium pressure P, respectively. From equation (22) it follows that $a(T)$ does not correspond to a_0, determined by the equation $\varphi'(a_0)=0$ (a_0 is the equilibrium interatomic distance of a one-dimensional chain) and since $\Delta\mu<0$, always $a<a_0$, i.e. from equation (22) it follows that for the clathrate phase stabilization it is necessary that the guest molecules should be situated in a clathrate in a compressed (in contrast to ideally stretched) state. Thus, the theory made possible to explain not only the dense packing, but even the shortening of the guest molecules, discovered experimentally, though only the supposition of their regular arrangement was included.

The model makes possible the explanation of partial filling the thiourea canal space by the guest molecules at low temperatures (140-160K), observed experimentally [10]: the temperature increase results in the fully occupied metastable host framework becoming labile at the expense of anharmonicity.

ACKNOWLEDGMENT. The authors gratefully acknowledge the work of N.V.Udachina in translating this report into English.

REFERENCES

1. J.H.van der Waals and J.C. Platteeuw:Adv.Chem.Phys.,2,1 (1959).
2. V.R. Belosludov, Yu.A. Dyadin, O.A. Drachiova and G.N. Chekhova:Izv. Sib.Otd.Akad.Nauk SSSR,4,9,60 (1979),C.A.,91,163770u (1979).
3. V.R. Belosludov, Yu.A. Dyadin G.N. Chekhova, and S.I. Fadeev:J.Incl. Phenom.,1, 251 (1984).
4. Yu.A. Dyadin, V.R. Belosludov:Izv.Sib.Otd.Akad.Nauk SSSR,5 72 (1986).
5. N.G. Parsonage,and R.C. Pemberton:Trans.Faraday Soc.,63,311'(1967).
6. G. Leibfried:Gittertheorie der mechanischen und thermischen Eigenschaften der Kristalle,Springer-Verlag,Berlin (1955).
7. G.N. Chekhova,and Yu.A.Dyadin:Izv.Sib.Otd.Akad.Nauk SSSR,5,66 (1986).
8. F. Laves, N.Nikolaides and K.C.Peng.Z.Kristallogr.,121,258 (1965).
9. N.U.Lenne, H.Ch.Mez,and W.Schlenk:Ann., 732, 70 (1970).
10. G.B.Sergeev, V.S. Komarov, and A.V. Zvonov:Zh.Obshch.Khim., 54, 985 (1984).

CLATHRATE FORMATION IN WATER-TETRAALKYL AMMONIUM IODIDE SYSTEMS AT HIGH PRESSURE

Yu.A. Dyadin, F.V. Zhurko, E.Ya. Aladko, Yu.M. Zelenin, L. A. Gaponenko
Institute of Inorganic Chemistry of the USSR Academy of Sciences, Siberian Branch, Novosibirsk, 630090, USSR

ABSTRACT. The phase diagrams of aqueous binary systems with $PrBu_3NI$(I), Bu_4NI(II), $i-AmBu_3NI$(III) and $i-Am_4NI$(IV) were studied at atmospheric and high pressures by DTA method. In systems III-H_2O and IV-H_2O at atmospheric pressure we observed polyhydrates melting incongruently at 7.1 and 14.7°, correspondingly. In systems I-H_2O and II-H_2O hydrates form at higher pressure only, there are $PrBu_3NI$ (15-25)H_2O at $P \geq 0.13$ kbar, Bu_4NI (25-35)H_2O at $P \geq 0.4$ kbar. In water systems with II-IV at pressure 1.2, 1.4 and 0.26 kbar correspondingly polyhydrates with smaller hydrate number form. Formation of hydrates in solution in the II-H_2O system does not occur at pressure greater than 7 kbar. The summarized P, T, X-phase diagram is discussed.

1. INTRODUCTION

The analysis of the thermal stability of clathrate polyhydrates in the tetrabutylammonium (Bu_4N) halogenide series has led us to believe that Bu_4NI clathrate hydrate must melt at positive temperatures /1/. This hydrate does not reveal itself, because the Bu_4NI crystal structure is stronger than that of other halogenides. This and a low hydration ability of the iodide ion are responsible for the low Bu_4NI solubility. In other words, the crystallization field of the salt itself overlaps that part of the phase diagram in which the crystallization field of the polyhydrate might be expected. A similar comparison /2/ of the thermal stability of the polyhydrates of $PrBu_3N$, $i-AmBu_3N$ and $i-Am_4N$ halogenides has also revealed that the greater the cation ability to stabilize the water framework the weaker the dependence of polyhydrate thermal stability upon the anion nature: $i-Am_4N$ fluoride, chloride, bromide hydrates have practically the same melting points and there is no reason to believe that the melting point of iodide hydrate (hydrates) would differ greatly, if it could melt congruently.

All this has caused us to believe that peralkylammonium iodides either form hydrates at the atmospheric pressure and they should be searched for more thoroughly, or, even if they are metastable, this metastability is not great and the conditions can be met under which they can become stable.

For the hydrate formers (cations) considered that typical polyhydrate structures are pressure stabilized structures /3/. Ice is known to be destabilized by pressure up to 2 kbar, so under pressure its crystallization field decreases allowing place for the crystallization field of polyhydrates. The pressure effect up to 2-5 kbar on the crystallization field of salts must be weaker than in the case of hydrates and ice. Therefore, using pressure we meant to obtain stable hydrates of the iodides concerned.

2. REAGENTS AND EXPERIMENTAL

The synthesis and purification of peralkylammonium iodides are described in /4/. The minimum content of the basic substance in the salts was not less than 99.5%. The analysis of the salts was carried out by potentiometric titration of tetraphenyl borate with a cation-selective indicator electrode /5/. The P,T-curves were determined by DTA technique in high pressure cells described in /2,6/. In the $PrBu_3NI-H_2O$ system thin wall pressure-transmitting teflon ampoules with a capacity of 0.12 ml were filled with the solution of the appropriate composition at 35°C (see the phase diagram, fig.1). In the rest of the systems the homogeneous solution with the appropriate composition cannot be obtained under the conditions convenient for carrying out work, that is why the ampoule was filled with the paste prepared of water and fine salt powder.

Exfoliation under pressure was studied using glass fiber light conductor introduced into the high pressure region as described in /6/.

3. RESULTS

Phase diagrams of the systems under consideration at atmospheric pressure based on data /4,7/ taking into account the results obtained in the present work are given in figure 1. Incongruently melting hydrates $i-AmBu_3NI \cdot (30\pm5)H_2O$ (7.1°C) and $i-Am_4BI \cdot (36\pm3)H_2O$ (14.7°C*) form in the last two systems. In the $PrBu_3NI-H_2O$ system the retrograde salt solubility in the temperature range 0–8°C and the existence of the low critical solvation temperature (LCST) /6/, are indicative of the conditions close to those necessary for clathrate formation. It is noteworthy that this is the only water-salt system with a closed exfoliation region we know, its temperature range being one of the smallest (LCST = 58.8°C; UCST = 68.4°C). The Bu_4NI-H_2O system is monotectic.

Figure 2 shows a fragment of the P,T-diagram of the $PrBu_3NI-H_2O$ system. If crystallization is carried out at atmospheric pressure, curve 3 corresponding to the incongruent melting (with the decomposition into salt and solution) of the $PrBu_3NI \cdot mH_2O$ hydrate versus pressure is obtained.

---

* The hydrate has been obtained. At 0°C ρ = 1.0099 g/cm^3. It is orthorhombic. Fmmm, a=12.1(1), b=21.6(2), c=49.9(5). According to data /7/ it melts incongruently at 13.3°C.

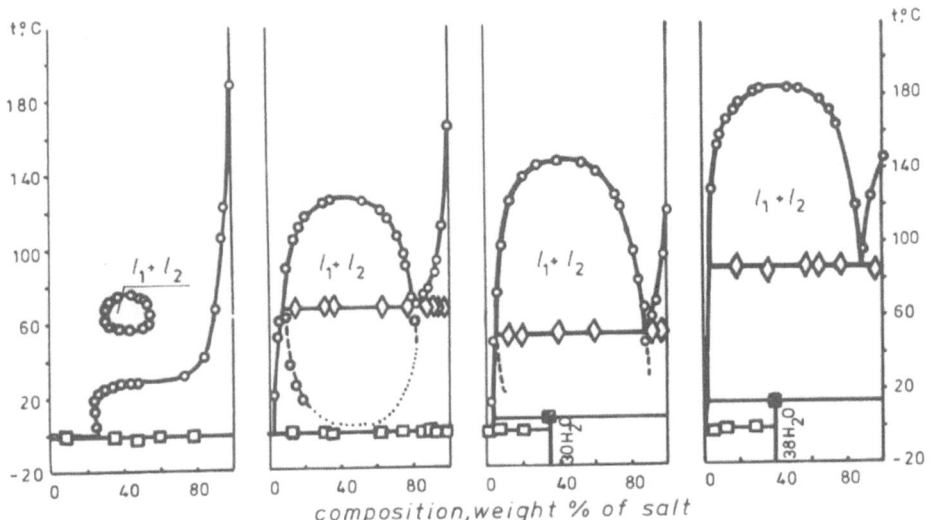

Figure 1. Phase diagrams of the tetraalkyl ammonium iodide - water systems at atmospheric pressure.

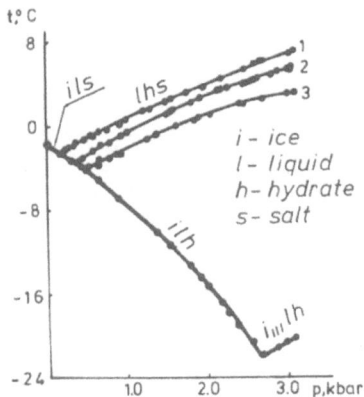

Figure 2. Fragment of P,T-phase diagram of the $PrBu_3NI-H_2O$ system.

The value of m found to be 15-25 was determined considering the presence or absence of heat melting effects of water eutectics. To determine the hydrate composition accurately by constructing Tamman's triangle was not possible, because it was difficult to make clathrate formation reaction go to completion.

Thus, with this type of reaction, clathrate formation begins at 0.5 kbar. However, if the samples were exposed to pressure of 1.5-2.4 kbar for 1-2 hours at $-15 \div -20°C$, the warming curves showed effects at temperatures described by curves 1 and 2. More stable hydrates (with the composition within the same limits) or polyamorphous modifications are likely to form under these conditions. The most stable of them (curve 1) has a stable crystallization field already at P ≥ 0.13 kbar.

Interesting is the behaviour of the eXfoliation region (fig.3). As follows from the supposition that LCST results from the destruction of clathratelike structures in the liquid phase /4/, first it increases with pressure up to 63.2°C at 0.36 kbar, then it decreases to 42.4°C at 2.45 kbar. UCST monotonously increases up to 130°C at 2.45 kbar. The composition of LCST changes so that it becomes richer in water.

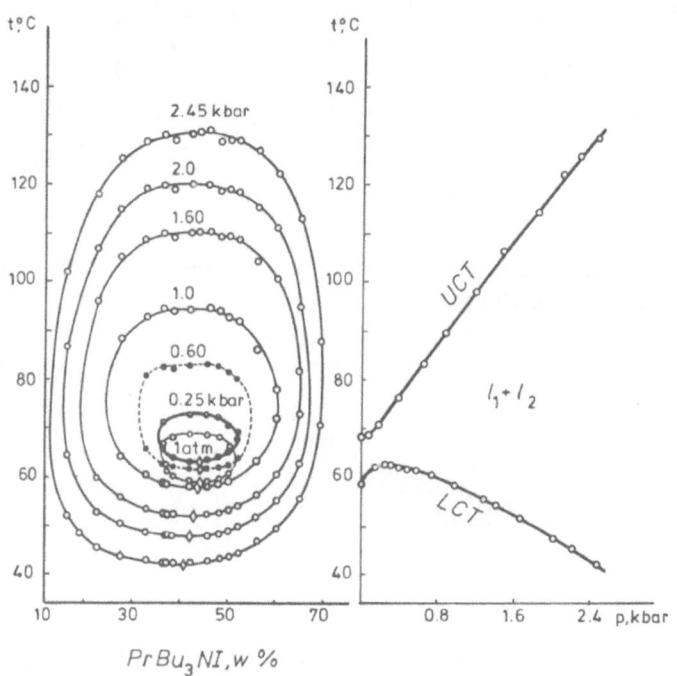

Figure 3. The exfoliation in the $PrBu_3NI-H_2O$ system under pressure.

The Bu_4NI-H_2O system in which no hydrates have been discovered at atmospheric pressure has been studied at pressures up to 9 kbar. At P = 0.4 kbar (fig.4) hydrate h_1 forms whose composition 1:(25-35) was determined as in the previous case. At P = 1.2 kbar one more hydrate h_2 with a smaller hydrate number forms which is evident due to the appearance of water eutectics, when pressure exceeds the transition pressure. At P = 7 kbar, which is not shown in the figure, no clathrate formation occurs in the system, at least, from the solution.

The $i-AmBu_3NI-H_2O$ system is shown in figure 4. In this system two hydrates form as in the previous case. One of them with the composition 1:(25-35) is stable both at high and at atmospheric pressure at temperatures up to +7.1°C at which it melts incongruently to give salt and

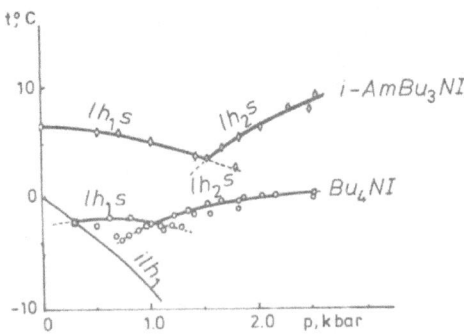

Figure 4. Fragment of the P,T-phase diagrams of the Bu_4NI-H_2O and i-$AmBu_3NI$-H_2O systems.

solution (fig.1). The formation of the hydrate requires some supercooling. This seems to be the reason why we have not found it earlier while studying the i-$AmBu_3NI$-H_2O system /4/. With increasing pressure the hydrate incongruent melting point decreases slightly (see fig.4). At P ≥1.4 kbar the second hydrate with a smaller hydrate number forms. This hydrate has a positive value of dt/dP. It should be noted that the analogous picture (but for the congruent melting of hydrates) was also observed in the i-$AmBu_3NBr$-H_2O system /8/. This shows that the formation of the salt during the clathrate decomposition into a separate phase and its behaviour at high pressure do not determine the behaviour of the system under pressure. The latter mainly depends on the structure and the specific volume of clathrate phases determined by the structure.

In the i-Am_4NI-H_2O system (fig.5) clathrate formation is rapid and occurs at atmospheric pressure /7,8/. For this reaction dt/dP =4.5 °C/kbar (1 atm). At 0.26 kbar one more hydrate (h_2) forms with a smaller hydrate number, which is shown by the appearance of the melting effects of water eutectics in the samples in which the salt content slightly exceeds that in the low pressure hydrate. We have managed to identify not only the incongruent melting lines of the hydrates, but also the monovariant equilibrium h_1+h_2= L line. The heat effect of this transition is negligible and it is approximately 100-200 times weaker than that of incongruent melting. The attempt to obtain the equilibrium h_2+h_1= S line by DTA technique was a failure. In figure 5 it is schematically depicted by a dashed line in the region allowed by the Schreinemakers rule /9/.

4. DISCUSSION

On the basis of the results obtained clathrate formation evolution in the water-iodides binary systems can be represented by the summarized P,T,X-diagram, whose P,T-projection and a number of characteristic isobaric sections are shown in figure 6.

First we will describe the case where at atmospheric or close to atmospheric pressure (P_1 in fig.6) the hydrate is metastable with

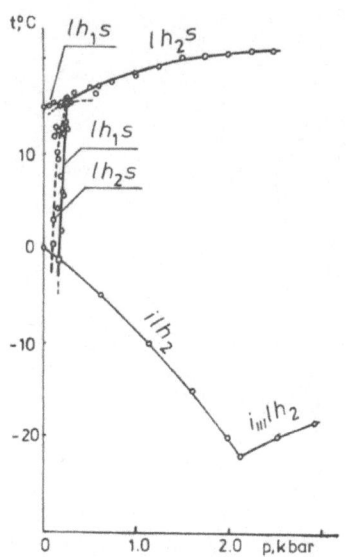

Figure 5. P,T-phase diagram of the $i\text{-}Am_4NI\text{-}H_2O$ system.

respect to both crystal salt and ice when the composition of mixtures is similar to that of clathrates, but the difference in the energy state of these phases is not great. Since with increasing pressure the melting point of ice Ih decreases, and the melting point of some clathrate polyhydrates generally increases /3/ (sometimes it can decrease but much more slowly than that of ice Ih) at certain pressures the hydrate becomes more stable than ice (P_2, fig.6). Further pressure increase results in the hydrate h_1 crystallization field overlapping the limits of that of ice and salt (P_4, fig.6) via nonvariant state at P_3*. Hydrate h_1 has upper and low stability limits drawing nearer under pressure (P_5* and P_6) and at P_7* it ceases to exist. Hydrate h_2 that appeared at comparatively low pressures (curve 7 on the P,T-projection) at this pressure begins to coexist with a liquid phase until it reaches the nonvariant state at P_9, where ice, stabilizing more rapidly under pressure (according to the scheme ice VI) depresses the crystallization field of hydrate h_2.

All types of diagrams in figure 6 have been obtained at the atmospheric pressure with iodides whose cations possess different ability to stabilize water-anion framework. The $PrBu_3NI\text{-}H_2O$ system at the atmospheric pressure with the hydrate whose melting is depicted by curve 3 in figure 2 can be attributed to the type at P_1 shown in figure 6. The cation of this salt (of all the salts considered) has the least ability to form clathrates. However, when with a stable hydrate (see fig.2, curve 1) this system just as $Bu_4NI\text{-}H_2O$ belongs to the type shown in figure 6 at P_2. The $i\text{-}AmBu_3NI\text{-}H_2O$ and $i\text{-}Am_4NI\text{-}H_2O$ systems belong to the type at P_4.

The probability of discovering the nonvariant state depicted by the isobar at P_3 in figure 6 at the atmospheric pressure is insignificant both because of a small probability of the appearance of such type of a

CLATHRATE FORMATION IN WATER-TETRAALKYL AMMONIUM IODIDE SYSTEMS 209

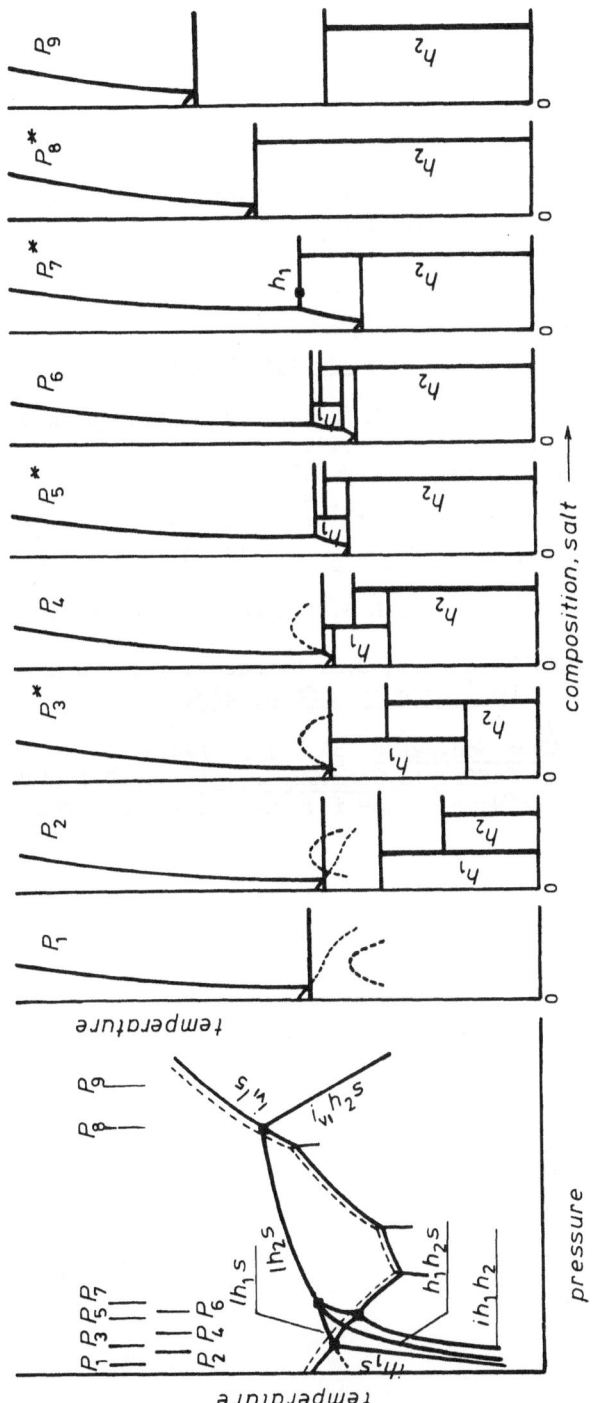

Figure 6. The summarized P,T,X-phase diagram of the tetraalkyl ammonium iodide - water system.

diagram at atmospheric pressure and considerable difficulties in the interpretation of the experimental data if they are obtained at one pressure only.

The P_{10} type of isobar via P_9 is discovered in the Bu_4NI-H_2O system at $P > 7$ kbar.

ACKNOWLEDGMENT

The authors gratefully thank N.V.Udachina for kind and disinterested translating this article into English.

REFERENCES

1. Yu.A. Dyadin and I.S. Terekhova:Izv.Sib.Otd.Akad.Nauk SSSR, 9, 88 (1980); C.A., 95, 13591e (1981).
2. Yu.A. Dyadin, F.V. Zhurko, Yu.M. Zelenin:Izv.Sib.Otd.Akad.Nauk SSSR, 2, 479 (1983); C.A., 98, 154144K (1983).
3. Yu.A. Dyadin, P.N. Kuznetsov and I.I. Yakovlev:Izv.Sib.Otd.Akad. Nauk SSSR, 7, 30 (1974); C.A., 81, 96833V (1974).
4. Yu.A. Dyadin Yu.M. Zelenin, L.S. Kiselyova (Zelenina), I.I. Yakovlev:Izv.Sib.Otd.Akad.Nauk SSSR, 12, 40 (1973); C.A., 31, 115682X (1973).
5. B.S. Smolyakov, Yu.A. Dyadin, L.S. Aladko:Izv.Sib.Otd.Akad.Nauk SSSR, 6, 66 (1980).
6. A.V. Nikolaev, P.N. Kuznetsov, D.S. Mirinsky, Yu.A. Dyadin, I.I. Yakovlev, N.S. Patrin:Dokl.Akad.Nauk SSSR, 223, 1, 101 (1975); C.A., 83, 137642U (1975).
7. H. Nakayama:Bull.Chem.Soc.Japan, 54, 3717 (1981).
8. R.K. McMullan, G.H. Jeffrey:J.Chem.Phys., 31, 5, 1231 (1959).
9. F.A. Schreinemakers:'Non-, mono- i divariantnye ravnovesiya!M.,(1948).

CATENATED AND NON-CATENATED INCLUSION COMPLEXES OF TRIMESIC ACID

F.H. Herbstein, M. Kapon and G.M. Reisner
Department of Chemistry, Technion-Israel Institute of Technology,
Haifa, Israel 32000.

ABSTRACT. Trimesic acid (benzene-1,3,5-tri-carboxylic acid; TMA) can in principle form two-dimensional hydrogen-bonded hexagonal networks in which central holes of the network have net diameters of 14 Å. Although such holes would be expected to be natural locations for guest molecules, non-catenated single networks have not been found in any of the crystals containing TMA studied in the last sixteen years. Instead, anhydrous α-TMA, TMA pentaiodide (TMA.I_5) and (so-called) γ-TMA have mutually triply-catenated structures in which triplets of networks are interlaced [3,4,5], while the hydrated complexes are based on non-catenated nets of composition TMA.H_2O [6]. We have now found conditions under which single networks are preserved without catenation, the cavities being occupied by guests such as n-tetradecane, n-heptanol, n-octanol, n-decanol, octene, cyclooctane and isooctane. The structures of 2TMA.n-tetradecane and 2TMA.isooctane have been solved and refined to R=13.0% and R=11.3%, respectively, disorder of the guest molecules having prevented further refinement of the room-temperature data. Determination of the crystal structures of the other complexes, which are isostructural with 2TMA.n-tetradecane, is now in progress. We are also investigating other potential guests.

Introduction

Trimesic acid (TMA; benzene-1,3,5-tricarboxylic acid) would be expected to be a favorable host for the formation of inclusion complexes since pair-wise hydrogen bonding between the carboxyl groups would lead to the formation of an infinite, two-dimensional "chicken-wire" network (Fig. 1). This network, based upon a hexagonal unit formed by six hydrogen-bonded TMA molecules, has a central hole having a net diameter (i.e. after taking into account the van der Waals radii of the TMA molecules) of 14 Å. A proper arrangement of these "chicken-wire" networks would lead to the formation of cylindrical channels characterized by a diameter of 14 Å. Such a situation should be contrasted with that found in urea channel inclusion complexes (net channel diameter ~ 4.5 Å) [1], those of thiourea (~ 5.5 Å) [1] and perhydrotriphenylene (~ 7 Å) [2]. Thus, TMA offers in potential much

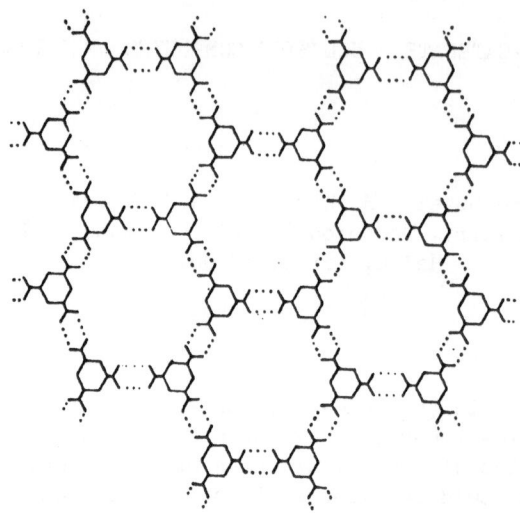

Figure 1: Idealised single "chicken-wire" network. The channels have net diameters of about 14 Å.

larger channels than those encountered in other channel inclusion complexes and may consequently act as host for guest molecules rather different from those forming complexes with urea or thiourea.

However, various attempts during the last 17 years to prepare crystals having <u>empty</u> "chicken-wire" networks and large guest molecules in the postulated cylindrical channels have failed. The various TMA polymorphs and complexes examined so far were found to be characterized by structures in which the channels formed by the TMA sheets are not available for accommodating guest molecules due to the remarkable phenomenon of mutual interlacing or triple catenation of the TMA networks [3,4,5].

Non-Catenated Complexes.

Recently we have succeeded in preparing TMA complexes in which the "chicken-wire" networks are preserved without catenation and so arranged as to form channels accommodating various guests. The guests used so far are: n-tetradecane, n-heptanol, n-octanol, n-decanol, octene, cyclooctane and isooctane. The crystals are colourless needles with pronounced cleavage normal to the needle axis and lose guest fairly easily to the atmosphere. Although complexes have been found with three different space groups, nevertheless comparison of cell dimensions suggests that the complexes are all isostructural.

INCLUSION COMPLEXES OF TRIMESIC ACID

The crystal structures of 2TMA.n-tetradecane (trigonal, space group $P3_1$ or $P3_2$, $\underline{a}=\underline{b}=16.500(6)$, $\underline{c}=10.071(4)$ Å, $Z=3$, $D_m=1.290$ g cm^{-3} $D_{calc}=1.297$ g cm^{-3}) and of 2TMA.isooctane (monoclinic, space group C2/c, $\underline{a}=28.599(9)$, $\underline{b}=16.596$ $\underline{c}=6.927(4)$ Å, $\beta=102.55(2)°$, $Z=4$, $D_{calc}=1.107$ g cm^{-3}) have been determined in outline. In both cases data collection was performed on a Philips PW 1100/20 four circle diffractometer (graphite-monochromated MoKα), structures solved by MULTAN [7] and refined using SHELX 77 [8]. Refinement converged to agreement factors of $R_F=0.130$ and $R_F=0.113$ for the n-tetradecane and isooctane

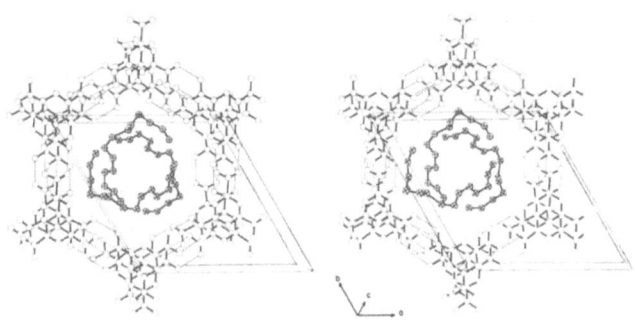

Figure 2: ORTEP [9] stereodiagram of the 2TMA.n-tetradecane complex viewed down the c axis.

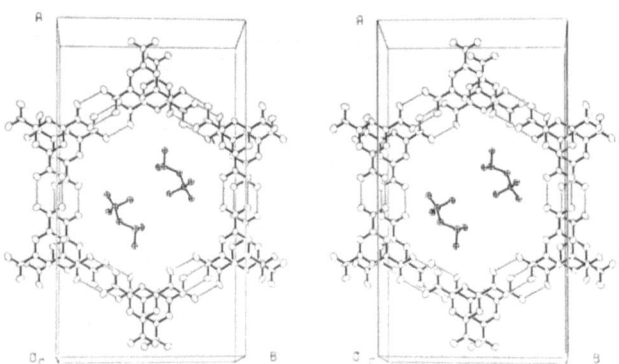

Figure 3: Stereodiagram of the C2/c polymorph of 2TMA.isooctane viewed down the c axis; guest molecule representations in both diagrams are schematic.

complexes, respectively, the disorder of the guest molecules having prevented further refinement of the room temperature data. The structure of 2TMA.n-tetradecane is shown in Fig.2. The guest molecules are disordered and their representation in the channels is schematic only. There are three TMA networks along the channel axis.

2TMA.isooctane has been found in two polymorphic forms; one is isomorphous with the tetradecane complex while the second is shown in Fig. 3. There are two networks in one period along the channel axis. The networks are somewhat non-planar in both crystals.

The 2TMA.n-octanol complex, whose structure has not yet been determined, should have, considering the length of the c axis, five TMA networks in one period along the channel axis.

In the future we plan to prepare and solve the structures of a variety of non-catenated TMA inclusion complexes with different types of guest molecules to test the chemical and steric limits to the formation of these complexes. It is our hope that the 14 Å diameter channels will provide a means for separating between large molecules of different shapes and sizes, some of which will be accommodated in the channels while others will not fit.

Acknowledgements: Technion Fund for Promotion of Research and Vice-president for Research Fund. U.S.-Israel Binational Science Foundation (earlier work).

References

1. K. Takemoto and N. Sonoda, 'Inclusion Compounds of Urea, Thiourea and Selenourea' in Inclusion Compounds: Structural Aspects of Inclusion Compounds Formed by Organic Host Lattices. Vol. 2, pp. 47-67 (1984), edited by J.L. Atwood, J.E.D. Davies and D.D. MacNicol, Academic Press, London etc.
2. M. Farina, ibid, pp. 69-95.
3. D.J. Duchamp and R.E. Marsh, Acta Cryst., B25, 5 (1969).
4. F.H. Herbstein, M. Kapon and G.M. Reisner, Proc. Roy. Soc. Lond., A376, 301 (1981).
5. F.H. Herbstein, M. Kapon and G.M. Reisner, Acta Cryst., B41, 348 (1985).
6. F.H. Herbstein and R.E. Marsh, Acta Cryst., B33, 2358 (1977).
7. P. Main, M.M. Woolfson, L. Lessinger, G. Germain and J.P. Declercq, "MULTAN 77. A system of computer programmes for the automatic solution of crystal structures for X-ray diffraction data". Universities of York, England and Louvain, Belgium (1977).
8. G.M. Sheldrick: "SHELX 77. A programme for crystal structure determination". University of Cambridge, England (1977).
9. Johnson, C.K. ORTEP. Report ORNL-3794 Oak Ridge National Laboratory, Tennessee (1965).

SELECTIVITY OF THE HOST $Ni(4\text{-mepy})_4(NCS)_2$ TOWARDS AROMATIC GUESTS

H. L. Wiener, L. Ilardi, P. Liberati, L. Dengler,
S. A. Jeffas, S. Saba, and N. O. Smith
Department of Chemistry
Fordham University
Bronx, New York 10458-5198
U. S. A.

The host $Ni(4\text{-mepy})_4(NCS)_2$, $\underline{I}$, forms isomorphous 1:1 inclusion compounds with many monocyclic aromatic guests. In continuing attempts to determine the basis for the selectivity shown by $\underline{I}$ towards pairs of such guests, distribution data between solid and liquid phases were reported for seven ternary systems at room temperature. These consist of $\underline{I}$, p-xylene, and each of the following in turn: p-bromotoluene, p-chlorotoluene, p-fluorotoluene, p-dichlorobenzene, 4-methylpyridine and benzene, as well as the system $\underline{I}$—p-chlorotoluene—p-dichlorobenzene. The results, as well as those already published, were reviewed.

The distribution data were given in the form of $\underline{R}_L$ and $\underline{R}_S$, where $\underline{R}_L$ is the mole ratio of the guests in the liquid and $\underline{R}_S$ that in the coexisting solid phase. Since $\ln \underline{R}_L$ appears to be a linear function of $\ln \underline{R}_S$ the data were fitted to

$$\ln \underline{R}_S = \underline{m} \ln \underline{R}_L + \underline{b}$$

giving the values of the parameters $\underline{m}$ and $\underline{b}$ shown in Table I. As $\underline{R}_S/\underline{R}_L$ is not, in general, a constant for any one system, its value when $\underline{R}_S = 1$ was chosen, somewhat arbitrarily, as a measure of the selectivity in each system. This quantity is also included in Table I and listed in the order of increasing value. For those guest pairs in which p-xylene is one member, the order given in the table is therefore the apparent order of preference of the other guests in the hierarchy of selectivity; namely, p-xylene-d_6 > p-xylene-d_{10} > p-bromotoluene > p-chlorotoluene > p-dichlorobenzene....... > benzene. However, this order bears no obvious relation to any molecular property of these guests.

Table I. Parameters for $\ln \underline{R}_S = \underline{m} \ln \underline{R}_L + \underline{b}$.

Guest pair	$\underline{m}$	$\underline{b}$	$\underline{R}_S/\underline{R}_L$ ($\underline{R}_S = 1$)
p-xylene/p-xylene-d$_6$	0.97	-0.12	0.88
p-xylene/p-xylene-d$_{10}$	.97	- .09	.91
p-xylene/p-bromotoluene	1.00	- .03	.97
(p-xylene/p-xylene	1.00	0	1.00)
p-xylene/p-chlorotoluene	0.999	.007	1.01
p-dichlorobenzene/p-chlorotoluene	1.00	.02	1.02
p-xylene/p-dichlorobenzene	.95	.15	1.18
p-xylene/p-fluorotoluene	.64	.99	4.63
ethylbenzene/toluene	.60	.94	4.75
p-xylene/ethylbenzene	1.16	2.3	7.56
p-xylene/toluene	.64	1.72	14.4
p-xylene/4-methylpyridine	.75	2.1	16.5
p-xylene/benzene	.32	2.56	2790

In order to reveal such a relation a quasi-thermodynamic approach was presented, based on the equilibria:

Guest A in liquid ⇌ Guest A in solid
Guest B in liquid ⇌ Guest B in solid

If $\underline{f}$ is fugacity then $\underline{f}_A^L = \underline{f}_A^S$ and $\underline{f}_B^L = \underline{f}_B^S$, where L and S refer to coexisting liquid and solid phases, respectively. For $\underline{R}_S = 1$, and assuming that Raoult's law applies to the guests in the liquid phase, $\underline{f}_A^L = \underline{p}_A^\bullet \underline{x}_A^L$ and $\underline{f}_B^L = \underline{p}_B^\bullet \underline{x}_B^L$, where $\underline{p}_A^\bullet$ and $\underline{p}_B^\bullet$ are the vapour pressures of the pure liquid A and B. Furthermore, writing $\underline{f}_A^S = (1/\underline{k}_A)\underline{x}_A^S$ and $\underline{f}_B^S = (1/\underline{k}_B)\underline{x}_B^S$ when $\underline{R}_S = 1$, gives

$$\underline{k}_A/\underline{k}_B = (\underline{R}_S/\underline{R}_L)(\underline{p}_B^\bullet/\underline{p}_A^\bullet)$$

in which $\underline{k}_A$ and $\underline{k}_B$ are measures of the attraction of the host for the respective guests. Thus, multiplying the $\underline{R}_S/\underline{R}_L$ values in Table I by $\underline{p}_B^\bullet/\underline{p}_A^\bullet$ yields $\underline{k}_A/\underline{k}_B$, a quantity which compares the host-guest attractions for A and B. Values of $\underline{k}_A/\underline{k}_B$ are shown in Table II, in which, with two obvious exceptions, A represents p-xylene and B the guest paired with it. Assigning an arbitrary value of 100 to $\underline{k}$ for p-xylene yields the values of $\underline{k}_B$, also given in the table.

The $\underline{k}$'s, with the exception of that for 4-methylpyridine, vary smoothly with the estimated van der Waals length of the guest molecule. Thus it is the latter property which appears to determine primarily the selectivity, the longer of the two competing guest molecules being preferred over the shorter — after correcting for the differences in their vapour pressures. An explanation for this is attempted.

Table II. Correlation between corrected guest selectivities and van der Waals length of guest molecule

A/B	k_A/k_B	k_B	length of B (nm)
p-xylene/p-bromotoluene	0.153	656	1.00
p-xylene/p-dichlorobenzene	.256	390	0.98
p-xylene/p-chlorotoluene	.40	250	.97
p-xylene/p-xylene-d_6	.94	106	.96
p-xylene/p-xylene-d_{10}	.97	103	.96
p-xylene/p-xylene	1.00	(100)	.96
p-dichlorobenzene/p-chlorotoluene	1.86	–	–
p-xylene/ethylbenzene	8.2	12.2	.90
p-xylene/p-fluorotoluene	11.0	9.1	.89
ethylbenzene/toluene	14.1	–	–
p-xylene/toluene	46.3	2.2	.85
p-xylene/4-methylpyridine	9.4	10.6	.77
p-xylene/benzene	3.0×10^4	0.0033	.74

The behaviour noted with $\underline{I}$, where the fit of the guests in the host lattice is loose was contrasted with that observed with other hosts where the fit is tight.

The assumption of ideality in the liquid phase permits the calculation of the activity coefficients of the guests in the solid phases. These were also presented. For each guest pair the preferred member is the one with the smaller activity coefficient.

SOME PHYSICAL AND THERMOPHYSICAL PROPERTIES OF CLATHRATE HYDRATES

D.W. Davidson, M.A. Desando, S.R. Gough, Y.P. Handa,
C.I. Ratcliffe, J.A. Ripmeester and J.S. Tse
Division of Chemistry
National Research Council of Canada
Ottawa K1A 0R9

1. STRUCTURES OF HYDRATES WITH SMALL GUEST MOLECULES

Oxygen, nitrogen and air, like argon and krypton (1), are found to preferentially form gas hydrates of structure II, rather than structure I as previously expected for gas hydrates of small guest molecules. Lattice parameters from X-ray diffraction are given in the Table.

TABLE I. Hydrate Structures of Small Guest Molecules

Guest	Diameter	Structure	a/Å at 100K
Ar	3.83Å	II	17.07±0.04
Kr	4.04	II	17.08±0.08
O_2	4.01	II	17.07±0.01*
N_2	4.10	II	17.11±0.06
Air		II	17.24±0.06
CO	4.10	?	
CH_4	4.58	I	11.77±0.01*
Xe	4.57	I	11.84±0.02

*Neutron diffraction of D_2O hydrate at 5K.

Laboratory-synthesized "air hydrate", like naturally-occurring air hydrate from the Greenland ice-cap, contains air which is enriched in oxygen and argon. The stable clathrate structure changes back to II for molecules of the size of SF_6 (5.9Å) and larger.

2. CARBON MONOXIDE HYDRATE

This hydrate, whose presence in comets has been suspected for some years, has been prepared for the first time and found to have a dissociation pressure of 128 bars at 0°C (figure 1).

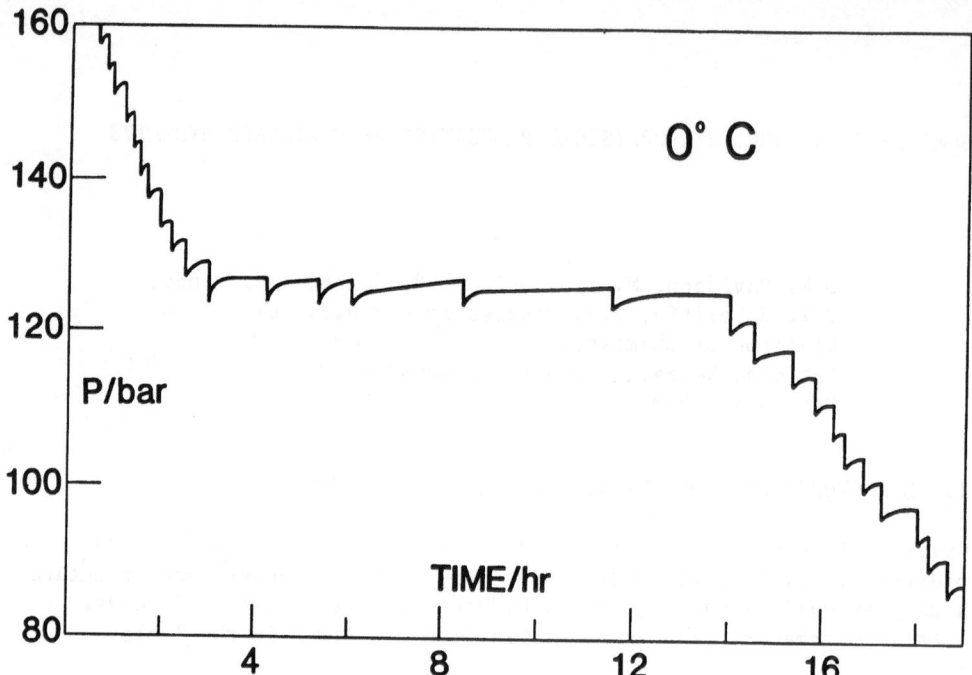

Fig. 1. The response of the pressure over a sample of CO hydrate to step-wise release of gas from the pressure vessel in which the hydrate was formed.

Dielectric absorption arising from reorientation of slightly dipolar CO molecules in the hydrate cages shows the barriers to reorientation to be extremely low (on average <0.5 kJ/mol) and widely distributed because of the frozen-in disorder of the orientations of the water molecules of the lattice. In solid CO itself the barrier is 6 kJ/mol.

3. THERMAL EXPANSIVITIES

Thermal expansivity measurements of ethylene oxide and tetrahydrofuran hydrates have been made by determining the temperature dependence of the spacings of several prominent lines in the powder X-ray diffraction patterns. These lines corresponded to Miller indices of 222, 123, 035 and 135 for ethylene oxide hydrate and 115+333, 135, and 066+228 for tetrahydrofuran hydrate. Expansivities at specific temperatures are given in Table II. These are seen to be much larger than for ice.

Table II. Linear Expansivities d ln a/d T (×10^6 K)

	100K	150K	200K	250K
Ice (average)	13	25	37	48
EO Hydrate (I)*	40	58	77	95
THF Hydrate (II)*	36	45	54	61
THF Hydrate**	28	42	52	62

*This work,
**Dilatometric results of Roberts et al. (1984) (2)

Compared to ice, the Grüneisen parameter is abnormally large for the clathrate hydrates at low temperatures. Molecular dynamics simulations using pair-wise additive intermolecular potentials show that the large anharmonicities in the forces acting between water molecules which are necessary to account for large expansivities occur because of the open structures of the lattices of clathrate structures and the presence of the guest molecules

4. CALORIMETRY OF CLATHRATE HYDRATES

Heat capacities above 90K and heats of dissociation of several of the most important gas hydrates have been measured with an automated Tian-Calvet heat flow calorimeter equipped for simultaneous measurement of the pressure. Hydrate compositions were determined by analysis of the decomposition products. The contribution of the water molecules to the heat capacity appears to be nearly the same as for the water molecules in ice.

The tabulated (table III) calorimetric values of the heats of dissociation of gas hydrate into ice and gas have been reduced to the standard state (0°C and 1 bar) by making use of small corrections based on heat capacities and compressibilities. These heats are in all cases more accurate than the heats which have been determined by application of the Clausius-Clapeyron equation to the temperature dependence of the dissociation pressures along the hydrate-ice-gas equilibrium line. The only previous data of comparable accuracy are the unpublished xenon hydrate dissociation pressure results of S.L. Miller at 16 temperatures between -28.5 and 0°C which give 25.32 ± 0.25 kJ/mol. This result is to be compared with the first entry in table III. The second entry refers to hydrate prepared under xenon pressure much higher than the three-phase equilibrium value. Combination of the two entries suggests that the heat of encagement of Xe depends on the degree of cage occupancy, at least for large cage occupancies.

Table III. Calorimetrically determined molar enthalpy changes for the dissociation of a hydrate into ice and gas at 273.15K and 1 bar

Hydrate	$\Delta H_m/\text{kJ mol}^{-1}$
$Xe \cdot 6.29 H_2O$	25.43 ± 0.17
$Xe \cdot 5.90 H_2O$	26.50 ± 0.17
$Kr \cdot 6.10 H_2O$	19.54 ± 0.24
$CH_4 \cdot 6.00 H_2O$	18.13 ± 0.27
$C_2H_6 \cdot 7.67 H_2O$	25.70 ± 0.37
$C_3H_8 \cdot 17.0 H_2O$	27.00 ± 0.33
$i\text{-}C_4H_{10} \cdot 17.0 H_2O$	31.07 ± 0.20

5. FREE ENERGY OF THE EMPTY STRUCTURE I LATTICE

The basic requirement for stability of a clathrate is that there be sufficient occupancy of the clathrate cages to reduce the free energy of the molecules making up the lattice to a value equal to or less than the free energy of the same molecules in the form (solid or liquid) stable in the absence of the guest species. In the case of gas hydrates of structure I, the minimum condition of stability is given by the ideal solution as

$$-\Delta\mu = kT \lceil \ln(1-\theta_S) + 3 \ln(1-\theta_L) \rceil / 23$$

where $\Delta\mu$ is the chemical potential of the water molecules in the empty hydrate lattice relative to ice (or liquid water) and θ_S and θ_L are the degrees of occupancy of the small and large cages which just stabilize the hydrate. We have determined $\Delta\mu$ for Xe hydrate (3) the most ideal of structure I hydrates.

The equilibrium hydrate at 0°C was found to have the composition $Xe \cdot (6.286 \pm 0.030)H_2O$. The ^{129}Xe proton-decoupled NMR spectrum gives two well-resolved peaks with relative areas which, allowing for the presence of three times as many large as small cages, give $\theta_S/\theta_L = 0.73 \pm 0.02$. Together these results determine $\Delta\mu = 1297 \pm 110$ J/mol. A nearly identical result was obtained from a recent study by Dharma-Wardhana et al. of type I cyclopropane hydrate with the assumption that cyclopropane is too large to occupy any of the small cages.

1. D.W. Davidson, Y.P. Handa, C.I. Ratcliffe, J.S. Tse, and B.M. Powell, Nature 311, 142 (1984).

2. R.B. Roberts, C. Andrikidis, R.J. Tarnish and G.K. White, Proc. 10th Int. Cryogenic Eng. Conf., Helsinki, p. 409, Butterworth (1984).

3. D.W. Davidson, Y.P. Handa and J.A. Ripmeester, J. Phys. Chem. (in press).

THREE-DIMENSIONAL METAL COMPLEX HOSTS BUILT OF α,ω-(LONG-CARBON-CHAIN)-DIAMINOALKANE LIGAND BRIDGING TWO-DIMENSIONAL CYANOMETAL COMPLEX NETWORK: HOFMANN-DIAMINOALKANE-TYPE CLATHRATES

Toschitake Iwamoto, Shin-ichi Nishikiori, and Tai Hasegawa
Department of Chemistry, College of Arts and Sciences,
The University of Tokyo
Komaba, Meguro, Tokyo 153, Japan

We have been developing systematically the chemistry of Hofmann-type and related clathrate compounds. Several modifications of the metal complex host structures have been derived from that of the Hofmann-type by replacing the ammine ligands by amines or diamines, and/or the square-planar tetracyanometallate by tetrahedral one [1]. However, their structures were substantially similar to that of the Hofmann-type: the volume of cavity was barely large enough to accommodate such small aromatic molecule as those enclathrated in the Hofmann-type host.

In order to accommodate a variety of guest molecules different in size and shape from those enclathrated in the Hofmann-type and the analogous clathrates previously derived, substitution of the ammine ligands in the Hofmann-type host by lipophilic amines or diamines have been attempted [2-6]. One of those attempts by us [6] and another group [4] was achieved by introducing α,ω-(long-carbon-chain)-diaminoalkanes into the host structures, where the diamine behaved as an ambidentate ligand to make a bridge between adjacent two dimensional cyanometal complex layers; the three-dimensional metal complex hosts thus derived may be called Hofmann-diam-type.

As listed in Table I more than 150 kinds of Hofmann-diam-type inclusion compounds have been prepared with the general formula $Cd[NH_2(CH_2)_nNH_2]Ni(CN)_4 \cdot xG$ for the hosts with $\underline{n} = 4 - 9$ and the guests with $\underline{x}=0.5-2$. They were characterized by chemical analyses, thermogravimetry, powder X-ray diffractometry, and vibrational (IR and Raman) spectroscopy; single crystal X-ray structure analyses have been carried out for the eight of them to demonstrate the characteristic features of the host structures [7-10]. In Table I the coefficient $\underline{x}$, i. e., the number of guest molecule per a formal unit of the host metal complex, the basal spacing $\underline{d}$ determined from the powder x-ray diffraction pattern, and the crystalline appearance are summarized; for some of them it was difficult to determine $\underline{x}$ and/or $\underline{d}$ as reproducible values but the formation of something comprised of the metal complex and the aromatic species was affirmed by the infrared spectrum. Those with the $\underline{x}$ of about 1 but not larger than 1 have probably the structures similar to those of the Hofmann-dabn-type (dabn = 1,4-diaminobutane) [10] or the

Table I. Hofmann-diam-type

n	4	5
Pyrrole	1.0; 7.84; A+	2.0; nd; C
Thiophene	F	F
Benzene	1.5; 9.73; B	1.0; 8.51; B
Phenol	F	NE
Aniline	1.5; 9.70; A+	1.0; nd; C
Toluene	1.5; 9.75; B	--------
Fluorobenzene	0.8; nd; C	1.0; nd; C
Chlorobenzene	1.5; 9.93; B	--------
Bromobenzene	1.3; nd; C	--------
Iodobenzene	1.3; nd; C	--------
Phenylacetylene	1.5; 9.69; B	1.5; 9.43; B
Styrene	1.5; 9.54; B	1.5; nd; B
Ethylbenzene	--------	--------
N,N-Dimethylaniline	1.0; 9.01; A+	1.45; 11.2; B
n-Propylbenzene	NE	NE
i-Propylbenzene	--------	--------
dl-1-Phenylethanol	--------	--------
o-Toluidine	1.5; 9.78; A	1.5; 10.2; A
m-Toluidine	1.28; 9.89; B	1.0; nd; C
p-Toluidine	1.5; 9.50; A	1.5; 11.0; C
o-Xylene	F	--------
m-Xylene	--------	--------
p-Xylene	--------	--------
o-Dichlorobenzene	1.5; 9.93; C	--------
m-Dichlorobenzene	--------	--------
p-Dichlorobenzene	--------	--------
m-Dibromobenzene	--------	--------
2,3-Xylidine	1.28; 10.0; A	1.29; 10.5; A
2,4-Xylidine	1.22; 9.45; A	--------
2,5-Xylidine	1.0; 9.48; A+	1.0; 10.2; A
2,6-Xylidine	1.25; nd; B	1.25; nd; C
3,4-Xylidine	1.25; nd; A	--------
3,5-Xylidine	1.18; 9.93; B	1.0; 10.8; B
1,2,3-Trimethylbenzene	--------	--------
1,2,4-Trimethylbenzene	--------	--------
1,3,5-Trimethylbenzene	--------	--------
2,4,6-Trimethylaniline	1.45; 10.3; B	1.34; 11.1; B
1,2,3,4-Tetramethylbenzene	NE	--------
1,2,3,5-Tetramethylbenzene	NE	NE
1,2,4,5-Tetramethylbenzene	NE	NE
Pentamethylbenzene	NE	NE
Hexamethylbenzene	NE	NE
Indene	2.0; 10.3; A	1.38; 10.5; B
Naphthalene	NE	NE
Phenanthrene	NE	NE

* $Cd[NH_2(CH_2)\underline{n}NH_2]Ni(CN)_4 \cdot \underline{x}$. The values of $\underline{x}$ and basal spacing $\underline{d}$ in guest G and the host of the $\underline{n}$ = 4, 5, 6, 7, 8, or 9. F: evidence of --------: examined but not yet obtained; nd: not determined; A+: single crystalline powder; D: fine powder; E: poor or unstable.

HOFMANN-DIAMINOALKANE-TYPE CLATHRATES

Compounds*			
6	7	8	9
0.9; nd; E	F	1.2; nd; C	NE
0.9; nd; C	0.3; nd; E	1.2; nd; D	NE
1.0; 9.19; E	1.0; 13.27; E	0.25; 10.73; D	0.45; nd; C
1.0; nd; D	NE	0.5?; 10.78; D	NE
1.0; nd; C	1.0; nd; C	1.2; nd; D	1.0; nd; C
1.0; 9.40; B	1.0; nd; B	0.5?; 11.05; A	0.5; 10.2; D
1.0; nd; C	0.8; nd; E	0.75; nd; C	NE
1.0; 9.21; C	1.0; 10.96; A	0.8?; nd; C	F
1.0; 9.49; C	1.0; nd; B	1.0; nd; C	NE
1.0; nd; C	1.0; nd; D	1.0; nd; D	NE
1.0; 9.59; B	1.0; nd; D	F	1.0; nd; B
0.9; 9.48; B	1.0; nd; C	F; 10.83; B	0.6; nd; D
0.7; 9.42; A	---------	1.0; 10.98; C	NE
0.8; 9.60; B	1.0; 10.64; C	F	2.5; nd; E
---------	---------	0.8; 10.96; C	NE
---------	---------	0.9; 11.33; C	NE
F	---------	F	NE
1.0; 9.42; A+	1.0; 9.63; A	2.5?; 10.92; B	2.8?; nd; C
1.0; 9.62; A+	1.0; nd; B	1.0; 11.01; C	2.3; nd; B
0.9; 10.56; A+	1.0; nd; B	1.0; 10.96; C	2.5; nd; B
F	1.0; nd; D	1.0; nd; C	NE
F	---------	1.0; 11.1; C	2.0; nd; C
0.7; 9.40; C	1.0; nd; C	1.0; 10.50; C	2.0; nd; C
1.0; nd; C	1.0; nd; B	1.0; nd; B	F
0.75; 9.45; B	---------	1.0; nd; C	F
0.8; 9.65; C	---------	1.0; nd; C	F
F	1.0; nd; C	F	NE
1.0; 9.82; A	1.0; nd; B	0.9; 10.81; A	3.0; nd; A
1.0; 9.47; A+	1.0; nd; C	0.9; 10.86; A	3.0; nd; A
1.0; 9.60; B	1.0; 10.35; B	0.9; 10.73; A	3.0; 13.78; A
0.7; nd; A	3.0?; nd; B	0.85; nd; A	0.5; 13.60; A
0.6; nd; B	0.5; nd; B	1.0; nd; A	2.5; nd; C
0.75; 9.61; B	0.8; nd; D	0.9; nd; A	2.5; nd; C
0.9; 9.98; C	---------	0.9; nd; C	2.0; nd; C
0.7; 9.98; C	---------	F	2.0; nd; C
---------	---------	1.0?; 11.14; D	2.0; nd; C
1.0; nd; B	0.9; nd; C	1.0; nd; C	2.5; nd; C
F	---------	0.6; nd; C	0.6?; nd; C
---------	---------	0.8; nd; D	0.6?; nd; C
0.7; 9.73; C	---------	0.5?; nd; D	NE
---------	---------	3.0; nd; C	NE
---------	---------	---------	---------
1.0; nd; A	1.0; nd; B	1.0; nd; B	2.5; nd; B
F	1.0; nd; C	F	F
2.22?; nd; E	F	NE	NE

d/A, and apparent crystalline state are shown for each compound with the formation has been obtained; NE: synthesis has not yet been examined; crystal data are available; A: good crystals; B: acceptable ones; C:

Hofmann-dahxn-type (dahxn = 1,6-diaminohexane) [8,9] ; their three-dimensional host structures are built of the stacking of two-dimensional cyanometal complex layers bridged by diaminoalkane ligands. Although the structure of the Hofmann-dabn-type aniline clathrate with $\underline{x}$ = 1.5 has been determined, it is rather unprobable that those with $1<\underline{x}<1.5$ are isostructural to the aniline one: the guests bulkier than aniline cannot be accommodated in the host with the ratio larger than 1 [10]. The products with the value of $\underline{x}$ larger than 1 are assumed to have structures conisderably different from those already determined.

During the research work on these clathrate compounds we discovered a rare case of conversion from an apparently more stable coordination complex to a less stable clathrate compound for the Hofmann-dahxn-type series. The $\underline{p}$-toluidine compound obtained under the experimental conditions similar to those for the $\underline{o}$- and the $\underline{m}$-toluidine clathrates was not a clathrate but a coordination complex $[Cd(\underline{p}-CH_3C_6H_4NH_2)_2-(dahxn)][Ni(CN)_4]$, which turned to a clathrate $Cd(dahxn)Ni(CN)_4 \cdot \underline{p}-CH_3C_6H_4NH_2$ isostructural to the $\underline{o}$- and the $\underline{m}$-toluidine clathrates on heating under ambient atmosphere: one mol of the $\underline{p}$-toluidine ligand was thermally liberated from the complex but the retained 1 mol was converted from a ligand to a guest in the solid state reaction [11].

Another notable fact is the accommodation of an aliphatic guest molecule in the Hofmann-daotn-type host (daotn = 1,8-diaminooctane). After screening more than thirty kinds of aliphatic molecules against the Hofmann-diam-type hosts, the Hofmann-daotn-type $\underline{n}$-hexanol clathrate was obtained. The guest is accommodated with the orientation of the aliphatic chain parallel to the all-trans daotn bridge in the host. The structure can be seen as a model of pillared intercalation compound for which few single crystal data have been available.

A phase transition probably to a superlattice structure has been observed for Hofmann-dahxn-type $\underline{o}$-toluidine clathrate at 200 K. Above this temperature the atomic temperature factors of the guest molecule increase steeply to suggest that the librational motion of the guest molecule is excited thermally. The details of the phase transition is still unknown, although it should be related to motion of the guest molecule.

1) T. Iwamoto, "The Hofmann-type and Related Inclusion Compounds," in Inclusion Compounds v. 1, J. L. Atwood, J. E. D. Davies, and D. D. MacNicol eds., Academic Press, London (1984), pp. 29-57.
2) S. Nishikiori and T. Iwamoto, Chem. Lett., 1035 (1982).
3) S. Nishikiori and T. Iwamoto, Chem. Lett., 1129 (1983).
4) J. E. D. Davies and A. M. Maver, J. Mol. Struct., **102**, 203 (1983).
5) S. Nishikiori and T. Iwamoto, Chem. Lett., 319 (1984).
6) T. Hasegawa, S. Nishikiori, and T. Iwamoto, J. Inclusion Phenom., **1**, 365 (1983/4).
7) S. Nishikiori and T. Iwamoto, J. Inclusion Phenom., **2**, 341 (1984).
8) T. Hasegawa, S. Nishikiori, and T. Iwamoto, J. Inclusion Phenom., **2**, 351 (1984).
9) T. Hasegawa, S. Nishikiori, and T. Iwamoto, Chem. Lett., 1659 (1985).
10) S. Nishikiori and T. Iwamoto, Inorg. Chem., **25**, 788 (1986).
11) T. Hasegawa, S. Nishikiori, and T. Iwamoto, Chem. Lett., 793 (1986).

MOLECULAR DETERMINANTS OF A NEW FAMILY OF HELICAL TUBULAND HOST DIOLS

Roger Bishop, Ian G. Dance, Stephen C. Hawkins, and
Marcia L. Scudder,
School of Chemistry,
The University of New South Wales,
P.O. Box 1, Kensington,
New South Wales 2033, Australia.

ABSTRACT. The alicyclic diols (1-5) constitute the first members of a family of novel helical tubuland hosts crystallising in space group $P3_121$ but possessing quite different canal shapes and dimensions. Consideration of their structural data has revealed two distinct sub-classes of these materials. The molecular features necessary for a diol to crystallise with the helical tubuland structure are defined and discussed.

1. INTRODUCTION

We have reported previously[1] that the racemic diol (1) forms stable crystalline inclusion complexes with a variety of solvent molecules. A network of host diol molecules (space group $P3_121$) is maintained by continuous helical spines of hydrogen bonds. Other diol molecules radiate from and interconnect these spines enclosing parallel open canals containing the disordered guest molecules. This tubuland structure[2] constitutes an especially interesting example since in each crystal these canals are surrounded by a double helical array of host diol molecules of the same chirality.

Following this initial discovery we embarked on a program of systematic synthesis in order to demonstrate that additional materials of similar structure could be obtained. Recently we have described the syntheses and crystal structures[3,4] of further alicyclic diols (2-4) of this type which also adopt the crystal space group $P3_121$. Their structural characteristics have been analysed and reported in detail[5]. The previously unreported diol (5) also belongs to this new family of helical tubuland host diols.

Comparative projections along c are shown to the same scale for one canal only of diols (1-5). The hydrogen bonded spines are circled and significant hydrogen atoms drawn as filled circles. Bond thickening indicates depth in individual molecules only because the helical characteristic is absent in these projections. The unobstructed canal cross-sections are drawn using projected van der Waals radii of the hydrogen atoms lining each canal.

2. STRUCTURAL ANALYSIS

Examination of the various crystal structures reveals two sub-classes of this helical tubuland family. That including (1), (3) and (5) has more efficient hydrogen bonding and smaller canal dimensions while the other, including (2) and (4), has weaker hydrogen bonding and larger canals. For (3) and (5) the canals are constricted so as to produce cages but because of the helical structure considerable void space is still present in these assemblies. Both isomeric diols, one of each sub-class, are prepared from the common diketone intermediate.

Five of the properties allowing distinction between the sub-classes are detailed here, namely: the melting point (M.P.); the hydrogen bond O.....O distance (O.....O); the unobstructed cross-sectional area of the canal (U.C.A.); the unit cell volume (U.C.V.); and the cross-ring O—C ---- C—O torsion angle (Torsion angle).

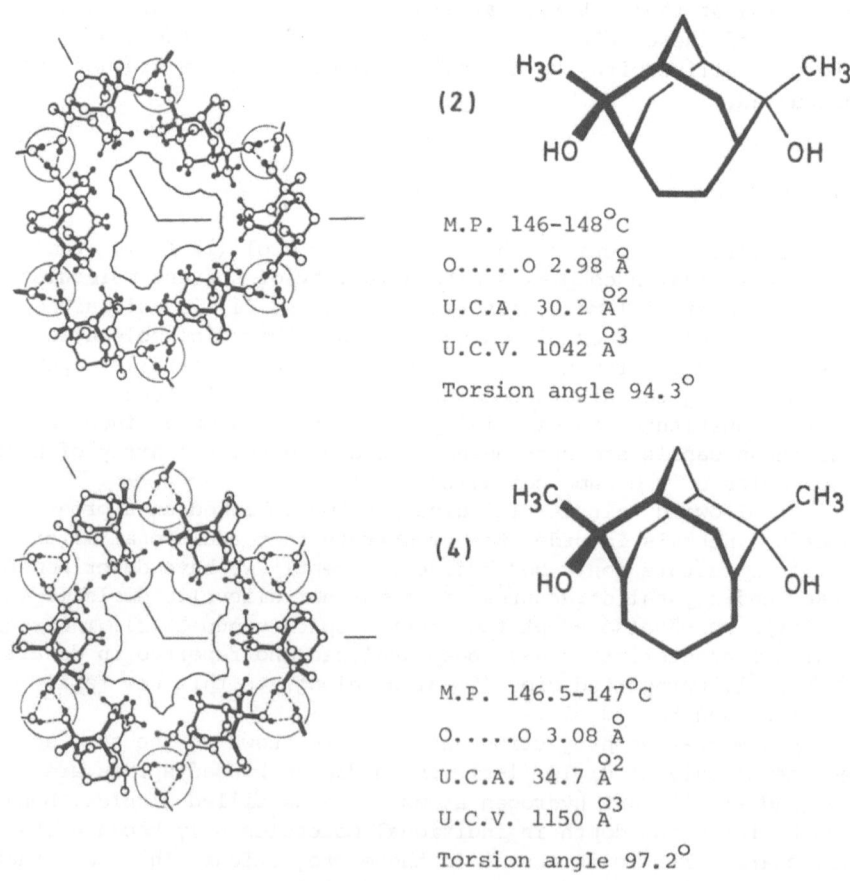

(2)

M.P. 146-148°C
O.....O 2.98 $\overset{o}{A}$
U.C.A. 30.2 $\overset{o2}{A}$
U.C.V. 1042 $\overset{o3}{A}$
Torsion angle 94.3°

(4)

M.P. 146.5-147°C
O.....O 3.08 $\overset{o}{A}$
U.C.A. 34.7 $\overset{o2}{A}$
U.C.V. 1150 $\overset{o3}{A}$
Torsion angle 97.2°

A NEW FAMILY OF HELICAL TUBULAND HOST DIOLS

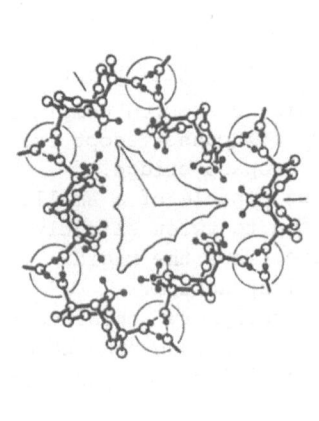

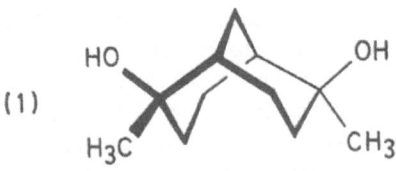

(1)

M.P. 189–191°C
O.....O 2.81 Å
U.C.A. 22.4 Å^2
U.C.V. 897 Å^3
Torsion angle 73.5°

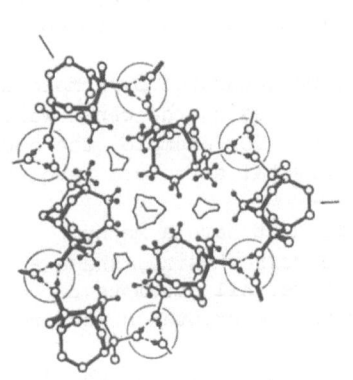

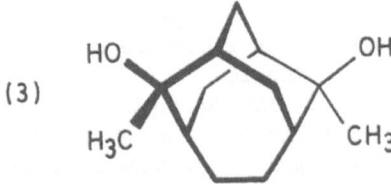

(3)

M.P. 245–247°C
O.....O 2.84 Å
U.C.A. 4.7 Å^2
U.C.V. 858 Å^3
Torsion angle 79.0°

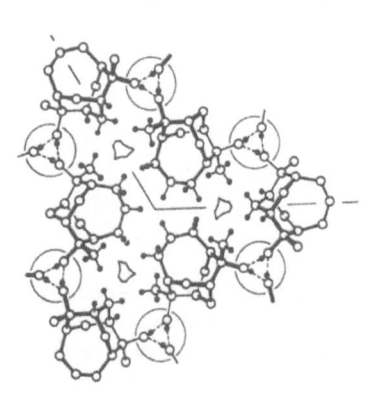

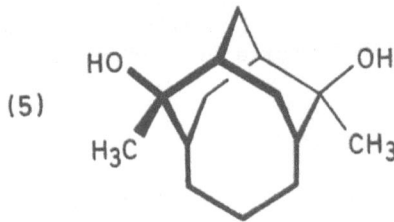

(5)

M.P. 249–250°C
O.....O 2.83 Å
U.C.A. 2.7 Å^2
U.C.V. 901 Å^3
Torsion angle 71.3°

3. MOLECULAR DETERMINANTS

Our program of synthesis has been planned to define the molecular features required in a host diol molecule for it to crystallise with the helical tubuland structure. The following molecular determinants have been found to be necessary.

(i) The diol molecules must have C_2 rotational symmetry in solution. However it is not necessary that this be adopted completely in the crystal. For example the diols (4) and (5) cannot adopt exact twofold symmetry in the solid because of the propano bridge.

(ii) The alicyclic structure must be capable of a small degree of flexibility. This allows the skeleton to twist slightly aiding the conformation imposed by the lattice. Thus the rigid adamantane analogues adopt a different crystal structure.

(iii) Substituent groups around the periphery appear so far to be deleterious. Polar groups may disrupt the hydrogen bonding of the host, while substituents in some positions will prevent the diol packing in a helical fashion.

(iv) A bridge on the opposite side to the hydroxy groups is optional. It can therefore be removed or modified in size to control the canal dimensions.

(v) The two hydroxy groups must be separated by a molecular bridge. This performs a key function in buttressing the canal walls against collapse to a denser structure. Thus, for example, the double epimer of (1) adopts a totally different crystal structure involving hydrogen bonded sheets.

(vi) The tertiary alcohol groups must have a methyl substituent. This appears to have just the correct size, shape and rigidity to support the canal wall structure. All attempts to replace these groups with others have so far led to new crystal structures being produced.

Although further factors are probably involved, the discovery of these structural requirements means that, within certain limits, new members of the helical tubuland family can be predicted with a reasonable degree of confidence.

4. REFERENCES

1 R. Bishop and I.G. Dance: *J. Chem. Soc., Chem. Commun.*, 992 (1979).
2 E. Weber and H.-P. Josel: *J. Incl. Phenom.*, **1**, 79 (1983).
3 R. Bishop, I.G. Dance, S.C. Hawkins and T. Lipari: *J. Incl. Phenom.*, **2**, 75 (1984).
4 I.G. Dance, R. Bishop, S.C. Hawkins, T. Lipari, M.L. Scudder and D.C. Craig: *J. Chem. Soc., Perkin Trans. 2*, in press, (1986).
5 I.G. Dance, R. Bishop and M.L. Scudder: *J. Chem. Soc., Perkin Trans. 2*, in press, (1986).

We wish to thank D.C. Craig for X-ray diffractometry and to acknowledge financial support through the Australian Research Grants Scheme.

SYNTHESIS AND STRUCTURE OF HEXAKIS(p-HYDROXYPHENYLOXY)BENZENE: A VERSATILE ANALOGUE OF THE HYDROGEN-BONDED HEXAMERIC UNIT OF β-HYDROQUINONE

D.D. MacNicol, P.R. Mallinson, A. Murphy, and C.D. Robertson
Department of Chemistry,
University of Glasgow,
Glasgow G12 8QQ, Scotland

ABSTRACT. The title hexaphenol (1), a direct analogue of the β-hydroquinone hexameric unit, has been prepared by six-fold demethylation of hexakis(p-methoxyphenyloxy)benzene (2) with BBr_3. Host 1 forms a trigonal adduct, space group $R\bar{3}$, a = 22.088(3), c = 12.232(3) Å, containing 6 molecules of pyridine per host molecule: a detailed X-ray study of this inclusion compound reveals a true clathrate structure, the closed cages of which accommodate a small, non-stoichiometric amount of water.

1. INTRODUCTION

The hexa-host analogy, the fundamental relationship between a hexa-substituted benzene and the pre-existing hydrogen-bonded hexameric unit of a number of important phenolic hosts, has led to the discovery of many new clathrate hosts.[1] The key hydrogen-bonded hexameric unit consolidating the host structure in the β-hydroquinone clathrates was first defined[2] during the pioneering X-ray studies of Powell and co-workers in the 1940s. The present work is concerned with the synthesis

(1)

(2)

and inclusion properties of hexakis(p-hydroxyphenyloxy)benzene (1) which corresponds directly to the hexameric unit of the β-hydroquinone clathrates. The hexaphenol 1, which has six hydroxyl groups round the molecular periphery, has a functional array suggesting the possibility of formation of a "hybrid" clathrate, partaking of hexahost character, in which exactly half of the [OH]$_6$ rings of the β-hydroquinone structure are replaced by permanent, hexa-oxygen-substituted benzene units. We describe the first synthesis of hexa-phenol 1 and give detailed structural information concerning one of its inclusion compounds.

2. EXPERIMENTAL

2.1 Synthesis of hexakis(p-hydroxyphenyloxy)benzene (1)

Hexaphenol 1 was prepared from the known[3] hexaether 2 by overnight exposure of the latter to excess BBr$_3$ (ca. 36 mol equiv.) in freshly distilled CH$_2$Cl$_2$, affording after work-up, 1 (>80%) as a fine white powder. Compound 1 has: ^{1}H n.m.r. resonances (δ in d^6-DMSO) at 9.02 (s) (6H), and 6.55 (s) (24H); and m/e [M$^+$] 726.1713 amu, C$_{42}$H$_{30}$O$_{12}$ requires m/e 726.1737 amu.

The compound is soluble in pyridine, and sparingly so in acetone. Recrystallisation from either DMSO or pyridine gives an inclusion compound with six solvent-guest components per host, the former being rapidly lost on standing in air. A second type of adduct containing about one mole of solvent (methanol and water) per host is obtained on recrystallisation from hot methanol in a sealed tube. Whilst dry hexakis(p-hydroxyphenyloxy)benzene quickly chars on heating, crystals of its pyridine inclusion compound darken in a sealed tube at 270 °C, and those of the methanol adduct can be taken to 320 °C on a hotplate with no apparent change.

2.2 Crystal data
C$_{42}$H$_{30}$O$_{12}$(1).6C$_5$H$_5$N.xH$_2$O, Formula weight = 1219.32 for x taken as 1, trigonal, R$\bar{3}$, a = 22.088(3), c = 12.232(3) Å, V = 5168(2) Å^3, Z = 3, D$_c$ = 1.18 g cm^{-3}, μ = 0.74 cm^{-1} for Mo-K$_\alpha$ radiation, λ = 0.7107 Å.

Number of independent reflections: 2503 from hexagonal needle, 0.6x0.2 mm. T = 293K. Final $\underline{R}$ for 537 reflections with $\underline{F}^2 > 2\sigma(\underline{F}^2)$: 0.094, $\underline{R}'$ 0.109.

X-ray intensity measurements for all possible reflections with sin $\theta/\lambda < 0.64$ Å^{-1} were made by $\theta-\omega$ scan on a Nonius CAD4 diffractometer. The small proportion of significant reflections, not caused by the use of an insufficiently large crystal, limits the precision of this analysis. The structure was solved by the MITHRIL computer program[4] and refined using the GX package.[5] During the anisotropic least-squares refinement, all aromatic hydrogen atoms of the host and pyridine guest molecules were placed in theoretical positions and allowed to ride on their attached carbon atoms. The hydroxyl hydrogen atom was not located. Two independent electron density peaks observed on and near the three-fold axis were ascribed to two types of statistically-disordered water molecule, the former oxygen site being of lower occupancy.

3. DISCUSSION

Figure 1(a) illustrates the individual hexakis(p-hydroxyphenyloxy)-benzene (1) host molecule in its pyridine adduct. Comparison [Figure 1(b)] with the hydrogen-bonded hexameric unit of β-hydroquinone (empty cage form) reveals a close parallel. Both units are located on a point of crystallographic $\bar{3}$ symmetry, with corresponding alternation of hydroxyl-containing moieties above and below the central core. A significant change is, however, found between the two units in that the torsion angle τ [denoted by the dotted line in Figure 1(b)], -58°, is significantly smaller in magnitude compared to the corresponding torsion angle O(1*)-O(1)-C(2)-C(7), -105°, for 1. In contrast to the situation for β-hydroquinone where each hydroxyl group is involved in forming hydrogen-bonded hexamers, for 1, as shown in Figure 2, six pyridine molecules are hydrogen-bonded to each host molecule.

Units comprising one host molecule and six pyridine molecules are stacked along the $\underline{c}$-axis, parts of two neighbouring infinite columns being shown in the stereoview at right angles to the $\underline{c}$-axis in Figure 3. Hydrogen bonds, length 2.71(2) Å, linking oxygen and nitrogen atoms are denoted by broken lines. The molecular packing is further illustrated in the view down the $\underline{c}$-axis, Figure 4.

The central benzene ring of 1 does not deviate significantly from planarity; the attached oxygen atoms, however, are disposed alternately 0.15(1) Å above and below this plane. The ether oxygen O(1) is displaced slightly from the plane of the outer benzene ring, by 0.06(1) Å.

An interesting feature of the molecular packing described above is that large centrosymmetric voids possessing 3-fold symmetry are formed between adjacent host molecules stacked along $\underline{c}$. The top and bottom of each such void are formed by the hexa-substituted benzene rings of 1, while the walls consist of p-hydroxyphenyl moieties, each hydrogen-bonded to a pyridine molecule. Figure 5 shows illustrative cavity contours drawn at right angles to the $\underline{c}$ axis at indicated fractions of the $\underline{c}$-spacing, showing the shape and indicating the closed nature of the clathrate cage. Residual electron density, corresponding to atoms

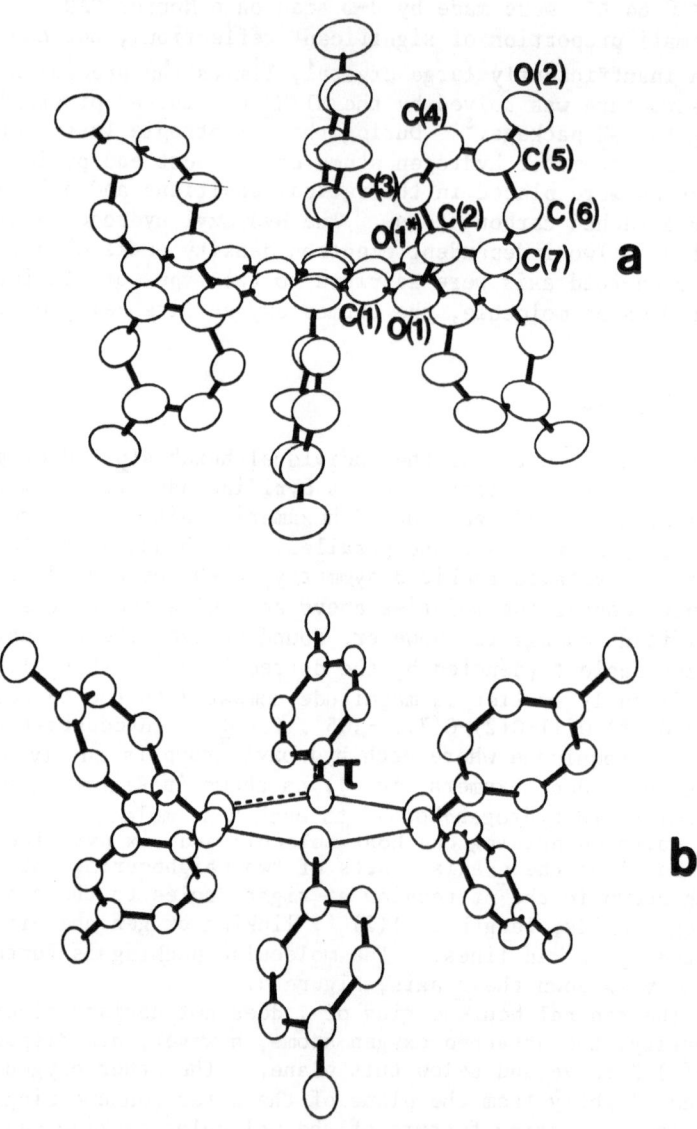

Figure 1. A comparison of (a) the molecule of hexakis(p-hydroxylphenyloxy)benzene (1) in its pyridine adduct with (b) the hydrogen-bonded hexameric unit of β-hydroquinone.

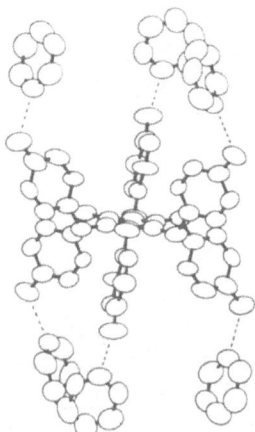

Figure 2. A view of the host molecule 1 with its six associated pyridine molecules. All hydrogen atoms have been omitted, and hydrogen bonds between oxygen and nitrogen are denoted by broken lines.

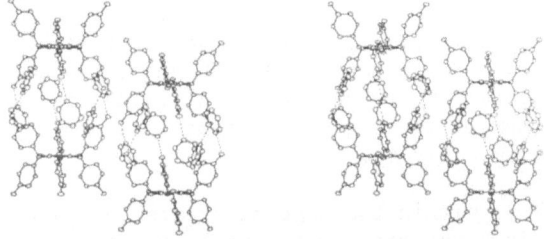

Figure 3. A stereoview normal to the c-axis illustrating the inter-column packing in the pyridine adduct of host 1 (as in Figure 4, included water is not shown).

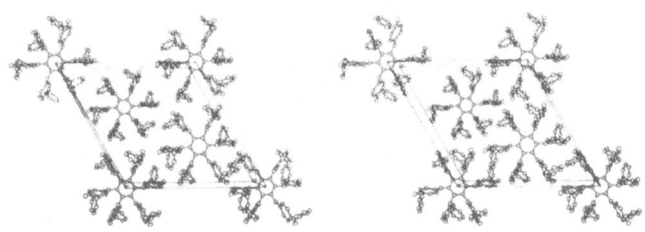

Figure 4. A stereoview looking down the c-axis showing the molecular packing in the pyridine complex of 1.

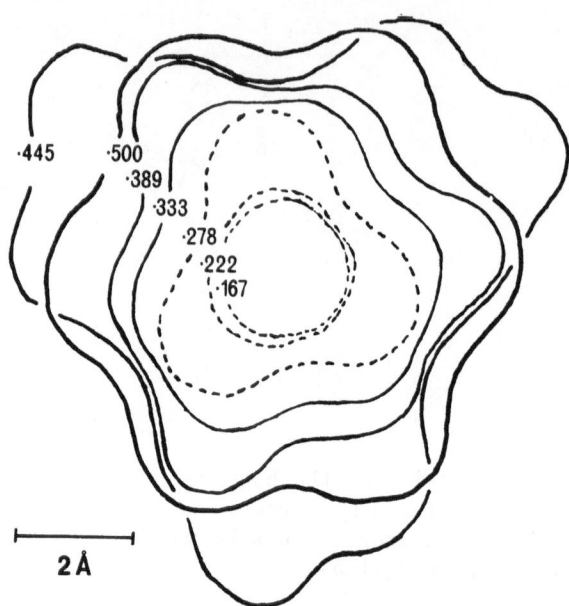

Figure 5. Cavity contours drawn at right angles to the c-axis at various fractional heights, indicating the free space available in the cage after allowing for the van der Waals volumes of the cage-wall atoms (drawn with the VDW computer program, ref. 6).

O(W1) and O(W2), within the cage may represent partial occupancy by water, this being consistent with small amounts of water (less than mole H_2O per **1**) detected by ^1H n.m.r. analysis.

References

[1] See, for example, D.D. MacNicol, in Inclusion Compounds. (Eds. J.L. Atwood, J.E.D. Davies, and D.D. MacNicol), Academic Press, London, Vol. **2**, Chapter 5, pp. 123-168 (1984).

[2] D.E. Palin and H.M. Powell: Nature (London) 156, 334 (1945); for other references see, for example, D.D. MacNicol, in Inclusion Compounds (Eds. J.L. Atwood, J.E.D. Davies, and D.D. MacNicol), Academic Press, London, Vol. **2**, Chapter 1, pp. 1-45 (1984).

[3] C.J. Gilmore, D.D. MacNicol, A. Murphy, and M.A. Russell: Tetrahedron Lett., **24**, 3269 (1983).

[4] MITHRIL, A Computer Program for the Automatic Solution of Crystal Structures from X-ray Data, C.J. Gilmore: J. Appl. Crystallogr., **17**, 42 (1985).

5 The GX Crystallographic Program System, P.R. Mallinson and K.W. Muir: J. Appl. Crystallogr., 18, 51 (1985).

6 J.C. Hanson, personal communication.

FREE RADICALS AS HOST MOLECULES

J. Veciana[a], J. Carilla[a], C. Miravitlles[b], E. Molins[b]
Departamento de Materiales Orgánicos Halogenados[a]. Centro de Investigación y Desrrollo (C.S.I.C.) C./ Jorge Girona Salgado, 18-26, 08034 Barcelona, Spain. Instituto de Materiales[b] (C.S.I.C.) C./ Alcarria s/n. P.O. 30102 Barcelona, Spain.

ABSTRACT. Perchlorotriphenylmethyl radical (1) and some of its radical derivatives form inclusion compounds with cyclic molecules. Thermal stabilities and crystal studies of some clatrates of 1 are presented and discussed.

1. INTRODUCTION

The design of new host molecules is currently an object of increasing interest. The general structure features of many clathrands are molecular bulkiness and limited conformational flexibility[1]. Molecular symmetry (specially three-or two-fold symmetry) also plays an important role in determining the inclusion ability of a host compound; providing a new principle for the design of novel host molecules[2].

Perchlorotriphenylmethyl radical (PTM, 1) is an exceptionally stable carbon free radical with a considerably bulky molecular geometry[3]. On the basis of space filling models it has been concluded that PTM radical has a propeller-like conformation (D_3 symmetry)[3a]. It has also been suggested that such conformation must be very rigid (high enantiomerization barrier for the reversal of propeller helicity) due to the intramolecular congestion of the three pairs of voluminous ortho-chlorine atoms[4]. Those clathratogenic features encouraged us to start a systematic study of the inclusion properties of PTM radical, functionalized derivatives and other polychlorinated triarylmethyl radicals[3].

$$C_6Cl_5 \diagdown \overset{\bullet}{C} - C_6Cl_5 \qquad \underline{1}$$
$$C_6Cl_5 \diagup$$

2. CLATHRATES AND THERMAL STABILITIES

PTM radical forms beautiful crystalline inclusion compounds with benzene, fluorobenzene, chlorobenzene, toluene, 1,4-dioxane, tetrahydrofurane, cyclohexane and cyclohexene; the host guest ratio being 1:1 for most of them. Those compounds are stable towards vaccumm drying at room temperature but on heating a release of the guest molecules takes place at different temperatures.

The thermal stability of some of the PTM inclusion compounds has been studied for the solid state declathration:

$$PTM \cdot Guest\ (s) \xrightarrow{\Delta} PTM\ (s) + Guest\ (g)$$

using DSC and TG techniques, under both non-isothermal and isothermal conditions.

The loss of the guest molecules corresponds to endothermic processes with low enthalpic values ($\Delta H_{dec} \sim 13-30$ KJ mol^{-1}). The rate constants for such processes were evaluated for each compound at several temperatures, by fitting isothermal TG curves to different kinetic physical mechanisms of solid state reactions (diffusion, nucleation, growth, nucleation-growth and homogeneous)[5]. The kinetic parameters (Ko, Ea) were calculated from an Arrenhius plot of the rate constants. The declathration physical mechanisms were assigned on the basis of agreement between these calculated kinetic parameter and those determined from non-isothermal TG curves by mean of Coats-Redfern method.

On the basis of both thermodynamic and kinetic parameters and the assigned physical mechanisms two types of thermal behaviour can be differentiated. One preferred for smaller guests (C_6H_6 and C_6H_5F; $\Delta H_{dec} \simeq 15$ KJ mol^{-1}, Ea>100 KJ mol^{-1} and Growth mechanism) and the other for the larger ones (C_6H_5Cl, C_6H_5Br and $C_4H_8O_2$; $\Delta H_{dec} \simeq 30$ KJ mol^{-1}, Ea<100 KJ mol^{-1} and Nucleation-Growth mechanism).

3. CRYSTAL STRUCTURES

The crystals of the inclusion compounds belong to the triclinic system, P$\bar{1}$ space group, with two PTM molecules and two molecules of guests in the unit cells. In PTM·1C_6H_6 the host molecules adopt a non-symmetrical propeller conformation with fixed benzene molecules accommodated in channels.

The packing of six neighbour host molecules through their coplanar phenyl groups (apparently a π-π-type interaction) create the channels were the guests are located. This structural arrangement is similar to that observed for the xylene clathrate of tris(1,8-naphtalenedioxy)cyclotriphosphazene[6] and differs substantially from those observed in most of the families of hosts with trigonal symmetry (triphenylmethane, tri-o-thymotide, cyclotriveratrylene, hexahosts and perhydrotriphenylene)[2].

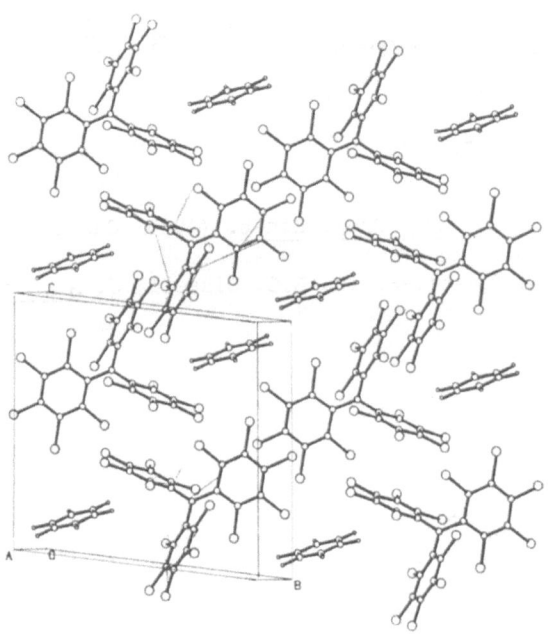

Figure. Packing in the PTM·1C$_6$H$_6$ clathrate

4. CONCLUDING REMARKS

Other polychlorinated triphenylmethyl radicals also present similar inclusion properties even with linear hydrocarbons. The introduction of functional groups (hydroxyl, phosphonium, etc) at the para position(s) of PTM radical permits to engineer new clathrate hosts with improved thermal stabilities, due to the higher energies of host-host interactions (hydrogen bonding, ionic). The PTM radical may be considered only as a first representative of a series of host compounds having radical character and therefore providing a stable paramagnetic contour with posible useful applications (radical intermediates, polymerizations).

REFERENCES

1. F. Vögtle, H-G.Löhr, J. Franke, D. Worsch, Angew. Chem. Int. Ed. Engl. 24 (1986) 727.

2. D.D. Mac Nicol, in J.L. Atwood, J.E. Davies, D.D. Mac Nicol (Eds): Inclusion Compounds. Vol. 2, Academic Press, London 1984, P. 123-168.

3. a) M. Ballester, J. Riera, J. Castañer, C. Badía, J.M. Monsó, J. Am. Chem. Soc. 93 (1971) 2215. b) M. Ballester, J. Riera, J. Castañer, A. Rodriguez, C. Rovira, J. Veciana. J. Org. Chem. 47 (1982) 4498. c) M. Ballester, J. Veciana, J. Riera, J. Castañer, C. Rovira, O. Armet. J. Org. Chem. 51 (1986) 2472.

4. K.S. Hayes, M. Nagumo, J.F. Blount, K. Mislow, J. Am. Chem. Soc. 102 (1980) 2773.

5. D.H. Bamford, Ed. Compr. Chem. Kinet. (1980) Chapter 3, p. 22.

6. H.R. Allcock, M. Teeter-Stein, E.C. Bissell, J. Am. Chem. Soc. 96 (1974) 4795.

^{35}Cl Nuclear Quadrupole Resonance studies of CCl_4 as a guest molecule in various clathrates.

L.Pang and E.A.C.Lucken
Physical Chemistry Department, Sciences II,
University of Geneva
30, Quai E.Ansermet,
1211 Geneva 4
Switzerland

Nuclear Quadrupole Coupling constants measure the electric field gradient tensor at a nucleus in a molecule. The field gradients in free molecules are determined by the electronic structure, but, in the solid state, are affected by environmental factors; hence their application to the study of inclusion complexes.

Thus the resonance frequencies of a guest molecule in an inclusion complex provide information on the following points.

Site Symmetry

If a guest molecule contains more than one equivalent nucleus, e.g. CCl_4 several distinct resonances indicate that the site symmetry is less than that of the free molecule.

Example : Dianin's compound is known to form cavities having a three-fold symmetry axis. The NQR spectrum of the ^{35}Cl nucleus in CCl_4 included in Dianin's compound shows two resonances, one from the single chlorine on the three-fold axis and another more intense resonance from the three remaining nuclei.

The complex CCl_4/Fe(III)tris-(AcAc) shows similar behaviour, indicating that, here too, the cavities have three-fold symmetry.

Site multiplicity

A guest molecule showing more distinct resonances than equivalent nuclei provides clear evidence of multiple inclusion sites.
Examples :

CCl_4/Ni(exan)$_2$	(2,2'-bipyridyl)
39.561	40.061
40.260	40.352
40.627	40.708
40.763	40.890

trans CHCl=CHCl/Fe(III)tris-acac

34.722	34.845
35.093	35.289

Chlorobenzene/Ni(4-picoline)$_4$(SCN)$_2$

34.475	34.670

Temperature Dependence

NQR resonance frequencies show a marked temperature dependence whose principal cause is the averageing of the field-gradient tensor brought about by molecular libration. For a nucleus such as ^{35}Cl in an organochlorine molecule, where the principal field-gradient axis lies along the C-Cl bond, libration about the bond-axis has no effect on the field-gradient, while libration about an axis perpendicular to the bond decreases the field-gradient by an amount proportional to the square of the librational amplitude. For CCl$_4$ in a trigonally-symmetric environment there is one librational mode about the three-fold axis and a pair of degenerate modes perpendicular to it. The temperature dependence of the NQR frequency of the axial chlorine atom is thus expected to be less than that of the remaining three, whose field-gradients are affected by all three librations.

The results for the two trigonal complexes are shown below. The fade out of the resonances indicates the onset of rapid reorientation and the surmounting of the librational barrier. This is particularly clear for the "equatorial" chlorine atoms in the Dianin complex, and is consistent with the egg-shaped form of the cavity in this compound.

The <u>Bayer-Kushida</u> theory of the temperature-dependence of NQR frequencies has the form :

$$\nu = \nu_0 - \nu_0 \sum_i \frac{A_i}{f_i \left[\exp f_i/kT - 1 \right]}$$

where :

ν_0 = NQR frequency at 0K

$A_i = \dfrac{3h \sin^2 \alpha_i}{8\pi^2 I}$

I = The molecular moment of inertia about the libration axis

α_i = The angle between the bond axis and the libration axis

$f_i = f_{0i} (1 - g_i T)$

f_{0i} = The libration frequency at 0K

g_i = A semi-empirical factor which includes the temperature dependence and anharmonicity of the librational frequency.

The temperature dependence of the NQR frequencies of ^{35}Cl in CCl_4 in the two trigonally-symmetric hosts have been fitted to this equation with the following results.

Host : Dianin's compound

ν_0^{ax} = 40.613 MHz ν_0^{eq} = 40.957 MHz

Axial libration :
f = 37.37 cm^{-1},   g = $2.047*10^{-3}$

Doubly-degenerate libration :
f = 47.04 cm^{-1},   g = $0.518*10^{-3}$

Host : Fe(III) tris-(acetylacetonate)

ν_0^{ax} = 41.088 MHz ν_0^{eq} = 40.529 MHz

Axial libration :
f = 43.78 cm^{-1},   g = $1.005*10^{-3}$

Doubly-degenerate libration :
f = 61.37 cm^{-1},   g = $1.070*10^{-3}$

The low librational frequency of 37 cm^{-1} and the high value of the parameter, g, is further evidence of the ease of rotation about the threefold axis in the Dianin complex. Indeed in the sulphur analogue of Dianin's compound, where the cavities are presumably slightly bigger, the resonance from the "equatorial" chlorines has already disappeared at 77K. In contrast to this the complexes between Dianin's compound and $BrCCl_3$ or CF_3CCl_3, where the more bulky axial substituent causes a tighter fit in the cavity, the "equatorial" resonances persist to 114K and 166K respectively.

These studies are being extended to 4K and complemented by structural and spectroscopic studies.

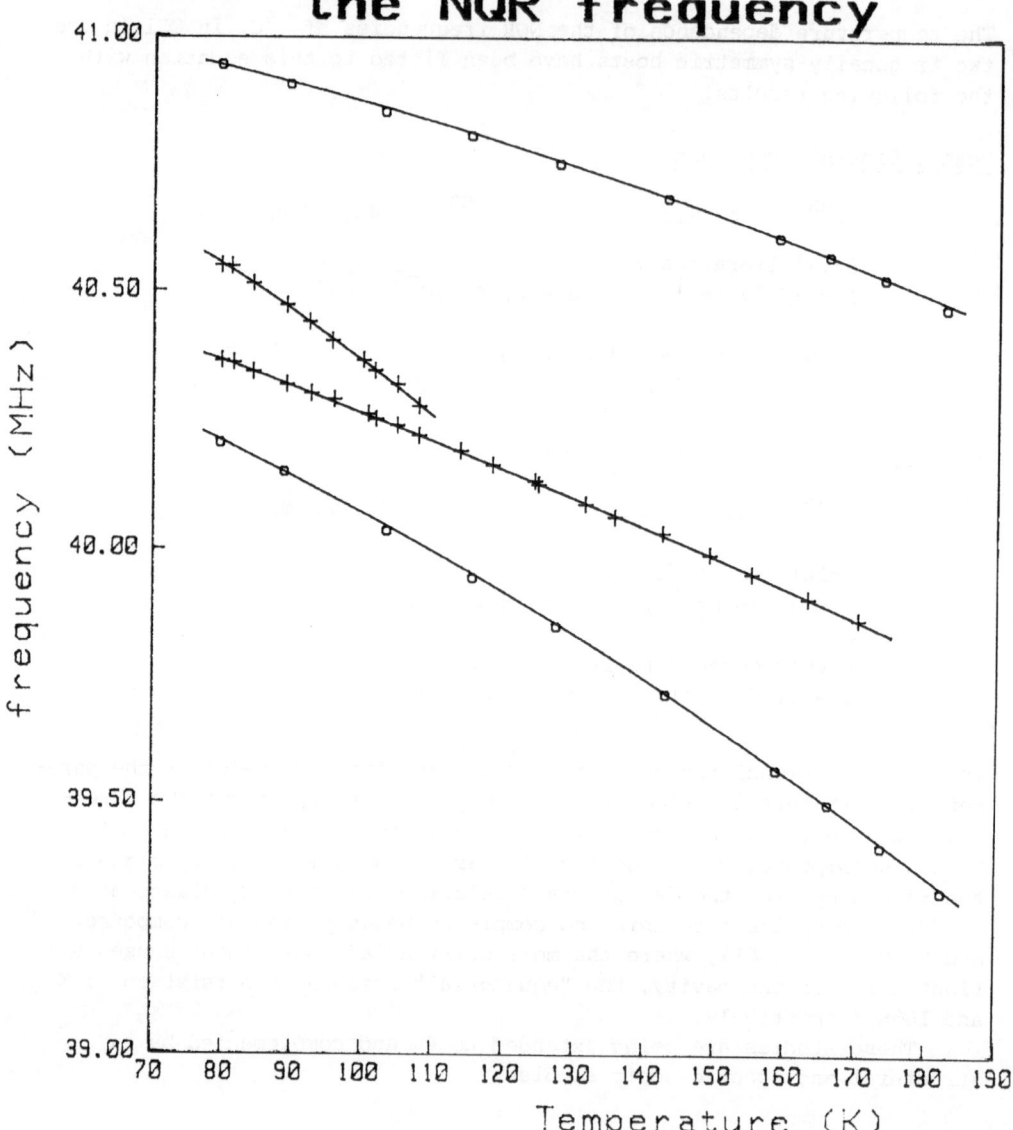

INCLUSION POLYMERIZATION OF DIENE AND DIACETYLENE MONOMERS
IN DEOXYCHOLIC ACID AND APOCHOLIC ACID CANALS [†]

Mikiji Miyata, Fusaharu Noma, Ken Okanishi,
Hiromori Tsutsumi, and Kiichi Takemoto

Department of Applied Fine Chemistry, Faculty of Engineering,
Osaka University, Suita, Osaka, 565 Japan

Inclusion polymerization is a unique one-dimensional polymerization which proceeds in a canal (channel) of inclusion compounds, and an excellent way for giving stereoregular polymers as well as composite materials at molecular level [1,2]. Such materials have a potential possibility to show some characteristic electrical, optical and magnetic properties in the inclusion state due to their one-dimensional structure.

We found earlier that deoxycholic acid (3α, 12α-dihydroxy-5β-cholan-24-oic acid ; DCA) and apocholic acid (3α, 12α-dihydroxy-5β-chol-8(14)-en-24-oic acid ; apoCA) can provide tunnel-like spaces, called canals or channels, which are suitable for one-dimensional polymerization [3-8].

In this paper we summarize our recent study on inclusion polymerization of various kinds of diene and diacetylene monomers in DCA and apoCA canals.

1. Characteristics of DCA and apoCA hosts

Both hosts form hydrogen-bonded bilayers stacked in an array which leaves hydrophobic canals between them [9]. Since the bilayer structure is thermally stable, we can study inclusion polymerization in a wide range of temperatures in DCA and apoCA canals as compared with other hosts such as urea and thiourea.

Deoxycholic acid(DCA)
[Apocholic acid(apoCA)
  8(14)en]

The hosts have slightly different canals in the sizes and shapes, but form inclusion compounds with the same monomers. So we can get significant information regarding effects of canal sizes and shapes on the inclusion polymerization.

The hosts form inclusion compounds with a great variety of organic substances with polar, nonpolar as well as bulky groups. Therefore we can try to polymerize more comprehensive monomers than before. That involves cyano, carbomethoxy or propyl group.

[†]Functional Monomers and Polymers 146

2. ESR observation of propagating radicals in the canals

Conventional free-radical initiators such as di-t-butyl peroxide can serve as the effective initiators for the inclusion polymerization of diene monomers in DCA and apoCA canals [10]. During the course of the polymerization we succeeded in observing ESR spectra of living-like propagating radicals of allylic type in the canals. This result confirms the existence of the polymerization reaction via free-radical mechanism.

3. Stereoregular polymerization

In case of a pair of DCA and apoCA hosts, we can study the inclusion polymerization of the same monomers in slightly different canals in size and shape at various temperatures ranging from -20°C to 140°C. This study enables us to find a dependence of the microstructures of the resulting polymers on polymerization temperatures and canal sizes. In case of butadiene derivatives, the following microstructures of the polymers should be evaluated : (1) 1,4-trans, 1,4-cis, 1,2; (2) head-to-tail, head-to-head(tail-to-tail); (3) erythro, threo; (4) asymmetry.

1,4-trans : In case of butadiene derivatives, the resulting polymers usually prefer 1,4-trans unit to 1,4-cis and 1,2 units. Inclusion polymerization of 2,3-dimethyl-1,3-butadiene in a DCA canal yielded almost completely 1,4-trans unit in a wide range of temperature from -20°C to 140°C. The microstructure, however, decreased remarkably over 80°C in an apoCA canal. In case of 1,3-butadiene the resulting polymer was composed of a mixture of those units even at low temperatures. These results are considered to reflect the difference of canal sizes of the hosts [11].

head-to-tail : In case of nonsymmetric monomers such as 1- or 2-alkyl-1,3-butadiene, the microstructure of head-to-tail or head-to-head (tail-to-tail) can be estimated. While the poly(1,3-pentadiene)s obtained always had completely head-to-tail structure, the polymers obtained from isoprene showed a dependence of the microstructure on polymerization temperatures and canal sizes. The fraction of head-to-head (tail-to-tail) unit of the latter polymer increased as decreasing the polymerization temperature from 50°C to -20°C in a DCA canal, while it was constant in an apoCA canal [11].

erythro, threo : In case of 1,4-disubstituted butadienes such as 2,4-hexadiene we can distinguish between erythro and threo structure. 2,4-Hexadiene was found to polymerize in DCA and apoCA canals on heating over 100°C for 10 to 20 days after γ-ray irradiation. Particullary trans,trans-2,4-hexadiene polymerized in an inclusion state via radical mechanism for the first time. The polymers from the monomer prefer erythro structure to threo structure in a DCA canal, while they do slightly threo to erythro in an apoCA canal. It is considered that the polymerization proceeded preferentially in the canals through trans opening to yield erythro diisotactic structure in a DCA canal [11].

Similarly many other monomers with bulky groups polymerized in DCA and apoCA canals on heating over 100°C for a long time on the basis of propagating radicals with very long life-times.

4. Asymmetric polymerization

DCA and apoCA can serve as effective host components for asymmetric inclusion polymerization of prochiral monomers such as 1-substituted butadienes. We reported previously the preparation of optically active polymers with extremely high specific optical rotation of arbitrary sign from (E)- or (Z)-2-methyl-1,3-pentadiene by inclusion polymerization in the canals [7,12-14]. Moreover we have found that butadiene derivatives with polar groups such as cyano or carbomethoxy group can be polymerized to yield optically active polymers. The $[\alpha]_D$ values of the resulting polymers were much higher than those of polymers obtained by other known polymerization method.

$$CH_2=\overset{R_1}{C}-CH=\overset{R_2}{CH} \longrightarrow \{CH_2\overset{R_1}{C}=CH\overset{R_2}{CH}\}_*$$

5. Inclusion of highly conjugated polymers

Electrically conductive polymers such as polyacetylene have recently received much attention as representative low-dimensional materials. This highly oriented assembly of the polymers is very attractive, because such chain alignment should improve their physical and chemical properties. It is expected that the incorporation of such polymers into canals will yield a novel type of functional composite materials at molecular level with unique electrical and optoelectronic properties.

$HC\equiv C-C\equiv CH$: It was found that the simplest diacetylene, butadiyne, can be polymerized spontaneously and mildly in DCA and apoCA canals at temperatures ranging from -20°C to 30°C. The polymerization was accompanied by a sequence of colour change from colorless through violet and brown to finally yield a black material with a metallic luster. The resulting polymers are characterized by infrared and Raman spectroscopies as having a polyconjugated main chain by 1,4-addition. The electric conductivity of the polymers in the inclusion state and of the separated polymers were about 10^{-7} to 10^{-9} S/cm under doping with iodine at room temperature [15].

$H_2C=CH-CH=CH-X$ (X : Cl, Br) : 1-Chloro- or 1-bromo-1,3-butadiene can be polymerized in DCA and apoCA canals to yield polymers with 1,4-trans structure. Since the separated polymers from the hosts are soluble in chloroform, they serve as precursors for polyacetylene. The dehydrohalogenation from the precursor polymer occurred slowly at room temperature and rapidly over 150°C in an atmosphere of dry nitrogen, accompanying a colour change from light yellow through brown to finally yield a black material with a metallic luster. The electric conductivity was about 10^{-3} S/cm under doping with iodine [16].

$H_2C=CX(CN)$ (X : Cl, Br) : The polymerization of 2-chloroacrylonitrile in DCA and apoCA canals was carried out in a way similar to that in case of butadiene derivatives. The dehydrochlorination of the polymer in the canals occurred gradually on heating at 140°C without adding an acceptor. The conversion in a DCA canal amounted to 70 % after one day. The electric conductivity of the polymer in the inclusion state was about 10^{-6} S/cm under doping with iodine at room temperature [17].

Ferrocene : We have recently found that DCA forms inclusion compounds with ferrocene and its derivatives. Particularly DCA-ferrocene inclusion compound is very easy to get a large crystal enough to analyze the crystal structure by X-ray diffraction method. It was found from the analysis of a single crystal that ferrocene molecules are tightly accommodated into a DCA canal in an array different from those of crystals of ferrocene itself [18]. It was also ascertained that DCA-ferrocene inclusion compounds are doped with iodine. On the other hand, apoCA does not form an inclusion compound with ferrocene itself, but with its derivatives.

The authors wish to thank the Radiation Laboratory, the Institute of the Scientific and Industrial Research, Osaka University, for allowing them to use the ^{60}Co facilities.

References

1. K.Takemoto and M.Miyata : *J.Macromol.Sci.Rev.Macromol.Chem.* C18, 83 (1980).
2. M.Farina : *Inclusion Compounds*, Vol.3, p.297, J.L.Atwood,J.E.D. Davies,D.D.MacNicol Eds., Academic Press, London, 1984.
3. M.Miyata and K.Takemoto : *J.Polym.Sci.,Polym.Lett.Ed.* 13,221(1975).
4. M.Miyata and K.Takemoto : *J.Polym.Sci.,Polym.Symp.* 55,279(1976).
5. M.Miyata,K.Morioka and K.Takemoto : *J.Polym.Sci.,Polym.Chem.Ed.* 15, 2987(1977).
6. M.Miyata and K.Takemoto : *Makromol.Chem.* 179,1167(1978).
7. M.Miyata,Y.Kitahara and K.Takemoto : *Polym.Bull.* 2,671(1980).
8. M.Miyata,Y.Kitahara,Y.Osaki and K.Takemoto : *J.Incl.Phenom.* 2,391 (1984).
9. E.Giglio : *Inclusion Compounds*, Vol.2, p.207, J.L.Atwood, J.E.D. Davies, D.D.MacNicol Eds., Academic Press, London, 1984.
10. M.Miyata,F.Noma,Y.Osaki,K.Takemoto and M.Kamachi : *J.Polym.Sci., Polym.Lett.Ed.* in press.
11. M.Miyata,F.Noma,S.Akizuki,T.Tsuzuki and K.Takemoto : to be published.
12. M.Miyata and K.Takemoto : *Polym.J.* 9, 111 (1977).
13. M.Miyata,Y.Kitahara and K.Takemoto : *Polym.J.* 13, 111 (1981).
14. M.Miyata,Y.Kitahara and K.Takemoto : *Makromol.Chem.* 184, 1771(1983).
15. H.Tsutsumi,K.Okanishi,M.Miyata and K.Takemoto : *International Conference on Science and Technology of Synthetic Metals* in Kyoto,Japan 1986 ; *Synthetic Metals* in press.
16. H.Tsutsumi,K.Okanishi,M.Miyata and K.Takemoto : to be published.
17. M.Miyata,K.Okanishi and K.Takemoto : *Polym.J.* 18, 185 (1986).
18. K.Miki,N.Kasai,H.Tsutsumi,M.Miyata and K.Takemoto : *J.Chem.Soc. Chem.Commun.* in contribution.

Journal of Inclusion Phenomena 5 (1987), 253–257.
© 1987 *by D. Reidel Publishing Company.*

CORRELATION BETWEEN LAYER CHARGE AND ACTIVATION ENERGY OF THERMALLY
INDUCED DEINTERCALATION IN ORGANO-LAYER SILICATES

J.H. CHOY, C.E. KIM*, K.W. HYUNG*, J.C. PARK
Department of Chemistry, Seoul National University, Seoul 151
*Department of Ceramics, Yonsei University, Seoul 131, Korea

The intercalation reaction of organic molecules with montmorillonite has been widely studied [1-3]. Some studies on the thermal behavior of n-alkylammonium-montmorillonite complexes has also been reported [4-5], but no research on the reaction kinetics of desintercalation.

The montmorillonite is a three layer lattice type of clay mineral having layer charge due to the isomorphous substitution in tetrahedral or octahedral sites where Si^{4+} or Al^{3+} are replaced by other cations of lower valence [6]. To compensate the layer charge in their natural state, the equivalent amount of cations must be introduced into the interlayer positions. These cations are exchangeable with other inorganic ions in aqueous solution or with organic cations, particularly with surface active agents such as n-alkylammonium cations. Since excess negative charge is delocalized over all oxygens in the lattice, n-alkylammonium cations are held to the interlayer surface by ionic bonding and van der Waals attraction. The bonding force of intercalated n-alkylammonium ions in the interlayer space is strengthend depending upon the extent of surface charge density of silicate.

The thermally stimulated alkylammonium complexes are desintercalated at a specific temperature domain of 320–470°C. Therefore, we have attempted to find the correlation between layer charge due to the isomorphous substitution of layer silicate and activation energy which represents the energy barrier to be surmounted of thermally induced desintercalation reaction of its decylammonium complex.

The preparation and characterization of n-decylammonium-montmorillonite complexes have been systematically studied in this work with the variation of charge density in the natural montmorillonite. The layer charge was estimated by n-alcohol method as follows [2-3]. At first, n-decylammonium complexes are synthesized by ion exchange reaction and by successive molecular intercalation of primary n-alcohol (ROH, where R = $C_{10}H_{21}$, $C_{12}H_{25}$, $C_{14}H_{29}$, etc....). The layer charge can be calculated from the basal spacings of n-decylammonium derivatives under n-alcohol by the following equations (1) and (2);

$$N_{CH_2} = [(x + y)n_c + \{2.0 - (x + y)\} n_A] \quad \cdots\cdots\cdots\cdots\cdots \quad (1)$$

for $n_A > n_C$

$$d = 1.4\, N_{CH_2} + 9.6\, \text{Å} \quad \ldots \ldots \ldots \ldots \ldots (2)$$

where n_A is no. of carbon atoms in n-alcanol, n_C is no. of carbon atoms in n-alkylammonium, N_{CH_2} is total no. of carbon atoms in $(Si,Al)_4O_{10}$ unit, (x+y) corresponds to layer charge (ξ) in formula unit. The obtained basal spacings, layer charge and interlayer C.E.C are listed in Table I.

TABLE I. Basal spacings (pm) of n-decylammonium-montmorillonites after swelling under n-alcanols and calculated layer charges and interlayer C.E.C.

MONT.	OBSERVED BASAL SPACINGS [d(ool)] (pm)			$\xi(x+y)$	IL-C.E.C. ($m_{eq.}$/100g)
	C_{10}-$\bar{C}_{10}$ mont.	C_{10}-$\bar{C}_{12}$ mont.	C_{10}-$\bar{C}_{14}$ mont.		
A	3843±25	4215±5	5667±5	0.38±0.01	102.57±0.5
B	3864±15	4200±5	4642±10	0.44±0.01	118.24±0.5
C	3837±5	4175±5	4618±5	0.49±0.02	138.27±1.0

C_{10} : n-decylammonium $\quad\quad \xi$: layer charge
$\bar{C}_{10}, \bar{C}_{12}, \bar{C}_{14}$: n-alcanol $\quad$ IL : Interlayer

For the kinetic study of desintercalation reaction along with the estimation of activation energy, Kissinger's [7] and Ozawa's [8] methods were employed by using a differential scanning calorimetry (DSC). As suggested by Kissinger, the activation energy of first order process may be estimated from the variation of the temperature at maximal intensity of DTA peak (Tm) with various heating rates. The kinetics of the thermal decomposition of volatile products can be described by the equation (3).

$$-\frac{dX}{dt} = DX^b \quad \ldots \ldots \ldots \ldots \ldots \ldots \ldots \ldots (3)$$

where $D = Z\exp(-E/RT)$, X is the fraction of the sample not yet reacted ($0 < X < 1$), b is the order of kinetics, E is the activation energy, R is the gas constant and Z is the pre-exponential factor. Assuming first order kinetics and simplifying eq. (3) gives

$$\frac{d(\ln(\phi/Tm^2))}{d(1/Tm)} = -\frac{E}{R} \quad \ldots \ldots \ldots \ldots \ldots (4)$$

The activation energy can be therefore obtained from the slope, by plotting the $\ln(\phi/Tm^2)$ vs $1/Tm$. Ozawa also proposed that the activation energy may be estimated from the shifting of DSC curve as the heating rate is changed. It is based on the general equation (5),

$$-\frac{dx}{dt} = Z \cdot f(x)\exp(-E/RT) \quad \ldots \ldots \ldots \ldots (5)$$

where f(x) can be general function of X. Assuming a constant heating rate and simplifying eq. (5) gives

$$\log \phi + 0.4567 \, E/RT = \text{constant.} \quad \ldots\ldots\ldots\ldots \quad (6)$$

Thus, plotting log ϕ versus $1/T_m$ gives the slope of which is equal to $-0.4567 \, E/R$ and the activation energy can be obtained.

In this study, the temperature of maximal peak intensity of desintercalation was detected with the variation of heating rate of 5, 7, 10, 15 and 20 deg./min. in an inert atmosphere. The estimated activation energies by Kissinger's and Ozawa's methods are shown in Fig. 1, 2 and 3.

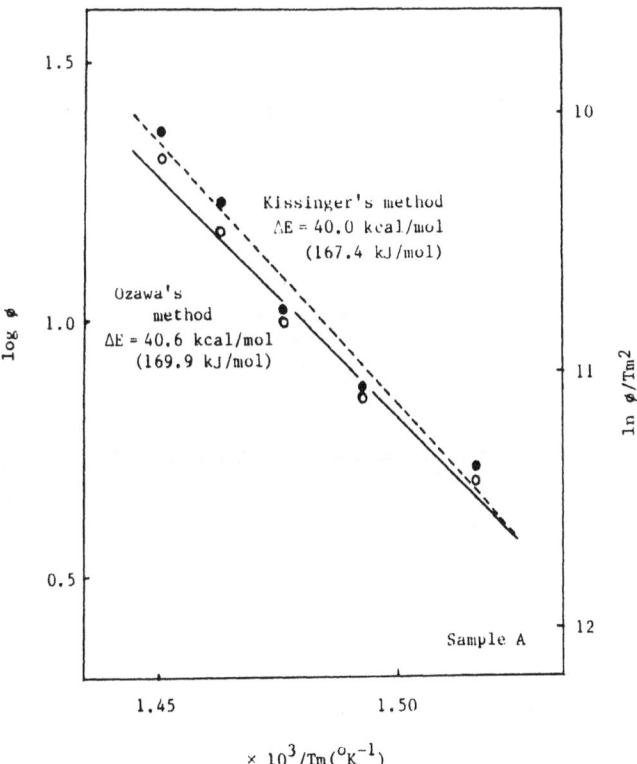

Figure 1. Plots of log ϕ vs. $1/T_m$ and ln ϕ/T_m vs. $1/T_m^2$ in montmorillonite A. ($\xi=0.38$)

The values obtained by Ozawa method are slightly higher than those of Kissinger's, but they have a good coincidence within the limit of experimental error.(Table 2) The activation energy in desintercalation process seems to be increased as the layer charge increases. Such a phenomenon might be well understandable by considering the enhanced bonding between silicate surface and decylammonium ion.

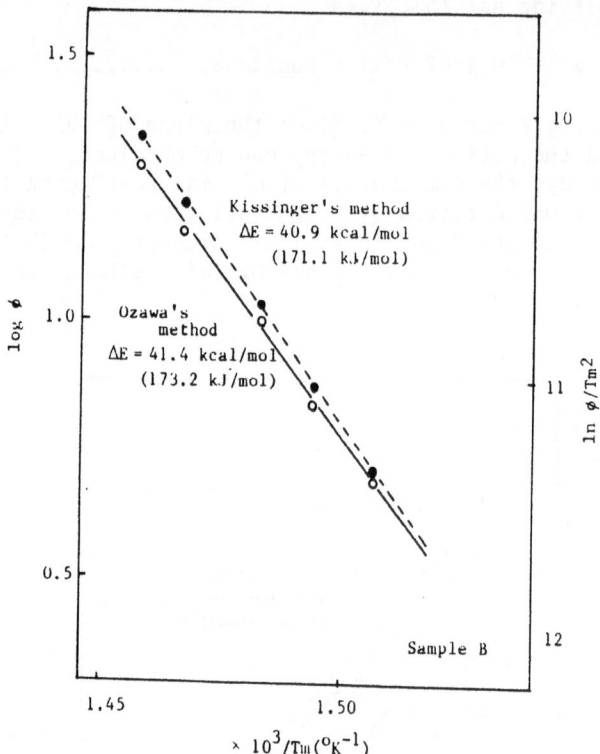

Figure 2. Plots of log ϕ vs. $1/T_m$ and ln ϕ/T_m vs. $1/T_m^2$ in montmorillonite B. ($\xi=0.44$)

TABLE II. Linear correlation between layer charge and activation energy (kj/mol)

MONT.	ξ	ΔE KISSINGER (kj/mol)	ΔE OZAWA (kj/mol)
A	0.38	167.4	169.9
B	0.44	171.1	173.2
C	0.49	214.6	214.6

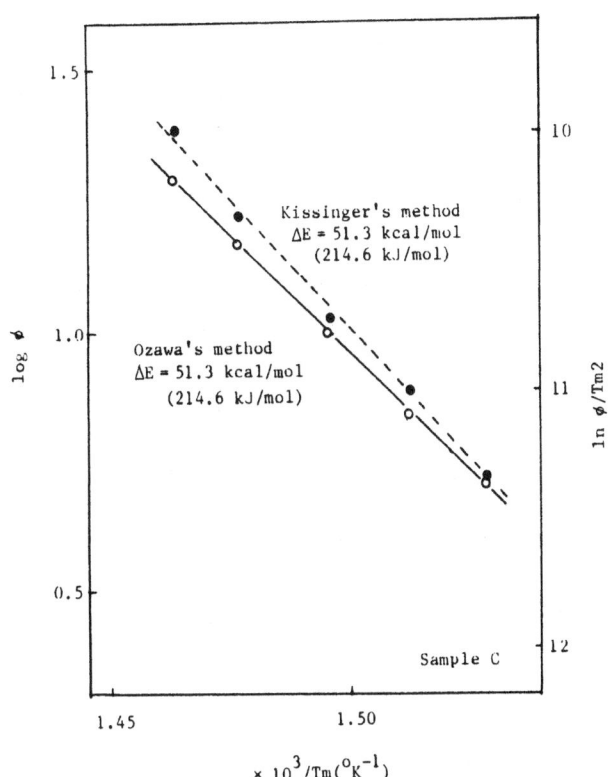

Figure 3. Plots of log ϕ vs. 1/Tm and ln ϕ/Tm vs. 1/Tm² montmorillonite C. (ξ=0.49)

The authors are grateful to the Korean Science & Engineering Foundation(KOSEF) for financial support.

References

1. R. Greene-Kelly: *J. Colloid Sci.* **11**, 77(1956).
2. G. Lagaly, A. Weiss: *Proc. Intern. Clay Conf., Tokyo,* 61(1969).
3. G. Lagaly: *Clay Minerals.* **16**, 1(1981).
4. A. Maes, A. Cremers: *J. Chem. Soc. Faraday Trans.* **1**, 77, 1553(1981).
5. B. Durand, et al.: *Clay and Clay Minerals.* **20**, 21(1972).
6. R.E. Grim: *Structure of the Clay Minerals*(Clay Mineralogy, Ed. R.E. Grim), p78(1968).
7. H.E. Kissinger: *Anal. Chem.* **29**, 1702(1957).
8. T. Ozawa: *Bull. Chem. Soc. Jpn.* **38**, 1881(1965).

INFRARED SPECTRAL INVESTIGATIONS OF ADSORPTION AND OXIDATION OF N,N-DIMETHYLANILINE BY SEPIOLITE, LOUGHLINITE AND DIATOMITE

S. AKYÜZ* and T. AKYÜZ†
* Hacettepe University, Department of Physics, Beytepe, Ankara, Turkey
† Mineral Research and Expolaration Institute of Turkey, (M.T.A.), Balgat, Ankara

ABSTRACT. The adsorption behaviour of sepiolite, loughlinite (Na-sepiolite) and diatomite for N,N-dimethylaniline (DMA) and its catalytically formed oxidation products are studied using IR spectroscopy. It is found that sepiolite and loughlinite adsorb DMA and then react with it more easily than diatomite. The adsorption capacity sequence of DMA for the minerals studied is sepiolite > loughlinite > diatomite.

I. INTRODUCTION

Sepiolite is a hydrated magnesium silicate having an internal structure of channels which can accomodate zeolitic water and other molecules. The magnesium ion in sepiolite crystal is exchangeable with various metal ions(1). Loughlinite has a similar structure to sepiolite and is known as natural Na-sepiolite(2). Sepiolite is an effective catalyst for hydrorefining of hydrocarbons, fuel oils(3) and conversion of ethanol into ethylene or bute-1,3-diene(1).

Diatomite is a siliceous rock made largely from the skeletons of aquatic plants (mainly planktonic) called diatoms. It is used as an absorber of dangerous chemicals in handling and storage, and for the refining of various acids(4).

Oxidative properties of clay minerals have been studied previously using simple organic molecules which are absorbed and then react at the surface yielding colored products(5,6).

In this study, the coloration of sepiolite, loughlinite and diatomite by the oxidation products of N,N-dimethylaniline (DMA) have been investigated using IR spectroscopy.

The oxidation of DMA by various oxidation agents (e.g. cupric sulphate) and the formation of methyl violet is a well known process used in the manufacture of the dye. The coloration and oxidation process is described by the following stages(6) : **(a)** sorption of DMA, **(b)** oxidation to N,N-dimethyl-N'-methylbenzidine and bis 4-(N,N-dimethylaminophenyl)-4-(N'methylaminophenyl) methane and **(c)** further oxidation to a blue quinoid cation or its protonated yellow component and finaly to methyl violet. The reaction can be formulated as follows :

$$C_6H_5N(CH_3)_2 + \{O\} \rightarrow C_6H_5NHCH_3 + HCHO \tag{I}$$

$$HCHO + C_6H_5NHCH_3 + C_6H_5N(CH_3)_2 \rightarrow HOH + (CH_3)_2NC_6H_4CH_2C_6H_4NHCH_3 \tag{II}$$

$$(CH_3)_2NC_6H_4CH_2C_6H_4NHCH_3 + C_6H_5N(CH_3)_2 + \{O\}$$
$$\rightarrow HOH + \{(CH_3)_2NC_6H_4\}_2CHC_6H_4NHCH_3 \tag{III}$$

Further oxidation of (II) and (III) in an acidic environment results in the formation of blue quinoid cation (B or IV) and methyl violet (V), respectively.

$$(CH_3)_2N^+ = \langle\!=\!\rangle = CH - \langle\;\rangle - NHCH_3 \tag{IV}$$

Blue Quinoid Cation (B)

$$C \begin{matrix} C_6H_4NHCH_3 \\ = \langle\!=\!\rangle = N^+(CH_3)_2 \quad \text{Methyl Violet (V)} \\ C_6H_4N(CH_3)_2 \end{matrix} \tag{V}$$

Yellow component is formed by protonation of B (IV) species :

$$(CH_3)_2N^+ \cdot C_6H_4 \cdot CH \cdot C_6H_4 \cdot NHCH_3 + H^+ \rightarrow (CH_3)_2N^+C_6H_4CHC_6H_4N^+H_2CH_3$$

Yellow (Y) component.

In order to verify that the oxidation reaction sequence is also valid for sepiolite, loughlinite and diatomite, the adsorption of DMA by these minerals have been performed.

2. EXPERIMENTAL

Sepiolite and loughlinite were obtained from the Mihalliccik region of Eskisehir (Turkey) and diatomite was obtained from the Kızılcahamam region of Ankara(Turkey). The samples were first investigated by X-ray diffraction, differential thermal analysis and elemental analysis. In loughlinite, analcime and dolamite were detected as significant impurities.

DMA-treated samples were prepared by immersing them in liquid DMA in sealed bottles for different periods of time at room temperature. They were then filtered and washed several times with benzene and dried.

IR spectra of the KBr disks of unwashed and benzene washed samples were recorded on Perkin-Elmer 621 and Nicolet MX-IE spectrometers.

3. RESULTS AND DISCUSSION

All the samples used in the present study were colored by DMA.

In the cases of sepiolite and loughlinite the samples turned to a green color immediately on immersion in liquid DMA. In longer treatments (1 week or more) the color turned to a dark bluish green. Sepiolite was found to be quickly affected by DMA vapour. The sample which was left in the open air near to a DMA bottle which was opened for few seconds also turned to a green color.

The IR spectrum of sepiolite which was treated with DMA for one week at room temperature and then washed with benzene several times is given in figure 1a. We do not observe vibrational bands of DMA(6,7), however we do observe vibrational bands of the V and Y components of the oxidation products of DMA(6). The IR spectrum of the same DMA-treated sepiolite sample, this time unwashed with benzene, is given in figure 1b. When the spectrum is compared with the IR spectra of DMA-treated laponite(6), it is found that the sample contains tetramethylbenzidine (tetrabase, TB) in addition to the V and Y components of the oxidation products of DMA. The IR spectrum of the loughlinite which was treated with DMA for two weeks and then washed with benzene several times is given in figure 1c. The vibrational bands of the V and Y components of the oxidation products of the DMA(6) are observed. We also observe the TB vibrational bands in the IR spectrum of the unwashed sample of the DMA-treated loughlinite.

The reaction of DMA with diatomite is found to be slower than in the cases of sepiolite and loughlinite. Diatomite turns slowly to a light blue color when it is contacted with DMA. If it is left under DMA vapour for a month at room temperature the color gets deeper. Figure 2 shows DMA-treated diatomite which is kept under DMA vapour for a month. We observe only the V component of the oxidation products of DMA. We do not observe any changes in the IR spectrum of DMA-treated diatomite after washing the sample with benzene. It is noted that the V and Y components of the oxidation products of DMA are not desorbed during the washing with benzene but the excess DMA and tetramethylbenzidine (TB) are washed out with benzene. IR spectroscopic analysis shows that sepiolite and loughlinite absorb DMA and then react with it more easily than diatomite. The adsorption capacity sequence of DMA for the minerals studied is found to be sepiolite > loughlinite > diatomite.

4. ACKNOWLEDGEMENTS

A part of the IR study was made at the Chemistry Department, University of Lancaster. The authors wish to thank Dr. J.E.D. Davies for allowing the use of the equipment in his laboratory. One of the authors (Dr. S. Akyüz) is also grateful to the British Council for financial support.

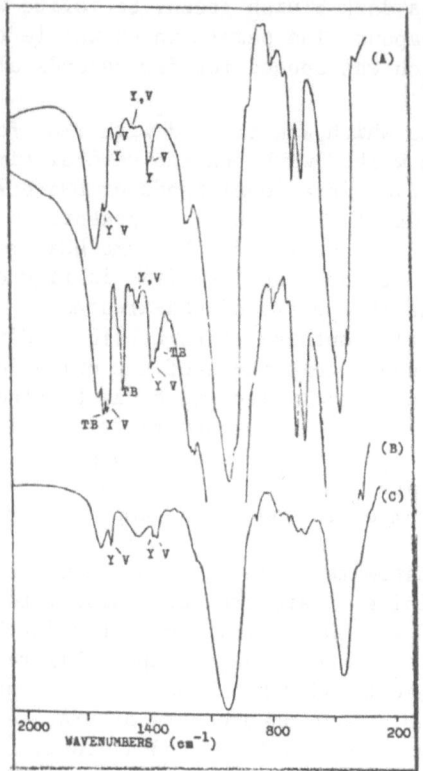
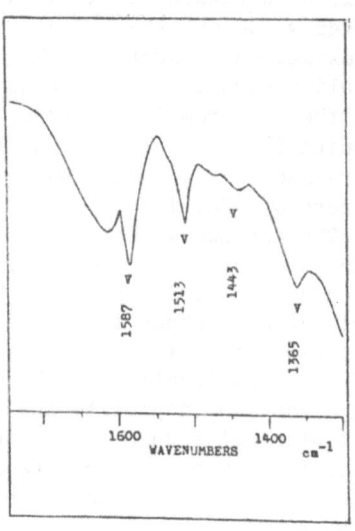

Figure 1. IR spectra of sepiolite and loughlinite samples treated with DMA. (a)(b) Sepiolite treated with DMA for one week, benzene washed (a), and unwashed(b). (c) Loughlinite treated with DMA for two weeks, benzene washed.

Figure 2. IR spectrum of diatomite treated with DMA for one month.

REFERENCES

1. Y.Kiyatama and A. Michishita, J.Chem.Soc.Chem.Comm., 401-2(1981)
2. W.Echle, Contrib.Mineral Petrology, **14**,86(1967).
3. A.J.Dandy and M.S. Nadiye-Tabbıruka, Clays and Clay Min.,**30**,347(1982).
4. T.Dickson, Industrial Minerals, **141**,33(1979).
5. B.K.G. Theng, Clay,Clay Min.,**19**,383(1971).
6. E.F.Vansant and S.Yariv, J.Chem.Soc.Far.,I,**73**,1815(1977).
7. P.N.Gates,R.A.R.Pearce and K.Radoliffe, J.Chem.Soc.Perkin Trans.II, **11**,1607(1972).

LITHIUM INTERCALATION CLUSTER COMPOUNDS

A.V. Mischenko, Yu.V. Moronov, P.P. Samojlov,
V.E. Fedorov
Institute of Inorganic Chemistry,
Siberian Branch, Academy of Sciences
630090 Novosibirsk
USSR

The important group of inclusion compounds based on the early transition metal layer and channel type chalcogenides has been investigated. The problems of the formation of Li intercalates by interacting n-bytyllithium hexene solutions with Nb, Mo, W and Re chalcogenides have been considered.

Solid original matrices of different structural type have been selected in order to show the influence of the matrices-"host" electronic and geometric structure on the stoichiometry of the phases formed.

It has been established that the known regularity, - viz the intercalation capacity decreases in the sulfide-selenidetelluride range for the present transition metal-was observed for ordinary layer dichalcogenides with quasi-two-dimensional van-der-Waals gaps. The intercalation capacity for that type of compound increases due to the appearance of localized metal-metal interactions i.e. by the metal atoms clustering in the "host" structure. For examle Li_3ReS_2 and $Li_{1.66}ReSe_2$ intercalates have been obtained for ReS_2 and $ReSe_2$ where rhombic Re_4 clusters are formed.

The other example of the metal-metal interactions influence on the intercalation capacity may be intercalates based on the Nb_3X_4 chalcogenides having separated lattice channels in the structure. These compounds form $LiNb_3S_4$ /1/, $Li_2Nb_3Se_4$, $Li_{0.4}Nb_3Te_4$ intercalates on the basis of Nb_3X_4 that is evidence of the influence of both electronic and lattice factors on the compounds intercalation capacity.

The intercalation capacity depends weakly on the chalcogen nature for Mo_6X_8 compounds with octahedral metal-clusters and $Li_{3.6}Mo_6S_8$ /2/, $Li_{3.3}Mo_6Se_8$, $Li_{3.3}Mo_6Te_8$ intercalates have been obtained. The obtained phases were identified and their crystal and electronic structure characteristics have been studied.

References

1. Schollhorn R. Angew.Chem.Int.Ed.Engl.,v.19,1980,p.983.

2. Schollhorn R.,Kumpers M.,Besenhord J.O. Mater.Res.Bull., v.12,1977,p.78.

METAL-CONTAINING CELLULOSE: SOME NOVEL MATERIALS

K F Gadd
School of Science,
Yeovil College,
Ilchester Road,
Yeovil,
Somerset BA21 2BA,
U.K.

ABSTRACT. This communication describes our preliminary studies of the preparation and characterisation of permeable cellulose films and filaments containing ca. 25% by weight metallic platinum, present as very small particles of colloidal dimensions dispersed throughout the cellulose matrix. The platinum (0)/cellulose displays high catalytic activity with respect to the decomposition of hydrogen peroxide.

1. INTRODUCTION

We have reported the preparation of small copper particles (3 to 30nm) dispersed throughout a permeable cellulose matrix[1]. The high loading (16% by weight metallic copper) means that the copper particles, while of colloidal dimensions, are far more closely packed than in conventional metal hydrosols.

An interesting feature of the copper(0)/cellulose is that particles of colloidal dimensions are easily accessible yet heterogeneous with respect to aqueous reaction mixtures. We were interested in extending our work to include the preparation of cellulose containing very small particles of the catalytically active noble metals; we describe here the preparation and properties of platinum-containing cellulose.

2. MATERIALS AND METHODS

Copper(0)/cellulose was prepared from solutions of cellulose dissolved in Cu/pn (a solution of copper(II) hydroxide in aqueous 1,3-diaminopropane[2]) as described previously[1].

Samples of copper(0)/cellulose, either washed but never dried or washed and dried, were immersed in ca. 0.1% by weight aqueous solutions of hexachloroplatinic(IV) acid of sufficient volume to ensure a large excess of platinum assuming that exchange occurred according to the equation:

$$2Cu(0)(cellulose) + Pt(IV) = 2Cu(II) + Pt(0)(cellulose)$$

Samples for spectroscopic examination were prepared by treating extremely thin copper(0)/cellulose films supported between Visking tubing (as described elsewhere[1]) with aqueous hexachloroplatinic(IV) acid; U.V./visible spectra were recorded on a Pye-Unicam SP8100 spectrophotometer.

The catalytic decomposition of hydrogen peroxide was monitored by measuring the evolution of oxygen with an electronic manometer. No attempt was made to maintain constant temperature in view of the vigour of the exothermic reaction, rather a temperature probe was employed to continuously record the temperature of the reaction mixture.

3. RESULTS AND DISCUSSION

Immersion of copper(0)/cellulose films or filaments in aqueous hexachloroplatinic(IV) acid resulted in an immediate darkening of the brown material; after ca. 20 minutes the platinum(0)/cellulose appeared almost black. The very thin samples sandwiched between Visking tubing were a transparent yellow-brown.

Three different reducing agents were used to prepare the copper(0)/cellulose films sandwiched between Visking tubing: sodium dithionite, hydrazine and sodium tetrahydridoborate(III) (all were 0.5 mol dm^{-3} in 1 mol dm^{-3} sodium hydroxide at 80°C). Spectral measurements (Figure 1) suggest that all reducing agents produce particles of the order 10 nm (undoubtedly a range of sizes as evidenced by our earlier transmission electronmicroscopic studies[1]) but that the average copper particle size increased with the following order of reducing agents under the conditions employed:

sodium tetrahydridoborate(III) < sodium dithionite < hydrazine

This may be related to two factors: the time required for the reduction to go to completion (sodium tetrahydridoborate(III) was the slowest reduction) and the rupturing of the cellulose structure by vigorously evolved nitrogen when hydrazine was used.

When these samples were exchanged with platinum the resulting sandwich films had very similar spectra to one another (Figure 2). The distinguishing feature of these spectra compared to that of a platinum sol (prepared in our laboratories by the citrate reduction method[3]) is the presence of a peak or shoulder at 250-300 nm. This feature was most prominent for films prepared from copper(0)/cellulose obtained by hydrazine reduction. The Mie theory suggests that the larger the platinum particles the greater should be the absorbance in this region[4] and this would be in accord with the pattern of particle sizes suggested for the copper(0)/cellulose precursors. At this stage a more detailed discussion of spectra is not warranted for the reasons outlined previously[1].

FIGURE 1

Absorption spectra of copper(0)/cellulose and platinum(0)/cellulose normalised at 650 nm and 450 nm respectively.

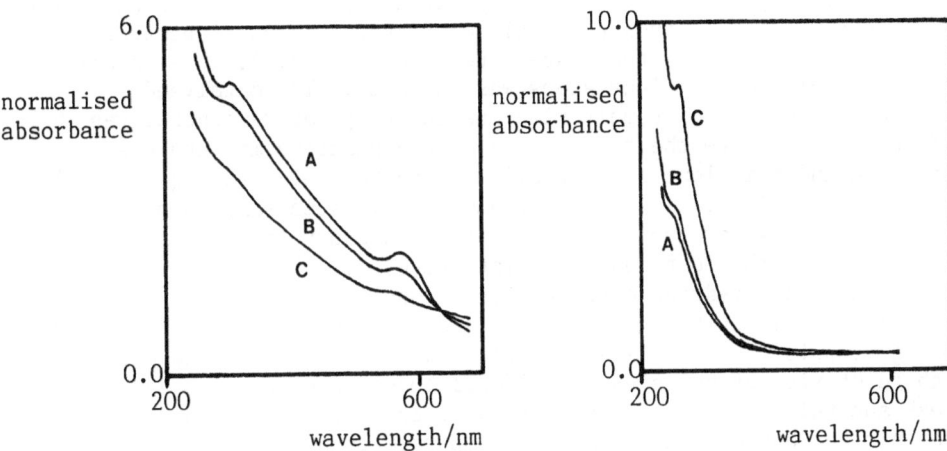

1.1 Cu(0)/cellulose; reducing agents:
A=NaBH$_4$ B=Na$_2$S$_2$O$_4$ C=N$_2$H$_4$

1.2 Pt(0)/cellulose from Cu(0)/cellulose in 1.1

FIGURE 2

Rate of oxygen evolution and temperature rise during the catalysed decomposition of 10 cm^3 hydrogen peroxide (ca. 15 volume).

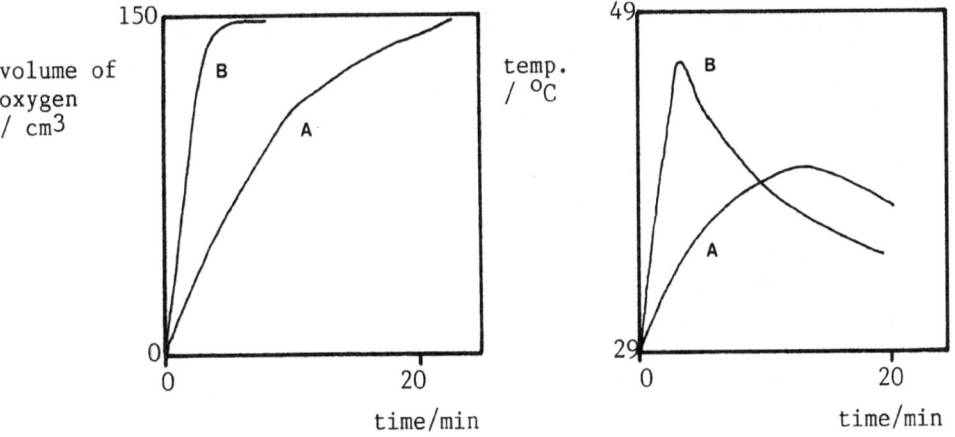

Catalysts used:
A: platinum hydrosol, B: ground, 'never-dried' platinum(0)/cellulose

Preliminary examination with a Philips 300 transmission electronmicroscope suggest platinum particles with overall dimensions of the order 5-50 nm but with a fine structure not apparent in the copper precursors: this might be anticipated in view of the probable growth pattern of the exchanging metal.

In order to assess the reactivity of the platinum(0)/cellulose, samples prepared by exchange with hydrazine reduced copper(0)/cellulose films, the catalytic decomposition of hydrogen peroxide was investigated. Comparison was made with a platinum sol prepared in our laboratories by the citrate reduction method (spectrum recorded above); in each case the number of moles of platinum atoms was the same. The catalysed decompositions were extremely vigorous and temperature control was difficult; therefore the temperature of the reacting mixture was monitored and this lent further evidence to the relative vigour of the catalysed reactions. All the platinum(0)/cellulose samples prepared in our laboratories were more effective than an equivalent amount of platinum sol. The most reactive platinum(0)/cellulose was obtained by exchanging a never-dried copper(0)/cellulose film with hexachloroplatinic(IV) acid, washing and grinding the product but not drying it prior to use. Results are shown in Figure 2. The extremely high reactivity of the platinum(0)/celluloses suggests high surface areas and this is as expected in view of the preparative method: crystal growth by exchange from parent copper particles of the order 3-30 nm in diameter.

We have described the preparation of a highly reactive form of platinum, small particles densely packed in a permeable cellulose support. Work is hand to characterise these materials more fully with respect to structure and catalytic activity. We hope to extend the work to other metals.

4. ACKNOWLEDGEMENTS

The efforts of the Yeovil College project group, in particular W. Murray, E. Fitzgerald and S. Mortimore, are gratefully acknowledged. The author also thanks the Royal Society and Professor D. H. Everett for their support through the Research in Schools scheme.

5. REFERENCES

1. E. Fitzgerald, K. F. Gadd, S. Mortimore and W. Murray, J. Chem. Soc., Chem. Comm., 1986, in the press.
2. K. F. Gadd, Polymer, 1982, 23, 1867.
3. G. C. Bond, Trans. Faraday Soc., 1956, 52, 1235.
4. D. Duff, private communication.

GENERATION AND MANAGEMENT OF THREE-DIMENSIONAL STRUCTURAL DIAGRAMS
FOR ZEOLITES ON STANDARD GRAPHICAL SUPPORT OF AN IBM-PC

G. Calestani, V. Sangermano, C. Rizzoli, G. Bacca and
G. D. Andreetti
Istituto di Strutturistica Chimica, Università di Parma
Viale delle Scienze, 43100 Parma
Italy

The crystal structure of zeolites is a complex arrangement of alumino-silicate tetrahedra, characterized by the presence of peculiar channels and cavities. Computer aided molecular graphics can represent an advantageous and powerful alternative to the classic solid model for the three-dimensional representation of this kind of structure, with the great advantage of a possible software device which just requires as input file the space group symbol and the few atomic coordinates of the independent atoms of the framework. The commonly used molecular graphics computer programs are however mostly unsuitable and some are cumbersome to use, since they have been designed mainly for organic structures and therefore limited to stick and ball or space filling models.
   The most significant representation of crystal structures of inorganic compounds is the idealized polyhedral drawing, with all the facilities of different colours or different shadowing. In an inorganic structure a polyhedron or a set of polyhedra can be considered as the equivalent of the unique-molecule in an organic lattice. Even when programs expressly written for inorganic structures are available, as far as the authors know, they are limited to the representation of simple fundamental polyhedra. Unfortunately zeolite polyhedral building blocks are much more complex than tetrahedra or octahedra, so that a convenient representation and manipulation of these models requires a more sophisticated approach.
   The basic idea of the present work is the implementation of a molecular graphics package, expressly   for inorganic structures and able to perform the two most significant representations of zeolite structures, i. e. framework model and polyhedral drawing.
   Since nowadays micro and personal computers offer low cost independent workstations, with relatively high graphic and computing performances, they represent a convenient substrate for this purpose. Moreover they are spreadly diffused and, in comparison with mainframes, they allow an easy direct access to system resources.
   VIDEOZEO is a completely original package of programs written and implemented for IBM-PC's or compatibles, running under MS-DOS operating system and able to represent zeolite structures as both polyhedral

blocks and framework models. In order to facilitate the access also to
non crystallographers, which represent the major users of molecular
graphics computing, the package has been made as much as possible user-
friendly. It is characterized by a menu-aided processing and supplies
an interactive run time user guide and error checking. The package
include a card image (80 columns wide) full screen file editor and the
most common MS-DOS commands allowing file management, leaving the
knowledge of the operating system out of consideration. Nevertheless a
full interfacing to MS-DOS is assured.

Peculiar features of the package are the possibility of perspective
view with default or user defined vantage point setting, rotation around
a user defined axis and removal or dashing of hidden lines in the
polyhedral representation.

VIDEOZEO has been developed in a modular way: a functional diagram
is shown in Fig. 1. The core of the package can be devided into four
principal blocks, namely the coordinates generator, stick modelling,
polyhedral modelling and picture display unit.

The first one is called when a new structure is considered. A
limited user's crystallographic skill is required only the space group
symbol and the independent coordinates are needed to generate the whole
unit cell content. Options are provided to link adjacent unit cells.

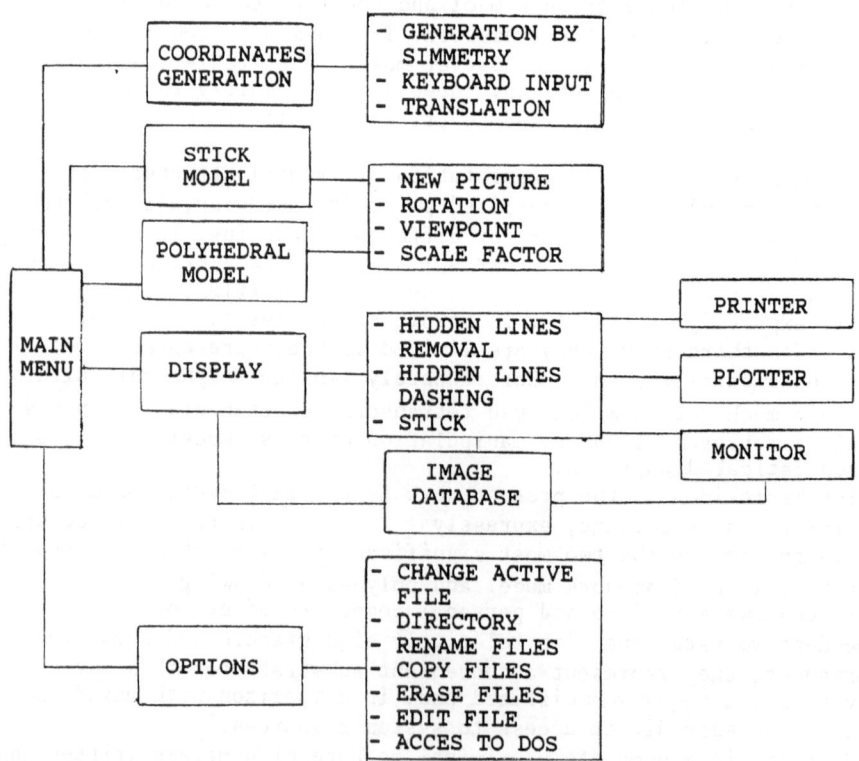

Figure 1. Functional diagram of the package

The second and third block provide respectively stick and polyhedral diagrams. In the latter case the program recognises automatically the faces of polyhedra, once the center are given. The algorithm is based only on the connectivity matrix and this permits the generation of non planar faces which are typical of some zeolite building blocks (e.g. the cancrinite 11-hedron, the chabazite 20-hedron and the gmelinite 14-hedron). The standard version of VIDEOZEO can recognize faces with up to eight edges and represent structures with up to 64 solids of up to 50 vertices. These limits can be however extended depending on the system memory configuration.

The fourth block performs output display for different graphic supports, such as IBM standard colour display, matrix printers (IBM-PC Graphic Printer or HP Laser Jet Plus) and plotters (CALCOMP M81 or M84). The pictures generated may be stored on hard or floppy disk forming an user's image data base, which can be managed inside the package. The high retrival and display rate of the stored figures allows to simulate animation. Some examples of zeolite framework as produced on a colour screen display are shown in Fig. 2.

The package is available on request from the Istituto di Strutturistica Chimica of the University of Parma.

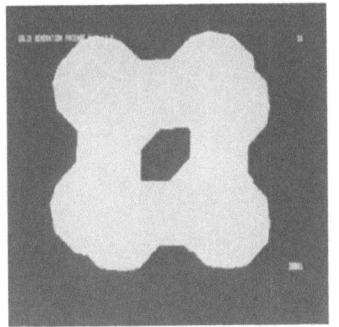
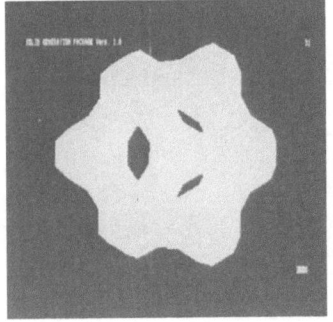
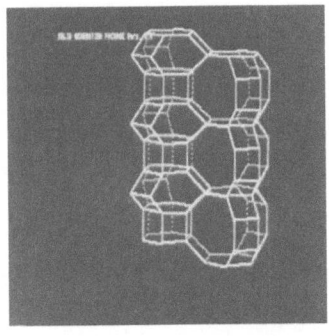
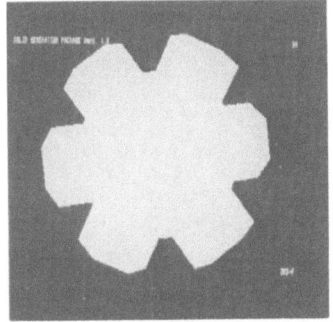

Figure 2. Examples of screen display of zeolite frameworks: a) Zeolite A; b) Zeolite X; c) Offretite; d) Zeolite ZK5

ESR and X-ray Diffraction Studies of Diacyl Peroxides in Urea and Aluminosilicate Hosts

Mark D. Hollingsworth, Kenneth D. M. Harris, William Jones and John M. Thomas
Department of Physical Chemistry, University of Cambridge, Lensfield Road, Cambridge, CB2 1EP, U. K.

Abstract. Electron spin resonance (ESR) spectroscopy was used to study the photodecomposition of long-chain diacyl peroxides trapped in channels within zeolites (silicalite and ferrierite) and urea clathrates. ESR spectra of radical pairs in single crystals of the urea clathrates of diundecanoyl peroxide (UP), lauroyl peroxide (LP) and bis(6-bromohexanoyl) peroxide (6-BrHP) show that the alkyl radicals respond to the CO_2 stress field by recoiling along the channel. In each clathrate, the inter-radical distance for the most relaxed pair is approx. 9.5Å, suggesting nearly complete relaxation of stress from the CO_2s. The rotational mobility and exceptional kinetic stability of the radicals is attributed to relaxation of stress and the lack of a convenient escape route for the CO_2s. X-ray diffraction indicates one-dimensional ordering of guests in 6-BrHP/urea and 3-dimensional ordering of guests in UP/urea. Solid state NMR experiments on LP/urea suggest high guest mobility under ambient conditions. When UP and 6-BrHP were intercalated into silicalite, photolysis yielded isolated radicals, but no radical pairs, even as low as 20K.

1. Introduction

The unique properties of radical pairs have allowed chemists to apply several convenient tools to study the effect of environment on the course of chemical reactions. Time-resolved CIDNP, laser flash photolysis, magnetic isotope effects and simple product studies all give information about the dynamics of radical pair reactions, but ESR spectroscopy is particularly powerful in providing both structural and dynamic information about the fate of the radical pairs. Over the past fifteen years, McBride and co-workers have used this technique to study photochemical reactions in single crystals of diacyl peroxides and azo compounds.[1] The detailed nature of their findings and the general principles that they have revealed about solid state reactivity have prompted us to apply ESR techniques to photochemical reactions of long-chain diacyl peroxides trapped in channels within zeolites (silicalite and ferrierite) and urea clathrates. For each of the guest molecules discussed here (diundecanoyl peroxide (UP), lauroyl peroxide (LP) and bis(6-bromohexanoyl) peroxide (6-BrHP)), the following reaction occurs,

$$R-C(O)-O-O-C(O)-R \xrightarrow{h\nu,\ \lambda>300\ nm} R\cdot\ CO_2\ CO_2\ \cdot R \longrightarrow Products$$

with decyl, undecyl and 5-bromopentyl radical pairs formed from the three guests, respectively. After briefly discussing our X-ray diffraction and solid state NMR studies of the peroxide/urea clathrates, we will present our preliminary ESR findings for the inclusion compounds of these guests in urea and zeolitic hosts.

2. Structural aspects of diacyl peroxide/urea inclusion compounds

The host structure in UP/urea, determined from single crystal X-ray diffraction data, consists of a hydrogen-bonded array of urea molecules, the packing of which is consistent with the space group $P6_1$. The guest molecules are located in the parallel,

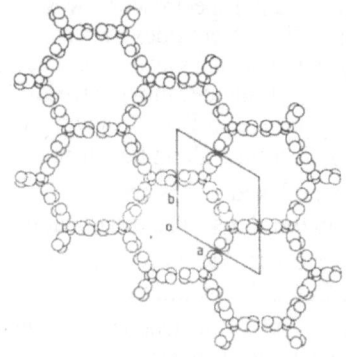

Fig. 1 Host structure of UP/urea viewed down z-axis with arbitrary atomic radii and no H atoms (guest molecules not included). Cell dimensions: a = b = 8.212(1)Å, c = 11.027(2)Å. Channel dimension ~5.25Å. (see also ref. 6)

linear, hexagonal channels shown in projection in Fig. 1. Although the urea structure is substantially the same for each of the systems studied, the structural characteristics of the guest depend critically on its identity. For example, the z-axis oscillation photographs of UP/urea and 6-BrHP/urea (Figs. 2a and 2b) each contain two sets of layer lines. The intense, widely spaced set is common to both photographs;

 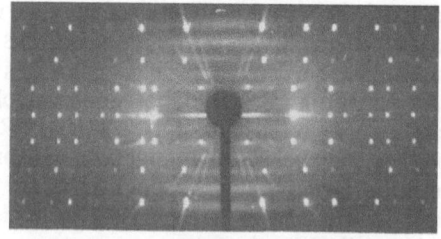

Fig. 2a z-axis oscillation photograph of UP/urea

Fig. 2b z-axis oscillation photograph of 6-BrHP/urea

the spacing of these lines corresponds to the z-axis repeat distance of the host structure. The less intense and more closely spaced set, which is attributed to diffraction by the guest lattice, differs for the two systems. For UP/urea these layer lines contain discrete spots, indicating some three dimensional ordering of the guest. However, for 6-BrHP/urea, the layer lines of the corresponding set appear as diffuse bands rather than discrete spots, suggesting that the guest molecules are ordered only within individual channels. A comprehensive crystallographic study of these and other diacyl peroxide/urea clathrates is in progress.

Solid state NMR experiments on LP/urea indicate a high degree of guest mobility under ambient conditions, a result consistent with similar studies of long-chain hydrocarbon/urea clathrates.[2,3]

3. ESR studies of radical pairs generated in urea clathrates.

From the relatively loose packing of several different long-chain compounds in urea channels,[4-6] one might expect radical motion to be very facile and radical pair collapse to occur at low temperatures in the diacyl peroxide/urea clathrates. We were therefore surprised to find that radical pairs are exceptionally persistent in crystals of UP/urea and 6-BrHP/urea. With UP/urea, collapse of decyl radical pairs occurs with a half-life of approximately 17 min at 163K.[7] That terminal bromine acts as a good "anchor"[9] is shown by the much slower decay rate of 5-bromopentyl radical pairs, the collapse of which occurs with a half-life of 26 min at 191K. These decay rates may be contrasted with those of decyl radical pairs in pure crystals of UP,[10] for which the half-life is 1 min at 133K. The 5-bromopentyl radical pair formed in 6-BrHP/urea is one of the most persistent unrearranged radical pairs ever generated by diacyl peroxide photolysis.[11]

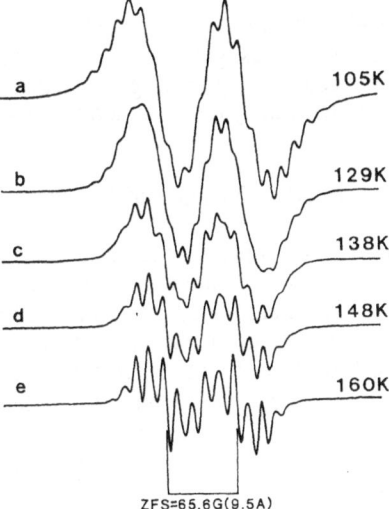

Figure 3 ESR spectra of decyl radical pairs generated by slight photolysis of a single crystal of UP/urea. With the z-axis parallel to the magnetic field, the zfs gives the component of the (symmetry averaged) inter-radical vector along the channel axis.

These kinetic anomalies are accentuated by the observation that in each system, alpha and beta hyperfine splittings are almost completely averaged above 160K (Fig. 3). This averaging is consistent with rapid rotation or large amplitude oscillations of the alkyl radical chains about the channel axis. Although detailed ESR studies with deuterated peroxides are needed before the structures and motions of these radical pairs can be fully assessed, our preliminary findings indicate that the absence of an escape route for the CO_2 molecules and the alleviation of the stress generated by them[12] allows the radicals to be kinetically stable, yet rotationally mobile, to very high temperatures.

Goniometric analysis of the zero-field splitting (zfs) of radical pairs in single crystals[13] indicates that in all three systems the radicals respond to the CO_2 stress field by recoiling along the channel before finally reacting. In each case, the radicals are separated by more than 8Å at temperatures as low as 20K, even though the incipient radical centers were separated by only ~5.7Å in the peroxide precursor.[10] In contrast, when generated in crystals of pure UP, decyl radicals are only 6.06Å apart at 20K, and only 7.68Å apart in the most relaxed pair at 125K.[10] For all three clathrate guests, the inter-radical distance for the most relaxed pair is approximately 9.5Å, suggesting nearly complete relaxation of stress from the CO_2 molecules.

From these results it seems very plausible that in the urea system, the tightly woven network of hydrogen bonds prevents the CO_2 molecules from escaping through the channel wall, while the radicals themselves hinder diffusion of CO_2 along the channel. At the same time, the radical chains (and presumably the adjacent peroxide molecules) are packed loosely enough to allow the radicals to recoil to a common relaxed position, where they can oscillate or rotate rapidly at higher temperatures. This system represents an extreme case in which the absence of a convenient escape route for the CO_2s and relatively loose packing of the guests allows fairly complete relaxation of stress without chemical reaction.[14] We are currently using radical pair ESR to investigate the role of terminal substituents, including those in adjacent molecules, by preparing mixed clathrates of terminally difunctionalized peroxides and terminally disubstituted alkanes.

4. ESR studies of radicals generated in silicalite and ferrierite

Intercalates of diacyl peroxides in silicalite and dealuminated ferrierite were formed by exposure of the zeolites to a solution of the peroxide in 2,2,4-trimethylpentane (2,2,4-TMP).[15] After treatment for 10 hrs in an ultrasonic bath at 20°C, the peroxide was removed from the outer surface by washing with large amounts of 2,2,4-TMP. When UP and 6-BrHP were intercalated into silicalite (channel dimensions 5.4Å by 5.6Å and 5.1Å by 5.5Å)[16] photolysis yielded isolated radicals, but no radical pairs, even as low as 20K. Similarly, with 6-BrHP/ferrierite (channel dimensions 4.3Å by 5.5Å)[16] only isolated radicals were generated at 20K. Three lines of evidence suggest that with silicalite,[17] the isolated radicals are generated within the channels and not on the surface:

1. Within a KBr pellet of unwashed UP/silicalite, the peroxide decomposes to give potassium carboxylate (indicated by IR bands at 1560 and 1410 cm^{-1}), whereas UP/silicalite washed with 2,2,4-TMP gives little or no such reaction.

2. Nitroso tert-butane (NtB) quenches those radicals formed by photolysis of unwashed UP/silicalite but not those formed in UP/silicalite that had been washed with 2,2,4-TMP.

3. When subjected to the intercalation and washing procedure described at the

beginning of this section, finely ground Spectrasil quartz (calcined at 500°C) and UP (or 6-BrHP) gives no signals from alkyl radicals after irradiation at 20K.

In the zeolite hosts, the high translational mobility of the radicals seems to be a consequence of several factors, including low loading levels of peroxide, the excess photolytic energy of 82 kcal/mol and the driving force generated by the incipient CO_2 molecules. Further ESR studies on these and other peroxide/zeolite systems are in progress.

5. Acknowledgement

We thank Mr. Wang-Nang Wang for his help with this work, which was supported by S.E.R.C. of Great Britain and by the North Atlantic Treaty Organization under a grant awarded to M.D.H. in 1985. K.D.M.H. thanks British Petroleum plc. for a studentship.

6. References

1. McBride, J. M., Accts. Chem. Res., **16**, 304 (1983) and refs. cited therein
2. Bell, J. D. and R. E. Richards, Trans. Farad. Soc., **65**, 2529 (1969)
3. Casal, H. L. , D. G. Cameron and E. C. Kelusky, J. Chem. Phys., **80**, 1407 (1984)
4. Griffith, O. H., J. Chem. Phys., **41**, 1093 (1964)
5. Chatani, Y., H. Anraku and Y, Taki, Mol. Cryst. Liq. Cryst., **48**, 219 (1978)
6. Smith, A. E., Acta Cryst., **5**, 224 (1952)
7. With UP/urea and 6-BrHP/urea, radical pair decay occurs with step-wise kinetics (see reference 8). The half-lives reported here represent the slower rates, which occur at low radical pair concentrations.
8. Whitsel, B. L., Ph. D. Thesis, Yale University, New Haven, CT. (1977)
9. see also Mills, D. E., Ph. D. Thesis, Yale University, New Haven, CT. (1986)
10. Segmuller, B. E., Ph. D. Thesis, Yale University, New Haven, CT. (1982)
11. Pairs of 1,1,2-triphenylethyl radicals formed in single crystals of bis(3,3,3-triphenylpropionyl) peroxide decay with a half-life of 1 min at 256K, but here, intramolecular rearrangement keeps the radical centers isolated from each other. Walter, D. W. and J. M. McBride, J. Am. Chem. Soc., **103**, 7069, 7074 (1981)
12. Hollingsworth, M. D., Ph. D. Thesis, Yale University, New Haven, CT. (1986)
13. McBride, J. M., M. W. Vary and B. L. Whitsel, ACS Symp. Ser., **69**, 208 (1978)
14. See Hollingsworth, M. D., and J. M. McBride, J. Am. Chem. Soc.,**107**,1792 (1985) for an related case in which a distant defect allows relaxation of stress.
15. Casal, H. L. and J. C. Scaiano, Can. J. Chem., **62**, 628 (1984)
16. Meier, W. M. and D. H. Olson, Atlas of Zeolite Structure Types, Structure Commision of the International Zeolite Association, Zurich (1978)
17. At the time of submission, the first two control experiments had not been done for 6-BrHP/ferrierite.

ALUMINATE SODALITES - A FAMILY OF INCLUSION COMPOUNDS WITH STRONG HOST-GUEST INTERACTIONS

W. Depmeier
Institut für Kristallographie
der Universität Karlsruhe (TH)
Kaiserstr. 12
D-7500 Karlsruhe, Bundesrepublik Deutschland

Aluminate sodalites belong to a structural family which, in principle, has been known for quite a long time and which is closely related to various zeolites. Naturally occurring sodalites belong to the class of aluminosilicates (e.g., sodalite in the proper sense has the idealized formula $Na_8[Al_6Si_6O_{24}]Cl_2$), whereas aluminate sodalites have a composition $M_8[Al_{12}O_{24}](XO_4)_2$ for a (pseudo-)cubic cell ($a_o \sim 9.5 Å$), with M=Ca,Sr, ... and X=S,Cr,Mo,W ...

In comparison with other inclusion compounds dealt with in these Proceedings, aluminate sodalites have a remarkable thermal stability. The usual preparation, by sintering appropriate mixtures of the corresponding oxides, requires temperatures of as high as $1350°$ C, whereas melting temperatures seem to be around $2000°$ C. Structural work and measurements of physical properties have not been possible until a technique of growing "single crystals" by the flux method had been developed. This technique works at about $1100°$ C and, because of the occurrence of phase transitions below these temperatures, the "single crystals" thus obtained are usually twinned.
The structure is characterized by corner-connected AlO_4 tetrahedra, the centres of which occupy the vertices of an - ideally - regular truncated octahedron. This is the so-called sodalite cage. The sodalite framework is then formed by joining together the sodalite cages in a body-centred arrangement, thereby filling the available space completely (the regular truncated octahedron is a space-filling polyhedron). The framework -it can be regarded as the host- is quite open and accommodates in its cavities the guests, viz. the so-called cage anions XO_4 and the cage cations M. The former occupy the centres of the sodalite cages, whereas the latter are situated near, but not necessarily at, the centres of the six-membered rings shared between two adjacent cages. Fig. 1 illustrates some particularities of the

sodalite structure.

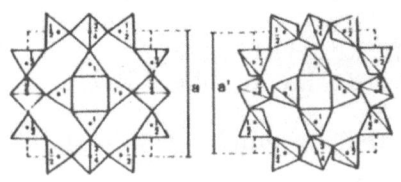

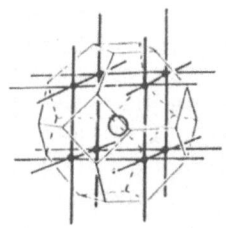

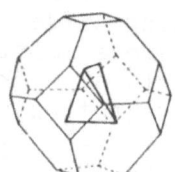 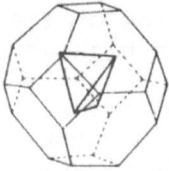

Fig. 1: Some particularities of the sodalite cage:
Top: Upper half of the sodalite (after ref. 9);
fully expanded (left) and partially collapsed
(right hand side).
Middle: The cage cations (small black circles) form
a framework (thick solid lines) which interpene-
trates the sodalite framework (truncated octahedron,
thin lines). The cage anions (big circle) occupy

the centre of the sodalite cage.
Bottom: Aluminate sodalites contain tetrahedral cage anions; these are oriented in such a way that their vertices (O atoms of the cage anions) point at the midpoints of <u>some edges of the truncated</u> octahedron (= framework O atoms). Two possible orientations are shown.

The AlO_4 tetrahedra are far from being ideal. The degree of distortion is, however, constant for the whole family of aluminate sodalites, as it is for any family of constant framework composition. An explanation has been proposed [1] for this intrinsic strain as being the result of an attempt of the framework to avoid an unfavourable conformation, required by its proper topology. The sodalite framework has the remarkable property of being quite flexible via cooperative rotations of the tetrahedra (so-called tilts, cf. fig. 1 and refs. 1,2). The flexibility enables the framework to adjust its size to that of the guest species via "partial collapse" [2]. Furthermore, the framework does not only adjust itself to the sizes, but also - in a certain sense, to the shape of the species [3]. This is shown schematically in Fig. 1. A tetrahedral cage anion within the sodalite cage is oriented in such a way that its O atoms at the vertices point at O atoms of the framework, rather than pointing into the direction of the threefold axes of the sodalite cages. The framework O atoms are repelled by this interaction and the cage is deformed. An important consequence is that the cubic symmetry of the structure is destroyed. By the repulsive interactions between the O atoms of the host and the guests several phenomena are produced which might be of interest for chemists and physicists as well:

i) <u>Bond Length and Angle Variations</u>, <u>Strain</u>
The present knowledge of the properties of the pure AlO_4 tetrahedron is surprisingly restricted. This is due, on the one hand, to the relatively small number of aluminate structures which have been determined with sufficient precision, and, on the other hand, a result of the fact that in aluminosilicates the behaviour of the AlO_4 tetrahedron is masked by that of the SiO_4 tetrahedron [4]. Thus, aluminate sodalites offer a good opportunity to study the Al-O bond and the Al-O-Al angle in their pure form and provide the theoretical chemist with empirical data to test his calculations (cf. ref. 4). Furthermore, the repulsive interactions produce strong local angular and also bond length distortions within the framework, whereby the interactions and corresponding distortions depend on the temperature (phase transitions, see below). Therefore, as structural work advances, a growing wealth of information on the pure AlO_4 tetrahedron should come to our disposal. The

study of the effects of intrinsic and extrinsic strains in the relatively simple sodalite structure should also facilitate studies in the more complicated zeolites, where these effects recently also began to attract attention (see, e.g. ref. 5).

(ii) Phase Transitions

The distortion of the framework means, in general, that the symmetry is lowered. The deviation from cubic symmetry can, however, be cancelled by order-disorder processes which might be brought about by either raising the temperature or by changing the composition of mixed crystals. All pure aluminate sodalites known so far are non-cubic at room temperature and undergo structural transitions at temperatures which range between ~300 K and ~700 K. The phase transitions are of the ferroic type; ferroelastic and ferroelectric species have been found so far [6]. The transitions occur at the boundary of the Brillouin zone; this fact accounts for the formation of complicated superstructures and for frequent pseudomerohedral twinning. First steps towards a phenomenological theory have been done; however, there are still many points which remain unclear; e.g., it is still a matter of question whether there is only one theory for all aluminate sodalites or more than one according to the various degrees of host - guest interaction. Very intriguing results come from studies on mixed crystals [7] which indicate the existence of tricritical points. Their study would be interesting in view of the six-component order parameter and because of possible implications for current theories of phase transitions [8].

References

[1] W. Depmeier, Acta Cryst. B40, 185 (1984)
[2] L. Pauling, Z. Kristallogr. 74, 213 (1930)
[3] W. Depmeier, Acta Cryst. C40, 226 (1984)
[4] K.L. Geisinger, G.V. Gibbs & A. Navrotsky, Phys.Chem.Minerals 11, 266 (1985)
[5] W. Depmeier, Acta Cryst. B41, 101 (1985)
[6] N. Setter, M.E. Mendoza-Alvarez, W. Depmeier & H. Schmid, Ferroelectrics 56, 49 (1984)
[7] N. Setter & W. Depmeier, Ferroelectrics 56, 45 (1984)
[8] E. Meimarakis & P. Tolédano, Jpn. J. Appl. Phys. 24, Suppl. 24-2, 350 (1985)
[9] D. Taylor, Miner. Magaz. 38, 593 (1972)

SYNTHESIS, CRYSTALLOGRAPHIC, AND THERMAL PROPERTIES OF A NEW POROUS SILICA

H. Gies
Mineralogisches Institut der Universität Kiel
2300 Kiel
FRG

ABSTRACT. A new porous tectosilicate has been synthesized in the presence of boric acid and 1,2,2,6,6-pentamethylpiperidine as guest molecules. It crystallizes in the monoclinic system with $a_o = 9.91(1)$Å, $b_o = 20.63(3)$Å, $c_o = 9.80(2)$Å, and $\beta = 99.7(2)°$. After heat treatment at 820°C for 1 hour the guest molecules are set free whereas the silica host framework is retained. From crystal morphology and thermal behaviour it is concluded that the new material possesses channel-like voids.

1. INTRODUCTION

A series of silica end-members of zeolites (zeosils) as well as clathrate compounds with framework composition SiO_2 (clathrasils) have been synthesized in the recent past [1, 2]. It has been concluded that the organic guest species alone act as templates for the formation of the open framework and that these guest molecules have a structure directing role. In a detailed study it has been shown that the size and dimensionality of the pore system in porous tectosilicates is dependent on the properties of the guest species used for synthesis [3].
In this paper we report on a new porous silica framework (porosil [1]) which has been crystallized in the presence of 1,2,2,6,6-pentamethylpiperidine and boric acid.

2. EXPERIMENTAL

The new porosil was synthesized from aqueous solutions of silicic acid and boric acid in the presence of 1,2,2,6,6-pentamethylpiperidine, ⟨structure⟩ , as guest molecule.
Tetramethoxysilane was hydrolyzed in 1M aqueous ethylendiamine to give a homogeneous 0.5M silica solution.

h	k	l	$d_{obs.}$	I/I_0	h	k	l	$d_{obs.}$	I/I_0
0	2	0	10.18	10	-1	4	2	3.464	1
					1	5	1		
1	0	0	9.72	5	3	0	0	3.255	1
0	1	1	8.715	1	-3	1	1	3.220	1
0	3	1	5.574	1	3	2	0	3.102	1
-1	3	1	5.102	1	-2	1	3	2.892	2
-2	0	1	4.695	1	1	1	3		
-1	1	2	4.527	2	-2	5	2	2.793	2
2	2	0	4.407	3	0	6	2		
0	4	2	3.531	3	-1	7	1	2.746	2
-2	2	2							

Tab. 1: X-ray powder data of the new zeosil.

characteristic morphology in sperulitic bundels (fig. 1). Single crystal diffraction studies revealed that the crystals were twinned with high disorder parallel to (100). From electron diffraction experiments the unit cell parameters have been obtained. The X-ray powder pattern could be indexed in the monoclinic system with a_o = 9.91(1)Å, b_o = 20.63(3)Å, c_o = 9.80(2)Å, and β = 99.7(2)° (tab. 1).

Thermogravimetric analysis was performed in the temperature range 30 - 820°C with 10°C/min heating rate (fig. 2). Continuous weight loss was observed in the temperature range from 150 - 820°C. At 500°C the sample turned black due to the decomposition of the organic guest molecules. The host framework is retained even after the heat treatment of the product at 820°C for 1 hour.

Using an energy dispersive system only Si as framework constituent has been detected. Boron analysis yielded about 1% B in the as synthesized sample, whereas after calcination only traces were detected.

From conditions of synthesis, thermal properties, and chemical analysis it is concluded that the new material belongs to the class of porosils having host framework composition of essentially SiO_2. The organic guest species 1,2,2,6,6-pentamethylpiperidine and the boric acid play an indispensable role as structure directing agents. From

To this boric acid (0.07M) was added. On mixing with boric acid a small amount of silica gel precipitated. After addition of the guest molecule (5-10vol%) the solution was sealed in silica tubes and kept in an oven in the temperature range 150 - 165°C. After about three months reaction time lath shaped crystals were obtained. The crystalline material was characterized by X-ray powder diffraction and electron diffraction. Silicon has been analyzed using an energy dispersive system. Boron analysis was performed with an Jobin Yvon JY38 plasma spectrometer (ICP) before and after calcining the sample. Thermogravimetric analysis was done using a DuPont 1090 Thermal Analyser.

3. RESULTS AND DISCUSSION

The new porosil framework typ could only be crystallized using 1,2,2,6,6-pentamethylpiperidine as guest molecule and in the temperature range 150 - 165°C and in the presence of boric acid. The variation of the concentration of the guest and of the ratio of guest/boric acid did not influence the formation of the product. However, at temperatures higher than 165°C and in the absence of boric acid at temperatures from 150 - 200°C only the clathrasil dodecasil 1H [4] was obtained.
The product appears as lath shaped transparent crystals of

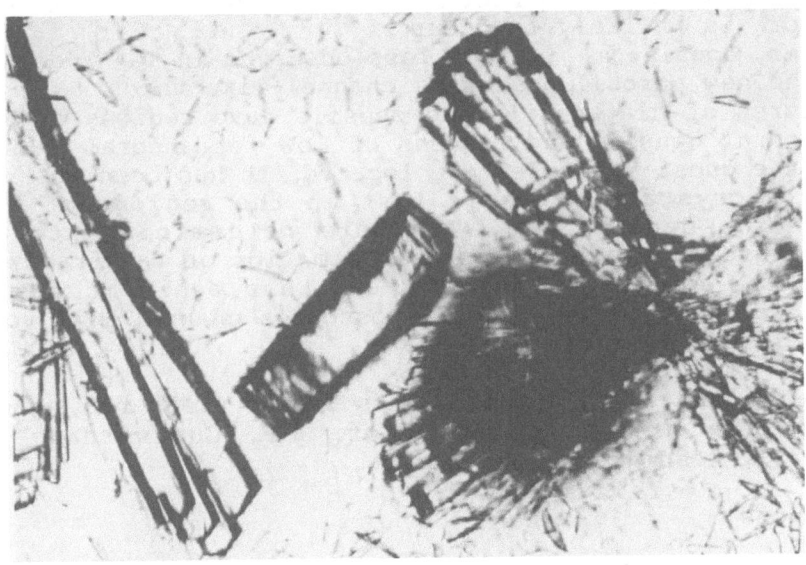

Figure 1: Micrograph of the new zeosil together with a dodecasil 1H crystal.

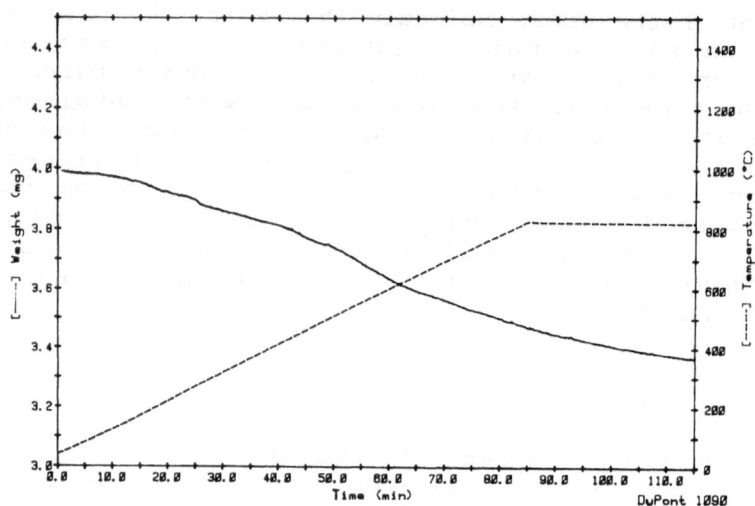

Figure 2: TGA curve of the new zeosil.

synthesis conditions it is evident that the organic guest species forms a Lewis acid-base complex with boric acid which probably acts as template. Therefore it is apparent that isomorphous replacement of Si by B in the host framework is unlikely to occur.
Moreover, continuous weight loss observed in TGA indicates that the new porosil possesses channel-like voids.
Structures of clathrasils and zeosils have regions of high tetrahedral density and regions of low tetrahedral density where the guest molecules are located. It has been observed that the crystal growth is fast in the regions of high tetrahedral density [5]. The rate determining step in crystal growth, however, is the formation of the framework around the guest species [6]. On this basis it may be suggested that the new porosil possesses channel-like voids parallel to the needle axis [001].

Acknowledgement. The author thanks Dr. M. Czank for support in electron microscopy and Prof. R.P. Gunawardane for critically reading the manuscript.

References

[1] R.M. Barrer in proceedings of the 7th International Zeolite Conference, Tokyo, 1(1986), Elsevier, New York, Amsterdam, Oxford.

F. Liebau, H. Gies, R.P. Gunawardane, and B. Marler, Zeolites, in press.
[2] H. Gies: Nachr. Chem. Tech. Lab. $\underline{33}$, 387(1985).
[3] R.P. Gunawardane, H. Gies, and B. Marler, in preparation for Zeolites.
[4] H. Gerke and H. Gies: Z. Kristallogr. $\underline{167}$, 11(1984).
[5] H. Gies: Fortschr. Miner. $\underline{64}$, Beiheft $\underline{1}$, 56(1986).
[6] H. Gies and R.P. Gunawardane, submitted to Zeolites.

The Effect of Water Vapour on the Cyclodextrin–Solute Interaction in Gas–Solid Chromatography

L. ANDĚRA and E. SMOLKOVÁ-KEULEMANSOVÁ*
Department of Analytical Chemistry, Charles University, Albertov 2030, 12840 Prague 2, Czechoslovakia

(Received: 25 September 1986)

Abstract. The effect of water vapour on the interaction between the solutes and a cyclodextrin was studied chromatographically in a gas/solid system. It has been shown that all the sorbates capable of forming inclusion complexes with α-CD are affected by the presence of water, in contrast to the sorbates that do not interact with α-CD, which are unaffected. The measurements were carried out with the cyclodextrin deposited on a solid carrier (Chromosorb W) and with water vapour contained in the mobile phase. Some advantages of this system have been demonstrated for an analytical application.

Key Words. Sorbate-cyclodextrin interaction, Effect of water vapour, Gas-solid chromatography, Analytical applications.

1. Introduction

Cyclodextrins (CDs) have recently found use as stationary phases in gas-solid chromatography (GSC) [1–8, 12–14] and in gas-liquid chromatography (GLC) [8–11], because of their selective separation capability. Their application to separations of stereoisomers (alkenes, pinenes) and positional isomers of aromatics (xylenes, trimethylbenzenes) has been found to be very advantageous. The inclusion process, which underlies selective separations, is, with cyclodextrins, also affected by the presence of water. It is well known that cyclodextrins form crystal hydrates and that the water of crystallization participates in the formation of inclusion complexes [15]. On the formation of an inclusion complex, the water molecules included in the CD cavity are liberated preferentially. This liberation is further enhanced under the dynamic conditions of gas chromatography. It can thus be assumed that water also plays an important role in the equilibrium processes between CD and a guest (sorbate) in the gaseous state.

To evaluate the intensity of the effect of water on the inclusion process, the conditions were modelled, so that the mobile phase consisted of an inert carrier gas saturated with water vapour, whereas a cyclodextrin represented a solid stationary phase. The use of mixed mobile phases consisting of water vapour in an inert gas has been known in gas chromatography for rather a long time [16, 17] and it is still utilized [18]. The water vapour present in the carrier gas affects the properties of the mobile phase (polarity, viscosity), as well as those of the stationary phase (interaction of the water vapour with the sorbent active sites). With cyclodextrins, the water

---

* Author for correspondence.

Presented at the Fourth International Symposium on Inclusion Phenomena and the Third International Symposium on Cyclodextrins, Lancaster, U.K., 20–25 July 1986.

enclosed in the cavities has also the function of a 'catalyst' of the inclusion complex formation. Therefore, it could be expected that the presence of water vapour in the carrier gas might not only affect the surface of the stationary phase, but also cause substantial changes in the formation of the inclusion complex, i.e. in the equilibration between the sorbate and the inclusion compound.

2. Experimental

α- and β-CD (Chinoin, Budapest) were deposited on Chromosorb W (60–80 mesh) from a dimethylformamide solution. The CD content in the stationary phase was determined from the results of organic elemental analysis and amounted to 4.75% wt. for α-CD and 7.32% wt. for β-CD. Columns 1200 mm long, 3 mm I.D., were used.

The chromatographic measurements were carried out on a CHROM 41 instrument (Laboratorní Přístroje, Prague, Czechoslovakia) with a flame ionization detector (FID). Mixed mobile phases were obtained by including a thermostatted water-saturator in the carrier gas stream. The measurements were carried out at water vapour pressures of 1.33, 2.97, and 6.50 kPa, corresponding to 0.8, 1.7, and 4.3% wt., respectively. The water vapour content in the carrier gas was determined gravimetrically, by absorption in $Mg(ClO_4)_2$. The carrier gas flow rate was 60 ml/min and the column temperature was 80 °C. Dilute vapours of the samples were injected with a Hamilton 10 µl syringe.

3. Results and Discussion

3.1. THE EFFECT OF WATER VAPOUR ON THE SORBATE RETENTION

As is demonstrated by the results summarized in Tables I and II, introduction of water vapour into the carrier gas led to a decrease in the retention times of all the sorbates, compared with those obtained in the dry carrier gas. An analogous decrease in the retention times when using water vapour in the carrier gas was also described in e.g. the paper by Guillemin and Millet [18]. The authors employed much higher water vapour contents (above 10%) and classical adsorbents (silica gel, alumina, porous polymers). With cyclodextrins, a perceptible decrease in the retention occurs even for 0.8% water vapour in the carrier gas. This phenomenon can be explained by partial adsorption of water molecules on the stationary phase surface, in a similar manner to common adsorbents. However, the experimental results, especially those obtained with α-CD as the stationary phase (Table I), indicate an effect on the equilibrium in the formation of the sorbate–cyclodextrin inclusion complex, in favour of the free guest. A great excess of water vapour over the sorbate apparently leads to competitive inclusion of water molecules and thus to faster desorption of the guest.

The dependence of the retention times of various sorbates on the water vapour pressure in the carrier gas, obtained for α-CD, can be formally classified into three groups, on the basis of the curve shape, characteristic of certain groups of substances:

1. Linear dependences with permanent decrease.
2. Hyperbolic dependences with a faster decrease at lower water vapour pressures.
3. Dependences with a very small dependence of the sorbate retention times on the water vapour concentration in the carrier gas.

WATER VAPOUR IN CYCLODEXTRIN GAS CHROMATOGRAPHY

Table I. Corrected retention times of sorbates on α-CD and their dependence on the water vapour in the carrier gas

Water vapour pressure (kPa)		0	1.33	2.97	6.50
Substance	b.p.(°C)	t'_R(s)			
n-Pentane	36.2	26	–	–	–
n-Hexane	69.0	205	123	72	25
n-Heptane	98.5	980	630	405	153
Cyclopentane	50.0	38	–	9	–
Cyclohexane	78.0	293	151	66	21
Methylcyclohexane	101.0	560	223	100	32
2,2,3,3-Tetramethylbutane	106.5	6	5.5	5	4.5
3-Methylheptane	115.5	1430	890	570	235
2,4-Dimethylheptane	133.5	170	83	44	15
2,2,4-Trimethylhexane	126.0	10	–	5	–
cis-3-Methyl-3-Hexene	94.0	278	168	118	52
trans-3-Methyl-3-Hexene	93.3	21	56	31	10
Cyclohexene	83.0	230	–	34	–
Benzene	80.1	66	38	22	6
Toluene	110.8	154	87	46	13
Ethylbenzene	136.2	685	426	249	84
n-Propylbenzene	159.2	–	–	838	325
Isopropylbenzene	152.4	160	69	36	24
sec-Butylbenzene	173.0	193	89	46	28
tert-Butylbenzene	169.0	35	27	21	18
o-Xylene	144.0	68	21	10	6
m-Xylene	138.8	137	63	33	15
p-Xylene	138.5	1080	605	356	126
1,2,3-Trimethylbenzene	176.1	27	20	16	12
1,3,5-Trimethylbenzene	164.7	20	16	12	9
1,2,4-Trimethylbenzene	169.4	136	56	30	17
o-Diethylbenzene	183.4	36	28	22	17
m-Diethylbenzene	181.0	776	437	263	87
m-Diisopropylbenzene	203.2	97	76	63	49
Diisopropylbenzene	210.3	128	95	75	56
Methanol	64.6	62	21	11	9
Ethanol	78.0	58	27	17	13
1-Propanol	97.0	134	66	43	28
2-Propanol	82.4	71	44	29	19
1-Butanol	117.7	520	251	172	97
Isobutanol	108.0	300	129	103	54
1-Pentanol	137.0	1550	1200	790	400

The dependences for alkanes and cycloalkanes are given in Figure 1 and are linear for n-alkanes; 2,4-dimethylheptane alternates between the first and second type, the second type is characteristic of cycloalkanes and the third type was found with 2,2,3,3-tetramethylbutane.

It has been known from the earlier papers [2, 6–8] that n-alkanes form very strong complexes with α-CD also under the conditions of gas chromatography, whereas voluminous molecules (2,2,3,3-tetramethylbutane) cannot interact with α-CD. Hence, it can be assumed that linear dependences are obtained for strongly interacting sorbates. With increasing concentration of water vapour in the carrier gas, inclusion

Table II. Corrected retention times of sorbates on β-CD and their dependence on the water vapour in the carrier gas

Water vapour pressure (kPa)		0	1.33	2.97	6.50
Substance	b.p.(°C)		t'_R (s)		
n-Hexane	69.0	9	–	3	–
n-Heptane	98.5	34	15	9	6
n-Nonane	150.5	467	195	124	82
Cyclohexane	78.0	37	22	16	11
Methylcyclohexane	101.0	60	32	22	15
2,2,3,3-Tetramethylbutane	106.5	255	112	74	63
3-Methylheptane	115.5	92	41	28	20
2,4-Dimethylheptane	133.5	193	84	55	40
2,2,4-Trimethylhexane	126.6	207	94	65	48
Cyclohexene	83.0	51	26	20	13
Benzene	80.1	59	33	22	14
Toluene	110.8	102	45	31	18
Ethylbenzene	136.2	282	118	77	45
n-Propylbenzene	159.2	880	373	231	133
Isopropylbenzene	152.4	626	304	187	118
sec-Butylbenzene	173.0	1240	600	369	234
tert-Butylbenzene	169.0	1370	735	485	315
o-Xylene	144.0	242	110	68	42
m-Xylene	138.8	202	82	52	32
p-Xylene	138.5	293	126	75	45
1,2,3-Trimethylbenzene	176.1	400	171	108	67
1,3,5-Trimethylbenzene	164.7	183	84	57	35
1,2,4-Trimethylbenzene	169.4	720	298	202	131
o-Diethylbenzene	183.4	536	234	139	98
m-Diethylbenzene	181.0	1150	495	304	183
Methanol	64.6	114	29	18	20
Ethanol	78.0	142	45	28	23
1-Propanol	97.0	518	182	122	82
2-Propanol	82.4	300	113	79	55
1-Butanol	117.7	1860	740	460	293
2-Butanol	98.0	877	336	219	138
Isobutanol	108.0	1380	595	400	274
tert-Butanol	83.0	593	232	159	110

of water is progressively more important and the retention of n-alkanes decreases. With the second group, the retention is strongly decreased even by low water concentrations. This concerns cycloalkanes, whose interaction with α-CD is complicated, in view of their conformational structures. It can be assumed from the experimental data that their interaction with the α-CD cavity is weaker and thus even low concentrations of water vapour in the carrier gas strongly shift the equilibrium in favour of the free sorbate. Sorbates with voluminous molecules that cannot be included into the α-CD cavity (2,2,3,3-tetramethylbutane) depend only negligibly on the water vapour content in the carrier gas and inclusion of water molecules has almost no effect on their retention.

Aromatic hydrocarbons (Figure 2), benzene, ethylbenzene, p-xylene, follow the first type of dependence, in agreement with the concept of their inclusion, while the

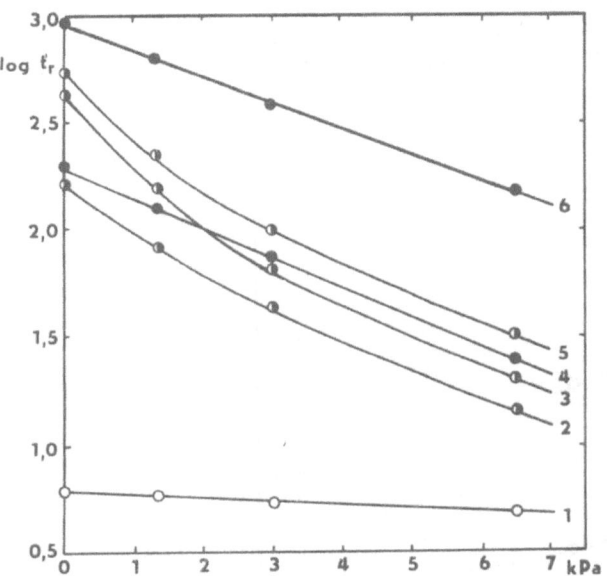

Fig. 1. A log t'_R dependence on the water vapour pressure (p) for alkanes and cycloalkanes on α-CD (● – first type, ◐ – second type, ○ – third type). 1 – 2,2,3,3-tetramethylbutane, 2 – 2,4-dimethylheptane, 3 – cyclohexane, 4 – n-hexane, 5 – methylcyclohexane, 6 – n-heptane.

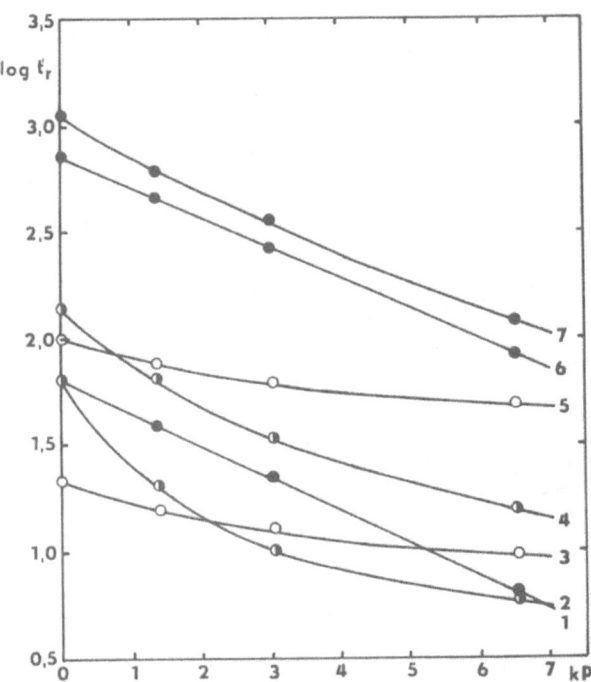

Fig. 2. A log t'_R dependence on the water vapour pressure (p) for aromatic hydrocarbons on α-CD. (● – first type, ◐ – second type, ○ – third type). 1 – benzene, 2 – o-xylene, 3 – 1,3,5-trimethylbenzene, 4 – m-xylene, 5 – m-diisopropylbenzene, 6 – ethylbenzene, 7 – p-xylene.

second type of dependence has been found for o- and m-xylene. The retention times of m-diisopropylbenzene and 1,3,5-trimethylbenzene are virtually independent of the water vapour content in the carrier gas.

The first type of dependence was obtained for the alcohols (Figure 3) 1-butanol

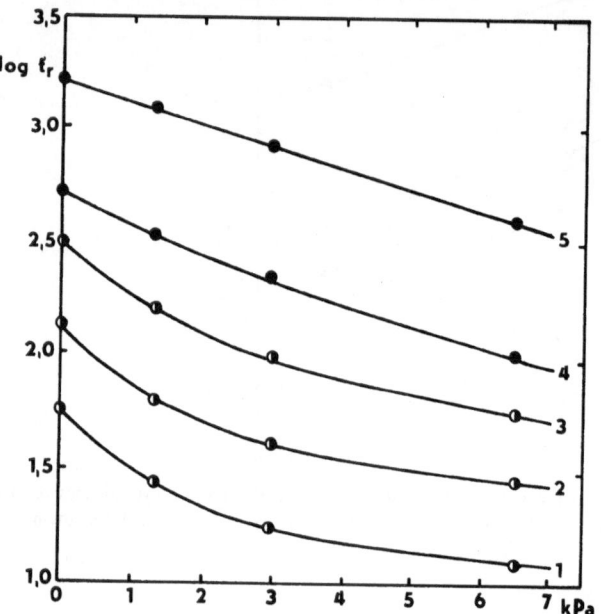

Fig. 3. A log t'_R dependence on the water vapour pressure (p) for alcohols on α-CD. 1 – ethanol, 2 – 1-propanol, 3 – isobutanol, 4 – 1-butanol, 5 – 1-pentanol (● – first type, ◐ – second type).

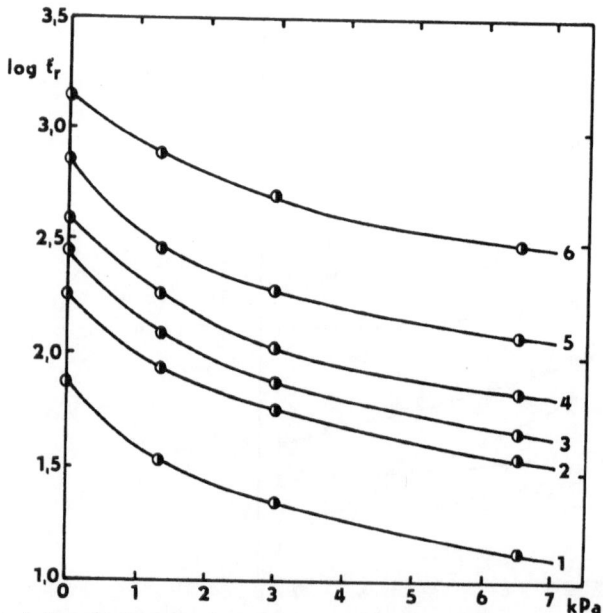

Fig. 4. A log t'_R dependence on the water vapour pressure for aromatic hydrocarbons on β-CD. (◐ – second type). 1 – benzene, 2 – 1,3,5-trimethylbenzene, 3 – ethylbenzene, 4 – 1,2,3-trimethylbenzene, 5 – 1,2,4-trimethylbenzene, 5 – tert-butylbenzene.

and 1-pentanol, whereas lower alcohols obeyed the second type. This finding is in agreement with the published data [19, 20] from which it follows that normal alcohols form progressively more stable complexes with increasing number of carbon atoms in the molecule (K_f for 1-pentanol is 210 M^{-1} and for 1-propanol only 20 M^{-1}).

The influence on the retention times of the test sorbates from water vapour cannot be explained in terms of the same concepts for β-CD. As follows from Table II, increasing water vapour pressure always leads to a decrease in the retention time, which, however, follows a different course from that obtained with α-CD. For positional isomers (xylenes, trimethylbenzenes) the elution order remains typical of interaction with CD over the whole range of the water vapour pressure. However, all the dependences (Figure 4) correspond to the second type, even for the substances with pronounced selectivity toward β-CD (aromatics) or for compounds with voluminous molecules, larger than the β-CD cavity.

3.2. THE EFFECT OF WATER VAPOUR ON THE SEPARATION PROPERTIES OF CD

A general phenomenon in the use of a carrier gas saturated with water vapour is shortening of the retention times of the sorbates and improvement in the shape of the elution curves. The shortening of the retention times of some little retained sorbates may adversely affect their separation. This is especially pronounced with substances that analogously interact with the CD cavity (benzene, toluene, cyclohexane, methylcyclohexane). A positive effect of water vapour can be observed in separations of

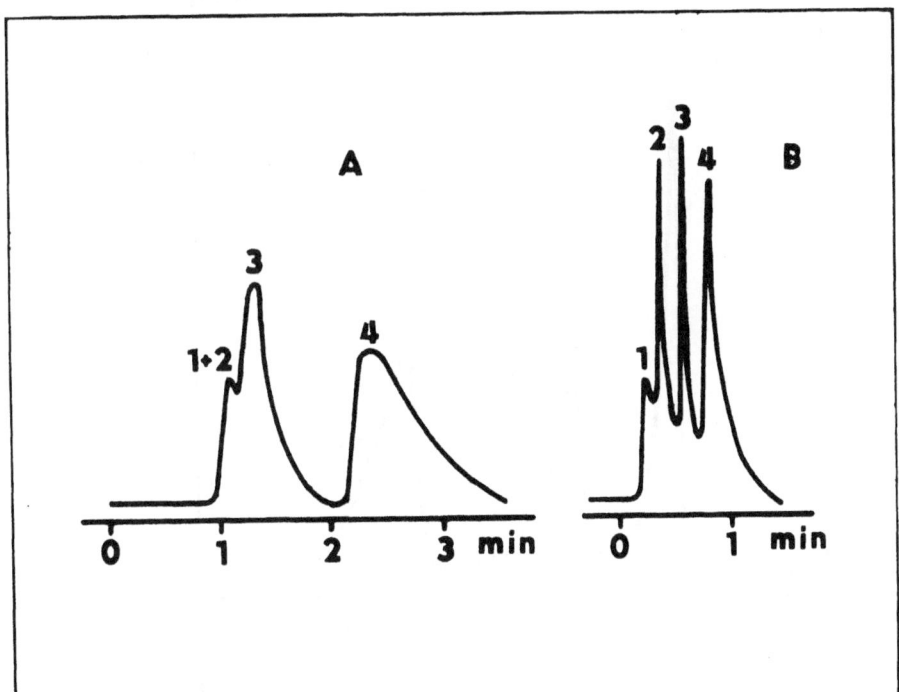

Fig. 5. Separation of alcohols on α-CD. Carrier gas flow rate, 60 ml/min. Column temperature, 80 °C. Fig. A – dry carrier gas (N_2); Fig. B – carrier gas with water vapour ($N_2 + H_2O$, $p_{H_2O} = 2.97$ kPa). 1 – methanol, 2 – ethanol, 3 – 2-propanol, 4 – 1-propanol.

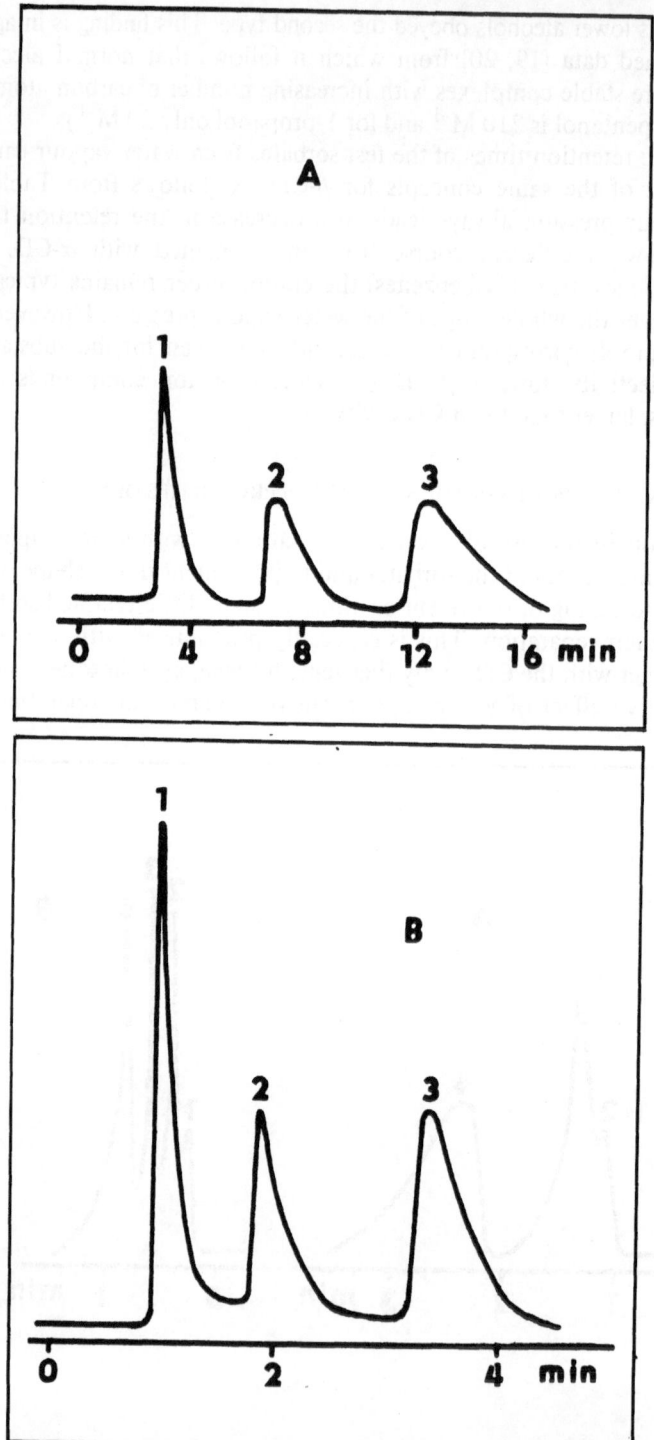

Fig. 6. Separation of trimethylbenzenes on β-CD. Carrier gas flow rate, 60 ml/min. Column temperature, 80 °C. Fig. A – dry carrier gas (N_2); Fig. B – carrier gas with water vapour ($N_2 + H_2O$), $p_{H_2O} = 2.97$ kPa. 1 – 1,3,5-trimethylbenzene, 2 – 1,2,3-trimethylbenzene, 3 – 1,2,4-trimethylbenzene.

substances that are strongly retained and thus produce broad, tailing elution curves (heptane, p-xylene, 1,2,4-trimethylbenzene).

The effect of water vapour in the carrier gas leading to shortening of the analysis and improvement in the separation can be demonstrated on a separation of lower alcohols on α-CD, depicted in Figure 5. It can be seen from the figure that saturation of the carrier gas with water vapour at a pressure of 2.97 kPa leads to a three-fold shortening of the analysis time and to separation of the two initial members of the homologous series.

With β-CD, advantages of saturation of the carrier gas with water vapour can be demonstrated on a separation of trimethylbenzenes (Figure 6). The three isomers are well separated even in the dry carrier gas, but the presence of water vapour shortens the analysis almost three-fold, while the separation efficiency is preserved.

4. Conclusion

The presence of water vapour in an inert carrier gas always affects the sorbate–cyclodextrin interaction. In contrast to analogous studies with sorbents such as alumina, silica gel and porous polymers, when the water vapour content in the carrier gas is 10% or more, a carrier gas (N_2) was used containing 1–5% water vapour. The general effect is a shortening of the retention times which occurs with CDs at water vapour contents as low as 0.5% and increases with increasing water vapour content. This effect is especially pronounced for sorbates whose structures correspond to the dimensions of the α-CD cavity. Analogous conclusions cannot be drawn for β-CD. A clarification of the contribution from the water vapour interaction with the cyclodextrin surface and the effect on its adsorption properties with respect to various types of sorbate is being studied further.

References

1. E. Smolková-Keulemansová: *J. Chromatogr.* **251**, 17 (1982).
2. E. Smolková, H. Králová, S. Krýsl, and L. Feltl: *J. Chromatogr.* **241**, 3 (1982).
3. M. Tanaka, S. Kanono, and T. Shono: *Fresenius Z. Anal. Chem.* **316**, 53 (1983).
4. Y. Mizobuchi, M. Tanaka, and T. Shono: *J. Chromatogr.* **194**, 153 (1980).
5. Y. Mizobuchi, M. Tanaka, and T. Shono: *J. Chromatogr.* **208**, 35 (1981).
6. J. Mráz, L. Feltl, and E. Smolková-Keulemansová: *J. Chromatogr.* **286**, 17 (1984).
7. E. Smolková-Keulemansová, L. Feltl, and S. Krýsl: *J. Incl. Phenom.* **3**, 183 (1985).
8. T. Kościelski, D. Sybilska, L. Feltl, and E. Smolková-Keulemansová: *J. Chromatogr.* **286**, 23 (1984).
9. D. Sybilska and T. Kościelski: *J. Chromatogr.* **261**, 357 (1983).
10. T. Kościelski, D. Sybilska, and J. Jurczak: *J. Chromatogr.* **280**, 131 (1983).
11. T. Kościelski and D. Sybilska: *J. Chromatogr.* **349**, 3 (1985).
12. S. Krýsl and E. Smolková-Keulemansová: 'A.J.P. Martin Honorary Symposium', Urbino (Italy) May 1985. *J. Chromatogr.* **349**, 167 (1985).
13. E. Smolková-Keulemansová, E. Neumannová, and L. Feltl: *J. Chromatogr.* **365**, 279 (1986).
14. E. Smolková-Keulemansová and L. Soják: *J. Am. Chem. Soc.* In press.
15. J. Szejtli: *Cyclodextrins and their Inclusion Complexes*, Akadémia Kiaďo, Budapest, 1982.
16. Ch. Dumazert and C. Ghiglione: *Bull. Soc. Chim. Fr.* **10**, 1770 (1960).
17. M. S. Vigdergaus, A. Gorusov, V. A. Ezraj, and V. I. Semkin: *Gazovaja Chromatografija s neidealnym eluentom*: Nauka, Moscow, 1980.
18. C. L. Guillemin, J. L. Millet, and E. Hauson: *J. Chromatogr.* **301**, 11 (1984).
19. M. Maeda and S. Tagaki: *Netsusokutei* **10**, 103 (1983).
20. R. Rymdén, J. Carlfors, and P. Stilbs: *J. Incl. Phenom.* **1**, 159 (1983).

The Inclusion Complex of Piromidic Acid with Dimethyl-β-cyclodextrin in Aqueous Solution and in the Solid State

NEVIN ÇELEBI*
Department of Pharmaceutical Technology, Faculty of Pharmacy, Gazi University, Etiler, Ankara, Turkey

OSAMU SHIRAKURA, YOSHIHARU MACHIDA, and TSUNEJI NAGAI
Department of Pharmaceutics, Faculty of Pharmaceutical Sciences, Hoshi University, Ebara, 2-4-41, Shinagawa-ku, Tokyo 142, Japan

(Received: 25 September 1986; in final form: 13 January 1987)

Abstract. Inclusion complex formation of piromidic acid (PA) with dimethyl-β-cyclodextrin (DM-β-CD) in aqueous solution and in the solid state was confirmed by the solubility method, differential scanning calorimetry (DSC) and proton nuclear magnetic resonance (^{1}H-NMR) spectroscopy. The apparent stability constant, K_c, of the complex was estimated to be 244 M^{-1}. The stoichiometry of the complex was given as the ratio 1:2 of PA to DM-β-CD. The dissolution rate of the PA/DM-β-CD complex was much greater than that of intact PA.

Key words: Piromidic acid/DM-β-CD inclusion complex, phase-solubility, DSC, ^{1}H-NMR, dissolution profile.

1. Introduction

The α-, β-, and γ-cyclodextrins are cyclic oligosaccharides consisting of six, seven, and eight glucose units respectively. Although they have found many applications [1] they have some undesirable properties. The limited application of cyclodextrins in the pharmaceutical field seems to be related to their relatively low aqueous solubility [2]. Recently, the chemically modified cyclodextrins have received considerable attention because their pharmaceutical properties and inclusion behaviors are different from those of natural cyclodextrins. For example, the methylated β-cyclodextrins, such as heptakis-(2,6-di-O-methyl)-β-cyclodextrin (DM-β-CD) and heptakis-(2,3,6-tri-O-methyl)-β-cyclodextrin (TM-β-CD), are extremely soluble in water (more than 30 w/v % at 25 °C), and they interact with a variety of drug molecules [3].

Piromidic acid (PA), 5,8-dihydro-8-ethyl-5-pyrrolidopyrido(2,3-d)pyrimidine-6-carboxylic acid is a pyrido pyrimidine derivative, a congener of nalidixic acid. PA is widely used for its antibacterial activity against several gram negative pathogens. PA is very slightly water soluble [4, 5].

Cyclodextrin complexation has been extensively applied to enhance the solubility, dissolution rate, membrane permeability and bioavailability of slightly soluble drugs [6–9].

---

* Author for correspondence.

Presented at the Fourth International Symposium on Inclusion Phenomena and the Third International Symposium on Cyclodextrins, Lancaster, U.K., 20–25 July 1986.

Thus, this investigation was carried out with the aim of improving the dissolution characteristics of piromidic acid.

2. Experimental

2.1. MATERIALS

Piromidic acid and the cyclodextrins (α, β, γ, DM-β-CD) were supplied from Dainippon Pharmaceutical Co., and Nippon Shokuhin Kako Ltd., respectively. All other chemicals and solvents were of analytical reagent grade.

2.2. METHODS

2.2.1. *Solubility Studies*

The phase solubility diagram was obtained according to the method described by Higuchi, *et al.*, [10]. Excess amounts of PA were added to buffer solutions at pH 7.4 (0.02 M Na-acetate) containing various concentrations of DM-β-CD in a measuring flask. These were shaken at 37 ± 0.5 °C for 4 days. Following equilibration, the contents of the measuring flask were filtered through a Toyo TM-2 membrane filter (0.45 µm). A portion of the sample was adequately diluted and analyzed using a Hitachi 323 spectrophotometer at 273 nm to determine the concentration of piromidic acid.

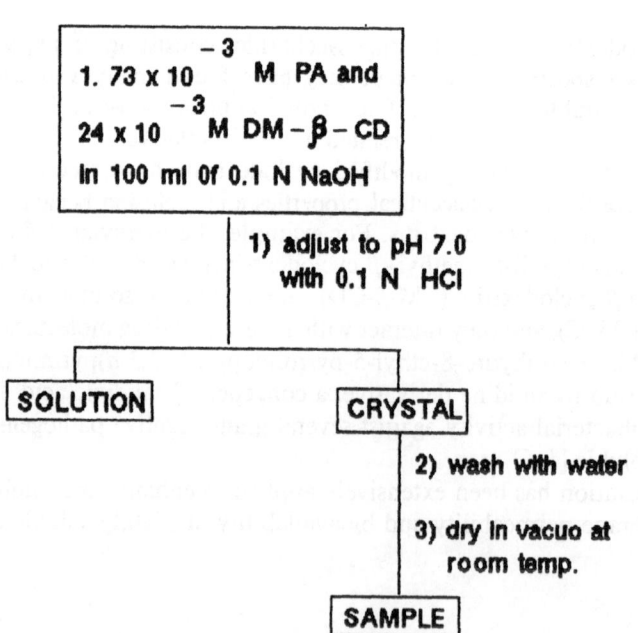

Chart 1. Method for mass preparation of inclusion complex.

2.2.2. Preparation of Inclusion Complex

The preparative method is shown in Chart 1. The stoichiometry of the complex prepared by the neutralization method was then analyzed and was found to be 1:2 (PA:DM-β-CD).

The physical mixture of PA with DM-β-CD in a 1:2 molar ratio was prepared by simple blending in a ceramic mortar (10 min).

2.2.3. Differential Scanning Calorimetry Study

The differential scanning calorimetry was carried out using a Perkin-Elmer model 1B differential scanning calorimeter at a scanning speed of 8 °C/min over the temperature range of 400 to 600 K.

2.2.4. ^{1}H-NMR Spectroscopy Study

The ^{1}H-NMR spectra were recorded in deuterium oxide and sodium hydroxide-d_1 solution (about 40% NaOD in D_2O) using a JNM-FX 100 spectrometer operating at ^{1}H–99.65 MHz. The ^{1}H-chemical shifts are given relative to external tetramethysilane within ±0.002 ppm.

2.2.5. Dissolution Studies

Dissolution rates of piromidic acid from the inclusion complex and the physical mixture were measured by the method of Nogami et al. [11]. The powdered sample (150 mesh) of the drug or its equivalent amount of the complex was put into 50 ml of buffer solutions at pH 7.4 in a dissolution cell which was kept at 37±0.5 °C and the dissolution medium was stirred. At appropriate intervals, 1 ml samples were filtered through a membrane filter (0.45 µm), diluted and assayed spectrophotometrically.

3. Results and Discussion

3.1. PHASE SOLUBILITY DIAGRAM

The formation of the complex between PA and DM-β-CD was studied by a solubility method. Figure 1 shows an equilibrium phase solubility diagram obtained for the PA/DM-β-CD system in buffer solution at pH 7.4. The plot shows a typical B_s-type solubility curve. The initial rising portion is followed by a plateau region and then decreases in total concentration of PA with precipitation of a microcrystalline complex at a high DM-β-CD concentration. The stoichiometry of the complex was found to be 1:2 (PA:DM-β-CD) from the PA content.

The apparent stability constant, K_c, of the DM-β-CD complex was calculated to be 244 M^{-1} from the initial straight line portion of the solubility diagram (Figure 1).

3.2. EVIDENCE OF INCLUSION COMPLEX FORMATIONS

As shown in Figure 2 the thermograms of PA and its physical mixture with DM-β-CD show an endothermic peak at around 316 °C corresponding to the melting

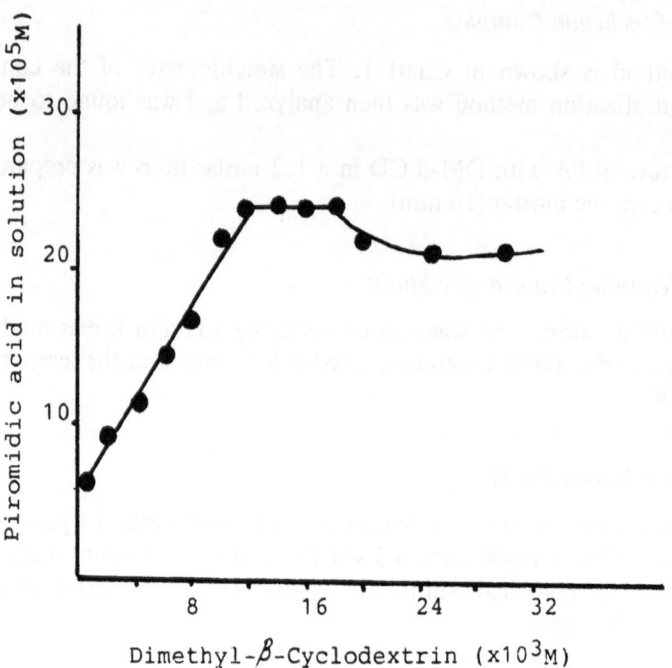

Fig. 1. Phase solubility diagram of piromidic acid/DM-β-CD system in buffer solution (0.02 M Na-acetate) pH 7.4 at 37 °C.

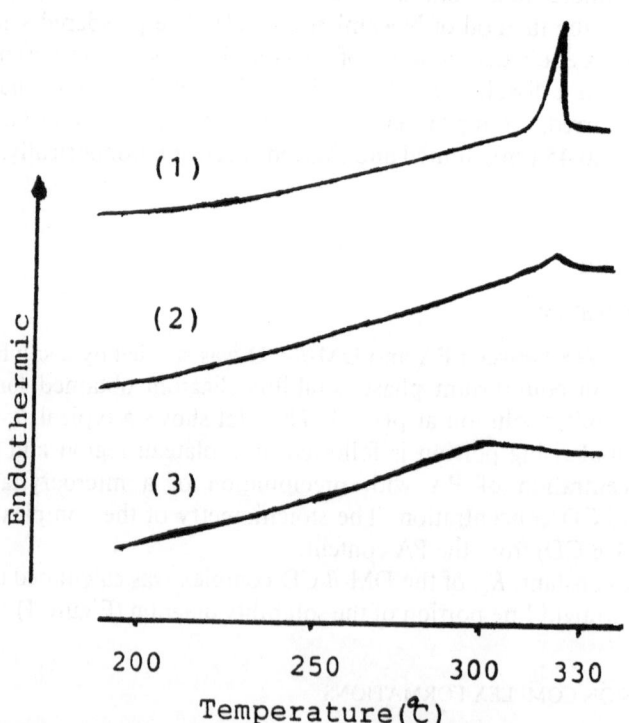

Fig. 2. Differential scanning calorimetry curves at a scanning speed of 8 °C/min. 1, Intact PA; 2, Physical mixture of PA and DM-β-CD; 3, PA/DM-β-CD inclusion complex (1:2 molar ratio).

THE DIMETHYL-β-CYCLODEXTRIN/PIROMIDIC ACID COMPLEX

point of the drug. In contrast, the endothermic peak was not observed in the case of the assumed inclusion complex.

Figure 3 shows the effect of DM-β-CD on the ^{1}H-NMR spectrum of PA in D_2O. All of the proton signals of PA shifted to lower field with increasing amounts of DM-β-CD. Similar chemical shift changes have been observed for other drug-CD systems [12, 13, 14]. The chemical shifts of PA to low field might be induced by the diamagnetic anisotropy of particular bonds of DM-β-CD and Van der Waals shifts [15].

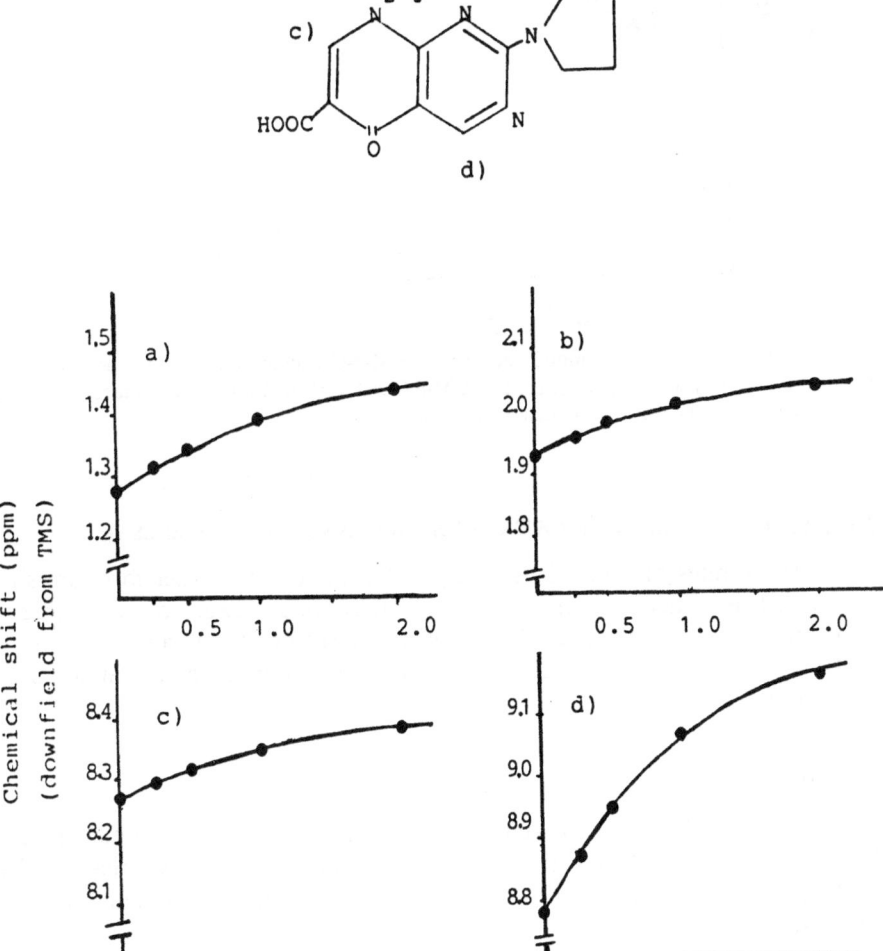

Fig. 3. Variation of ^{1}H chemical shifts of 0.01 PA with concentration of DM-β-CD in 0.1 N NaOD at room temperature.

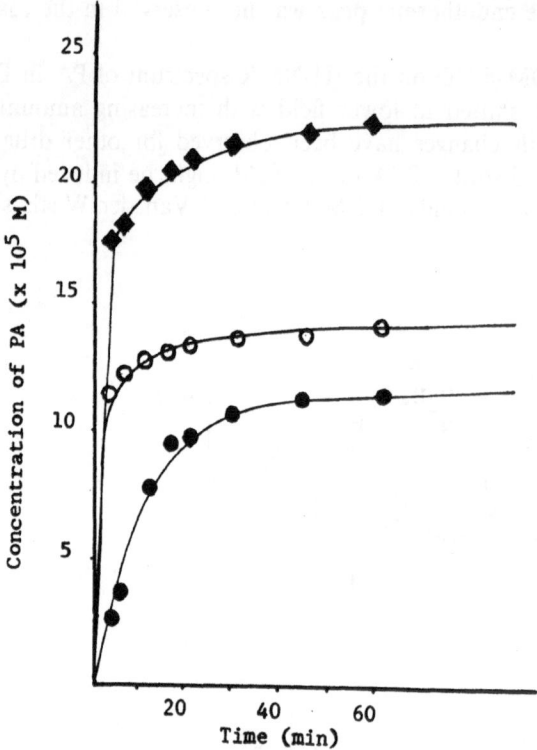

Fig. 4. Dissolution profiles of piromidic acid and its DM-β-CD complex in pH 7.4 buffer solution at 37 °C. ●, Intact PA; ○, Physical mixture of PA and DM-β-CD; ◊, PA/DM-β-CD inclusion complex (1:2 molar ratio). Each point is the mean of five determinations.

3.3. DISSOLUTION BEHAVIOR OF PIROMIDIC ACID AND ITS COMPLEXES

The relative rates of PA and its DM-β-CD complex in powder form are shown in Figure 4. It is evident that the PA/DM-β-CD complex dissolved much more rapidly than intact PA. The enhancement of the dissolution characteristics of PA by DM-β-CD inclusion complexation may be due to improvements in solubility and wettability.

4. Conclusion

The above results suggest that complex formation with DM-β-CD may enhance the bioavailability of PA. Thus, the increased dissolution rate suggests that the PA/DM-β-CD complex may have utility in the development of fast dissolving dosage forms with improved bioavailability.

References

1. W. Saenger: *Angew. Chem. Int. Ed. Engl.* **19**, 344 (1980).
2. J. Szejtli: *The Cyclodextrins and their Inclusion Complexes*, pp. 74, Akademiai Kiado, Budapest (1982).

3. K. Uekama: *Pharm. Int.* **6**, 61 (1985).
4. S. Minami, T. Shono, and J. Matsumoto: *Chem. Pharm. Bull.* **19**, 1426 (1971).
5. S. Minami, T. Shono, and J. Matsumoto: *Chem. Pharm. Bull.* **19**, 1482 (1971).
6. Y. Hamada, N. Nambu, and T. Nagai: *Chem. Pharm. Bull.* **23**, 1205 (1975).
7. K. Uekama, F. Hirayama, K. Esaki, and M. Inove: *Chem. Pharm. Bull.* **27**, 76 (1979).
8. M. Nakano, K. Juni, and T. Arita: *J. Pharm. Sci.* **65**, 709 (1976).
9. K. H. Froemming and I. Weyermann: *Arzneim. Forsch.* **23**, 424 (1973).
10. T. Higuchi and K. A. Connors: *Adv. Anal. Chem. Instrum.* **4**, 117 (1965).
11. H. Nogami, T. Nagai, and T. Yotsuyanagi: *Chem. Pharm. Bull.* **17**, 499 (1969).
12. W. G. Craig, P. V. Demarco, D. W. Mathieson, L. Sounders, and W. B. Whalley: *Tetrahedron* **23**, 2339 (1967).
13. H. Ueda and T. Nagai: *Chem. Pharm. Bull.* **28**, 1415 (1980).
14. M. Otagiri, K. Uekama, and K. Ikeda: *Chem. Pharm. Bull.* **23**, 188 (1975).
15. B. Howard, B. Linder, and M. T. Emerson: *J. Chem. Phys.* **3**, 485 (1962).

Improvement of Fat Digestion in Rats by Dimethyl-β-cyclodextrin

ANDREA GERLÓCZY*, LAJOS SZENTE, and JÓZSEF SZEJTLI
Biochemical Research Laboratory, Chinoin Pharmaceutical and Chemical Works, Endrődi Sándor u. 38/40, H-1026 Budapest, Hungary

ANNA FÓNAGY
Frédéric Joliot-Curie National Research Institute for Radiobiology and Radiohygiene, H-1775 Budapest, Hungary

(Received: 2 October 1986; in final form: 24 November 1986)

Abstract. Heptakis-(2,6-di-O-methyl)-β-cyclodextrin (DIMEB), a compound having a great water-solubility enhancing effect on lipids via inclusion complex formation was investigated as a potential bile-substituting agent *in vivo* in rats. The normal fat digestion was inhibited by ligating the bile duct. 3-H-Stearic acid or edible oil were administered orally to rats and the effect of simultaneously administered DIMEB on the lipid absorption was studied by measuring the blood radioactivity level or plasma triglyceride and free fatty acid concentrations. The lipid absorption was significantly improved by DIMEB. Accordingly, it seems to be a new fat digestion and absorption enhancing drug, i.e. a possible bile-substituting agent.

Key words: Dimethyl-β-cyclodextrin, inclusion complex, bile substituent, lipid absorption enhancement.

1. Introduction

A precondition for the digestion of fats — both by humans and animals — is the production and excretion of bile. Natural bile is a very complicated mixture, the main ingredients are the various cholic acid derivatives. The nutritional fats can be absorbed only when they are emulsified by the bile, and thereafter hydrolysed by the lipases. In some hepatic and biliary diseases either the production or the excretion of natural bile is impaired. In such cases the bile has to be substituted by drugs, which contain either cholic acid of animal origin, or other fat-emulsifying agents. According to preliminary studies, heptakis-(2,6-di-O-methyl)-β-cyclodextrin (DIMEB) may be such a fat digestion facilitating compound.

DIMEB is a compound prepared by the selective methylation of all C(2) secondary and C(6) primary hydroxyl groups of beta-cyclodextrin. DIMEB is soluble in organic solvents and it is very soluble in cold water. 25–30% solutions of increased viscosity can be readily prepared, while a syrupy 50% solution can be prepared by prolonged stirring and shaking. DIMEB has a great solubility enhancing effect on poorly soluble compounds, e.g. lipids via inclusion complex formation [1].

Oral administration of DIMEB to male and female mice resulted in no toxic symptoms up to 3000 mg/kg. Six hours after intravenous administration only traces of DIMEB can be detected in the blood, the majority of the substance is removed from the circulation within the first two hours [1].

---

* Author for correspondence.

2. Materials and Methods

2.1. MATERIALS

Heptakis-(2,6-di-O-methyl)-beta-cyclodextrin (DIMEB) (Chinoin Pharm.-Chem. Works): At least 90% of the substance corresponds to the above mentioned chemical name. The remaining 10% is a mixture of different, very closely related methylated derivatives. The heavy metal content is less than 10 ppm, the organic solvent residue is less than 100 ppm. $[\alpha]_D^{15}: 161°$ (0.1% H_2O). Sodium deoxycholate (Reanal); 9,10-3-H-stearic acid (Izocommerc), specific radioactivity: >1110 GBq/mmol (labelled stearic acid was diluted with appropriate amount of non-labelled stearic acid (Reanal) before the experiments); edible oil.

2.2. METHODS

2.2.1. *Experiments with 3-H-stearic acid*

Following a 24-hour starvation, the choledocus of CFY rats weighing 160–170 g was ligated in diethyl-ether narcosis. The absorption experiments were performed 48 hours following the operation. The animals were starved for 12 hours before the treatment and during the experimental period as well. In all of the experiments, one group of rats was administered orally 3-H-stearic acid (40 to 120 mg/animal dose), the other group was given 3-H-stearic acid (40 to 120 mg/animal dose) and DIMEB (12.5 to 50 mg/animal dose) simultaneously. The materials were suspended in 1% methylcellulose solution.

Blood samples of 50 µl were taken at predetermined time intervals from the tail vein. The samples were solubilized in 0.75 ml of 1:1 Soluene-100 (Packard Instruments):isopropanol solution (v/v) and were bleached with 0.25 ml 33% hydrogen peroxide for 30 min at room temperature. Then 10 ml of a 9:1 Insta Gel (Packard):0.5 mol/l HCl (v/v) solution was added and the samples were incubated in the dark for 24 hours at room temperature. The samples were assayed for radioactivity in a Searle Nuclear Chicago Mark III 6880 apparatus. The radioactivity (dpm) measured was related to 10 ml of blood and expressed as % of total administered radioactivity.

2.2.2. *Experiments with edible oil in intact rats*

CFY rats of 205–230 g weight were administered orally 2 ml edible oil or 2 ml edible oil + 50 mg DIMEB (dissolved in 0.5 ml distilled water). The aqueous solution of DIMEB was added to edible oil at physiological temperature, and following a 5-minute vigorous shaking it was administered immediately to the animals. Blood samples were taken at 0, 1, 2.5, 4 and 6 hours after treatment by the decapitation of rats. Plasma triglyceride and free fatty acid concentrations were determined according to Van Handel [2] and Duncombe [3], respectively.

3. Results

3.1. THE EFFECT OF DIMEB ON THE ABSORPTION OF ORALLY ADMINISTERED 3-H-STEARIC ACID IN BILE DEFICIENT RATS

The free fatty acid absorption enhancing effect of DIMEB was investigated in rats suf-

IMPROVEMENT OF FAT DIGESTION IN RATS

fering from an experimental bile deficiency. 3-H-Stearic acid or 3-H-stearic acid and DIMEB were administered orally to the animals and the blood radioactivity level was determined. The blood radioactivity values versus time are demonstrated in Figure 1.

Blood radioactivity level was considerably higher (2 to 15 fold) in DIMEB-treated groups than in those administered 3-H-stearic acid alone. The absorption enhancing effect of DIMEB was most apparent in the first two hours. On the contrary, a slow increase of blood radioactivity was observed in the case of animals treated with 3-H-stearic acid and sodium deoxycholate. In this case, a more apparent increase was observed only after the 6th hour.

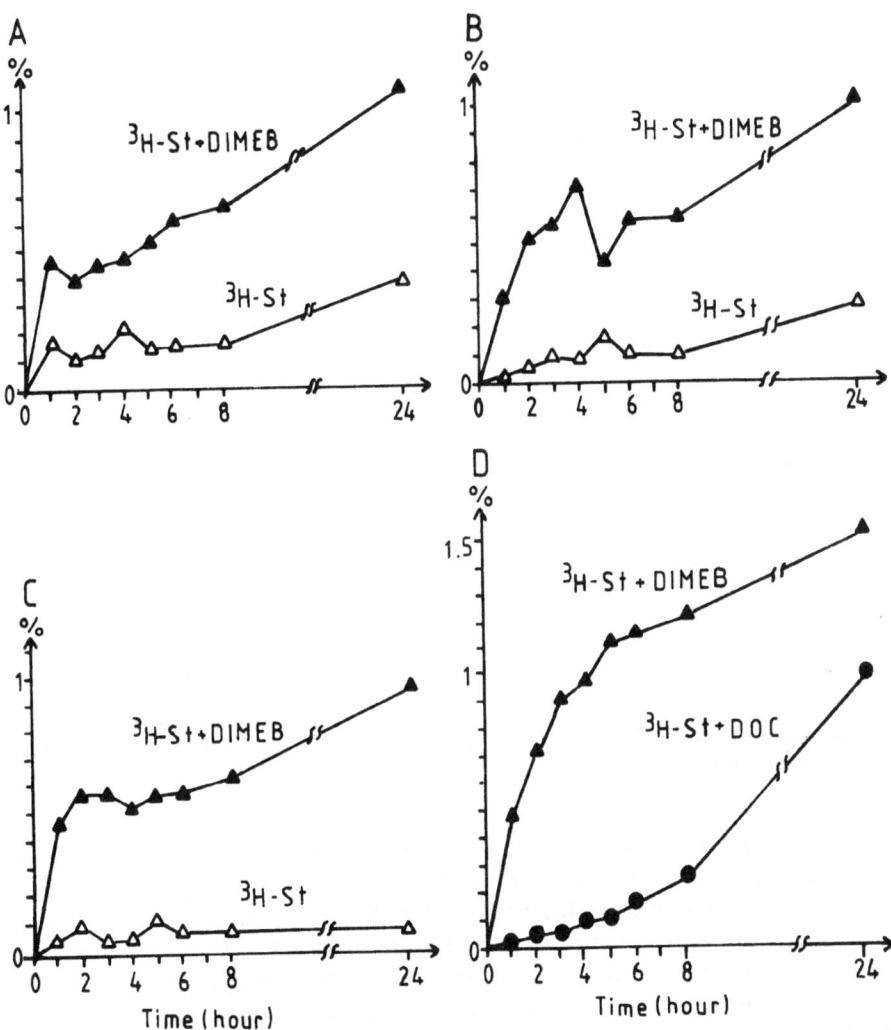

Fig. 1. Absorption of orally administered 3-H-stearic acid in the presence and absence of DIMEB in bile deficient rats. Blood radioactivity values are expressed in % of total administered radioactivity and related to 10 ml of blood. Key: (△──△)3-H-St: 3-H-stearic acid; (▲──▲) 3-H-St + DIMEB: 3-H-stearic acid + DIMEB; (●──●) 3-H-St + DOC: 3-H-stearic acid + sodium deoxycholate. Doses: A: 3-H-St: 40 mg/animal, 3-H-St + DIMEB: 40 mg + 12,5 mg/animal, resp. B: 3-H-St: 40 mg/animal, 3-H-St + DIMEB:: 40 mg + 25 mg/animal, resp. C: 3-H-St: 120 mg/animal, 3-H-St + DIMEB: 120 mg + 50 mg/animal, resp. D: 3-H-St + DIMEB: 40 mg + 25 mg/animal, resp., 3-H-St + DOC: 40 mg + 25 mg/animal, resp.

Therefore, the absorption of stearic acid was considerably enhanced due to the presence of DIMEB in bile duct-ligated rats. In all probability, the absorption enhancing mechanism of DIMEB differs from that of sodium deoxycholate.

3.2. THE EFFECT OF DIMEB ON THE ABSORPTION OF EDIBLE OIL IN INTACT RATS

Two ml edible oil was administered to intact rats orally in the presence or absence of DIMEB. The absorption process was followed by measuring the plasma triglyceride and free fatty acid concentration, as demonstrated in Figure 2.

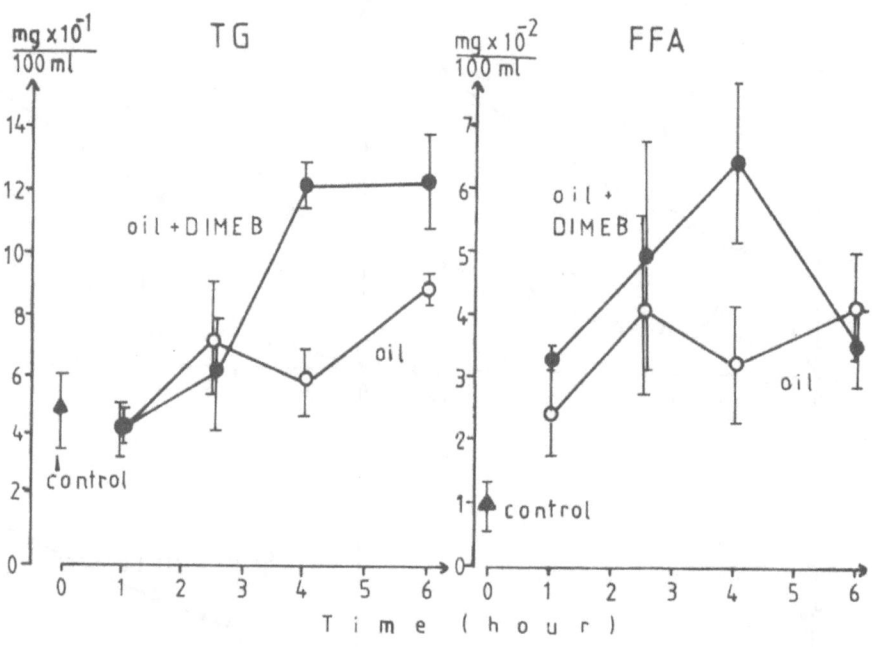

Fig. 2. Plasma triglyceride (TG) and free fatty acid (FFA) concentration of intact rats following oral administration of edible oil in the presence or absence of DIMEB. Key: (▲) control, animals were bled at the beginning of the treatment; (O——O) 2 ml edible oil/animal; (●——●) 2 ml edible oil + 50 mg DIMEB (dissolved in 0.5 ml water)/animal.

Four hours following the treatment there is no significant difference in the plasma triglyceride concentration of the control and the oil-treated group. At the same time, a significant increase can be observed in the DIMEB treated group. There is a significant difference between the oil-treated and the oil + DIMEB treated group even at the 6th hour. A similar tendency was observed by measuring the free fatty acid concentration. Therefore, DIMEB promoted the lipid absorption in healthy, intact rats.

4. Discussion

It is well known that the presence of bile is of capital importance in lipid absorption. In some hepatic and biliary diseases either the production or the excretion of bile is impaired. However, in some cases it is unadvisable to administer bile acids to the patients. Bile acids should be substituted by a compound being no bile-like regarding the chemical structure, but being capable of solubilizing lipids to a similar extent to bile.

DIMEB seems to be a suitable material for this purpose. It is poorly absorbed orally and is probably non-toxic [1]. The solubility of lipids is highly enhanced by DIMEB via inclusion complex formation [1]. According to rat experiments, DIMEB promoted the absorption of edible oil even in intact rats having a normal bile supply (Fig. 2). Plasma triglyceride and free fatty acid concentrations were significantly increased in the presence of DIMEB 4 hours following the treatment. In the case of bile-deficient, choledocus ligated rats, the absorption of a free fatty acid, stearic acid, was considerably increased (2 to 15 fold) by DIMEB mainly in the first two hours (Fig. 1). Comparing the effect of DIMEB and sodium deoxycholate in a preliminary experiment, it was found that the degree of absorption enhancement effect was similar but the mechanism of action is probably quite different (Fig. 1).

According to these results, DIMEB seems to be a possible bile-substituting agent.

Acknowledgements

Thanks are due to Dr Péter Szabó (Department of Zoology and Anthropology, Kossuth Lajos University, H-4010 Debrecen, Hungary) for his help and advice regarding the experiments on intact rats.

References

1. J. Szejtli: *J. Incl. Phenom.* **1**, 135, 1983.
2. E. Van Handel: *Clin. Chem.* **7**, 249, 1961.
3. W. G. Duncombe: *Biochem, J.* **88**, 7, 1963.

Reduction of Phytotoxicity of Nonionic Tensides by Cyclodextrins

K. BUJTÁS
Institute of Soil Science and Agricultural Chemistry of the Hungarian Academy of Sciences, Budapest, Hungary

T. CSERHÁTI
Plant Protection Institute of the Hungarian Academy of Sciences, Budapest, Hungary

and

J. SZEJTLI*
Biochemical Research Laboratory of Chinoin Pharmaceutical and Chemical Works, H-1026 Budapest, Hungary

(Received: 2 October 1986; in final form: 24 November 1986)

Abstract. The effect of nonionic tenside nonylphenylnonylglycolate and its α-, β-, γ-cyclodextrin, 2,6-di-*O*-methyl-β-cyclodextrin (DIMEB) and 2,3,6-tri-*O*-methyl-β-cyclodextrin (TRIMEB) complexes was tested on the potassium influx of wheat seedling roots. Tenside alone inhibited strongly the potassium influx. This noxious effect was alleviated by cyclodextrins. The alleviating effect increased with increasing cyclodextrin: tenside molar ratio, in the order: DIMEB > βCD > γCD > αCD ≈ TRIMEB.

Key words: Tenside, phytotoxicity reduction, cyclodextrin.

1. Introduction

Nonionic tensides are widely used for many purposes, as e.g. in pesticide formulations. The concomitant toxic phenomena are also varied. Although they are easily biodegradable they nevertheless decrease the performance of activated sludge systems [1, 2], damage pine trees by increasing the uptake of sodium ions [3, 4, 5], and by dissolving the surface wax layer of pine needles [6]. Tensides exert an inhibitory effect on the growth of *Spirodela polyrrhiza* (L.) Schleiden [7] and they are lethal to *Medaka Oryzias latipes* [8].

Their biological activity is based on their interaction with membrane phospholipids [8, 9] resulting in increased excretion of glutamate and in enhanced uptake of various substances through stomata. Besides the membrane damaging effect tensides influence markedly the activity of enzymes [10] and enhance the biological activity of insecticides [11] by improving their solubility.

Because cyclodextrins form inclusion complexes with tensides [12, 13, 14, 15] it was expected that they could modify their biological activity. The prognosticated use of cyclodextrins in pesticides formulations [16, 17] eventually together with nonionic tensides, rendered necessary the study of the damaging effect of some nonionic

---

* Author for correspondence.

tensides on the membrane function (K⁺ influx) in wheat seedling roots and to assess the preventative effect of cyclodextrins.

2. Material and Methods

Nonylphenylethyleneoxide polymers containing on average 4, 9 and 30 ethyleneoxide groups per molecule (further T_4, T_9 and T_{30}) were purchased from Hoechst (FRG) and were used without further purification. The cyclodextrins: alpha-cyclodextrin (αCD), beta-cyclodextrin (βCD), gamma-cyclodextrin (γCD), 2,6-di-O-methyl-beta-cyclodextrin (DIMEB) and 2,3,6-tri-O-methyl-beta-cyclodextrin (TRIMEB) were produced by Chinoin Pharmaceutical Works (Hungary) and were chromatographically pure.

Excised roots of 4-day old seedlings of winter wheat (*Triticum aestivum* L.cv. GK Szeged), grown in darkness at ambient temperature, under low salt concentration (0.5 mM $CaSO_4$) were used for the potassium influx experiments. Fragments of roots (0.8 g) were allowed to absorb potassium from 100 cm³ of 0.1 mM KCl (^{86}Rb) + 0.5 mM $CaSO_4$ + 50 mg/dm³ tenside for 2 hours then the non-absorbed ions were removed by rinsing the roots with 100 cm³ of non-labelled 0.1 mM KCl + 0.5 mM $CaSO_4$ solution. The radioactivity of roots was measured directly in a scintillation counter and influx values (μM potassium per g fresh weight/2 hours) were calculated. Cyclodextrins were added to the most active T_9 tenside in the molar ratio 5:1, 1:1, 0.5:1, 0.2:1 and 0.1:1 cyclodextrin:tenside. Due to its limited solubility in water the effect of βCD was not tested at the highest molar ratio. The activity of T_9 was checked also

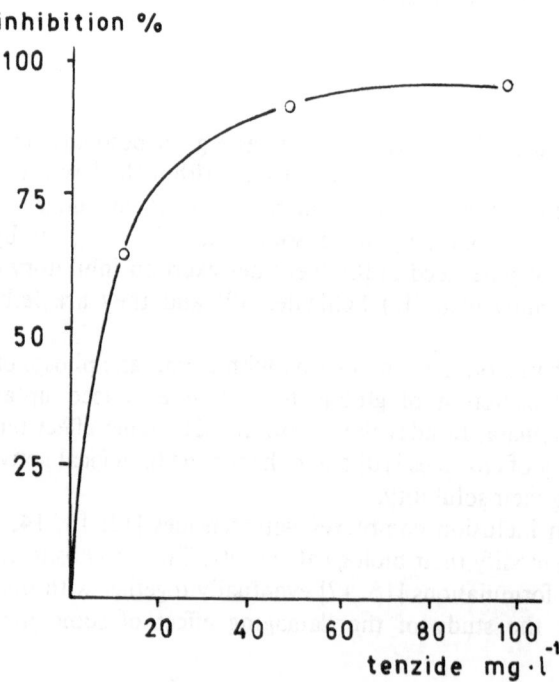

Fig. 1. Effect of nonylphenyl-nonylglycolate on the potassium influx of wheat seedling roots.

at 10 and 100 mg/dm^3 concentrations. The % decrease of potassium influx was considered as the measure of biological activity.

Each experiment was made in triplicate, corresponding to three separate experiments.

3. Results and Discussion

The dependence of the inhibition of K$^+$ influx on the concentration of T$_9$ is shown in Figure 1. The tenside inhibits the K$^+$ influx very strongly even at 10 mg/dm^3 concentration, the inhibition is nearly complete at 50 ppm. The probable explanation is that the tenside damages the membrane structure around the ion channels disrupting them partially or completely. The fact that this effect was the highest for T$_9$ (94,4% inhibition) and lower for T$_4$ (70,7%) and T$_{30}$ (25,0%) indicates that the noxious effect of tensides depends considerably on the length of the hydrophilic ethyleneoxide chain of the tensides. These observations are in good agreement with other results [18, 19, 20].

The alleviating effect of cyclodextrins is summarized in Figures 2 and 3. Each

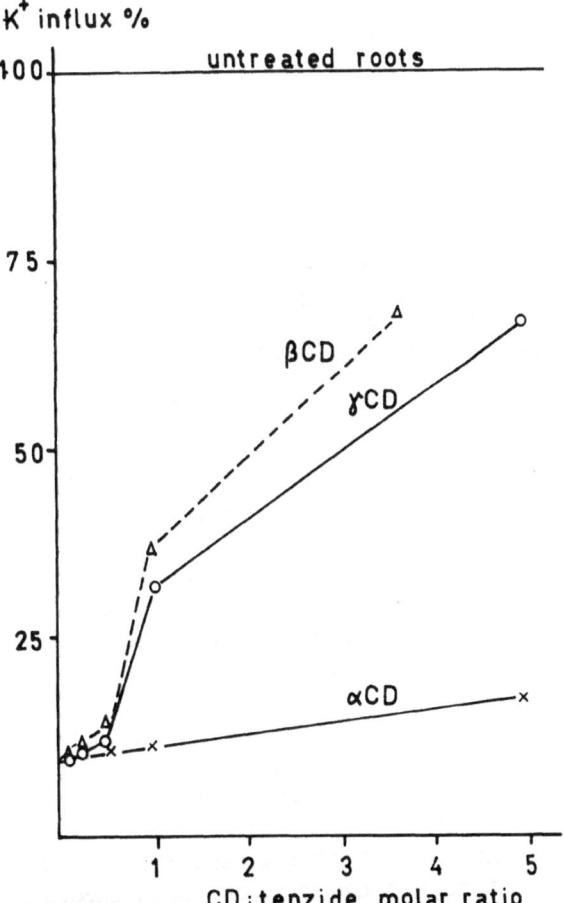

Fig. 2. Effect of nonylphenyl-nonylglycolate and its α-, β-, and γ- cyclodextrin complexes on the inhibition of K$^+$ influx of wheat seedling roots.

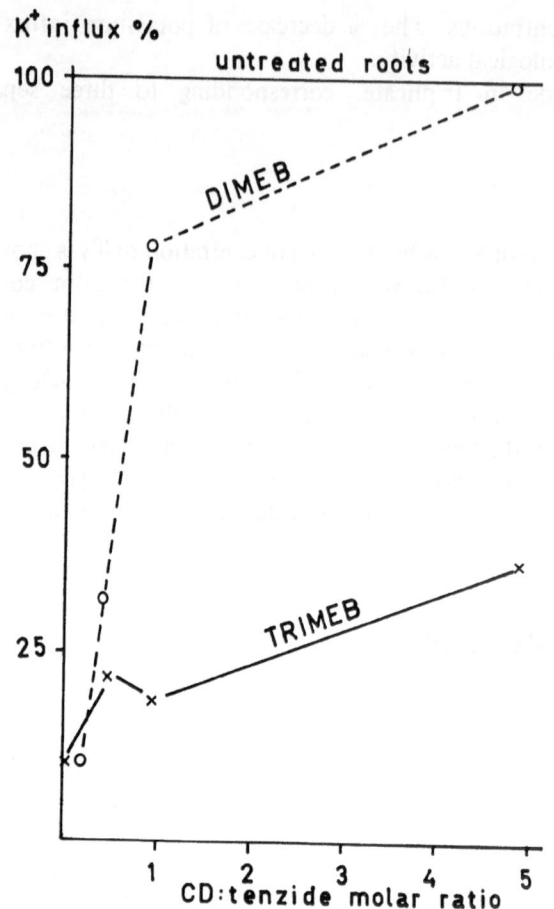

Fig. 3. Effect of nonylphenyl-nonylglycolate and its DIMEB and TRIMEB complexes on the inhibition of K⁺ influx of wheat seedling roots.

tenside reduces the inhibitory effect of T_9. This result is probably due to the fact that the cyclodextrins form inclusion complexes with the tensides, lowering the concentration of free tenside molecules responsible for the biological effect. This observation also indicates that the tenside-cyclodextrin complexes exert a negligible effect on the potassium influx of wheat seedling roots. Nonionic tensides also form complexes with membrane phospholipids [21], however our data suggest that the interactive forces between cyclodextrins and T_9 are higher than those between membrane phospholipids and T_9.

The alleviating effect of cyclodextrins differs considerably from each other. Among the nonmethylated derivatives βCD showed the highest, αCD the lowest preventive effect, the difference between the activity of βCD and γCD was lower than that between γCD and αCD (Figure 1). It means that T_9 forms inclusion complexes of commensurable stability with βCD and γCD but only a less stable one with the αCD. The cavity of αCD is probably not large enough to accommodate adequately the nonylphenyl hydrophobic moiety of T_9.

As the interaction between the nonmethylated cyclodextrins and phospholipids is rather weak and does not influence significantly the membrane permeability [22], the above data reflect entirely the effect of inclusion complex formation on the biological activity of T_9.

The alleviating effects of methylated beta-cyclodextrins deviate markedly from that of unmethylated βCD (Figure 3), the effect of DIMEB is higher, that of TRIMEB is lower than the effect of βCD. It suggests that their complex forming capacities are different. Methylation decreases the accessibility of the cyclodextrin cavity for T_9, moreover the increased lipophilicity of methylated cyclodextrins [23] favours the micelle formation like hydrophobic-hydrophobic interactions with T_9. In the case of DIMEB the favourable effect of hydrophobic-hydrophobic interactions predominates, the association with T_9 is stronger than with βCD. In the case of TRIMEB the complex formation is weaker than with βCD probably the steric hindrance caused by the numerous methyl groups overshadows the favourable effect of the outer sphere hydrophobic interactions.

The overal effectivity order of cyclodextrin derivatives was DIMEB > βCD > γCD > αCD ≈ TRIMEB.

References

1. T. Hashinaga, M. Dazai, and M. Uehara: *Rep. of Ferment Res. Inst. Japan* **62**, 55 (1984).
2. P. A. Gilbert and R. Pettigrew: *Int. J. Cosmetic Sci.* **6**, 149 (1984).
3. R. Gellini, F. Pantani, P. G. Rossoni, F. Bussotti, E. Barbolani, and C. Rinallo: *Eur. J. Forest Path.* **13**, 296 (1983).
4. R. Truman and M. J. Lambert: *Aust. J. Plant. Physiol.* **5**, 337 (1978).
5. H. G. M. Dowden, M. J. Lambert, and T. Truman: *Aust. J. Plant. Physiol.* **5**, 387 (1978).
6. R. Gellini, F. Pantani, P. Grossoni, F. Bussotti, E. Barbolani, and C. Rinallo: *Eur. J. Forest Path.* **15**, 145 (1985).
7. J. Buczek.: *Acta Societatis Bot. Poloniae* **53**, 551 (1984).
8. M. Kikuchi and M. Wakabayashi: *Bull. Jpn. Soc. Sci. Fisheries* **50**, 1235 (1984).
9. A. Helenius and K. Simons: *Biochim. Biophys. Acta* **415**, 29 (1975).
10. H. Nakahara, S. Okada, H. Ohmori and M. Masui: *Chem. Pharm. Bull.* **32**, 3803 (1984).
11. J. N. Mkhize and A. P. Gupta: *Insect. Sci. Applic.* **6**, 183 (1985).
12. J. Szejtli: *Cyclodextrins and their Inclusion Complexes*, Akadémiai Kiadó, Budapest, 1982.
13. J. Koch: in *Proc. of the First Int. Symp. on Cyclodextrins* (Ed.: J. Szejtli), Akadémiai Kiadó, Budapest, and D. Reidel Publ. Co. 1982, p. 487.
14. K. Králová and L. Mitterhauszová: in *Proc. of the First Int. Symp. on Cyclodextrins* (Ed.: J. Szejtli), Akadémiai Kiadó, Budapest, and D. Reidel Publ. Co. 1982, p. 217.
15. K. Králová, L. Mitterhauszová, and J. Szejtli: *Tenside Detergents* **20**, 37 (1983).
16. J. Szejtli: *Inclusion Compounds*, Vol. III. (eds.: J. L. Atwood, J. E. D. Davies, and D. D. MacNicol), Academic Press, London, 1984.
17. J. Szejtli: *Starch*, **37**, 382 (1985).
18. T. Cserháti, M. Szőgyi, B. Bordás, and A. Dobrovolszky: *Quant. Struct. Act. Relat.* **3**, 56 (1984).
19. T. Cserháti, M. Szőgyi, and B. Bordás: *Gen. Physiol Biophys.* **1**, 225 (1982).
20. M. Szőgyi, F. Tölgyesi, and T. Cserháti: in *Physical Chemistry of Transmembrane Ion Motions*. (Ed.: G. Spach), Elsevier, Amsterdam, 1983, p. 29.
21. T. Cserháti, M. Szőgyi, and L. Győrfi: *J. Chromatogr.* **349**, 295 (1985).
22. J. Szejtli, T. Cserháti, and M. Szőgyi: *Carbohydr. Polym.* **6**, 35 (1986).
23. T. Cserháti, L. Szente, and J. Szejtli: *J. High Res. Chromatogr. Chromatogr. Commun.* **7**, 635 (1984).

Water Soluble Cyclodextrin Polymers: Their Interaction with Drugs

J. SZEMÁN, E. FENYVESI, J. SZEJTLI*
Chinoin Pharmaceutical and Chemical Works Ltd., Budapest, Endrődi S- u. 38/40, Hungary, H-1026

and

H. UEDA, Y. MACHIDA, T. NAGAI
Faculty of Pharmaceutical Sciences, Hoshi University Ebara 2-4-41, Shinagawa-ku, Tokyo 142, Japan

(Received: 2 October 1986; in final form: 13 January 1987)

Abstract. The complex forming ability of a water-soluble β-cyclodextrin epichlorohydrin polymer (CDPS) and its different molecular weight fractions was studied and compared with the complexing properties of β-cyclodextrin (βCD) and dimethyl-βCD (DM-βCD). CDPS was separated into two main fractions. CDPS and its fractions formed well soluble inclusion compounds with the studied drugs. The low molecular weight fraction formed rather stable complexes with small guest molecules, the high molecular weight fraction was found to be more efficient in binding larger substrates. Structural studies of furosemide-CD complexes were attempted by NMR spectroscopy.

Key words: Cyclodextrin polymer, solubility method, complex stability, NMR spectroscopy.

1. Introduction

The water soluble cyclodextrin polymers [1, 2, 3] form amorphous, well soluble complexes with drugs [4, 5]. Sublingual/buccal administration of such steroid complexes resulted in improved absorption [6]. These CD derivatives have no known undesirable or toxic effects, and they neither enter nor damage oral tissue [6]. The simplest way to prepare such polymers is crosslinking CD molecules with epichlorohydrin. The degree of polymerization depends on the preparative procedure [3]. The present work deals with the complex forming ability of a water-soluble β-cyclodextrin epichlorohydrin polymer [CDPS] and its different molecular weight fractions with some drug molecules. The results have been compared to the efficacy of βCD and dimethyl-βCD (DM-βCD).

2. Materials and Methods

CDPS is a pilot product of Chinoin Pharm. Chem. Works Ltd. (Hungary) [3], a white powder with the following characteristics: βCD content – 52.6%; average molecular weight – 4150. The following other materials were used, βCD (Nihon Shokuhin Kako Co., Ltd.) DM-βCD (Toshin Chemical Co., Ltd.), butylparaben (Tokyo Kasei T.C.I.) hydrocortisone (Nakarai Chemicals Ltd.), cinnarizine (Eisae Co. Ltd),

---

* Author for correspondence.

furosemide (Hoechst Japan Co. Ltd), acetohexamide (Shionogi Pharm. Co. Ltd.), NaOD and D_2O (Merck).

Chromatographic separation: Column: Ultrogel AcA 54 gel, eluent: water containing 0.02% w/v NaN_3, the separation was monitored by Sepa-200 High Sensitive Polarimeter (Horiba). The chromatogram was evaluated on the basis of the relationship between the relative elution volumes and molecular weight [3]. Two main fractions were collected, dialyzed (1-7/8 DM cellulose tubing, Union Carbide Corp.) and freeze dried. The apparent βCD content of the samples was determined by the acidic hydrolysis method [1].

The observed stability constants were calculated from the phase solubility diagrams [7, 8], which were determined in water at 25 °C. ^{1}H-NMR spectroscopy was carried out in 0.02M NaOD solution at 24.5 ± 0.5 °C using a JEOL FX-100 spectrometer. ^{1}H chemical shifts were referred to tetramethyl silane external standard.

3. Results and Discussion

CDPS could be separated into two main fractions (Figure 1). The lower molecular weight fraction, CDPS-L is a mixture of different isomers of CD glyceryl ethers, its average molecular weight is about 1600, and its apparent CD content is 51.7%. The other peak represents the real polymer fraction (CDPS-H) which consists of 4-5 or more CD rings interconnected with longer or shorter glyceryl ether chains. The average molecular weight of this fraction, CDPS-H, is more than 9000, and its apparent βCD content is 51.4%. The weight ratio of separated fractions was found to be about 1:1.

The complex forming ability of CDPS and its fractions was studied with small and large drug molecules (Table I).

Butylparaben (BPB), a small guest molecule can fit well into the βCD ring. The βCD derivatives, DM-βCD, CDPS and its fractions formed more stable complexes with BPB than the parent βCD. CDPS and its fractions resulted in relatively low

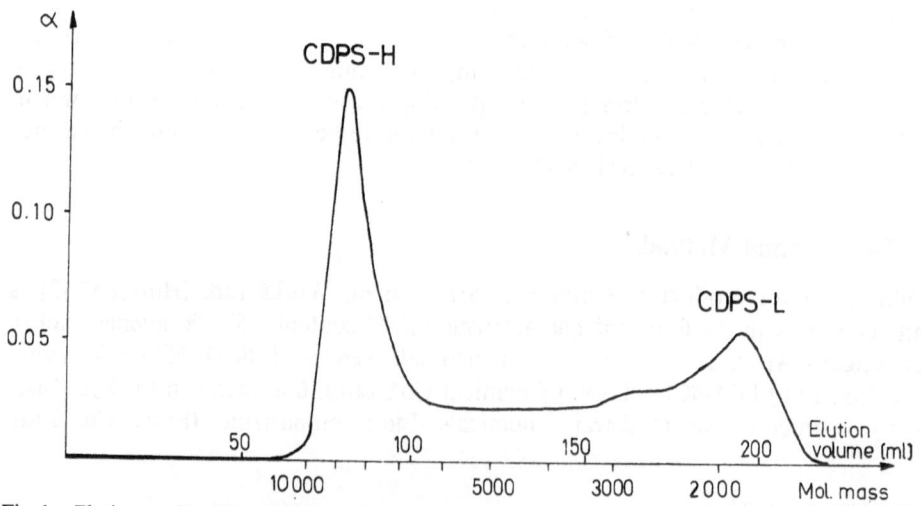

Fig. 1. Elution profile of the CDPS on Ultrogel ACA 54.

CYCLODEXTRIN POLYMER–DRUG INTERACTIONS

Table I. Apparent stability constants, type of solubility curves and increase of solubility of some drugs with βCD and CD derivatives in water at 25 °C

Drug	CD	Apparent stability constant (M^{-1}) $K'_{1:1}$	$K'_{1:2}$	Type of sol. curve	Increase of sol. c_x/c_0 [a]
Butylparaben	β-CD	2130	–	B_S	3.6
	CDPS	7260	–	A_N	90
	CDPS-H	8160	–	A_L	95
	CDPS-L	8160	–	A_L	95
	DM-β-CD	7260	–	A_N	70
Hydrocortisone	β-CD	4170	–	B_S	5.5
	CDPS	990	–	A_L	45
	CDPS-H	1250	–	A_L	50
	CDPS-L	1250	–	A_L	50
	DM-β-CD	5910	–	A_L	70
Cinnarizine	β-CD	4510	–	B_S	7.5
	CDPS	2490	3.4	A_P	300
	CDPS-H	5200	–	A_L	330
	CDPS-L	2190	85	A_P	500
	DM-β-CD	8640	13	A_P	2500
Tolnaftate	β-CD	7140	–	B_S	70
	CDPS	17000	–	A_L	3000
	CDPS-H	42000	–	A_L	4000
	CDPS-L	17000	–	A_L	3000
	DM-β-CD	17000	29.5	A_P	45000
Acetohexamide	β-CD	890	–	B_S	4.1
	CDPS	1900	–	A_L	190
	CDPS-H	890	–	A_L	90
	CDPS-L	890	–	A_L	90
	DM-β-CD	810	9.5	A_P	125
Furosemide	β-CD	62	–	B_S	8.8
	CDPS	590	–	A_L	45
	CDPS-H	330	–	A_L	32
	CDPS-L	330	–	A_L	32
	DM-β-CD	160	26	A_L	70

[a] c_0: water solubility of drug. c_x: solubility in 0.1M solution of the respective CD (molarity of CDPS relates to the actual CD content of the polymer) in βCD solutions the maximally obtained solubility.

stability constants with hydrocortisone (HC), probably because of the steric hindrance caused by the hydrophilic substituents on the βCD rings. In spite of this fact the increase of solubility is about 10 fold compared to βCD, because the CDPS and its complexes are much more soluble. In the case of cinnarizine (CN), which is also a relatively large guest molecule, CDPS and CDPS-L were less effective than βCD or DM-βCD. The observed stability constant (K') of the CN-CDPS-H system is higher than the K' of CN-βCD but lower than the K' of the CN-DM-βCD. A_p type solubility isotherms were obtained with DM-βCD, CDPS and CDPS-L because they are overlapped by a micelle formation.
The largest complex stability was observed for the high molecular weight CDPS-H

fraction and tolnaftate (TN) system, but as a means of increasing the solubility of TN, the DM-βCD was the most effective, because of its A_p type solubility curve. Acetohexamide (AH) formed complexes of similar stability with βCD and its derivatives, except the unfractionated CDPS, which resulted in the best solubilization of AH. Similar results were found with furosemide (FS). The highest complex stability was obtained with the CDPS, and both its fractions form more stable complexes with FS than the βCD or DM-βCD.

^{1}H-NMR spectra showed differences in the mode of inclusion between CDPS, βCD and DM-βCD (Figure 2.). The inclusion of the phenyl moiety was found to

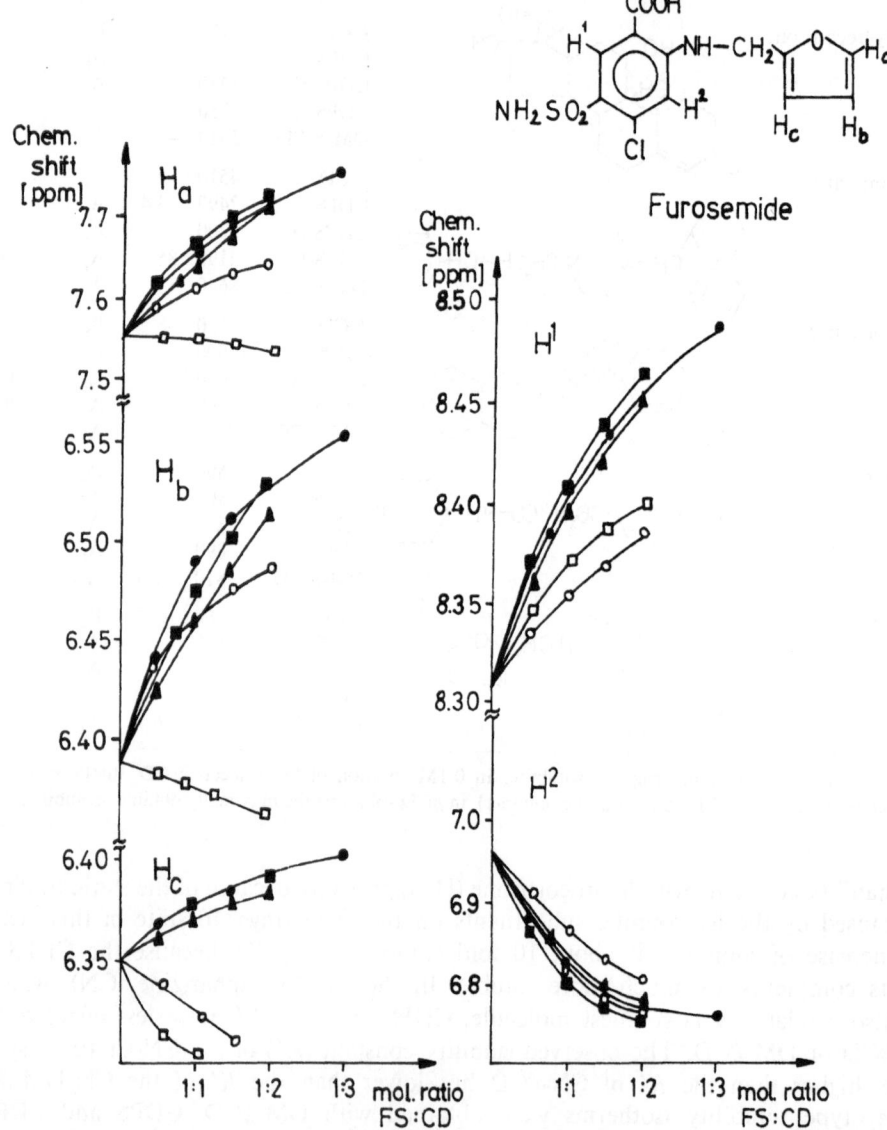

Fig. 2. Variation of chemical shifts of Furosemide with the concentration of βCD and its derivatives in NaOD solution at 24.5 ± 0.5 °C. βCD ○, DM-βCD □, CDPS ●, CDPS-H ▲, CDPS-L ■.

be similar with all CD-hosts. Both rings of furosemide were included in the CD cavity of CDPS and its fractions. Also βCD includes both rings, but the H_c proton of the furanyl group is located outside of the CD cavity. The H_a, H_b and H_c proton signals are shifted towards higher fields in the case of DM-βCD, the furanyl group of FS is not included in the CD cavity.

4. Conclusions

CDPS and its different molecular weight fractions form more or less stable inclusion complexes with several drugs. The water solubility of the CDPS complexes are considerably higher than that of the complexes of the parent βCD. The complex forming ability of CDPS and its fractions depends on the structure of the guest molecule, the hydrophilic substituents on the βCD rings may prevent or help the inclusion of guest molecules. The interaction of the studied drugs with CDPS-L was equivalent or somewhat weaker than that with CDPS. The solubilizing effect of CDPS-H was better with larger guest molecules (cinnarizine, tolnaftate), which probably can be explained by cooperativity in binding between the adjacent CD units of the polymer molecule [2].

A small fraction of CDPS was lost upon separation and dialysis. Perhaps this lost fraction plays some complementing role because the unfractionated CDPS resulted in larger stability constants with furosemide and acetohexamide, than the less heterogeneous fractions.

Mixing the separated CDPS-H and CDPS-L fractions in a 1:1 weight ratio resulted in about the same stability constant with FS, than that with the fractions.

Acknowledgement

This work was supported by the Otani Research Grant, Hoshi University, for which the authors are grateful.

References

1. N. Wiedenhof, J. N. J. J. Lammers, and C. L. van Panthaleon van Eck: *Stärke* **21**, 119 (1969).
2. A. Harada, M. Furue, and S. Nozakura: *Polymer J.* **13**, 777 (1981).
3. É. Fenyvesi, M. Szilasi, B. Zsadon, J. Szejtli, and F. Tüdős: *Proceeding of the Ist International Symposium on Cyclodextrins* (Ed.: J. Szejtli), pp. 345-356, D. Reidel Publishing Co., Dordrecht, Holland, 1982.
4. K. Uekama, M. Otagiri, T. Irie, H. Seo, and M. Tsuruoka: *Int. J. Pharm.* **23**, 35 (1985).
5. J. Pitha and Jan Pitha: *J. Pharm. Sci.* **47**, 987 (1985).
6. J. Pitha, S. M. Harman, and M. E. Michel: *J. Pharm. Sci.* **75**, 165 (1986).
7. T. Higuchi and K. A. Connors: *Adv. Anal. Chem. Instr.* **4**, 117 (1965).
8. K. A. Connors and T. W. Rosanske: *J. Pharm. Sci.* **69**, 173 (1980).

Cyclodextrins Lessen the Membrane Damaging Effect of Nonionic Tensides

M. SZÖGYI*
Semmelweis Medical University, Institute of Biophysics, P.O. Box 263, 1444 Budapest, Hungary

T. CSERHÁTI
Plant Protection Institute of Hungarian Academy of Sciences, Budapest, Hungary

J. SZEJTLI
Biochemical Research Laboratory of Chinoin Pharmaceutical and Chemical Works, Budapest, Hungary

(Received: 24 October 1986)

Abstract. Nonylphenyl-ethyleneoxide polymers containing 5, 9 and 30 ethyleneoxide groups per molecule build into the hydrophobic fatty acid chains of the cell membrane phospholipid dipalmitoyl-phosphatidylcholine (DPPC) resulting in a decreased main transition temperature, a decreased enthalpy of the main transition and in enhanced potassium permeability of DPPC liposomes. The α-, β- and γ-cyclodextrins form inclusion complexes with the tenzides lowering their free concentration. The complex formation lessens or sometimes totally prevents the membrane damaging effect of tensides. The effectivity order of cyclodextrins is $\beta CD > \gamma CD > \alpha CD$.

Key words: Liposomes, permeability, tenside cyclodextrin complex.

1. Introduction

Pesticide formulations generally contain nonionic tensides that improve the technical parameters (lower drop volume, higher suspension or emulsion stability, better spreading on the leaf surface etc.) of formulations [1]. Besides these effects the nonionic tensides can modify the biological efficiency of an active ingredient [2, 3] and even its selectivity [4]. Sometimes they themselves show marked microbicidal effect [5, 6] and enhance the phytotoxicity [7, 8, 9]. The mode of action of nonionic tensides has been explained by the fact that they interact with the membrane phospholipids [10] and increase the membrane permeability [11]. This effect depends on the lipophilicity and on the structural characteristics of the hydrophobic part [12]. Due to their capacity to form inclusion complexes with a large number of organic compounds cyclodextrins are finding growing acceptance and application in human therapy and in agrochemistry [13, 14]. As they form complexes also with tensides [15, 16, 17] it was assumed that the interaction between nonionic tensides and cyclodextrins can be used to lessen or to prevent the membrane damaging effect of tensides. The non-methylated cyclodextrins themselves exert only a negligible effect on membranes, however, the methylated derivatives show marked membrane damaging activity [18].

---
* Author for correspondence.

Presented at the Fourth International Symposium on Inclusion Phenomena and the Third International Symposium on Cyclodextrins, Lancaster, U.K., 20–25 July 1986.

2. Materials and Methods

The nonylphenyl-polyethyleneoxide nonionic tensides (Hoechst FRG) contain 5, 9 and 30 ethyleneoxide groups per molecule on average (abbreviated to T_5, T_9 and T_{30} respectively). The cyclodextrins: alpha-(αCD), beta-(βCD) and gamma-cyclodextrin (γCD) are produced by Chinoin (Hungary). Dipalmitoyl-phosphatidylcholine (DPPC) was used as purchased from the Sigma Chemical Co.

Differential Scanning Calorimetry (DSC) studies and the determination of the permeability of DPPC liposomes were carried out as described in [18] and [19]. The molar ratio of samples varied in the range DPPC:tenside:cyclodextrin 100:0–1:0–5. The main transition temperature (T_m °C) and the enthalpy of the main transition (ΔH mJ/mg) was determined from the DSC data. The permeability time constant ($P_t s^{-1}$) was calculated according to [20].

3. Results and Discussion

The results of DSC measurements are compiled in Figures 1 and 2. Each tenside considerably lessens the main transition temperature of DPPC (Figure 1). This observation suggests that the tensides – after binding to the bilayer surface – penetrate among the hydrocarbon chains of lipid molecules. This intercalation depends on

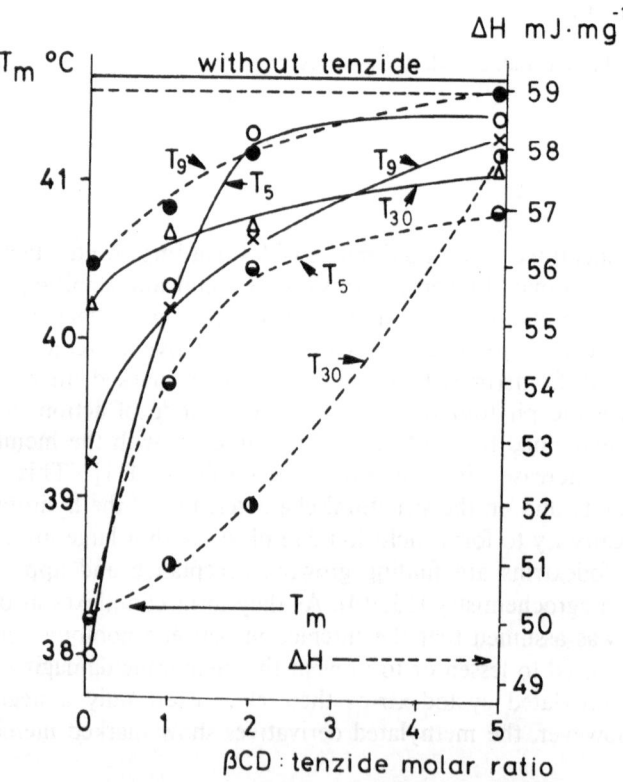

Fig. 1. Effect of tensides (T_5, T_9 and T_{30}) as a function of the βCD molar ratio on the main transition temperature (T_m) and enthalpy (ΔH) of DPPC. The DPPC:tenside molar ratio was 100:1.

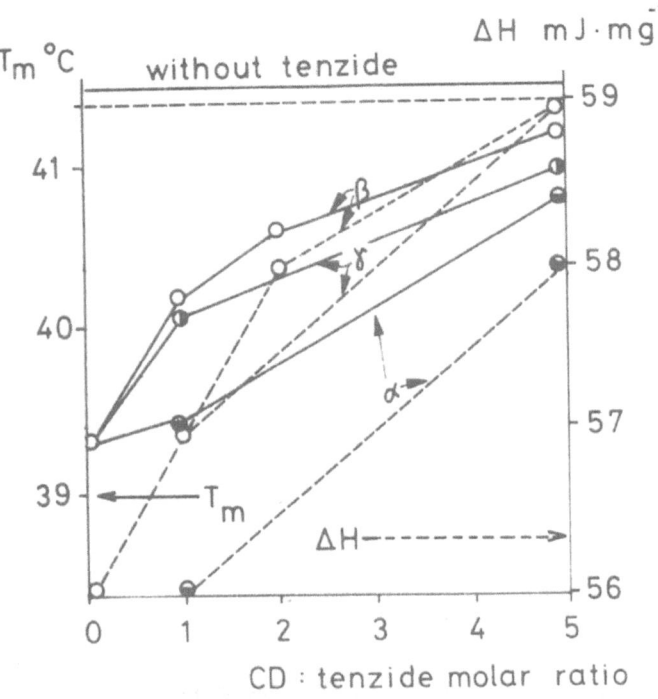

Fig. 2. Effect of T_9 in the presence of cyclodextrins on the main transition temperature (T_m) and enthalpy (ΔH) of DPPC. The DPPC:tenside molar ratio 100:1.

the ratio of hydrophobic and hydrophilic parts within the tenside molecule. The modifying molecules act as structural defects loosening the lipid packing density and lowering the transition temperature and enthalpy (Figure 1). βCD lessens or even annuls these effects in each case. The interaction between DPPC and the tensides decreases nonlinearly with the increasing molar ratio of βCD:tenside, indicating the formation of fairly stable inclusion complexes between tensides and βCD.

The fact that at higher βCD molar ratios the differences between the effects of the different tensides becomes rather small suggests that the bulky hydrophobic part of tensides (identical in each tenside) is responsible for the complex formation and the length of the hydrophilic ethyleneoxide chain (different in each tenside) has a negligible effect on the complex formation.

The various cyclodextrins exert different effects on the interaction between the T_9 tenside and DPPC (Figure 2). βCD exerts the highest and αCD the lowest preventive effect. This finding indicates that the dimensions of the βCD and γCD cavities are nearly equally adequate for the hydrophobic nonylphenyl moiety of the tensides. The lower effect of αCD is due to the smaller cavity.

The results of permeability determinations on liposomes support the conclusions formulated above (Figure 3). The intercalated tenside molecules disturb the organization of lipid bilayers. The efflux of potassium ions is higher through the partially disorganized, more loosely packed lipid layers. The cyclodextrins counteract the permeability increasing effect of T_9 and this effect increases with increasing CD:T_9

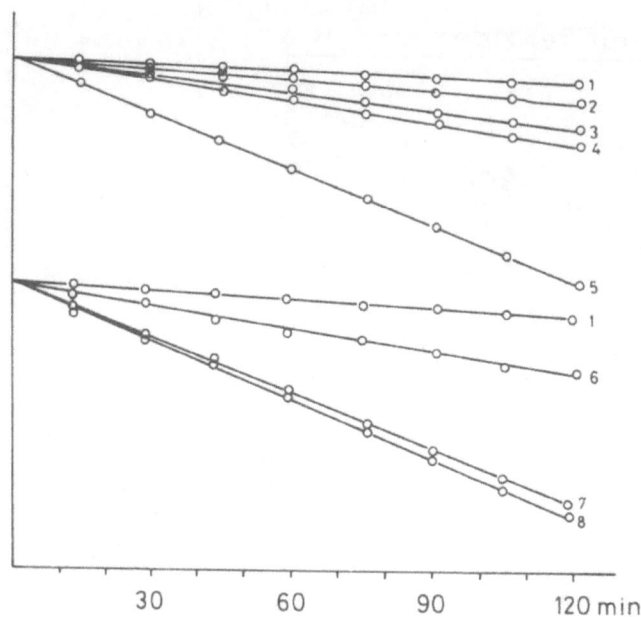

Fig. 3. Effect of T_9 in the presence of cyclodextrins on the ^{42}K efflux of DPPC liposomes. 1. DPPC; 2. DPPC:T_9:βCD 100:1:5; 3. DPPC:T_9:βCD 100:1:2; 4. DPPC:T_9:βCD 100:1:1; 5. DPPC:T_9 100:1; 6. DPPC:T_9:αCD 100:1:5; 7. DPPC:T_9:αCD 100:1:1; 8. DPPC:T_9 100:1.

molar ratio. αCD exhibits also in this case a lower preventive effect than βCD proving again the poorer stability of its inclusion complex with T_9. The permeability constant changes nonlinearly with the cyclodextrin:T_9 molar ratio (Figure 4). At higher cyclodextrin-tenside molar ratios the permeability increase caused by tenside is nearly completely supressed. The effectivity order of cyclodextrins is the same as in the DSC measurements: βCD > γCD > αCD.

Under such conditions in aqueous solutions, the majority of the tenside molecules are in the complexed form. The membrane damaging effect of free tenside at reduced concentration is of course lower.

Our data do not exclude the possibility that the cyclodextrin-tenside complexes may interact with the DPPC. However, the facts that the cyclodextrins do not show any membrane damaging effect and the tenside-cyclodextrin complex is highly hydrophilic and has considerable dimensions contradict this supposition.

References

1. W. Van Valkenburg: *Pesticide Formulations*, Marcel Dekker Inc., New York (1973).
2. P. J. Dunleavy, A. H. Cobb, K. E. Pallett, and L. G. Davies: *Proc. 1982 British Crop. Prot. Conf. Weeds* **1**, 187 (1982).
3. R. A. Spotts and B. B. Peters: *Plant Disease* **6**, 725 (1982).
4. R. Müller and U. Bueth: Abhandlungen der Akademie der Wissenshaften der DDR. Abteilung Mathematik, Naturwissenschaften. Technik N1. Jahrgang 1982. **1**, 315 (1983).

MEMBRANE DAMAGE OF NONIONIC TENSIDES

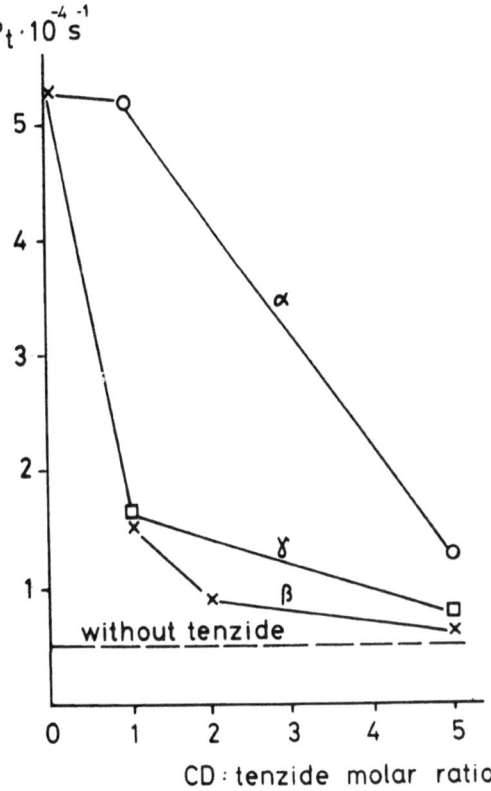

Fig. 4. Effect of T_9 in the presence of cyclodextrins on the permeability time constant of DPPC liposomes.

5. D. R. Clifford and E. C. Hislop: *Pestic. Sci.* **6**, 409 (1975).
6. T. Baicu and A. Jilaveau: *Acta Horticult.* **58**, 453 (1977).
7. A. G. T. Babiker and H. J. Duncan: *Pestic. Sci.* **6**, 655 (1975).
8. J. V. Parochetti, H. D. Wilson, and C. A. Beste: *Proc. Northeastern Weed Sci. Soc. Baltimore,* **31**, 105 (1977).
9. G. D. Leroux and R. G. Harvey: *Proc. North Central Weed Control Conf.* **36**, 40 (1981).
10. T. Cserháti, M. Szőgyi, and L. Győrfi: *J. Chromatogr.* **349**, 295 (1985).
11. M. Szőgyi, F. Tölgyesi, and T. Cserháti in: *Physical Chemistry of Transmembrane Ion Motions.* (ed.: G. Spach) pp. 29–34. Elsevier Sci. Publishers B.V., Amsterdam (1983).
12. T. Cserháti, M. Szőgyi, B. Bordás, and A. Dobrovolszky: *Quant. Struct. Act. Relat.* **3**, 56 (1984).
13. J. Szejtli: *Inclusion Compounds,* Volume 3 (Eds: J. L. Atwood, J. E. D. Davies, and D. D. MacNicol). Academic Press, London (1984).
14. J. Szejtli: *Cyclodextrins and their Inclusion Complexes.* Akadémiai Kiadó, Budapest (1982).
15. J. Koch in: *Proc. of the First Int. Symp. on Cyclodextrins* (Ed.: J. Szejtli) p. 487. Akadémiai Kiadó, Budapest and D. Reidel, Holland (1982).
16. K. Králová and L. Mitterhauszová in: *Proc. of the First Int. Symp. on Cyclodextrins* (Ed.: J. Szejtli). p. 217. Akadémiai Kiadó, Budapest and D. Reidel, Holland (1982).
17. K. Králová, L. Mitterhauszová, and J. Szejtli: *Tenside Detergents* **20**, 37 (1983).

Fig. 10.5.

Effect of Cyclodextrin Complexation on the Reduction of Menthone and Isomenthone

L. SZENTE and J. SZEJTLI*
Biochemical Research Laboratory of Chinoin Pharmaceutical and Chemical Works, Budapest, Hungary

LE TUNG CHAU
Institute of Materia Medica, Ha Noi 38, Quang Trung St., Vietnam

(Received: 2 October 1986; in final form: 6 January 1987)

Abstract. The β- and γ-cyclodextrin complexation of menthone and isomenthone was found to modify the ratio of epimeric menthol products formed upon a $NaBH_4$/MeOH reduction of these ketones.

Key words: Stereoselectivity, selective reduction, menthol menthone cyclodextrin complex.

1. Introduction

The presence of cyclodextrins in reaction mixtures often results in the alteration of well known chemical transformations, due either to their catalytic activities or by controlling the formation of certain reaction products. This latter may lead to the alteration of the ratio of products, formed, and to the improvement of the selectivities and yield of the reactions [1]. Characteristic examples are the selective chlorination of anisole [2] and the selective formylation of phenol [3].

The present work reports preliminary data on the $NaBH_4$ reduction of menthone and isomenthone in the free and cyclodextrin complexed form.

The reduction of these terpenketones results in a mixture of epimeric products which are difficult to separate [4, 5, 6] (Figure 1). A reduction method with improved selectivity would be of practical importance, especially in the production of menthol from menthone.

2. Experimental

The reduction of menthone and isomenthone or their CD-complexes with $NaBH_4$ was carried out in methanol according to Hedin [7]. The cyclodextrin complexes were only suspended. The process was followed by TLC on a silica layer using a benzene-ethylacetate 95:5 mixture.

The end product was analyzed by GLC. The β- and γ-cyclodextrin complexes of menthone and isomenthone were prepared by co-crystallization, as previously described [8, 9]. Menthone and isomenthone were purchased from Dragoco Co. (Holzminden, West, Germany).

---

* Author for correspondence.

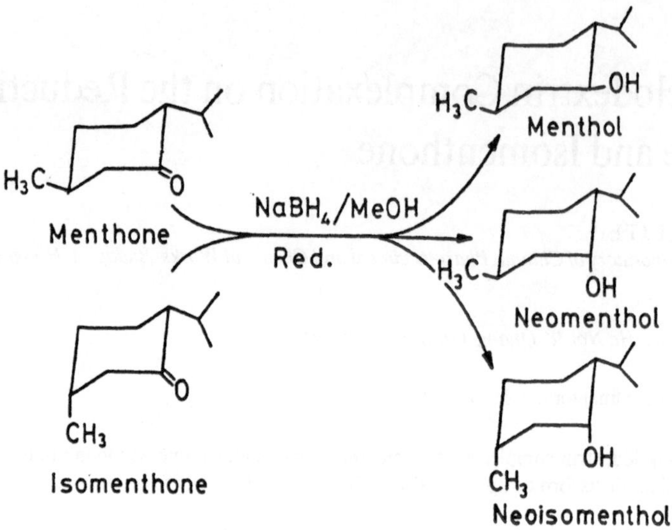

Fig. 1.

β-and γ-cyclodextrin are produced by Chinoin Pharm. and Chem. Works Ltd. (Budapest, Hungary).

3. Results and Discussion

It has earlier been reported, that both menthones and menthols form inclusion complexes with cyclodextrins [8, 9]. It is presumed that the ketones and the isomer alcohols formed upon reduction, show different affinities towards inclusion complexation by cyclodextrins.

3.1. REDUCTION OF FREE AND CYCLODEXTRIN COMPLEXED MENTHONE

The $NaBH_4$-reduction of menthone and its β- and γ-cyclodextrin complex suspended in methanol at 22 °C for 2 hours gave the reaction mixture of menthol epimeric products as listed in Table I, according to gas-chomatographic analysis.

Table I. Products obtained by reducing menthone or cyclodextrin-complexed menthone with $NaBH_4$

Ketone	Menthol (%)	Neomenthol (%)	Neoisomenthol (%)	Isomenthol (%)
Free Menthone	42.06	43.47	13.86	not detectable
Menthone-β-cyclodextrin	64.60	29.94	8.37	not detectable
Menthone-γ-cyclodextrin	68.58	23.50	7.01	not detectable

3.2. REDUCTION OF FREE AND CYCLODEXTRIN COMPLEXED ISOMENTHONE

Since it has been known for many years that the reduction of isomenthone gives all the four possible isomeric menthols [5] we compared the ratio of isomeric alcohols in the reaction mixture of free and complexed isomenthone after the $NaBH_4$ reduction.

Surprisingly no detectable amount of isomenthol was found by gas-chromatography in either reaction mixtures. Similarly only menthol, neomenthol and neoisomenthol were formed from both the free and the complexed isomenthone. The composition of products after its reduction is shown in Table II.

Table II. Products obtained by reducing isomenthone or cyclodextrin-complexed isomenthone with $NaBH_4$

Ketone	Menthol (%)	Neomenthol (%)	Neoisomenthol (%)	Isomenthol (%)
Free isomenthone	20.41	18.16	61.42	not detectable
Iso-menthone-β-cyclodextrin	46.38	11.34	36.19[a]	not detectable
Iso-menthone-γ-cyclodextrin	50.72	12.07	36.81	not detectable

[a] Two more unidentified products detected (~3–3% of each).

The following conclusions have been drawn from the above observations:

- Both β- and γ-cyclodextrin were found to influence the ratios of isomeric menthols formed in the reduction of menthone and isomenthone.
  A remarkable increase of menthol formation took place due to the presence of cyclodextrins, while the appearance of both epimers (neomenthol and neoisomethol) was hindered.
- No significant difference between the effect of β- and γ-cyclodextrin on the reduction was observed.

As for the practical usefulness of the above results, we note that the essential oil of the Vietnamese Murraya glabra plant contains 85–95% of menthone and isomenthone. This essential oil is available in huge amounts in Vietnam and could be used as starting material for an industrial menthol production, involving the reduction of the menthones of the oil. The aim of the present work was to study the effect of cyclodextrin complexation on the selectivity of menthone and isomenthone reductions.

The remarkable enhancement of menthol formation (and the significant decrease of the formation of menthol epimers) as a result of cyclodextrin complexation may be of practical importance.

References

1. I. Tabushi: *J. Synth. Chem. Jpn.* **10**, 892 (1982).
2. R. Breslow and P. Campbell: *J. Am. Chem. Soc.* **91**, 3085 (1969).
3. M. Ohara and J. Fukuda: *Pharmazie* **33** 467 (1978).
4. W. Hückel and S. Geiger: *Justus Liebigs Ann. Chem.* **624**, 235 (1959).

5. K. Stroh: *Angew. Chem.* **69**, 702 (1957).
6. W. Hückel and C. Z. Khan Cheema: *Chem. Ber.* **91**, 311 (1958).
7. I. A. Hedin: *Anal. Chem.* **44**, 1254 (1972).
8. J. Szejtli, L. Szente and E. Bánky: *Acta Chim. Acad. Sci. Hung.* **101**, 27 (1979).
9. J. Szejtli: *Cyclodextrins and their Inclusion Complexes*, Akadémiai Kiadó, Budapest, 1982, p. 269.

β-Cyclodextrin-Catalyzed Effects on the Hydrolysis of Esters of Aromatic Acids

DAO-DAO ZHANG*, NAI-JU HUANG, LING XUE, and YONG-MING HUANG
Department of Chemistry, Fudan University, Shanghai, People's Republic of China

(Received: 20 October 1986; in final form: 3 March 1987)

Abstract. In order to make the behavior of the hydrophobic cavity of β-cyclodextrin clear, we have studied β-cyclodextrin-catalyzed hydrolysis of a series of nitrophenyl esters of aromatic acids. We defined a new kinetic parameter to determine the structure of the inclusion compounds. The kinetic parameters obtained provide evidence that the aromatic acid moiety rather than the nitrophenyl moiety of the esters mainly enters the hydrophobic cavity of β-cyclodextrin to form the inclusion complex.

Key words. Cyclodextrin-catalyzed, hydrolysis, hydrophobic cavity, inclusion complex, kinetic parameter RMP.

1. Introduction

Cyclodextrin has attracted great attention as an enzyme model for serine acylase enzymes such as chymotrypsin. It has a hydrophobic cavity and shows hydrophilic characteristics by the hydroxyl groups on both sides of the torus. In cyclodextrin-catalyzed reactions, a substrate binds into the cyclodextrin molecular cavity and then undergoes reaction with one of the cyclodextrin hydroxyls. According to Bender's work [1], the cyclodextrins cause a markedly stereoselective acceleration of the release of phenols from substituted phenyl acetates, the rate accelerations with meta-substituted esters being larger than with the corresponding para-substituted esters, i.e., the rate accelerations are independent of the electronic nature of the substituents. In order to make clear the behavior of the hydrophobic cavity of β-cyclodextrin, we have studied the β-cyclodextrin-catalyzed hydrolysis of the esters containing both hydrophobic and hydrophilic groups, such as *m*- and *p*-nitrophenyl nicotinates and 3-β-pyridylacrylates [2], and the other *m*- and *p*-nitrophenyl esters of *p-t*-butylbenzoic, *p-t*-butylcinnamic, α-naphthoic, acetic and β-anthraquinone carboxylic acids.

2. Experimental

Aqueous buffers were prepared with deionized water. The dimethyl sulfoxide used in the kinetic runs was dried over potassium hydroxide and distilled. β-cyclodextrin was recrystallized from water and dried overnight at 80 °C (0.05 torr) just prior to use.

A Shimadzu UV 260 spectrophotometer was used both for recording UV/visible

---

* Author for correspondence.

absorption spectra and for the kinetic studies. pH measurements were accomplished with a PHM 84 research pH meter equipped with a GK2401C combined electrode.

2.1. KINETIC MEASUREMENTS

Reaction buffers were prepared by adding 4 volumes of 10 mM aqueous potassium dihydrogen phosphate/sodium hydroxide buffer (pH 6.8–7.6) to 6 volumes of dimethyl sulfoxide. The resulting solutions had pHs ranging from 10.6 to 11.5 as determined with the glass electrode. β-Cyclodextrin solutions (0.9–6.0 mM) were prepared with this buffer and stored under nitrogen. Substrate solutions (3–4 mM in dimethyl sulfoxide) were stored in the dark.

A kinetic run was initiated by equilibrating 3.00 ml of β-cyclodextrin solution to 30.0 ± 0.5 °C in the spectrophotometer chamber. A 30 µl sample of substrate solution was injected (to make the solution 30–40 µM in substrate) and the absorbance at 410 or 415 nm monitored as a function of time. After a suitable interval (10 half-lives), the final absorbance was measured. The pH of the solution throughout the run was found to remain constant to within ± 0.02 unit. In the case of substrates showing monophasic kinetics, data from the first 3–4 half-lives were fitted to a simple exponential by using a standard nonweighted least-squares routine. The resulting rate constants showed standard deviations of less than 1%.

The maximal rate constants (k_c) and binding constants (K_m) were extracted from rates measured at six different concentrations of β-cyclodextrin by using the method of Eadie [3].

2.2. PREPARATION OF MATERIALS

All substrates were prepared by condensation of carboxylic acids using dicyclohexyl-carbodiimide or acid chlorides with p- and m-nitrophenol. The esters were identified

Table I. The melting point of the esters used [a]

Substrates	m.p. (°C)	Ref.[b]
1. m-nitrophenyl acetate	55–56	
2. p-nitrophenyl acetate	80–81	
3. m-nitrophenyl nicotinate	134–135	
4. p-nitrophenyl nicotinate	172–173	
5. m-nitrophenyl 3-β-pyridylacrylate	163–164	
6. p-nitrophenyl 3-β-pyridylacrylate	186–187	
7. m-nitrophenyl p-t-butylbenzoate	85–86	[5]
8. p-nitrophenyl p-t-butylbenzoate	123–124	[6]
9. m-nitrophenyl p-t-butylcinnamate	82–83	[4]
10. p-nitrophenyl p-t-butylcinnamate	147–148	
11. m-nitrophenyl α-naphthoate	116–117	[7]
12. p-nitrophenyl α-naphthoate	142–143	
13. m-nitrophenyl β-anthraquinone carboxylate	212–213	[8]
14. p-nitrophenyl β-anthraquinone carboxylate	243–244	

[a] All melting points are uncorrected. Esters, except 1, 2, 3, 4, 8, and 12, are the unknown compounds.
[b] References to the synthetic methods for the esters or for the corresponding acids or the acid chlorides.

by elemental analysis and spectral measurements before use. All reported melting points are listed in Table I.

p-t-Butylcinnamic acid was prepared from p-t-butylbenzaldehyde and propanedioic acid [4], m.p. 198–199 °C. Anal. Calcd. (Found) for $C_{13}H_{16}O_2$: C, 76.44(76.22); H, 7.89(8.11).

3. Results and Discussion

According to Bender's work [1], the rate acceleration k_c/k_{un} (k_{un} is the observed first-order rate constant for hydrolysis of the ester in the absence of β-cyclodextrin and k_c is the maximal catalyzed rate due to decomposition of the fully complexed ester) with m-nitrophenyl acetate is larger than with p-nitrophenyl acetate, since the cyclodextrin forms a complex with the nitrophenyl group of the esters and the m-nitrophenyl group is more suitable than the p-nitrophenyl group for cyclodextrin catalysis. To express our views clearly, we define a new kinetic parameter, RMP, as the ratio of the rate acceleration for β-cyclodextrin-catalyzed hydrolysis of the m-nitrophenyl ester to that of the corresponding p-nitrophenyl ester, $(k_c/k_{un})_m/(k_c/k_{un})_p$. So the RMP values of the m- and p-nitrophenyl acetates is 10.5 (calculated using Bender's data [1]). If the β-cyclodextrin forms a complex with the acyl group of the esters, the complexes of m- and p-substituted esters are similar from the point of view of catalysis and the rate acceleration will be almost the same and the RMP values will be nearly unity. Otherwise similar results to Bender's work will be observed. Because the methyl group is too small to stay in the cavity of β-cyclodextrin and the anthraquinonyl group is too large to enter the cavity [9], the nitrophenyl groups of these two esters will enter the cavity and we can get a result similar to that from Bender's work.

The maximal rate constants k_c and binding constants K_m of β-cyclodextrin-ester complexes in 6DMSO–4H$_2$O solvent are shown in Table II. For nitrophenyl acetates the temperature and solvent we employed is different from Bender's [1], so the data obtained are different. From the data in Table II, differences are observed between the substrate pairs (1, 2) and (13, 14) and the others.

The RMP values of the substrate pairs (1, 2) and (13, 14) are 4.5 and 4.8 and are larger than those of the others (less than 1.5). For substrates 1, 2, 13 and 14, β-cyclodextrin forms complexes with the nitrophenyl groups and a markedly stereoselective acceleration is caused. For the other substrates β-cyclodextrin forms the complexes with their acyl groups and a similar complex is formed with the p- and m-nitrophenyl esters. So it will not cause a markedly stereoselective acceleration and the RMP values will be nearly unity.

On the other hand, from the data in Table II we found that there are differences in the values of K_m only between substrates 1 and 2 or 13 and 14. The values of K_m for the other substrates are almost the same. This can be explained by the fact that if the acyl group enters the β-cyclodextrin cavity, the structures of the inclusion complexes of the corresponding p- and m-substituted esters with β-cyclodextrin are similar and the values of K_m will be very close. If the p- and m-nitrophenyl groups enter the β-cyclodextrin cavity, the apparent differences in the values of K_m of the substrates 1 and 2 or 13 and 14 can be expected because of the difference of the steric effect between two nitrophenyl groups.

The k_c values of the m-nitrophenyl esters are larger than those of the p-nitro-

Table II. Kinetic and Binding Constants at 30.0 ± 0.5 °C

Sub-strate	k_{un}[b] ($s^{-1} \times 10^5$)	k_c[a] ($s^{-1} \times 10^3$)	K_m[c] ($M \times 10^3$)	pH	k_{un} corrected to pH 10.00 ($s^{-1} \times 10^6$)	k_c corrected to pH 10.00 ($s^{-1} \times 10^4$)	k_c/k_{un}	RMP[d]
1	20.9 ±0.1	12.6 ±0.42	20 ±1	11.45	7.42	4.46	60.1	
2	43.1 ±0.2	5.78±0.53	8.3 ±1.1	11.45	15.3	2.05	13.4	4.5
3	15.8 ±0.1	5.28±0.55	5.3 ±1.0	11.09	12.8	4.29	33.5	
4	17.6 ±0.1	5.02±0.75	6.6 ±2.1	11.05	15.7	4.48	28.5	1.2
5	3.51±0.01	5.10±0.65	6.7 ±1.4	11.00	3.51	5.10	145	
6	5.74±0.01	5.42±0.82	6.5 ±1.7	11.00	5.74	5.26	94.4	1.5
7	2.22±0.03	1.81±0.09	0.32±0.03	11.49	0.719	0.587	82	
8	3.80±0.02	4.70±0.12	0.35±0.02	11.49	1.23	1.52	124	0.66
9	2.51±0.01	3.45±0.16	0.39±0.03	11.37	1.07	1.47	137	
10	5.16±0.01	5.70±0.16	0.40±0.02	11.37	2.20	2.43	111	1.2
11	7.40±0.01	1.54±0.16	9.8 ±1.3	11.50	2.34	0.488	21	
12	5.79±0.01	1.83±0.10	9.5 ±1.0	11.50	1.83	0.578	35	0.60
13	49.8 ±0.2	18.5 ±1.9	11 ±4	10.66	109	40.5	71.7	
14	62.6 ±0.3	9.69±2.88	3.6 ±.2.0	10.66	137	21.2	15.5	4.8

[a] Pseudo-first-order rate constants for acylation of β-cyclodextrin by fully bound substrate in 6DMSO-4 H₂O at the 'pH' indicated, as measured with a glass electrode.
[b] Pseudo-first-order rate for hydrolysis of the substrate in the absence of β-cyclodextrin at the 'pH' indicated.
[c] Dissociation constant of the substrate-β-cyclodextrin complex.
[d] $(k_c/k_{un})_m/(k_c/k_{un})_p$.
[e] All errors are standard deviations.

phenyl esters for substrates 1, 2, 13, and 14 while the opposite trend is observed for the other substrates. The nitrophenyl groups do not enter the β-cyclodextrin cavity on complexation and the substrates, apart from 1, 2, 13, and 14 will display the electronic effect of the acylation reaction. The k_c value of the p-nitrophenyl ester is larger than that of the corresponding m-nitrophenyl ester. For example, the k_c value of substrate 8 is 1.52×10^{-4} and that of substrate 7 is 0.58×10^{-4}.
But for substrates 1, 2, 13, and 14, the nitrophenyl groups enter the β-cyclodextrin cavity and the same result is not observed (for substrate 1, k_c is 4.46×10^{-4} and for substrate 2, $k_c = 2.05 \times 10^{-4}$).

Complexes formed between β-cyclodextrin and the aromatic acid moiety of the esters, apart from 3, 4, 5, and 6 are more suitable for catalysis than those formed with the nitrophenyl group, since the aromatic acid moiety is larger than nitrophenyl. So it is reasonable to draw this conclusion.

The pyridyl group is less hydrophobic than the nitrophenyl group and the volume of the pyridyl group is a little less than that of the nitrophenyl group (from CPK models). Why does the end of the pyridyl group enter the β-cyclodextrin cavity more readily than the end of the nitrophenyl group? We consider that maybe there is a hydrogen bond between the N atom of the pyridyl group and one of the hydroxyl groups of the β-cyclodextrin in the inclusion complex, or that other, more complicated effects occurred. It is necessary to investigate this point further.

From the results and discussion above, we can get the following preliminary conclusion: it is the aromatic acid moiety rather than the nitrophenyl moiety of the esters that enters the hydrophobic cavity of β-cyclodextrin to form the inclusion complex.

Acknowledgement

The project was supported by the Science Fund of the Chinese Academy of Sciences. The authors gratefully thank Professor Xi-kui Jiang of Shanghai Institute of Organic Chemistry, Academia Sinica for useful suggestions.

References

1. R. L. van Etten, J. F. Sebastian, G. A. Clowers, and M. L. Bender: *J. Am. Chem. Soc.* **89**, 3242 (1967).
2. D. D. Zhang, D. K. Shao, Z. H. Xue, and J. P. Fang: *Acta Chimica Sinica*, submitted.
3. G. S. Eadie: *J. Biol. Chem.* **148**, 86–92 (1942).
4. A. I. Vogel: *Vogel's Textbook of Practical Organic Chemistry*, 4th edn., 767–768, Longman, London (1978).
5. C. J. O'Connor and T. D. Lomax: *Aust. J. Chem.* **36** (5), 917 (1983).
6. N. M. Cullinane and D. M. Leyshon: *J. Chem. Soc.* 2944 (1954).
7. M. Yukito, A. Yasuhiro, K. Masaaki, and N. Akio: *Bull. Chem. Soc. Jpn.* **50** (12), 3365 (1977).
8. P. Arjunan and K. D. Berlin: *Org. Prep. Proced. Int.* **13** (5), 368 (1981).
9. A. Kuboyama and S. Y. Matsuzaki: *J. Incl. Phenom.* **2**, 755 (1984).

Journal of Inclusion Phenomena 5 (1987), 449–458.
© 1987 by D. Reidel Publishing Company.

Asymmetric Halogenation and Hydrohalogenation of Styrene in Crystalline Cyclodextrin Complexes

HIDETAKE SAKURABA* and HIROKAZU ISHIZAKI
Department of Industrial Chemistry, Faculty of Engineering, Kanto Gakuin University, 4834 Kanazawa-Mutsuura, Yokohama, Kanagawa 236, Japan

and

YOSHIO TANAKA and TOSHIMI SHIMIZU
Research Institute for Polymers & Textiles, 1-1-4 Yatabe-Higashi, Tsukuba, Ibaraki 305, Japan

(Received: 20 October 1986; in final form: 16 February 1987)

Abstract. Asymmetric halogenation and hydrohalogenation of styrene in microcrystalline cyclodextrin complexes were studied in the gas-solid state, and compared with the homogeneous reactions in aqueous or dimethyl sulfoxide solutions. The gas-solid brominations in the α- and β-cyclodextrin complexes produced predominantly (−)-1,2-dibromo-1-phenylethane. The chiral induction for the reaction of the α-cyclodextrin complex rose to 9 times that of the β-cyclodextrin complex. Brominations in the homogeneous solutions containing the α- or β-cyclodextrin complexes gave no dibromide but racemic bromohydrin. In the gas-solid chlorination, the α-cyclodextrin complex gave (−)-dichloride, S-(+)-2-chloro-1-phenylethanol (14% ee) and (+)-1,2,2-trichloro-1-phenylethane, and the β-cyclodextrin complex produced (+)-dichloride, S-(+)-chlorohydrin (8% ee) and (+)-trichloride. The chiral induction of the gas-solid halogenation using the solid cyclodextrin complexes is attributed to the ability to hold rigidly a chiral conformation of the crystalline state. However, the gas-solid hydrohalogenation all gave racemic products.

Key words: Asymmetric halogenation, hydrohalogenation, styrene, cyclodextrin complex.

1. Introduction

Asymmetric synthesis through reaction in the solid state demands the formation of chiral crystalline structures having certain intramolecular or intermolecular features [1]. Penzien and Schmidt [2] reported the first example of absolute asymmetric synthesis where a single crystal of 4,4'-dimethylchalcone reacts with gaseous bromine. Since this approach, however, has to use compounds which form giant single crystals in a chiral space group, the number of molecules available for such reactions are quite limited. Chiral host species such as cyclodextrins (CDs) form chiral crystalline inclusion complexes with prochiral guest molecules by incorporating them within the relatively nonpolar cavity of the host molecules [3]. Therefore, these materials may produce asymmetric reactions in the solid state, which are topochemically controlled by the crystalline lattice of the inclusion complexes. Such reactions of CD complexes in the solid state, however, have not been accomplished, but asymmetric syntheses such as reduction [4], epoxidation [5] and sulfoxidation [6, 7] catalyzed by native CDs have been studied in solution.

* Author for correspondence.

Presented at the Fourth International Symposium on Inclusion Phenomena and the Third International Symposium on Cyclodextrins, Lancaster, U.K., 20–25 July 1986.

All the products gave low optical yields with 0–34% ee. Furthermore, asymmetric halogenation of olefins with CD as a chiral matrix has not been investigated in the past apart from our reports [8–10].

Previously [8], the authors succeeded in achieving a strong chiral induction (88–100% ee) for the chlorination of methacrylic acid in the crystalline CD complexes. Here, we report on the asymmetric addition of gaseous bromine, chlorine, hydrogen bromide and hydrogen chloride to styrene in the crystalline complexes of α-CD (cyclohexa-amylose) or β-CD (cyclohepta-amylose).

2. Materials and Methods

2.1. MATERIALS

α- and β-CDs were purchased from Sanraku-Ocean Co., Ltd., and recrystallized from water. Styrene was distilled immediately before use. Chlorine and hydrogen chloride were purchased from Komatsugawa Sanso and Tsurumi Soda Co., Ltd., respectively, and passed through a sulfuric acid trap prior to use. Hydrogen bromide was prepared by the procedure given in the literature [11]. All other chemicals were purified using standard methods [12].

2.2. PREPARATION OF INCLUSION COMPLEXES

To 500 ml of aqueous solutions of α-CD (1.7×10^{-1} mol/l) or β-CD (3.0×10^{-2} mol/l), equimolar amounts of styrene were added at 40 °C and dissolved by stirring. After stirring for 2 h at room temperature, the microcrystalline precipitates were filtered and dried in vacuo at room temperature for 1 day. Then the dried powders were washed with n-pentane to eliminate the guest molecule not included and dried again. The complex formation was confirmed by X-ray powder diffraction and TG-DSC techniques. Thus, the crystalline 2:1 and 1:1 complexes were obtained for α- and β-CDs with styrene in 90 and 80% yields, respectively. The contents of styrene in the complexes were estimated by ^{1}H NMR spectra in dimethyl-d_6 sulfoxide (DMSO-d_6).

2.3. HALOGENATION AND HYDROHALOGENATION OF THE CRYSTALLINE INCLUSION COMPLEXES

A typical experimental procedure was as follows. The solid α-CD inclusion complex of styrene (ca. 2 g, 1 mmol) was exposed to bromine vapour (10 mole % excess) in a desiccator (ca. 600 ml) in the dark under air at 0 °C. After an exposure of 2 h, the powder obtained was dissolved in water containing sodium thiosulfate as a reducing agent for the excess of bromine. The reacted and the unreacted guests were extracted repeatedly with diethyl ether from the aqueous solution until no aromatic compounds were detected in the aqueous layer by UV spectroscopy in the region of 230 to 350 nm. Thus, 95–98% of the extract was recovered from the water layer, and chromatographed with dichloromethane on silica gel (Wakogel C-300). The products were identified as 1,2-dibromo-1-phenylethane and 2-bromo-1-phenylethanol by ^{1}H NMR and IR spectra. Chlorination, hydrobromination and hydrochlorination of the CD complexes of styrene were carried out by a similar procedure to this bromination.

The homogeneous bromination was carried out by dissolving the α- and β-CD complexes (containing 1 mmol of styrene) in water (200 ml) or dry dimethyl sulfoxide (DMSO, 5 ml), to which an equimolar amount of bromine was added, at 25 °C for 10 min. Then the reaction mixture was poured into 100–200 ml of 15% aqueous sodium chloride containing sodium thiosulfate (1 mmol), followed by extraction with diethyl ether. The extract was recovered in 98% yield and chomatographed as described above.

2.4. ANALYTICAL METHODS

Optical rotations were measured in various organic solvents on a Union Giken PM-101 polarimeter using 1 dm cells. The other spectroscopic measurements were carried out by a JEOL-PMX 60 for ^{1}H NMR at 60 MHz and Hitachi IR-285 spectrometers for IR spectra. The X-ray powder diffraction patterns of the solid samples were taken in the region of 5 to 35° by a Rigakudenki Model 2037 X-ray diffractometer using Ni-filtered Cu-K_α radiation. The thermal behavior of the specimens was observed with a Rigakudenki TG-DSC standard analyzer at a fixed heating rate of 10 °C/min.

3. Results and Discussion

3.1. INCLUSION COMPLEXES

The inclusion complexes between styrene and α- or β-CDs were obtained as microcrystalline precipitates from the aqueous solutions in good yields. The molar ratios of styrene to CDs were found to be 0.5 for the α-CD complex and 1.0 for the

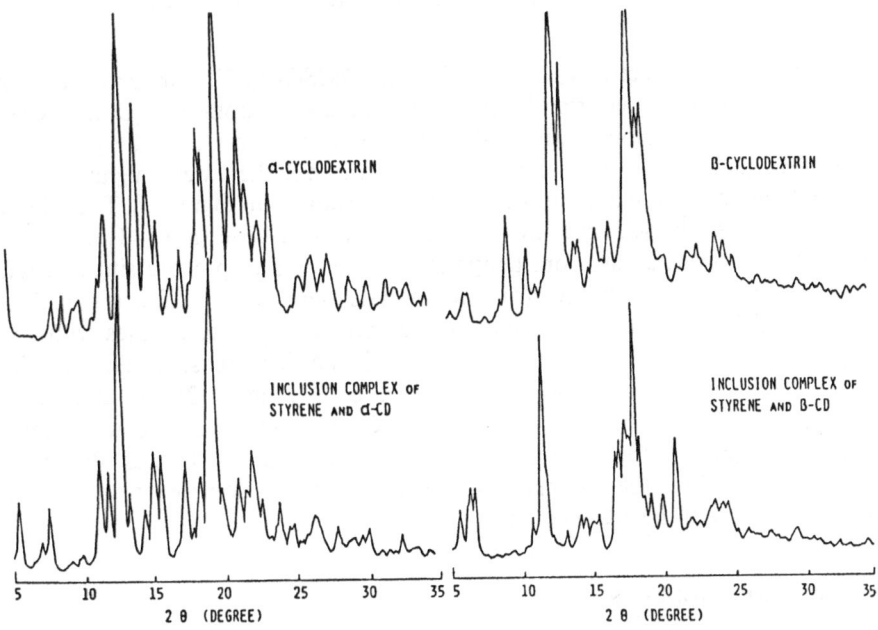

Fig. 1. X-ray diffraction patterns of α- and β-CDs, and their inclusion complexes with styrene.

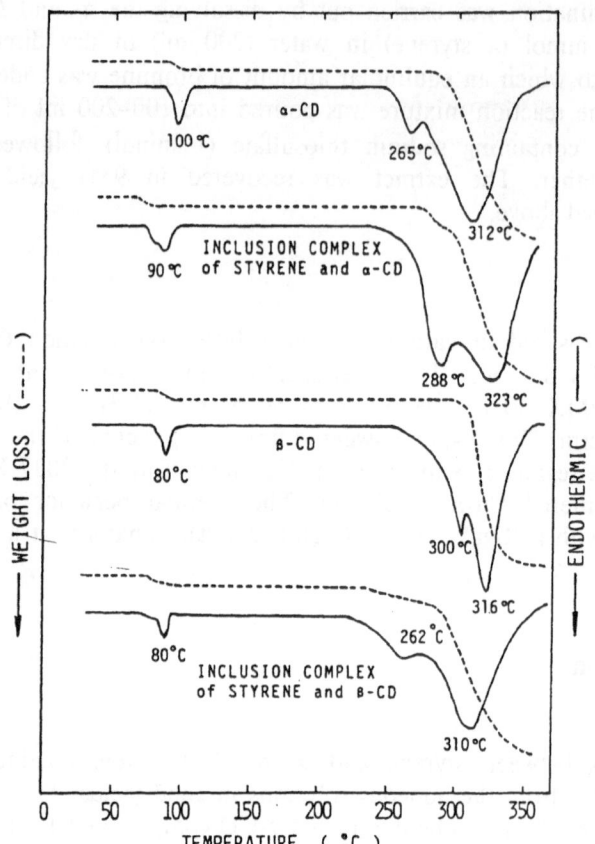

Fig. 2. TG-DSC curves of α- and β-CDs, and their inclusion complexes with styrene.

β-CD complex, respectively, by ^{1}H NMR in DMSO-d_6. The water content of the CD complexes, measured from ^{1}H NMR spectra ($\delta = 3.4$ ppm, s, H$_2$O, in DMSO-d_6), was about 7 in the molar ratio of the water molecules to CDs. The powder X-ray diffraction patterns of these complexes showed that they were highly crystalline, and did not correspond to those of CDs, as shown in Figure 1. The diffraction diagram of the physical mixtures of CDs and oily styrene (mp -31 °C), however, could not be obtained at room temperature. Attempts to prepare single crystals of the CD complexes for X-ray structure analysis were unsuccessful. Figure 2 shows the thermal behavior (TG-DSC curves) for CDs and their complexes with styrene. α- and β-CDs seem to dehydrate at 80–100 °C and decompose at 290–300 °C. The α-CD complex does not lose styrene (bp 146 °C) even at 250 °C and the β-CD complex also does not until reaching 230 °C, as a result of the formation of a complex. Although the decomposition temperature of α-CD was slightly lower than that of β-CD, the thermal stability of the α-CD complex was found to be higher than that of the β-CD complex.

3.2. REACTION PRODUCTS

The gas-solid halogenation of styrene **1** in the crystalline CD complexes produced

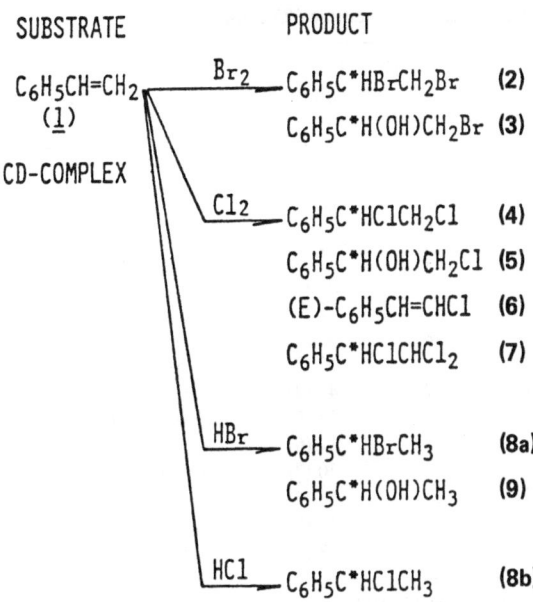

Fig. 3. Products from the gas-solid halogenation and hydrohalogenation of styrene in the crystalline cyclodextrin complexes.

optically active products, but the hydrohalogenation gave racemic products. As shown in Figure 3, the bromination of the CD complexes produced (−)-1,2-dibromo-1-phenylethane (2) and racemic 2-bromo-1-phenylethanol (3). In the chlorination, the CD complexes gave (+)-1,2-dichloro-1-phenylethane (4), (+)-2-chloro-1-phenylethanol (5) and (+)-1,2,2-trichloro-1-phenylethane (7), and accompanied with (E)-1-chloro-2-phenylethane (6). The CD complexes produced racemic 1-bromo-1-phenylethane (8a) and 1-phenylethanol (9) in hydrobromination and racemic 1-chloro-1-phenylethane (8b) in hydrochlorination. The ^{1}H NMR and IR spectra of all products were virtually identical with those of the authentic samples given in the literature, e.g., 6 and 7 [13]. 6: NMR (CCl$_4$) δ 6.62 (1H, d, $J = 14.2$ Hz, β-CH), 6.92 (1H, d, $J = 14.2$ Hz, α-CH), 7.28–7.75 (5H, m, aromatic H); IR (neat) 1620 cm^{-1}, (Found: C, 69.5; H, 5.2%. C$_8$H$_7$Cl requires C, 69.3; H, 5.1%). 7: NMR (CCl$_4$) δ 5.20 (1H, d, $J = 6.0$ Hz, α-CH), 6.00 (1H, d, $J = 6.0$ Hz, β-CH), 7.45 (5H, s, aromatic H), (Found: C, 46.0; H, 3.4%. C$_8$H$_7$Cl$_3$ requires C, 45.9; H, 3.4%).

3.3. GAS-SOLID HALOGENATION

Table I shows the results for the gas-solid bromination and chlorination of styrene in the crystalline CD complexes. However, the maximum values of the specific optical rotation cannot be found in the literature for the optically pure halides apart from halohydrin derivatives of styrene. The optical yield of chlorohydrin (5) was calculated from the maximum value $[\alpha]_D^{25}$ −48.1° (c, 1.73, cyclohexane)

Table I. Asymmetric halogenation of styrene in crystalline CD complexes

Host	Gaseous reagent	Temp. °C	Time h	Yield[a] %	Product[b] compn./%	$[\alpha]_D^{25}/°$ (c = 0.5, solvent)	ee %[c] (Config.)
α-CD	Br$_2$	0	2	90	96 (2)	−47.0 (CH$_2$Cl$_2$)	–
					4 (3)	,0 (CHCl$_3$)	0
β-CD	Br$_2$	0	2	95	100 (2)	−5.5 (CH$_2$Cl$_2$)	–
α-CD	Cl$_2$	−20	48	0	–	–	–
α-CD	Cl$_2$	0	48	25	12 (4)	−4.1 (CH$_2$Cl$_2$)	–
					36 (5)	0 (C$_6$H$_{12}$)	0
					52 (6)	–	–
α-CD	Cl$_2$	25	2	38	43 (5)	+6.6 (C$_6$H$_{12}$)	14 (S)
					57 (7)	+6.5 (C$_6$H$_{12}$)	–
β-CD	Cl$_2$	−20	48	47	43 (4)	+20.5 (CH$_2$Cl$_2$)	–
					15 (5)	0 (C$_6$H$_{12}$)	0
					12 (6)	–	–
					30 (7)	+20.5 (C$_6$H$_{12}$)	–
β-CD	Cl$_2$	25	2	22	68 (4)	+6.5 (CH$_2$Cl$_2$)	–
					14 (5)	+3.8 (C$_6$H$_{12}$)	8 (S)
					18 (7)	+6.3 (C$_6$H$_{12}$)	–

[a] Based on styrene consumed as estimated from ^{1}H NMR analysis.
[b] Based on the total amount of isolated products obtained by silica gel chromatography. Percentages are normalized to 100%.
[c] Enantiomeric excess (ee) values based on $[\alpha]_D^{25}$ − 48.1° (C$_6$H$_{12}$) for R-5 from [14].

for R-5 [14]. We tried to determine the enantiomeric excess (ee) of the other products by means of ^{1}H NMR analysis with chiral shift reagents such as Eu(hfc)$_3$ (tris[3-(heptafluoropropylhydroxymethylene)-d-camphorato]europium(III)) or Eu(fod)$_3$, or by other methods such as liquid chromatography with various chiral stationary phases, e.g., Daicel, Chiralcel, OB or OC, but could not succeed in the optical resolution of the products. Thus, it is not clear in the present experiment how much of the chiral induction on the gas-solid halogenation of styrene is due to the use of CDs.

3.3.1. Gas-Solid Bromination

As shown in Table I, when styrene in the α-CD complex was brominated at 0 °C for 2 h, the levorotatory dibromide (2) ($[\alpha]_D^{25}$ −47.0°) and racemic bromohydrin (3) were isolated in 90% yield (2:3 = 96:4). Bromination of the olefin in the β-CD complex gave no 3 but 2 ($[\alpha]_D^{25}$ −5.5°) in 95% yield at the same reaction condition. The chiral induction for the reaction of the α-CD complex rose to 9 times that of the β-CD complex. The same sign of the specific rotations of 2 shows that styrene forms complexes with α- and β-CDs such that the access of bromine to the olefinic plane occurs into the same enantiotopic face in the two cases, and this face may be slightly less blocked by the inclined plane in both the asymmetric cavities of CDs. A detailed mechanism, however, cannot be described at the present time, because neither crystalline nor molecular structures were determined for the solid CD complexes. No bromination of styrene in the CD complexes occurred at a temperature of −10 °C or below because the vapor pressure of bromine is not enough to sustain the reaction: bromine solidifies at −7.3 °C.

HALOGENATION OF STYRENE IN CYCLODEXTRIN COMPLEXES

Next, the chiral induction of the gas-solid bromination was compared with that of the homogeneous reactions in DMSO or aqueous solutions. The homogeneous reactions in the presence of CDs gave no **2** but **3** in 60–75% yields at 25 °C for 10 min. No chiral induction was observed in the homogeneous reactions. The racemic bromohydrin (**3**) is produced non-enantioselectively through the reaction of the bromonium cation with water contained in the CD complexes in dry DMSO solution, and not by the hydrolysis of **2** under the same condition. Thus, it is clear that the observed chiral induction in the gas-solid reaction is due to the ability to hold rigidly the chiral conformation of the CD complex in the crystalline state. This solid state of the inclusion complex is essential but complex formation occurs in solution.

Figure 4 shows the changes of the specific rotations of **2** during the course of the gas-solid bromination with the α-CD complex. The values were constantly up to 90% conversion of styrene, but decreased with an increase of reaction time after 2 h. Furthermore, after exposure for 40 h the value decreased to 44% of that for a 2 h exposure. When the optically active **2** ($[\alpha]_D^{25}$ −47.0°) included in the crystalline β-CD complex, not done in α-CD, was exposed to bromine vapour at 0 °C for 20 h, the optical purity of recovered **2** ($[\alpha]_D^{25}$ −33.2°) also decreased to 29.4%. In contrast, the optically active **2** crystal (mp 72 °C) did not racemize under bromine vapour. These results show that the racemization was catalyzed by CDs through exposure of bromine vapour for a long time.

In an additional result, the optical resolution of racemic **2** by the method of Cramer [15] did not recover the optically active **2** from the β-CD inclusion complex, and the racemic **2** did not form the inclusion complex with α-CD having at narrower cavity. Therefore, the formation of the β-CD complex with the racemic **2** was completely non-stereospecific.

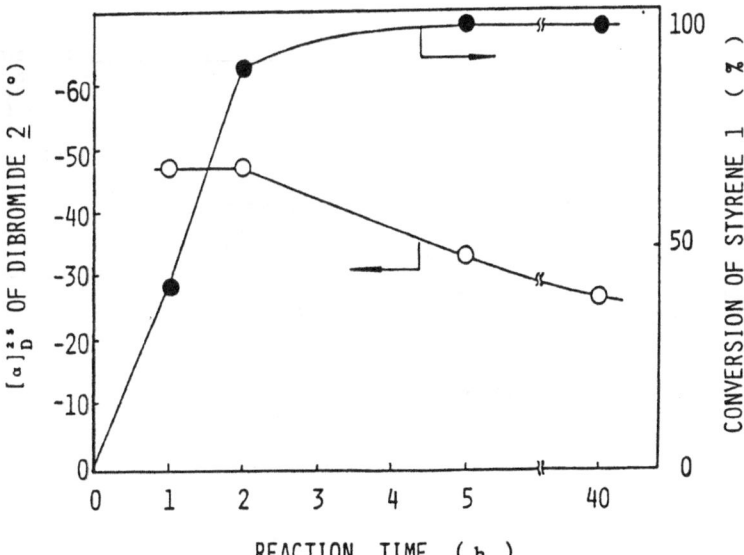

Fig. 4. Changes of specific rotations of dibromide during the course of the gas-solid bromination of styrene in the α-CD complex at 0 °C. ○, Change of specific rotations of dibromide **2**, measured in dichloromethane; ●, change of conversions (%) of styrene **1**.

3.3.2. *Gas-Solid Chlorination*

The gas-solid chlorination of the α-CD complex did not proceed at -20 °C. Dichloride **4** ($[\alpha]_D^{25}$ $-4.1°$) was obtained at 0 °C, but repeating the experiments two more times lacked reproducibility for the yields (0 ~ 3%) and the specific rotations ($[\alpha]_D^{25}$ 0 ~ $-2°$) of **4** under the same condition. Chlorination at a higher temperature such as 25 °C gave no dichloride **4** but *S*-chlorohydrin **5** (14% ee) and trichloride **7** ($[\alpha]_D^{25}$ $+6.5°$), as shown in Table I. Chlorination of the β-CD complex, however, proceeded at -20 °C for 48 h, gave the optically active products, **4** and **7** (the same specific rotation of $[\alpha]_D^{25}$ $+20.5°$), racemic **5** and (*E*)-1-chloro-2-phenylethene **6** in 47% yield (**4**:**5**:**6**:**7** = 43:15:12:30). The reaction at 25 °C gave no chloroolefin **6** but three optically active products, **4**, **5** and **7** ($[\alpha]_D^{25}$ $+4$ ~ $+7°$). The chlorination afforded different product species from those obtained in the bromination. The formation of **7** as a typical different product may proceed through the addition of chlorine to **6**, which is formed by dehydrochlorination between a free chloride anion and an open carbonium intermediate (A in Figure 5) involving the C_α–C_β rotation of the latter [16]. In the case of the bromination, the reaction proceeds through a strong bridged ion intermediate, so that the C_α–C_β rotation of the intermediate is difficult to occur, thus showing no dehydrobromination. Furthermore, since the trichloride **7** was the optically active product, both the dehydrochlorination and the subsequent chlorination of **6** should proceed in the chiral cavities of CDs without an escape of the guest molecule from the cavities. The chlorohydrin **5** was also produced by nucleophilic attack of water to a carbonium cation intermediate (C in Figure 5) similar to the formation of **3** on bromination, but was the optically active product (8 ~ 14% ee) different from the racemic **3** formed by bromination. In the chlorination of anisole in aqueous solution, Breslow *et al.* [17] suggested the covalent participation by the CD: the chloronium cation transferred first to a

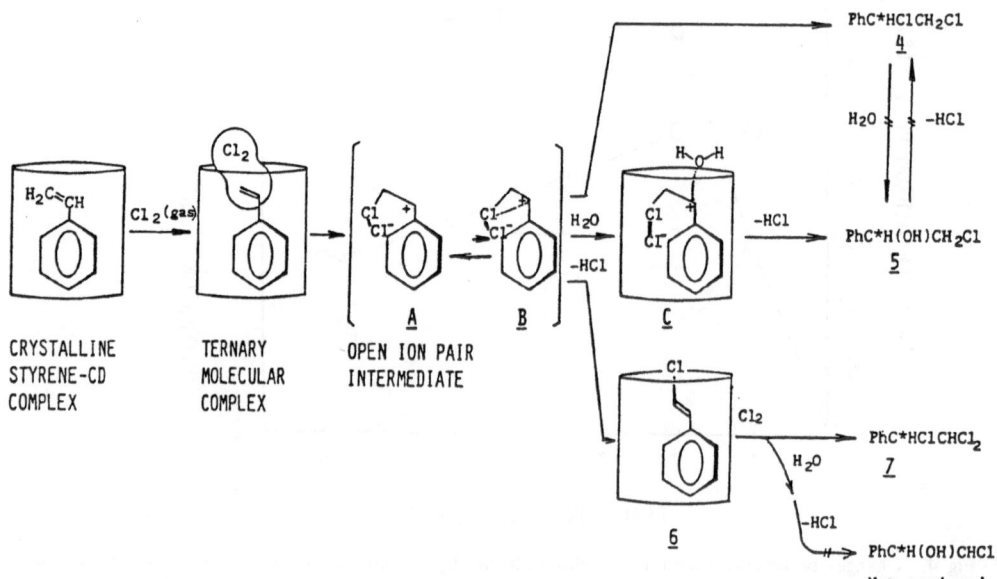

Fig. 5. Reaction mechanism of the gas-solid chlorination of styrene in the CD complexes.

hydroxyl group of the CD, then to the guest. Such a scheme also seems to occur in this reaction. However, no chloride-substituted products (e.g., p-chlorostyrene derivatives) and chlorinated CD were detected by ^{1}H NMR analysis of the recovered extract and CD in 95~98% yields after reaction. Thus, in the solid state, the hydroxyl groups of the CD should be less reactive than those in solution [17].

In the chlorination of the β-CD complex, the optical yields of **4** and **7** increase about three times on lowering the temperature from 25 to −20 °C. Judging from the result that the chiral induction and reactivity for the gas-solid chlorination of styrene included in β-CD depends on the temperature, the motion of the guest molecule within the host molecule should be restricted by decreasing temperature. This result, however, is different from that of the gas-solid chlorination of (E)-cinnamic acid in the β-CD cavity, where the optical yields of products were nearly constant over the temperature range −25 to 50 °C as reported previously [9].

Enantioselectivity for the chlorination with both the α- and β-CD complexes provided the products, **5** and **7** but no **4**, with the same configuration similar to that of the bromination.

3.4. GAS-SOLID HYDROHALOGENATION

Table II shows the results for the gas-solid hydrohalogenation of the α-and β-CD complexes with styrene at 25 °C. Both the CD complexes produced racemic **8a** and **9** in 30~50% yields (**8a**:**9** = 1:1) under hydrobromination for 3 h, and no **9** but racemic **8b** in 30~40% yields under hydrochlorination for 20 h.

Table II. Gas-solid hydrohalogenation of styrene in crystalline CD complexes at 25 °C

Host	Gaseous reagent	Time h	Yield[a] %	Product[b] compn./%	ee %
α-CD	HBr	3	28	52 (**8a**)	0
				48 (**9**)	0
β-CD	HBr	3	52	51 (**8a**)	0
				49 (**9**)	0
α-CD	HCl	20	29	100 (**8b**)	0
β-CD	HCl	20	37	100 (**8b**)	0

[a] Based on styrene consumed estimated from ^{1}H NMR analysis.
[b] Based on the total amount of isolated products obtained by silica gel chromatography. Percentages are normalized to 100%.

The hydrohalogenation shows a remarkable decrease in the enantioselectivity of halide anions attacking the carbonium cation intermediate, formed by first addition of the acids to the olefin [18]. This nonchiral induction suggests that the rotation of the groups on C_α of the intermediate occurs even in the crystalline CD complexes.

4. Conclusion

Asymmetric bromination and chlorination of styrene are achieved in microcrystalline cyclodextrin complexes. It is attributed to the ability to hold rigidly a chiral conformation of the crystalline state. The homogeneous reaction shows no chiral induction. The gas-solid hydrohalogenation, however, gave racemic products.

References

1. B. S. Green, M. Lahav, and D. Rabinovich: *Acc. Chem. Res.* **6**, 191 (1979).
2. K. Penzien and G. M. Schmidt: *Angew. Chem., Int. Ed. Engl.* **8**, 608 (1969).
3. M. L. Bender and M. Komiyama: *Cyclodextrin Chemistry*, Springer-Verlag, Berlin (1978).
4. N. Baba, Y. Matsumura, and T. Sugimoto: *Tetrahedron Lett.* 4281 (1978).
5. S. Banfi, S. Colonna, and S. Julia: *Synth. Commun.* **13**, 1049 (1983).
6. A. W. Czarnik: *J. Org. Chem.* **49**, 924 (1984).
7. J. Drabowicz and M. Mikolajczyk: *Phosphorus and Sulfur* **21**, 245 (1984).
8. Y. Tanaka, H. Sakuraba, and H. Nakanishi: *J. Chem. Soc., Chem. Commun.* 947 (1983).
9. H. Sakuraba, T. Nakai, and Y. Tanaka: *J. Incl. Phenom.* **2**, 829 (1984).
10. Y. Tanaka, H. Sakuraba, Y. Oka, and H. Nakanishi: *J. Incl. Phenom.* **2**, 841 (1984).
11. O. Shimamura and M. Takahashi: *Bull. Chem. Soc. Jpn.* **22**, 60 (1949).
12. J. A. Riddick and W. B. Bunger: *Organic Solvents*, Wiley-Interscience, New York (1970).
13. M. C. Cabaleiro, M. D. Johnson, B. E. Swedlund, and J. E. Williams: *J. Chem. Soc. B*, 1022 (1968).
14. L. C. J. van der Laan, J. B. N. Engberts, and T. J. de Boer: *Tetrahedron* **27**, 4323 (1971).
15. F. Cramer and W. Dietsche: *Chem. Ber.* **92**, 378 (1959).
16. R. C. Fahey and C. Schubert: *J. Am. Chem. Soc.* **87**, 5172 (1965).
17. R. Breslow and P. Campbell: *J. Am. Chem. Soc.* **91**, 3085 (1969).
18. F. Freeman: *Chem. Rev.* **75**, 439 (1975).

… *Journal of Inclusion Phenomena* 5 (1987), 459–468.
© 1987 by D. Reidel Publishing Company.

^{13}C Nuclear Magnetic Resonance Spectra of Cyclodextrin Monomers, Derivatives and their Complexes with Methyl Orange

MIYOKO SUZUKI*, YOSHIO SASAKI
Faculty of Pharmaceutical Sciences, Osaka University 1-6, Yamadaoka, Suita, Osaka, 565, Japan

JÓZSEF SZEJTLI, ÉVA FENYVESI
Chinoin Pharmaceutical and Chemical Works, Tó u. 1-5, H-1045 Budapest, Hungary

(Received: 20 October 1986; in final form: 17 March 1987)

Abstract. Low molecular mass fractions of water soluble α-, β-, and γ-cyclodextrin epichlorohydrin polymer products (cdx-Ep) were characterized by ^{13}C nuclear magnetic resonance. The derivatives proved not to be polymers, but substituted cdx having one or two glyceryl groups per one glucose at the C-2, C-3 and C-6 positions. Spectra of analogous hydroxy-propyl β-cdx indicate that the degree of substitution is rather higher at the C-6 position. Methyl orange (MO) was included into nine kinds of cdx having different inner diameters and hydrophobic torus heights; α-, β-, and γ-cdx monomers, 2, 6-dimethyl and 2, 3, 6-trimethyl β-cdx, water soluble α-, β-, and γ-cdx-Ep and ethyleneglycol-bis(epoxy-propyl) ether products. The inclusion shifts were compared with each other and with the dioxane-induced solvent shift of MO. The N, N-dimethyl-aniline side of MO shifted to a higher field site with the increase of the inner diameter in cdx. By substituting cdx with ether groups of different length, the mechanism of inclusion formation remains substantially the same, but by lengthening the hydrophobic cavity, the hydrophobic interaction becomes stronger, as a better resemblance of inclusion shifts and solvent shifts can be observed.

Key words: Alpha, beta, gamma, cyclodextrin, epichlorohydrin, carbon NMR, methyl-β-cyclodextrin, methyl orange, inclusion shift, solvent shift.

1. Introduction

Water soluble cyclodextrin polymers are useful in the pharmaceutical area because of their high solubility and nontoxity. The characterization of the above products is worth studying because they can be studied in solutions and can be a clue to characterize the insoluble polymers. In the previous paper [1], the circular dichroism (CD) spectra of methyl orange (MO) were recorded in the presence of separated fractions of water soluble polymers prepared by reacting cdx with epichlorohydrin (cdx-Ep) or with ethyleneglycol-bis(epoxypropyl) ether (cdx-DiEp) as well as in the presence of the parent monomers, 2, 6-dimethyl β-cdx (DMβ) and 2, 3, 6-trimethyl β-cdx (TMβ). The complexes with α-cdx-Ep and α-cdx-DiEp showed an exciton splitting in the induced π-π* band of the N=N group. The above phenomenon disappeared when larger host molecules and azo dyes were used. In the present work, ^{13}C nuclear magnetic resonance (NMR) spectra were recorded on all the above compounds. Specific emphasis was placed on the following points;

---

* Author for correspondence.

1. positions and degrees of substitution and polymerization of the products obtained.
2. whether the increase of the inner diameter in cdx and the lengthening of the hydrophobic torus by substituting with the linking agents and methyl groups affect the inclusion properties.
3. whether the included geometry may be deduced by compáring the inclusion shift with the chemical shift of the guest molecule in dioxane.

2. Materials and Methods

Cdx-Ep and cdx-DiEp products were prepared according to [1]. DMβ and TMβ were prepared according to [2]. Hydroxypropyl β-cdx was a generous gift from Dr J. Pitha. Other compounds used are the same as described previously [1, 3]. ^{13}C NMR spectra were recorded on a Hitachi R-22 CFT spectrometer (22.5 MHz) and a JEOL JNM-GX 500 FT NMR (125 MHz) at a temperature of 307 °K. Host molecules of the order of ~0.05M were added to ~0.05M guest molecules. Chemical shifts were measured in ppm downfield from external tetramethylsilane (TMS). A positive sign indicates a low field shift.

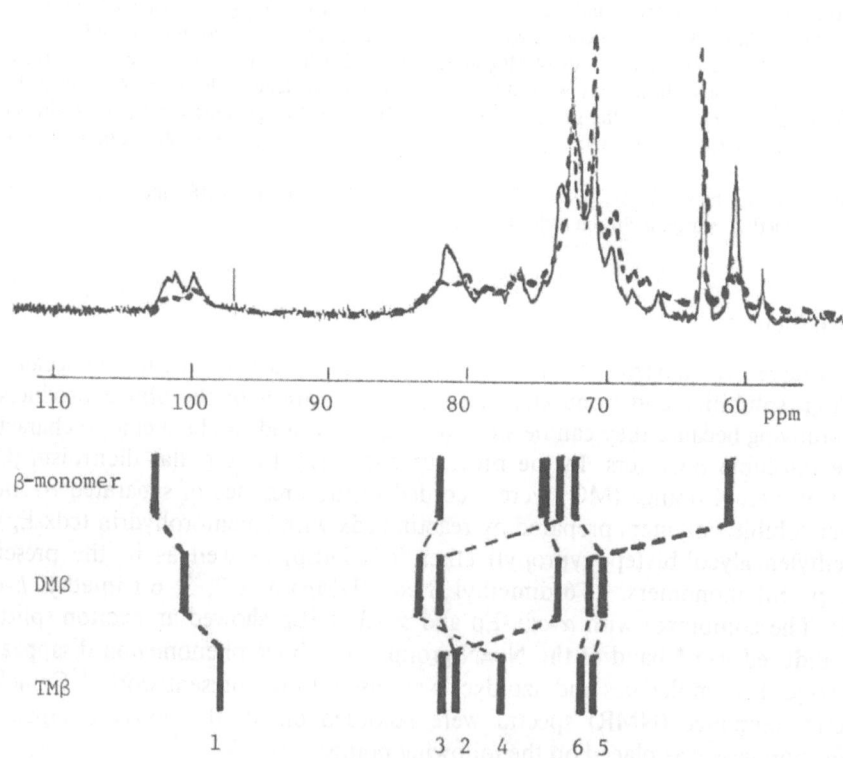

Fig. 1. ^{13}C NMR spectra of β-cyclodextrin epichlorohydrin products (125 MHz). ——— Low molecular mass; ----- high molecular mass.

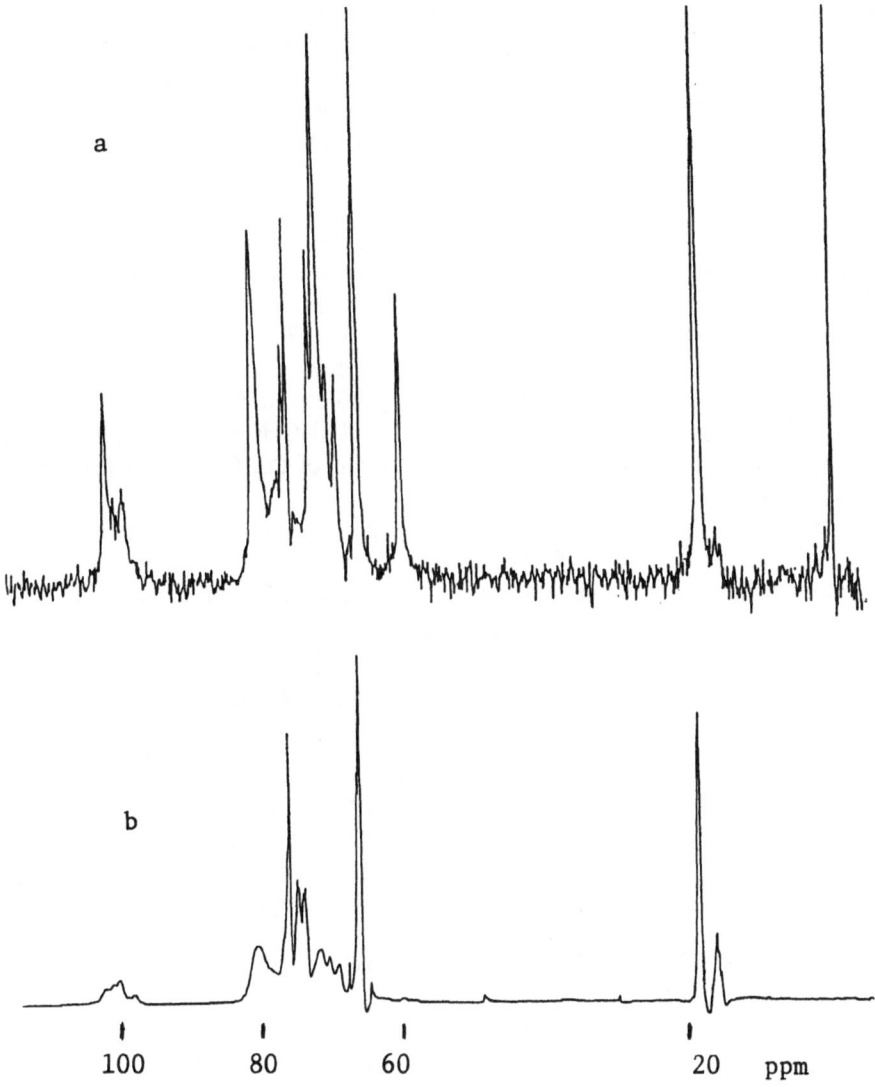

Fig. 2. ^{13}C NMR spectra of hydroxypropyl β-cyclodextrin. (a) Degree of substitution = 7.2; (b) Degree of substitution = 18.4.

3. Results and Discussion

3.1. ASSIGNMENT OF CDX-EP

When cdx is treated with Ep, substitution and crosslinking take place at the same time. So the product is a mixture of cdx glyceryl ethers of different degree of substitution (D.S.) and so called polymers containing 2 or more cdx rings.

$$\text{cdxOH} + \text{CH}_2\text{CHCH}_2\text{Cl} \longrightarrow \text{cdxOCH}_2\text{CHCH}_2 \longrightarrow \text{cdxOCH}_2\text{CHOHCH}_2\text{OH}$$
$$\phantom{\text{cdxOH} + \text{CH}_2}\underset{O}{\diagdown\diagup}\phantom{\text{H}_2\text{Cl} \longrightarrow \text{cdxOCH}_2}\underset{O}{\diagdown\diagup}$$

$$\text{cdxOCH}_2\text{CHCH}_2 \longrightarrow \text{cdxOCH}_2\text{CHOHCH}_2\text{Ocdx}$$
$$\phantom{\text{cdxOCH}_2}\underset{O}{\diagdown\diagup}$$

The above reactions resulted in rather low molecular mass products in the case of α-cdx and β-cdx (α-cdx-Ep and L-β-cdx-Ep, cdx contents 59 and 68%, mass average molecular mass 1400 and 2200, average D.S. = 9 and 7, respectively). In the case of γ-cdx, the product prepared by further crosslinking was a colloidal solution of higher molecular mass (γ-cdx-Ep, cdx content 59%, mass average molecular mass 7500, D.S. = 12).

The ^{13}C NMR spectrum of L-β-cdx-Ep was compared with those of β-cdx, DMβ and TMβ (Figure 1 and Table I). The C-1 signal gives a doublet-like pattern; C-1 and C-1' (101.4 and 100.4 ppm). The C-1' appears as a new band due to the substitution at C-2. The C-4 signal (81.8 ppm) becomes broad due to overlapping by the C-2' and C-3' signals. The new C-4' signal (77.5 ppm) results from the 2, 3, 6-trisubstituted derivative. Three new sharp peaks appear at 72.8, 71.1 and 63.4 ppm. They were assigned to cdxOCH$_2$, CHOH and CH$_2$OH of the substituent CH$_2$CHOHCH$_2$OH and the linking agent OCH$_2$CHOHCH$_2$O. The substitution at C-6 induces the decrease in the peak height of the C-6 signal (61.0 ppm) and the appearance of the C-5' signal (69.9 ppm).

The spectral pattern becomes broader in the high molecular mass β-cdx-Ep (H-β-cdx-Ep, cdx content 59%) prepared by further crosslinking. The solution becomes colloidal, so it was impossible to separate fractions of different molecular

Table I. Chemical shifts of β-cyclodextrin derivatives[a] (ppm from TMS)

	β	DMβ	TMβ	L-β-Ep	D.S. of Hydroxypropyl β-cdx			
					2.5	5.1	7.2	18.4
1	102.3	100.4	97.5	101.4	102.4	102.4	102.2	102.4
1'				100.4			100.9	101.3
							100.0	100.4
1''								98.6
2	72.7	82.0	80.6		72.5	72.6	72.5	
2''				~82				
3	73.7	73.2	81.5	73.6	73.7	73.7	73.6	72.5
4	81.6	82.9	77.5	81.8	81.7	81.7	81.5	81.3
4'				77.5				
5	72.3	70.4	70.9		72.5	72.6	72.5	
5'				69.9	69.8	70.1	70.0	70.0
6	61.0	71.3	71.4	61.0	60.9	60.9	61.0	
6'				71.2	71.3	71.4	71.1	71.3
2-Me		60.1	58.6					
3-Me			60.2					
6-Me		58.6	58.9					
CH$_2$OR				72.8	77.6	77.7	77.5	76.7
					76.8	76.9	76.8	75.7
								74.5
CHOH				71.1	67.1	67.2	67.2	67.3
					66.8	66.8	66.8	66.7
								68.4
Me					19.0	19.1	19.0	19.0
					18.7	18.8		16.5
CH$_2$OH				63.4				

[a] Assigned by selective decoupling, off resonance and peak height.
' Indicates new shifts induced by the substitution.

Table II. Methyl orange-induced ^{13}C chemical shifts of cyclodextrin (ppm)

	α	α-Ep	β	β-Ep	DMβ	TMβ	γ	γ-Ep
1	0.6	0.6	1.0	0.3	0.0	2.7	0.7	
1'		0.6		0.1				
2	0.2		0.1		-0.2	1.0	0.0	
3	0.1		0.3		0.0	0.6	0.1	
4	-0.2	1.5	-0.1	-0.4	0.0	4.1	0.1	0.0
5	-0.1		0.3		0.1	-0.3	-0.1	
6	-0.5	-0.2	-0.4	-0.6	-0.4	-0.1	-0.6	-0.2
6'				-0.4				0.3
2-Me					-0.2	-0.1		
3-Me						1.4		
6-Me					-0.1	0.0		
CH$_2$OR		0.0		-0.3				0.2
CHOH		0.1		-0.2				0.1
CH$_2$OH		0.0		-0.3				0.0

mass by gel chromatography to calculate mass average molecular mass. The C-3 signal (73.6 ppm) disappears. The intensities of the C-2' and C-5' signals (81 ~ 80 ppm and 69.9 ppm) increase and those of the C-4 and C-6 signals (81.6 and 61.0 ppm) decrease. The degree of the substitution increases. Comparison of the peak heights of cdxOCH$_2$ and CH$_2$OH in the glycerylether group may indicate the degree of polymerization. Now the CH$_2$Ocdx signal in H-β-cdx-Ep is better separated from those of the unreacted carbons of cdx (C-2, C-3, C-5) than that in the case of L-β-cdx-Ep in the measurement at 125 MHz. Qualitatively, in the case of H-β-cdx-Ep, the peak height CH$_2$OH is rather higher than that of cdxOCH$_2$. This fact means that the soluble fraction measured in H-β-cdx-Ep does not contain the polymer part, but only an increase in the degree of the substitution. The spectra of α-cdx-Ep and the soluble fraction in γ-cdx-Ep were almost the same as that of L-β-cdx-Ep, though the latter obviously contains the fraction above 10 000 molecular mass weight.

Hydroxypropyl β-cdx of different D.S. will give clearer spectra, because the CH$_3$ and CHOH signals in the substituent are separated from those of cdx. It is therefore possible to examine the behavior of the substituent (Figure 2 and Table I). In the spectrum of D.S. = 2.5, new signals appear; C-6' and C-5' (71.3 and 69.8 ppm), CH$_2$Ocdx (77.6 and 76.8 ppm), CHOH (67.1 and 66.8 ppm), CH$_3$ (19.0 and 18.7 ppm). With increasing D.S. values, the above peaks increase in intensity, but remains split. In the spectrum of D.S. = 7.2, the C-1 signal begins to separate (102.2, 100.9 and 100.0 ppm). The neighborhood of the C-4 signal becomes more intense and broader due to the overlapping by the C-2', and C-3'? signals. In the spectrum of D.S. = 18.4, the C-1 signal becomes broader (102.4, 101.5, 100.4 and 98.6 ppm) and the C-6 signal disappears. The former may explain the substitution to C-3. All signals in the substituent group split. This fact must be due to the substitution to the different positions of β-cdx. Judged from the pattern of the C-1 signal, the substitution at C-3 cannot be perceived in the D.S. = 2.5 spectrum. So the splitting of the signal in the substituent may be ascribed to the substitution at C-2 and C-6; 77.6 and 67.1 ppm are ascribed to cdx-C-2-OCH$_2$CHOHMe and cdx-C-2-OCH$_2$CHOHMe, 76.8 and 66.8 ppm are ascribed to cdx-C-6-OCH$_2$CHOHMe and cdx-C-6-OCH$_2$CHOHMe. From comparison of the above peak heights, it is possible to estimate the ratio of the substitution (C-6/C-2): it is roughly 3/2. The ratio is unchanged in the spectrum of D.S. = 7.2 (Figure 2b).

The glyceryl ether parts of α-, β-, and γ-cdx substituted at the 2, 3 and 6 positions at which D.S. are above seven and the ratio of substitution (C-6/C-2) may be rather higher at the C-6 position have been used in the further studies.

3.2. INCLUSION SHIFTS OF MO IN COMPLEXES BETWEEN α-, β-, AND γ-CDX AND MO

Figure 3 gives the α-, β-, and γ-cdx-induced chemical shifts of MO. Inclusion shifts in the α-cdx-complex gather together to low field. When the inner diameter of the cavity becomes large for the guest molecule, inclusion shifts spread to high field, especially on the N, N-dimethyl aniline side. By CPK molecular model, MO does not show any compression with β-cdx, but it shows the compression in the neighborhood of N=N with α-cdx. The fact that inclusion shifts gather together to low field may be due to some strain of the substituents. Both complexations consist of a 1 : 1 molar ratio [3]. γ-cdx-induced chemical shifts of MO gather together to high field. Plots of a molar ratio of γ-cdx/MO vs. change in chemical shifts of MO in Figure 3c are almost zero below a molar ratio of 0.5 for γ-cdx/MO and increase gradually after that (Figure 3c). Job plots give a 1 : 1 molar ratio (Figure 4), but the shape is distorted. These facts suggest the co-existence of the 2 : 1 (MO : γ-cdx) complex. In the investigation of this complex by CD spectra [1], the red shift and the absence of the exciton splitting of the maximum peak in the π-π^* region of the N=N group suggest the parallel planar arrangement of the two MO molecules with respect induced to the

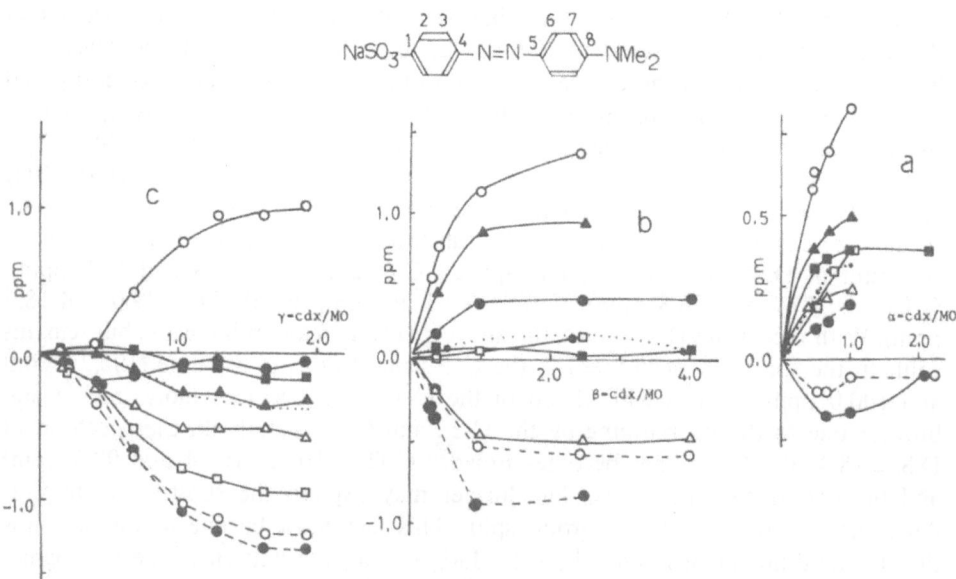

Fig. 3. γ-, β-, α-Cyclodextrin-induced shifts of methyl orange plotted as a function of the molar ratio of γ-, β-, α-cyclodextrin to methyl orange: C-1, ○——○; C-2, ●——●; C-3, ■——■; C-4, □——□; C-5, ▲——▲; C-6, △——△; C-7, ○- - -○; C-8, ●- - -●; Me, ... α-Cyclodextrin complex plots decrease above a molar ratio = 1 for α-cyclodextrin/methyl orange, because a precipitate appears.

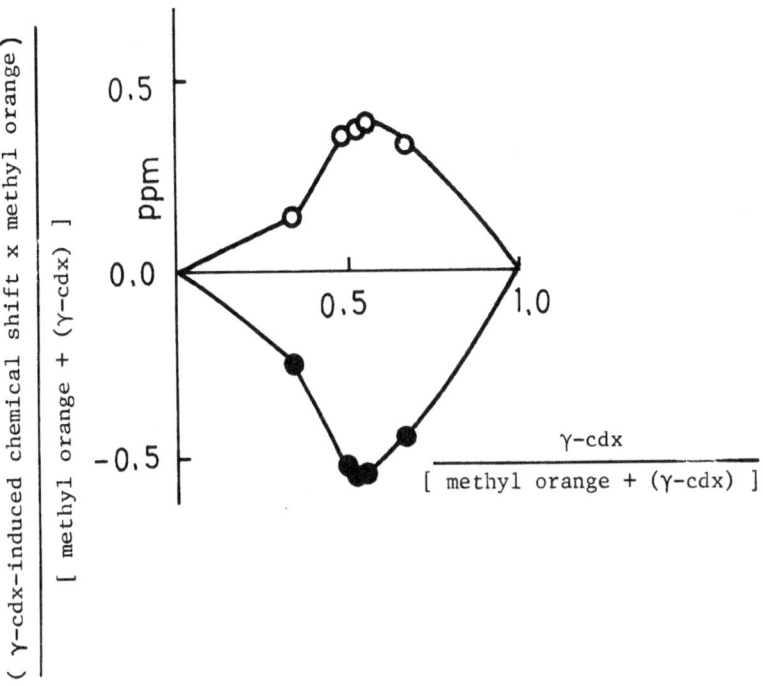

Fig. 4. Continuous variation plots of γ-cyclodextrin-induced chemical shifts of methyl orange at 53°C. C-1, ○; C-8, ●.

annular axis of γ-cdx. In the present ^{13}C NMR work, such a situation induces a ring current effect of MO on each other and may increase the high field shift.

3.3. SOLVENT SHIFTS OF AZO DYES

In general, ^{13}C NMR inclusion shifts may be divided into hydrophobic, van der Waals, dipolar, ring current and steric compression interactions with the various substrate species [4]. In the spectroscopic experiments, the complex formation makes the absorption spectrum of the substrate almost the same as that in dioxane [5], and strengthens the fluorescence degree of 1-anilino-8-naphthalene sulfonate in the fluorescence measurement [6]. In ^{13}C NMR spectroscopy [7], inclusion shifts of the substrate in the strainless host-guest complexes give shifts similar to that in hydrophobic dioxane. Now, to check the role of the hydrophobic interaction in the inclusion shifts, ^{13}C NMR of MO in dioxane was measured (Figure 5). When Figure 5 was compared with Figure 3, the former resembles Figure 3b the most; the inclusion shifts of the β-cdx complex may be mainly due to hydrophobic interaction. On the other hand in the case of the p-(2-hydroxy-1-naphthylazo) benzenesulfonic acid sodium salt (orange II) which has a larger width, the solvent shifts agree with the γ-cdx-induced shifts rather than those of β-cdx [7]. Concerning longer guest molecules (for example, NaSO$_3$—◯—N=N—◯—NH—◯), the above resemblance was not observed [8]. Thus, the inner size of cdx and the width and the length of the guest molecule sensitively affect the resemblance between both shifts.

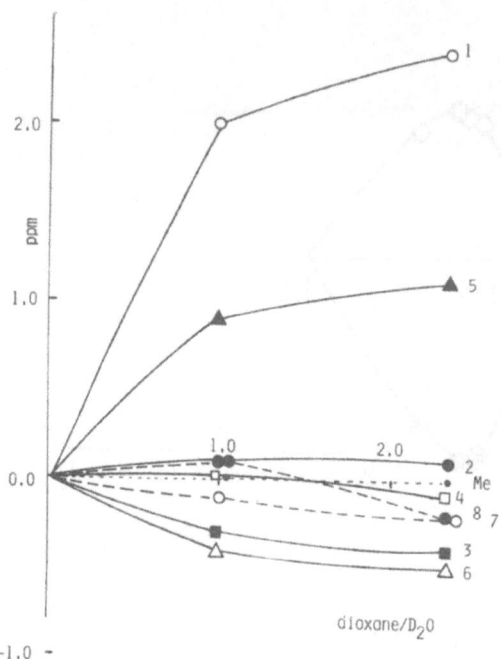

Fig. 5. Solvent shifts of methyl orange in dioxane + D_2O at 53 °C.

3.4. NINE KINDS OF CDX-INDUCED ^{13}C CHEMICAL SHIFTS AND THE DIOXANE-INDUCED SOLVENT SHIFTS OF MO

From the comparison between the inclusion shifts and the solvent shifts, it may be possible to deduce roughly the situation of the included molecule. The lengths of the skeleton carbons on MO and orange II are ~9Å. The torus heights of the monomers and DMβ are ~8Å [9] and ~11Å [2b], respectively. α- and β-cdx-Ep have a longer torus. Thus, nine kinds of cdx having different inner diameters and torus heights were used to include MO. The inclusion shifts were compared with each other and with the dioxane-induced solvent shifts (Figure 6).

Figure 6 shows the cdx-induced chemical shifts of MO with a molar ratio of one for cdx/host molecules. The behavior of the carbons at both ends of MO coincides with those of the parent cdx; in the α-cdx series the C-8 signals move to low field and in the β- and γ-cdx series move to high field. The lengthening of the torus gives the following change in the β-cdx series; inclusion in DMβ and β-cdx-Ep causes the general pattern of MO better to resemble that of the solvent shifts and inclusion in TMβ causes the change to the N, N-dimethylaniline side of MO.

Table II shows the MO-induced chemical shifts of cdx molecules with a molar ratio of one for cdx/host molecules. The values are smallest in DMβ and largest in TMβ. TMβ induces the largest distortion of all the cdx at the wide rim side, and the smallest one at the narrow rim side.

Figure 6 and Table II suggest that DMβ serves as a long and strainless hydrophobic environment (like dioxane solvent) to MO.

CARBON NMR OF CYCLODEXTRIN DERIVATIVES

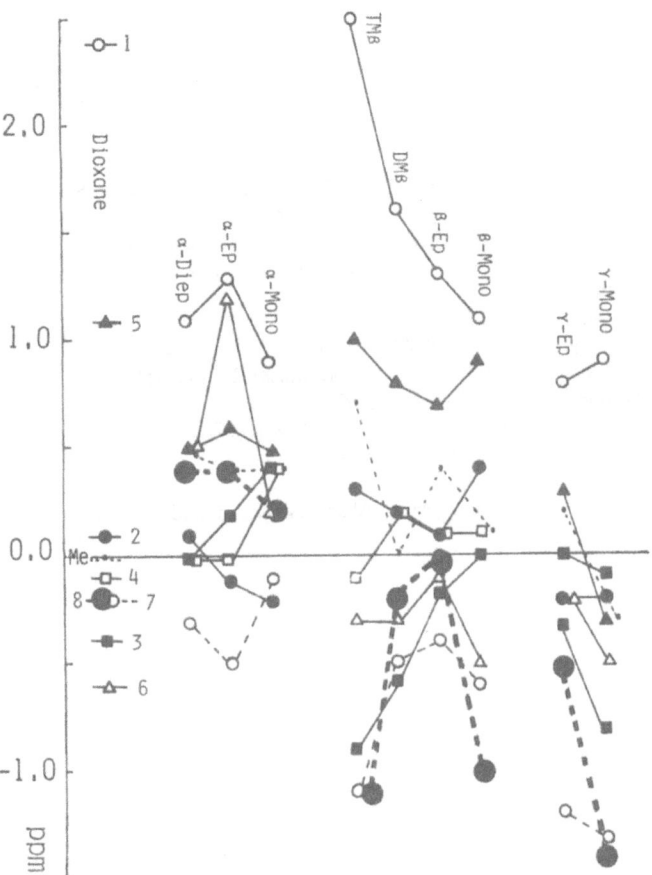

Fig. 6. Nine kinds of cyclodextrin-induced ^{13}C chemical shifts and the dioxane-induced solvent shifts of methyl orange.

4. Conclusion

1. Water soluble α-, β-, and γ-cdx-Ep polymer products proved to be not polymers, but substituted cdx having one or two glyceryl groups per one glucose at the C-2, C-3 and C-6 positions. The degree of the substitution may be rather higher at the C-6 position. 2. Comparison between nine kinds of cdx-induced inclusion shifts and dioxane-induced solvent shifts proved that the hydrophobic interaction plays an important role in inclusion. Moreover, the inclusion shifts in the α-cdx-MO series induce van der Waals and/or compression effects at the N, N- dimethylaniline side and those in γ-cdx-MO series induce the anisotropic ring current effect by the parallel planar arrangement of the two MO molecules included parallel to the annular axis of γ-cdx. 3. By substituting cdx with ether groups of different length, the mechanism of inclusion formation remains substantially the same, but the hydrophobic interaction becomes stronger.

Acknowledgement

The authors are grateful to Dr J. Pitha of NIH/National Institute on Aging Gerontology Research Center for the generous gift of hydroxypropyl β-cdx.

References

1. This part is Part VI of Cyclodextrin and Azo Dyes Part V. M. Suzuki, E. Fenyvesi, M. Szilasi, J. Szejtli, M. Kajtár, B. Zsadon, and Y. Sasaki: *J. Incl. Phenom.* **2**, 715 (1984).
2. J. Szejtli: *Proceeding of the 1st Symposium on Cyclodextrins*, Akademiai Kiadó, Budapest, (1982).
   (a) A. Lipták, P. Fügedi, Z. Szirmai, J. Imre, P. Nánasi, and L. Szejtli: p. 275.
   (b) John J. Stezowski, M. Czugler, and E. Eckle; p. 151.
3. M. Suzuki and Y. Sasaki: *Chem. Pharm. Bull.* **27**, 609 (1979).
4. R. Bergeron and M. A. Channing: *Biorg. Chem.* **5**, 437 (1976).
5. (a) Y. Shibusawa, T. Hamayori, and R. Sasaki: *Nihon Kagaku Kaishi*, 2121 (1975).
   (b) Y. Shibusawa and Y. Hirose: *Sen-i Gakkaishi* **29**, 1 (1973).
6. F. Cramer, W. Saenger, and H.-Ch. Spats: *J. Am. Chem. Soc.* **89**, 14 (1967).
7. M. Suzuki and Y. Sasaki: *Chem. Pharm Bull.* **32**, 832 (1984).
8. In preparation.
9. J. Szejtli: *Cyclodextrins and their Inclusion Complexes*, Akademiai Kiadó, Budapest (1982), p. 25.

Journal of Inclusion Phenomena 5 (1987), 469–472.
© 1987 by D. Reidel Publishing Company.

Complexes of Na-, Ca-, and Zn-Montmorillonites with an Aminated Cyclodextrin

TSUYOSHI KIJIMA*, SATOSHI TAKENOUCHI
National Institute for Research in Inorganic Materials, Sakura-mura, Niihari-gun, Ibaraki, 305 Japan.

and

YOSHIHISMA MATSUI
Department of Agricultural Chemistry, Shimane University, Nishikawazu, Matsue, 690 Japan

(Received: 26 September 1986; in final form: 1 December 1986)

Abstract. The uptake of mono-(6-β-aminoethylamino-6-deoxy)-β-cyclodextrin (CDen) by Na-, Ca- and Zn-montmorillonites has been examined at 25 °C. Each of the first two minerals forms only one intercalated phase in which the CDen molecules are intercalated as a monolayer, while the third yields a mixture of two intercalated phases whose interlayer spaces are occupied by mono- and bilayers of the guest molecules. Intercalation proceeds by ion exchange with the interlayer cations for the Na-complex and by forming metal-aminoethylamino chelate complexes for the others.

Key words: Montmorillonite, aminated cyclodextrin, intercalation.

1. Introduction

Recently we prepared a complex of layered Cu(II) montmorillonite with mono-(6-β-aminoethylamino-6-deoxy)-β-cyclodextrin (CDen) [1]. It was also found that the CDen molecules in their neutral or cationic forms are taken up by the formation of complexes with the interlayer Cu(II) ions and by displacing protons at the clay surface [2]. These findings aroused our interest in a comparative study on montmorillonites with other interlayer cations.

An attempt was thus made to examine the intercalation properties of CDen toward Na-, Ca-, and Zn-montmorillonites.

2. Experimental

The Na-montmorillonite sample was supplied by Kunimine Industry Co. Ltd. The Ca and Zn exchanged minerals were prepared as described previously [3]. The compositions and characteristics of these clay minerals are given in Table I. CDen was prepared in the same manner as described previously [1, 2].

Each mineral sample was soaked in an aqueous solution containing 1.0 and 2.0 mmol of CDen per gram of clay at 25 °C for 10 days, centrifuged and air-dried at 40 °C. The X-ray diffraction patterns were taken with a Nippon-Denshi diffractometer at a rate of 1/2° min^{-1} using FeKα radiation. Silicon was used as an external standard.

* Author for correspondence.

Table I. Compositions and interlayer spacings of Na-, Ca- and Zn-montmorillonites

Sample[a]		Interlayer spacing (Å)
Na-mont	$Na_{0.34}K_{0.008}Ca_{0.03}(Al_{1.60}Fe_{0.10}Mg_{0.32})[Si_{3.83}Al_{0.17}]O_{10}(OH)_2 \cdot 6\ H_2O$	12.6
Ca-mont	$Ca_{0.278}(Al_{1.60}Fe_{0.10}Mg_{0.32})[Si_{3.83}Al_{0.17}]O_{10}(OH)_2 \cdot 10\ H_2O$	14.9
Zn-mont	$Na_{0.005}Zn_{0.250}(Al_{1.60}Fe_{0.10}Mg_{0.32})[Si_{3.83}Al_{0.17}]O_{10}(OH)_2 \cdot 11\ H_2O$	12.3

[a] mont = montmorillonite

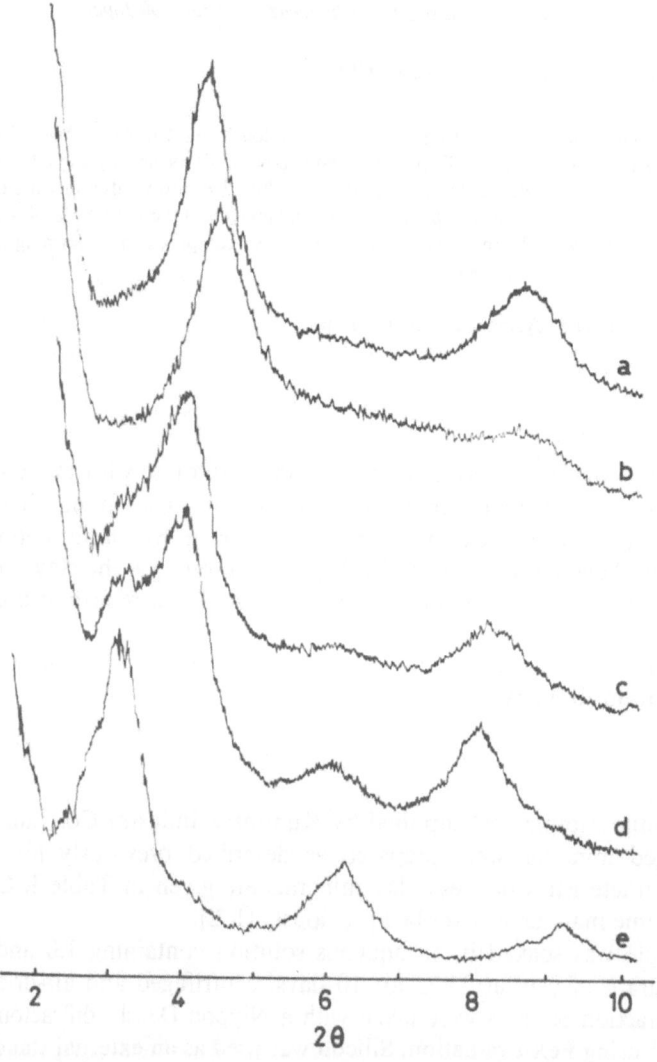

Fig. 1. X-ray diffraction patterns of Na-(a), Ca-(b), Zn-(c,d) and Cu-(e) montmorillonite complexes with CDen (FeKα radiation). CDen addition levels (mmol g^{-1}): 1.0 (a–c), 2.0 (d), 1.4 (e). The data for the Cu-complex are from [1].

Thermogravimetry was performed at a heating rate of 10 °C min^{-1}. The amount of CDen taken up was determined as the difference between the weight losses for the resulting and the starting solids in TG. The amounts of interlayer cations released to solution were determined by atomic absorption analysis.

3. Results and Discussion

Figure 1 shows the X-ray diffraction patterns of the resulting solids, along with that of the CDen-Cu(II)-montmorillonite complex which has an interlayer spacing of 33.4 Å [1]. Table II summarizes the intercalation parameters for the complexes obtained. Here the Δ value indicates the thickness of the intercalated layer determined by subtracting 9.5 Å for the thickness of the silicate layer [1] from the observed spacing. On the assumption that the CDen molecules are placed with their cavity axes perpendicular to the inorganic sheets and hexagonally close-packed to form mono- and bilayers in the interlayer space, the CDen content can be evaluated as 0.30 and 0.60 mmol per g of clay, respectively [2]. These packing models, in combination with an assumption for the location of interlayer cations, also enable us to calculate the Δ value for each complex.

Table. II. Characterization of montmorillonite-CDen complexes with various interlayer cations. The data for the Cu(II)-complex are from [2].

Interlayer cation	Interlayer spacing (Å)	Δ value (Å) obs.	Δ value (Å) cal.	Amount of CDen/ m mol g^{-1} clay^c obs.	Amount of CDen/ m mol g^{-1} clay^c cal.	Amount of water/ m mol g^{-1} clay^c obs.	Amount of interlayer cation/m mol g^{-1} clay^c desorbed	Amount of interlayer cation/m mol g^{-1} clay^c remaining	Type of arrangement
Na	25.6	16.1	15.8	0.35	0.30	6.0	0.70	0.22	I
Ca	24.0	14.5	13.5	0.32	0.30	7.7	0.007	0.736	II
Zn	26.9	17.4	13.5	0.69^b	0.30	11	0.046	0.645	II
	34.7	25.2	22.0^a		0.60				III
Cu	23.2	13.7	13.5	0.28	0.30	5.5	0.01	0.681	II
	33.4	23.9	22.0^a	0.58	0.60	6.7	0.144	0.547	III

^{a} tail-to-tail dimer model ^{b} mixture ^{c} anhydrous form of clay

From the above data, some characteristics are noticed for the intercalation behavior of CDen towards montmorillonites. Each of the Na- and Ca-montmorillonites forms only one intercalated phase with an interlayer spacing of 25.6 or 24.0 Å, while Zn-montmorillonite yields a mixture of two intercalated phases whose interlayer spacings are 26.9 and 34.7 Å. It is also inferred that the 34.7 Å phase for Zn-montmorillonite is a bilayered complex with structure III in Figure 2 and that the other three phases can be ascribed to monolayered phases with structures I and II. Except for the Na-complex, the amounts of cation desorbed during intercalation are much less than those of CDen taken up. This suggests that CDen molecules are taken up by ion exchange with the interlayer cation in the Na-complex and by forming metal chelate complexes in the others. It is well known that the stability constants for metal–ethylenediamine complexes increase in the order of Na<Ca<Zn<Cu [4]. Thus, CDen is likely to form a bilayered intercalate only with montmorillonites

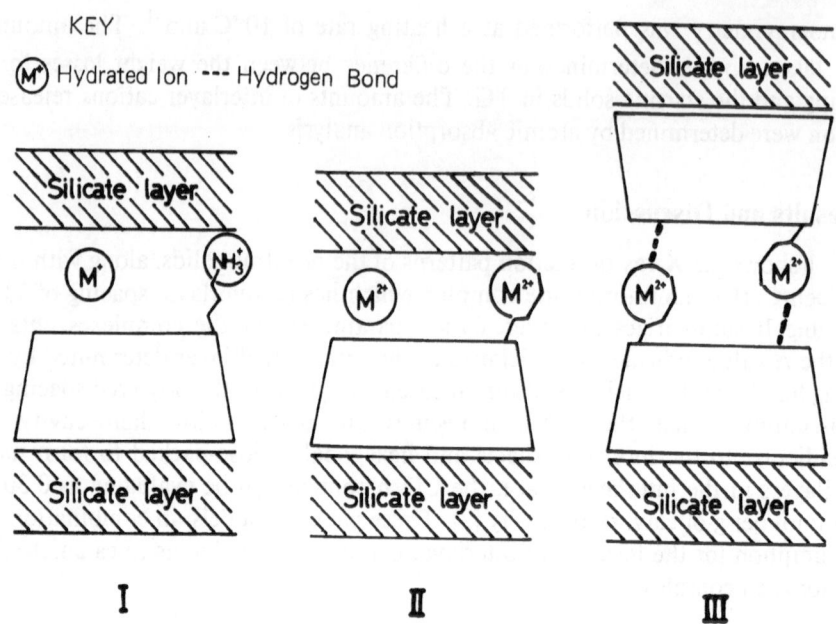

Fig. 2. Models proposed for CDen-montmorillonite formed by ion exchange of the terminal $-NH_3^+$ group with the interlayer monovalent cation (I) and by coordination of the aminoethylamino group to the interlayer divalent cation (II and III); M = Na, Ca, Zn, and Cu(II).

bearing interlayer cations with a high stability constant for the aminoethylamino group. In the previous report on the bilayered CDen-Cu-montmorillonite complex, we presumed that pairs of 1:1 complexes of Cu(II) with CDen are dimerized with either a tail-to-tail or, less likely, a head-to-head arrangement in the interlayer space [1, 2]. The present observation that Ca-montmorillonite forms only a monolayered complex supports the former arrangement because, if CDen molecules were dimerized head-to-head with hydrogen bonds between their secondary hydroxyl groups in the interlayer space of Cu(II)-or Zn-montmorillonite, such an arrangement would also be observable for the Ca exchanged form which takes up CDen by the same mechanism as the other two divalent forms.

References

1. T. Kijima, J. Tanaka, M. Goto, and Y. Matsui: *Nature* **310**, 45 (1984).
2. T. Kijima, M. Kobayashi, and Y. Matsui: *J. Incl. Phenom.* **2**, 807 (1984).
3. T. Kijima: *J. Incl. Phenom.* **4**, 333 (1986)
4. O. Yamauchi and M. Chiguma: *Inorganic Biochemistry* (ed. O. Yamane, H. Tanaka, and N. Kidani), Nankodo, Tokyo (1980) chap. 3.

Synthesis of an Intercalated Compound of Montmorillonite and 6-Polyamide

YOSHIAKI FUKUSHIMA* and SHINJI INAGAKI
Toyota Central Research and Development Labs., Inc. 41-1 Yokomichi Nagakute Aichi-gun Aichi-ken 480-11 Japan

(Received: 7 October 1986; in final form: 16 December 1986)

Abstract. Natural montmorillonite, fractionated from bentonite produced in Yamagata, Japan, was ion-exchanged for NH_3^+—$(CH_2)_{11}$—$COOH$, NH_3^+—$(CH_2)_5$—$COOH$, Al^{3+}, Cu^{2+}, Mg^{2+}, Co^{2+}, Li^+, K^+ and H^+. The mixtures of the ion-exchanged montmorillonite and ε-caprolactam were heated at 263 °C in glass ampoules for various periods. The intercalated compounds before and after the heating were examined by X-ray powder diffraction, DSC and GPC. Although ε-caprolactam was not polymerized without montmorillonite, it was polymerized at 263 °C in the presence of montmorillonite. The polymerization rate varied with the interlayer cations in the order of NH_3^+—$(CH_2)_{11}$—$COOH > Al^{3+} > NH_3^+$—$(CH_2)_5$—$COOH > H^+ > Cu^{2+} > Mg^{2+} > Co^{2+} > Li^+ > K^+$. After heating at 263 °C for 5 h, the mean number-average molecular weight was about 1.5×10^4. Although the interlayer distance of NH_3^+—$(CH_2)_{11}$—$COOH$ type montmorillonite/ε-caprolactam compound increased from 2.85 nm to 4.90 nm by heating at temperatures above the melting point of ε-caprolactam, those of other compounds were not changed. After heating at 263 °C, an intercalated compound of montmorillonite and 6-polyamide, whose interlayer distance was more than 10 nm, was obtained. It is concluded that montmorillonite acts as a Brönsted acid and initiates the open ring polymerization of ε-caprolactam and that the driving force of swelling is the polymerization energy.

Key words: Intercalation, clay mineral, montmorillonite, ε-caprolactam, 6-polyamide, XRD, DSC, GPC, polymerization, catalysis, swelling.

1. Introduction

Intercalated compounds of clay minerals are formed by the introduction of organic and inorganic molecules between the silicate layers. Various applications of the clay/organic complexes, such as absorbents [1], additives for controlling rheological properties of organic solvents [2], separators for optically active metal complexes [3], antioxidants for rubber materials [4], curing agents for synthetic resins [5] etc., have been reported. A composite material of a clay and a polyamide might be a useful material, because the mechanical properties of a polyamide are strongly affected by inorganic additives. Although many studies on clay/organic polymer complexes [6] have been reported, clay/polyamide intercalation has not been studied extensively.

Smectite clay minerals, such as montmorillonite, saponite, nontronite, hectorite, stevensite, vermiculite and haloysite have been known to act as hosts of intercalated compounds [7] and to function as catalysts for various organic reactions [8]. The properties of smectites depend upon the interlayer cations [9], which are easily exchangable for inorganic or organic cations.

---

* Author for correspondence.

Presented at the Fourth International Symposium on Inclusion Phenomena and the Third International Symposium on Cyclodextrins, Lancaster, U.K., 20–25 July 1986.

To understand the catalytic effect of smectites and their interlayer cations on the polymerization of amide molecules and the swelling behaviour of layered silicates by organic polymers, the intercalated compound of cation exchanged montmorillonite and ε-caprolactam and the polymerization of ε-caprolactam in the presence of the montmorillonite have been studied in this work.

2. Experimental

2.1. CATION EXCHANGE OF MONTMORILLONITE

The interlayer cations of high purity natural montmorillonite (particle size < 2 μm) fractionated from bentonite produced in Yamagata, Japan, by suspending in water, (available from Kunimine Industries Co., Ltd.) were exchanged for NH_3^+—$(CH_2)_{11}$—COOH, NH_3^+—$(CH_2)_5$—COOH, Al^{3+}, Cu^{2+}, Mg^{2+}, Co^{2+}, Li^+, K^+, and H^+.

One liter 1.0N aqueous solution of the nitrate of one of the inorganic cations was mixed with 200 ml of a cation-exchanged resin (DuPont 50W-X8), whose cation exchange capacity and particle sizes were 20 meq./ml and 20–50 mesh, respectively. After washing with 2.0 liters of deionized water five times, the cation exchanged resin was mixed with 3 wt % of an aqueous suspension of montmorillonite followed by settling for 24 h at 25 °C. The exchange with organic ions was accomplished by mixing 30 g of montmorillonite powder with one liter of 1.0N aqueous solutions of a chloride of one of the cations, followed by repetitive washing by deionized water and filtration until the chloride ions could not be detected by reaction with $AgNO_3$. The cation exchanged montmorillonites were frozen in liquid nitrogen and vacuum dried at 25 °C. The cations in the montmorillonite were analyzed by atomic absorption after extraction using 1.0N aqueous solution of ammonium acetate. The results of the analysis are shown in Table I.

Table I. The results of the atomic absorption analysis of interlayer inorganic cations of cation exchanged montmorillonite

Type of montmorillonite	Al^{3+}	Cu^{2+}	Mg^{2+}	Co^{2+}	Li^+	K^+	Na^+	Ca^{2+}	Total
	(meq./100 g clay)								
Al^{3+}	31.8	0	1.4	0	0	0.2	4.4	0	37.9
H^+	7.2	0.4	5.0	0	0	0.3	8.0	4.0	24.9
Cu^{2+}	0	53.6	9.0	0	0	11.9	8.9	0.6	84.0
Mg^{2+}	0	1.8	88.8	0	0	0.8	8.1	0.4	99.5
Co^{2+}	0	0.8	19.2	72.0	0	0.6	11.4	1.8	105.8
Li^+	0	0.2	1.8	5.4	90.0	0.4	7.0	0	104.9
K^+	0	0	20.8	0	0	52.9	13.6	0	80.5
$NH_3^+(CH_2)_{11}COOH$	0	0	1.0	0	0	0.1	0.4	0	1.5

2.2. PREPARATION OF INTERCALATED COMPOUNDS

Mixtures of 2.0 g of the cation exchanged montmorillonite powders and 8.0 g of ε-caprolactam were heat treated for 2 h in air at 80 °C, which is about 10 °C higher than the melting point of ε-caprolactam. These mixtures were vacuum dried at

25 °C for 2 h followed by sealing in glass ampoules, and were heat treated at 263 °C for 1, 2, 3 and 5 h. The amount of 6-polyamide was estimated by the heat of fusion of ε-caprolactam, which was measured by DSC on a Seiko DSC-10, using a 15 ml sealed capsule made of aluminum, and an aluminum metal disk as a reference sample. About 10 mg of the sample was used for the DSC measurement, and the heating rate was 5 °C/minute. ε-Caprolactam without montomorillonite was dried, heated and analyzed by DSC in a similar manner.

2.3. MEASUREMENT OF MOLECULAR WEIGHT OF POLYAMIDE

The molecular weight of the organic components in the mixtures of Cu^{2+}, NH_3^+—$(CH_2)_{11}$—COOH, H^+ or Mg^{2+} type montmorillonites and ε-caprolactam heat treated at 263 °C for 5 h were analyzed by gel permeation chromatography (GPC).

The organic components were extracted with m-cresol at 100 °C for 1 h. The molecular weights of the extracted molecules were analyzed using a Waters 150-C GPC and a column of polystyrene gel (Shodex AD-20 μ/s from Showadenko Co., Ltd.) heated at 100 °C, and at a 1.0 ml/minute flow of the 0.25 wt/vol % m-cresol solution. The molecular weight of common 6-polyamide (Amilan CM1017 from Toray Co. Ltd.) was analyzed in a similar manner.

2.4. X-RAY POWDER DIFFRACTION (XRD) STUDY

XRD data on powdered samples were taken using a conventional Bragg-Brentano type diffractometer, where Fe-filtered Co-K_α ($\lambda = 0.1789$ nm) was used as the X-ray source, the tube voltage was 30 kV, the tube current was 20 mA, 1°-0.15 mm-1° ($2° < 2\theta < 10°$) or 0.167°-0.15 mm-0.167° ($1° < 2\theta < 3°$) slits were used and the scanning speed was 2°(2θ)/ minute.

XRD data of the mixtures of liquid ε-caprolactam and NH_3^+—$(CH_2)_{11}$—COOH, NH_3^+—$(CH_2)_5$—COOH, Mg^{2+} or Cu^{2+} type montmorillonites were taken at 80 °C using a diffractometer designed for the examination of liquids and an aluminum sample holder with a silicone plate heater, a thermocouple and an aluminum foil cover. The conditions of the XRD for liquid samples were as follows; Ni-filtered Cu-K_α ($\lambda = 0.1542$ nm) was used. Tube voltage was 30 kV, tube current was 20 mA, 0.5°-0.15 mm-0.5° slits were used and scanning speed was 1.0°(2θ)/minute.

The diffractometers were calibrated by using a standard reference fluorophlogopite mica powder [10] from NBS for the XRD d-spacing.

3. Results

3.1. MOLECULAR WEIGHT OF POLYAMIDE

The results of the GPC analyses are listed in Table II, which shows that ε-caprolactam was polymerized to 6-polyamide, whose mean number-average molecular weight; $\overline{Mw}(N)$, was about 1.5×10^4. The molecular weight, however, was smaller than those of common 6-polyamide, whose $\overline{Mw}(N)$ is about 2×10^4. A small peak or shoulder, which suggests the existence of oligomers, whose $\overline{Mw}(W) \approx 800$, was also observed.

Table II. Mean number-average molecular weight; $\overline{Mw}(N)$, mean weight-average molecular weight; $\overline{Mw}(W)$ and distribution of molecular weight; $\overline{Mw}(W)/\overline{Mw}(N)$ of the 6-polyamide, polymerized at 263 °C for 5 h in the presence of montmorillonites

Type of montmorillonite	$\overline{Mw}(N)$	$\overline{Mw}(W)$	$\overline{Mw}(W)/\overline{Mw}(N)$
$NH_3^+(CH_2)_{11}COOH$	1.4×10^4	1.1×10^5	7.9
H^+	1.4×10^4	1.2×10^5	8.6
Cu^{2+}	1.0×10^4	1.0×10^5	10.0
Mg^{2+}	1.8×10^4	1.0×10^5	5.5

3.2. POLYMERIZATION RATE OF ε-CAPROLACTAM

DSC curves for the mixture of ε-caprolactam and the Al^{3+} type montmorillonite after various periods of treatment at 263 °C are shown in Figure 1. The area of the endothermic peak between 40 and 80 °C, corresponding to melting of ε-caprolactam, decreased and that between 150 and 180 °C, corresponding to melting of 6-polyamide, increased with increasing reaction time. These results also suggest the polymerization of ε-caprolactam.

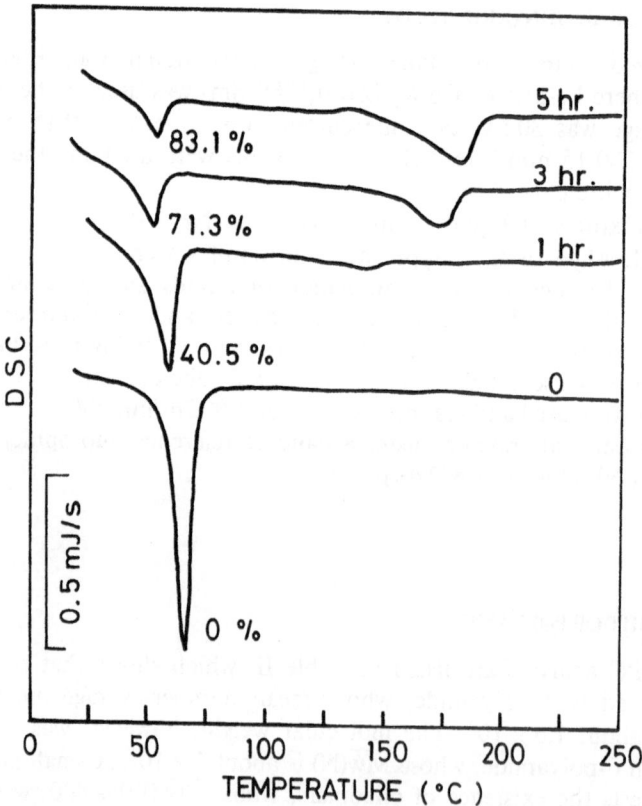

Fig. 1. DSC curves for the mixture of ε-caprolactam and Al^{3+} type montmorillonite after the various periods of treatment at 263 °C.

MONTMORILLONITE-6-POLYAMIDE INTERCALATE

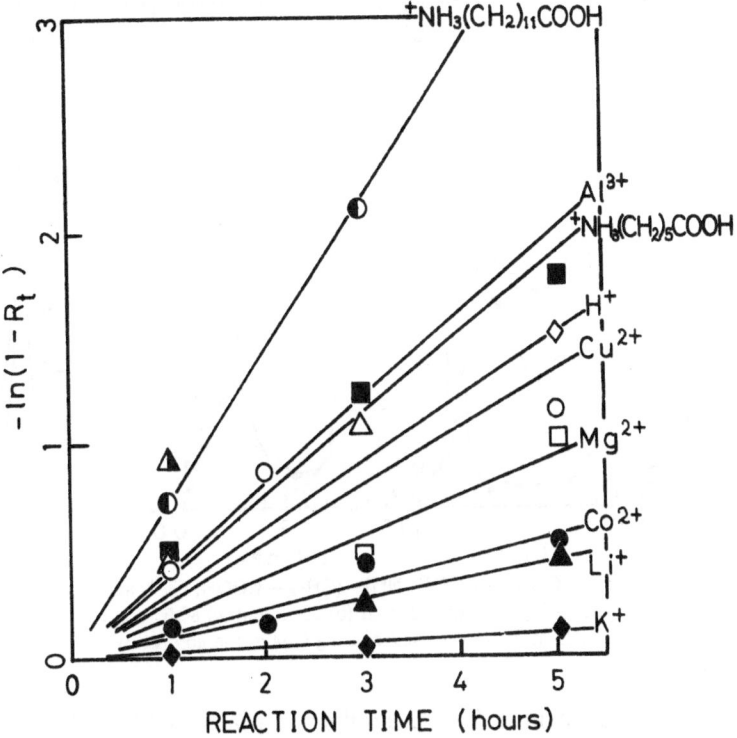

Fig. 2. First order reaction plot for the polymerization of polyamide in the presence of various types of montmorillonite. R_t is the degree of polymerization estimated by Eq. (1).

The degree of polymerization after t hours of treatment at 263 °C, R_t, was estimated using the heat of fusion of ε-caprolactam, according to the equation,

$$R_t = \left(1 - \frac{\Delta H_t}{\Delta H_0}\right), \quad (1)$$

where ΔH_t and ΔH_0 are enthalpy changes calculated by using the endothermic peak between 40 and 80 °C after and before the heat treatment, respectively. The R_t values for the polymerization are shown in Figure 2. The polymerization rate decreased with the interlayer cations of montmorillonite in the order of $NH_3^+-(CH_3)_{11}-COOH > Al^{3+} \geqslant NH_3^+-(CH_3)_5-COOH > H^+ > Cu^{2+} > Mg^{2+} > Co^{2+} > Li^+ > K^+$. However, ε-caprolactam was not polymerized in the absence of montmorillonite and the DSC curve was not altered after 5 h treatment at 263 °C. Even in the presence of montmorillonite, no changes in DSC curves were observed by treatment at a temperature less than 100 °C.

3.3. RESULTS OF XRD

The XRD patterns in the low angle region of $NH_3^+-(CH_2)_{11}-COOH$, $NH_3^+-(CH_2)_5-COOH$, Cu^{2+}, and Mg^{2+} type montmorillonites and the mixture with ε-caprolactam before and after 5 h treatment at 263 °C are shown in Figure 3. Although the (001) peak of Mg^{2+} type montmorillonite with ε-caprolactam was

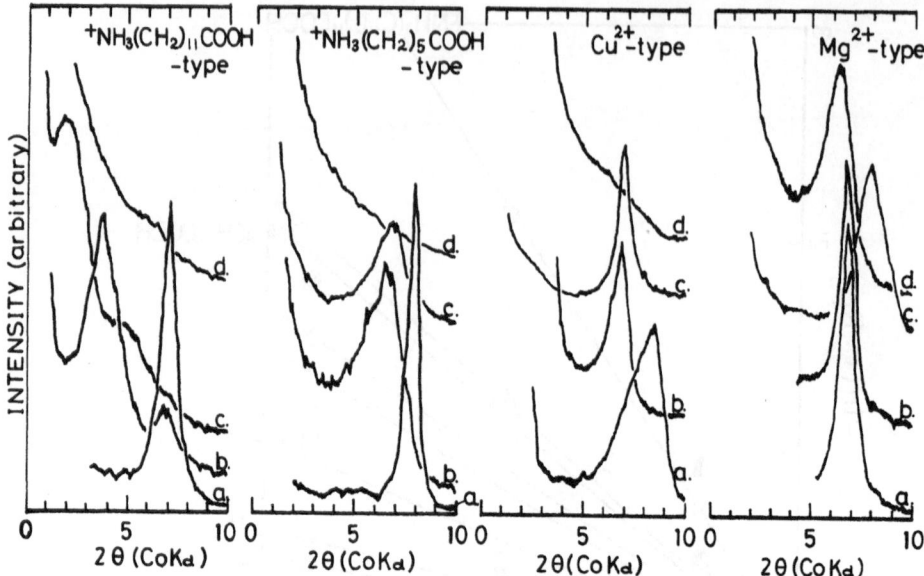

Fig. 3. XRD patterns of (a) NH_3^+—$(CH_2)_{11}$—COOH, NH_3^+—$(CH_2)_5$—COOH, Cu^{2+}, and Mg^{2+} type montmorillonite, and the mixtures with ε-Caprolactam taken at (b) 25 °C and (c) 80 °C and (d) those after 5 h treatment at 263 °C. (Although XRD at 80 °C were taken using CuKα, the abscissa was normalized as if CoKα was used.)

still observed after 5 h treatment at 263 °C, no distinct peaks could be observed but only an increase of the background level was observed for the NH_3^+—$(CH_2)_{11}$—COOH, NH_3^+—$(CH_2)_5$—COOH and Cu^{2+} type montmorillonites with ε-caprolactam. Changes in XRD patterns of the Cu^{2+}-montmorillonite with ε-caprolactam after various periods of treatment at 263 °C are shown in Figure 4. Increasing the heating time, that is increasing the polymerization rate, the (001) peak of montmorillonite was decreased and the background in small angle regions was increased. In Figure 5, XRD patterns of the complexes with various 6-polyamide/NH_3^+—$(CH_2)_{11}$—COOH type montmorillonite ratios are shown. With an increase in the 6-polyamide content, peaks in the low angle region, $d \approx 2.0$ nm, 5.4 nm and 8.4 nm and an increased background level in small angle regions were observed. This result shows that the increase in the background in small angle regions for the mixture of 2 g of montmorillonite and 8 g of ε-caprolactam, which was shown in Figure 3, suggests that the stacking of silicate layers became disordered and/or the mean interlayer distances in these mixtures were more than 10 nm. The mean interlayer distances of montmorillonites, estimated by (001) peak positions are listed in Table III. The interlayer distances increased with the intercalation with ε-caprolactam and the interlayer distances were not changed by heating at temperatures above the melting point of ε-caprolactam, except for the mixture of NH_3^+—$(CH_2)_{11}$—COOH type montmorillonite and ε-caprolactam. For the mixture of NH_3^+—$(CH_2)_{11}$—COOH type montmorillonite and ε-caprolactam, the interlayer distance was increased from 2.85 nm to 4.90 nm by heating to 80 °C and decreased to 2.85 nm with solidification of ε-caprolactam by cooling. After the polymerization, the interlayer distances of NH_3^+—$(CH_2)_{11}$—COOH, NH_3^+—$(CH_2)_5$—COOH, Al^{3+}, H^+, and Cu^{2+} type montmorillonites with ε-caprolactam were increased beyond 10 nm. However, those of the Mg^{2+}, Co^{2+}, Li^{2+} and K^+ type montmorillonite were not changed by the 5 h treatment at 263 °C.

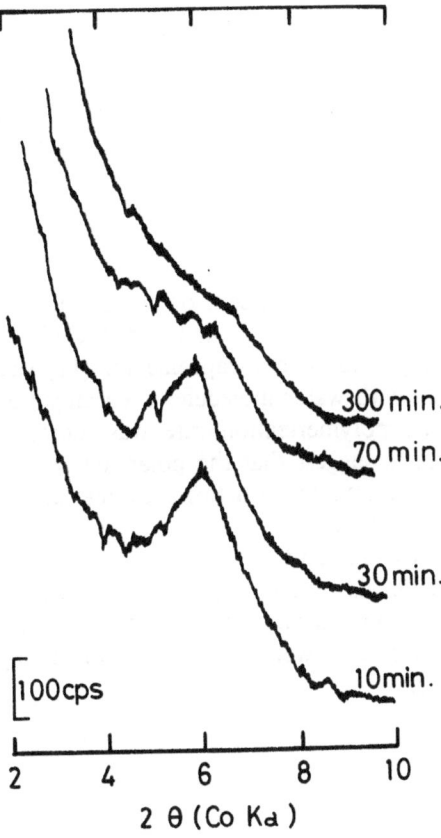

Fig. 4. XRD patterns of Cu^{2+} type montmorillonite/ε-caprolactam mixture after various periods of treatment at 263 °C.

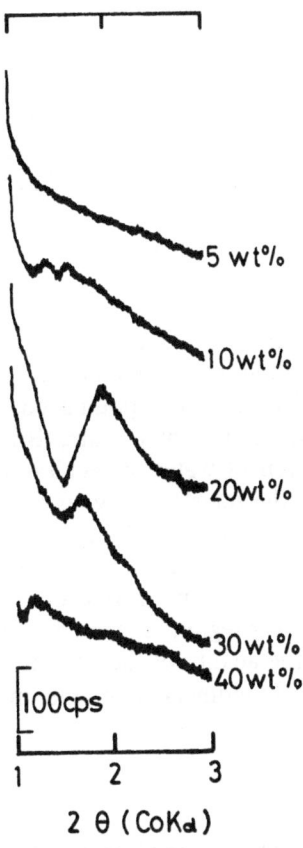

Fig. 5. XRD patterns of complexes with various 6-polyamide/NH_3^+—$(CH_2)_{11}$—COOH type montmorillonite ratio.

Table III. Interlayer distances of montmorillonites and that of the montmorillonite/ε-caprolactam and montmorillonite/6-polyamide complexes

Type of montmorillonite	Interlayer distance (nm)			
		With ε-caprolactam		
		25 °C	80 °C	after × 5 h
$NH_3^+(CH_2)_{11}COOH$	1.65	2.85	4.9	Ind.
Al^{3+}	1.59	1.71	1.7	Ind.
$NH_3^+(CH_2)_5COOH$	1.32	1.47	1.5	Ind.
H^+	1.32	1.51	—	Ind.
Cu^{2+}	1.22	1.51	1.5	Ind.
Mg^{2+}	1.51	1.51	1.5	1.60
Co^{2+}	1.49	1.51	—	1.86
Li^+	1.21	2.00	—	2.00
K^+	1.22	1.51	—	1.53

Ind. = Tendancy to indefinite swelling.

4. Discussion

4.1. ROLE OF MONTMORILLONITE IN THE POLYMERIZATION

It is well known that the smectite clays such as montmorillonite act as Brönsted or Lewis acids and the acidity varies with the interlayer cation. The polymerization rate constant, k, was calculated by using the formula for first order reactions,

$$-\ln(1 - R_t) = kt. \qquad (2)$$

As shown in Figure 6, the polymerization rate is dependent on the strength of the electric field around the interlayer cations. An EXAFS study on the Cu^{2+} type montmorillonite and its intercalated compound with ε-caprolactam [11] suggests that Cu^{2+} in these compounds coordinated with water molecules even after vacuum drying at room temperature. Besides, the polymerization rate was increased by the addition of water molecules. These facts suggest that the polarized water molecules in the interlayer region of montmorillonite act as a Brönsted acid and opens the ε-caprolactam rings to initiate the polymerization.

The results of the previous EXAFS study [11] and of this work suggest that the ring-opening polymerization of ε-caprolactam is initiated by the cations existing in the interlayer region of layered silicates. The polymerization would be controlled not only by the acidity of the cations, but also by the swelling behaviour of the layered silicate by monomers of polymer molecules.

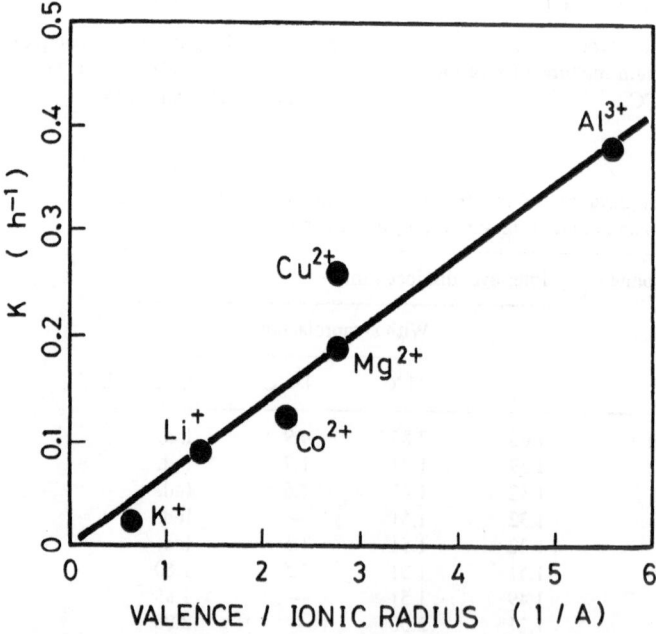

Fig. 6. Relationship between the polymerization rate constant, k, defined in Eq. (2) and the electric field strength around the cations.

4.2. SWELLING BEHAVIOUR OF MONTMORILLONITE

The swelling phenomena of montmorillonites are considered to be a result of a balance between the interlayer cohesive force and the force required to increase the interlayer distance. The latter force in this system would be the attractive force between interlayer cations and ε-caprolactam molecules, which would be increased with the strength of the electric field around the cations, as discussed in the previous section. Since the cohesive force would be the van der Waals or ionic interaction, the cohesive force is expected to decrease with increasing interlayer distance.

As the attractive force between ε-caprolactam molecules and polarized water molecules or cations would be strong enough to increase the interlayer distance, the intercalated compound of montmorillonites and ε-caprolactam was formed. By heating at 80 °C, the attractive force would be enough to increase the interlayer distance of the NH_3^+—$(CH_2)_{11}$—COOH type montmorillonite from 2.85 nm to 4.90 nm, because of the weak interlayer cohesive force due to the large interlayer distance. However, in other montmorillonites, the interlayer cohesive force is strong enough to prevent further swelling. These results correspond well to the result of the study on the swelling behaviour of $CH_3 \cdot (CH_2)_{n-1} N^+(CH_3)_3$ type montmorillonite by liquid toluene, where only the montmorillonites with $n \geq 12$ were swollen by toluene [12].

When the mixtures of montmorillonites and ε-caprolactam are heated at 263 °C in ampoules, the driving force of swelling may be the polymerization energy of ε-caprolactam. Since the polymerization energy was enough to increase the interlayer distance, the interlayer distance was increased remarkably in the system in which a high polymerization rate could be obtained.

Acknowledgements

The authors are grateful to Dr M. Mizuno, Government Industrial Research Institute, Nagoya for his assistance in the XRD study. Discussions with Dr O. Kamigaito and Mr A. Usuki, Toyota R & D Center, were also very helpful for this work.

References

1. P. J. Dodson and P. Somasundaran: *J. Colloid Interface. Sci.* **97** (1984), 481–487.
2. J. T. A. M. Weltzen, H. N. Stein, J. M. Stevels, and C. A. M. Siskens: *J. Colloid Interface. Sci.* **81** (1981), 455–467.
3. A. Yamagishi: *J. Chromatogr.* **262** (1983), 41–60.
4. Y. Fukushima, K. Mori, and A. Murase: *J. Incl. Phenom.* **2** (1984), 305–315.
5. N. Adachi, M. Koizumi, and F. Kanamaru: *Am. Mineralogist*, **60** (1975), 650–658.
6. G. Lagaly and K. Beneke: *Am. Mineralogist*, **60** (1975), 650–658.
7. D. M. C. MacEwan and M. J. Wilson: *Crystal Structures of Clay Minerals and their X-ray Indentification* (Ed. G. W. Brindley and G. Brown) pp. 197–248, Mineralogical Society, London (1980).
8. J. M. Thomas: *Intercalated Chemistry* (Ed. M. S. Whittingham and A. J. Jacobson) pp. 55–100, Academic Press (1982).
9. K. Norrish: *Disc. Faraday Soc.* **18** (1954), 120–134.

10. C. R. Hubbard: *Adv. X-Ray Anal.* **26** (1983), 45.
11. Y. Fukushima, S. Inagaki and T. Okamoto: *Nendo Kagaku* (*J. Clay Sci. Japan*, in Japanese), **26** (1986), 187.
12. Y. Fukushima, to be published in the Proc. of 5th International Symposium on Surfactants in Solution, Bordeaux, 1984.

Isomorphous Substitution Effects on the Thermally Induced Interlayer Reaction in N-Hexylammonium Layered Aluminosilicates

J. H. CHOY* and Y. J. SHIN
Department of Chemistry, Seoul National University, Seoul 151, Korea

G. DEMAZEAU and P. HAGENMULLER
Laboratoire de Chimie du Solide du CNRS – Talence, France

(Received: 24 October; in final form: 3 March 1987)

Abstract. It was found that the variation in the thermally evolved gases obtained by decomposition of n-hexylammonium layered aluminosilicates is mainly due to the difference between octahedral and tetrahedral coordination of aluminium in the lattice and also to the contribution of the excess negative layer charges.

ESCA, TG-DSC, GC and MS results indicate that the layer charge originated from the octahedral substitution induces only desintercalation of n-hexylamine around 250–360 °C, whereas that from the tetrahedral substitution induces the catalytic decomposition reaction involving the cleavage of C—N and C—C bonds at 350–450 °C with the evolution of ammonia, ethylene, pentene and hexene.

It is therefore concluded that the former reaction step is a simple desintercalation, but for the latter one a Brönsted acid catalytic mechanism is proposed.

Key words. N-hexylammonium, montmorillonite, catalytic decomposition reaction, intercalate, layer charge.

1. Introduction

The layer charge in the 2:1 type lattice expanding layered aluminosilicates is known to result from isomorphous substitutions of tetrahedral Si^{4+} or octahedral Al^{3+}, Fe^{3+} and Mg^{2+} by cations of lower charge [1].

The isomorphous substitution site distribution has been studied recently by employing MAS-NMR [2], X-ray photoelectron diffraction [3], IR [4], etc. These studies suggest that the character of the silicate surface should be differentiated according to the site where the isomorphous substitution occurred.

It has been also reported that the layer silicate intercalates can undergo an organic solid-state reaction within the interlayer region, such a reaction can be considered as highly regiospecific due to the distinct surface properties such as surface acidity or layer charge [5–7]. Moreover, size- and shape-selective layer silicates have been prepared [8, 9] and the catalytic properties investigated in recent years [10].

However, up to now, the possibility has not been well studied whether the difference in the surface character attributed to each type of isomorphous substitution can affect the chemical reaction. In this report, we have, therefore, attempted to con-

---

* Author for correspondence.

firm this possibility for the first time and to determine the pathway of a specific de-decomposition reaction of n-hexylammonium derivatives of silicates with different layer charges.

2. Experimental

The natural montmorillonite (Junsei Chem. Co., Japan) and the synthetic saponite (Hoechst Lab., BRD) were used as starting materials. The sodium exchanged form of montmorillonite was prepared by treating with 1.0 N NaCl, centrifuging and discarding the supernatant liquid. The particle size was controlled below 0.2 μm according to the sedimentation rule. The approximate chemical composition of montmorillonite is $Na_{0.70}[Al_{3.92}Mg_{0.08}][Si_{7.38}Al_{0.62}]O_{20}(OH)_4$ and that of saponite is $Na_{0.8}[Mg_{6.0}][Si_{7.2}Al_{0.8}]O_{20}(OH)_4$. The latter shows only tetrahedral substitution in the lattice.

These two silicates were converted into their n-hexylammonium forms with an aqueous solution of 2.0N n-hexylammonium chloride at 65 °C.

The interlayer cation exchange capacity or the layer charge was estimated by the n-alkylammonium method [11].

X-ray diffractograms were obtained on a Jeol instrument using Cu$K\alpha$ radiation with a Ni-filter; samples were examined as oriented aggregates in order to enhance the basal reflections.

ESCA spectra were recorded on a Perkin-Elmer PHI-558 with unmonochromatized Mg$K\alpha$ radiation (1253.6 eV). The separation of an asymmetric peak into two symmetric ones was performed by using a reiterative curve fitting computer program provided by Perkin-Elmer.

The thermally induced interlayer reaction of n-hexylammonium silicate complexes were studied by TG and DSC with a Dupont 1090 Thermal Analyzer. The dehydroxylation of lattice —OH groups, desintercalation and calalytic decomposition of n-hexylammonium in interlayer silicates were measured in a flowing atmosphere of N_2 with a flow rate of 1.0 cm^3/s. The heating rate was 10°/min. for both TG and DSC.

The evolved gas products at various temperatures were identified by GC using a Yanaco 180 G instrument and partly by MS using a Hewlett-Packard 5985B GC-MS system.

3. Results and Discussion

The layer charges of the montmorillonite and the saponite were determined by the n-alkylammonium method [11] as 0.80 and 0.71 per unit cell composition, respectively.

The ion exchange reactions for n-hexylammonium silicate complexes can be formulated as follows:

$$Na^+_{0.71}\text{—Montmorillonite} + 0.71\ n\text{-}C_6H_{13}NH_3^+Cl^- \longrightarrow (n\text{-}C_6H_{13}NH_3^+)_{0.71}\text{—}$$
[A]
Montmorillonite + 0.71 NaCl (1)
[B]

$$\text{Na}^+_{0.80}\text{—Saponite} + 0.80\ n\text{-}C_6H_{13}NH_3^+Cl^- \longrightarrow (n\text{-}C_6H_{13}NH_3^+)_{0.80}\text{—Saponite} +$$
[C]   [D]

0.80 NaCl   (2)

From the X-ray diffraction data, the basal spacings of [B] and [D] were estimated to be ~13.6 Å in both cases, which can be interpreted as the sum of the van der Waals interval occupied by alkyl chains, 4.0 Å, and of that of the silicate layer, 9.6 Å, respectively. It shows that the intercalated cations are oriented with the flat-lying monolayer structure in the interlamellar space of the silicate. The interlayer surface area of an unit cell can be calculated as $a \times b = 5.14\ \text{Å} \times 9.00\ \text{Å} = 46.3\ \text{Å}^2$ for montmorillonite and $5.28\ \text{Å} \times 9.18\ \text{Å} = 48.5\ \text{Å}^2$ for saponite respectively. Since the effective surface area of one n-hexylammonium ion is 45.16 Å^2 [11], the silicate surface is covered by n-hexylammonium cations up to about 69% in the montmorillonite and to 75% in the saponite.

From the ESCA spectra, the binding energies of Al-$2p$ and -$2s$ orbital electrons of both silicates have been obtained. The asymmetric line shape of the Al-$2p$ signal, as shown in Fig. 1, was observed because of the octahedral and tetrahedral coordinations of aluminium in the lattice and it was separated into two peaks with the help of a curve fitting program. One of the separated peaks at 73.2 eV is attributable to Al-$2p$ of a tetrahedrally coordinated aluminium atom, because it corresponds exactly to the Al-$2p$ line of synthetic saponite, where all the aluminium atoms occupy the tetrahedral sites. The major peak at 74.3 eV is consequently assigned to the Al-$2p$ of octahedrally coordinated atoms. The peak shift of about 1.0 ± 0.1 eV

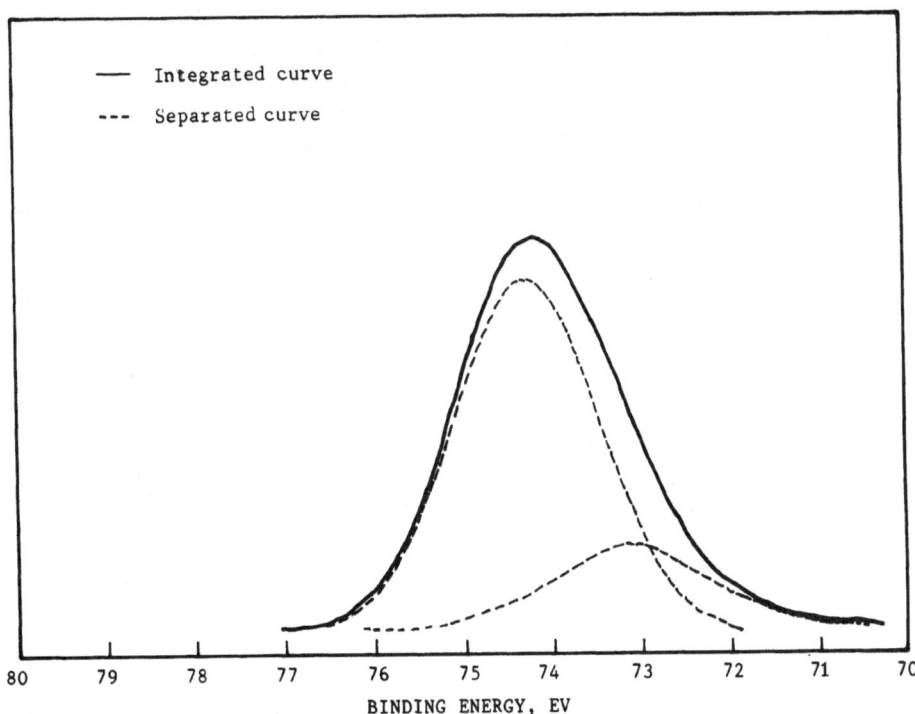

Fig. 1. ESCA spectrum of Al $2p$-orbital electrons of the montmorillonite.

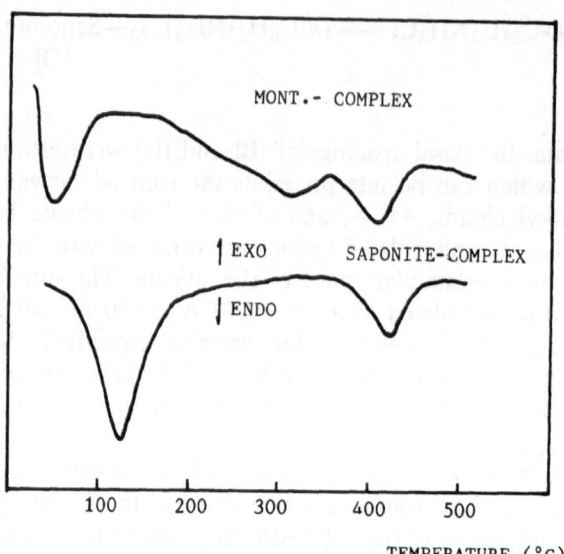

Fig. 2. DSC curves for n-hexylammonium complexes of (a) montmorillonite, and (b) saponite under a N_2-flowing atmosphere with the temperature elevating rate of 10 °C/min.

is associated with the difference of the coordination number of aluminium [12] and also with formation of the excess negative layer charge by isomorphous substitution.

Figure 2 shows the DSC curves of the silicate complexes. The thermogram of the saponite complex indicates only a single endothermic peak at 350–450 °C, whereas an additional peak appears at 250–350 °C in the montmorillonite complex. Correlating the endothermic peaks with the substitution site, it is strongly believed that the peak around 250–350 °C is due to the octahedral substitution and that of 350–450 °C results from the tetrahedral one. This consideration is reasonable since the ionic bonding character between n-hexylammonium ion and the surface oxygen can be considered to be stronger in the tetrahedral substitution than in the octahedral one.

The evolved gas of the first step around 300 °C of the montmorillonite complex was identified as n-hexylamine by GC-analysis (Fig. 3), which indicates the simple desintercalation of amines. According to the TG-analysis of the first step, approximately 25% of the total weight decrease was measured, and this indicates that the layer charge originating from the octahedral substitution might be 0.18 per unit cell composition. On the basis of this result, along with that of the layer charge estimation by the n-alkylammonium method, the chemical equation of the first step can be formulated as follows;

$$(n\text{-}C_{16}H_{13}NH_3^+)_{0.71}^{Oh+\ Td}\text{--Montmorillonite} \xrightarrow{250-350\ °C} (n\text{-}C_6H_{13}NH_3^+)_{0.53}^{Td}\text{--Montmorillonite} + 0.18\ n\text{-}C_6H_{13}NH_2$$

The protons left on the silicate surface after desintercalation of the first step seem to be combined with the —OH group of the octahedral sheet to form H_2O, because

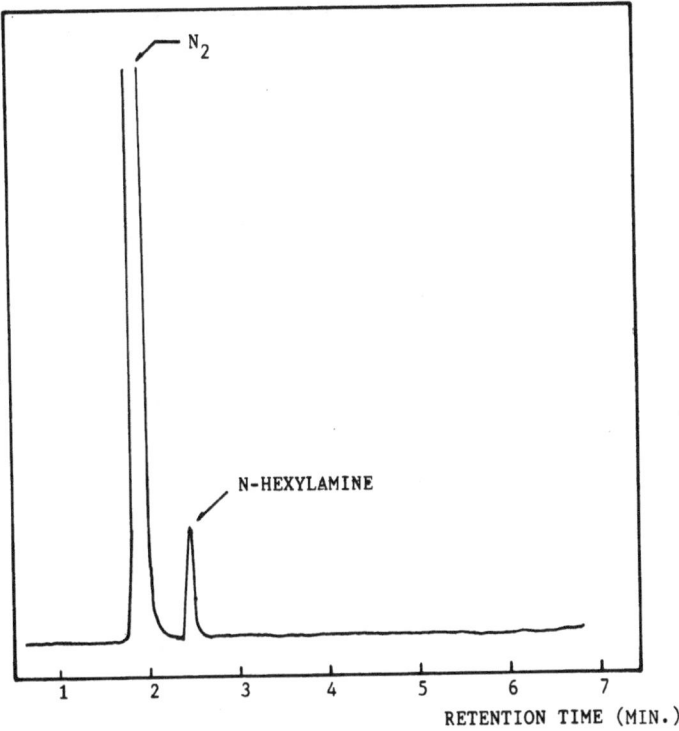

Fig. 3. G.C. results for the evolved gas from the *n*-hexylammonium montmorillonite complex at 300 °C.

the interlayer oxygen atoms affected by the octahedral substitution were apparently not negative enough to trap protons [4].

The results of TG analysis of montmorillonite and its hexylammonium derivative support this concept; the weight loss in the latter case in the temperature range 550–710 °C, (where the former undergoes dehydroxylation), reaches only 52% of that of the former, however, the expelled H_2O was not detected in GC analysis. As a consequence, the Brönsted acid sites which might catalyze the interlayer reaction of *n*-hexylamine diffusing out from the silicate lattice cannot be formed on the surface.

On the other hand, GC results on the decomposition of the saponite complex at 350, 380 and 420 °C indicate that catalytic alkylchain cleavage is involved in this step, in which the tetrahedral substitution participates (Fig. 4). It is noteworthy that deintercalation of amine occurs as the first step, even though the amount of evolved *n*-hexylamine is almost negligible. In this case, the protons left on the surface are attached to the oxygen atoms bridging Si and Al atoms in the tetrahedral sheet, on which the negative charges are strongly localized, to form silanol groups, namely, Brönsted acid sites on the silicate surface. These proton sites are considered to catalyze the evolved substrate diffusing out from the interlayer space.

It is also interesting to note that the ratios of evolved gas components varies as the temperature increases. The relative content of *n*-hexylamine and hexene has increased while the other components such as pentene, ammonia and ethylene have decreased. These results can be understood as following a mechanism including a carbocation intermediate (Fig. 5).

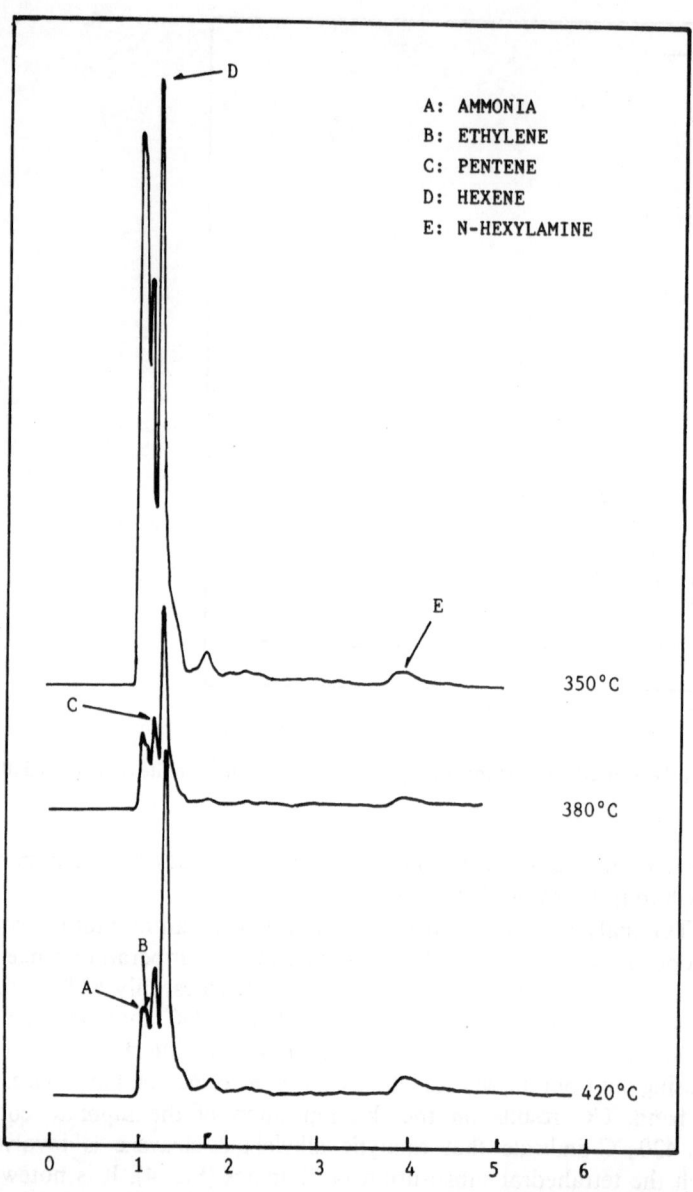

Fig. 4. GC results for the evolved gas from the *n*-hexylammonium saponite complex at 350 °C, 380 °C and 420 °C. (The identifications were also confirmed by mass spectroscopy.)

4. Conclusions

(1) Deintercalation of amine occurs as the first step. The *n*-hexylamine molecules then diffuse out through the pathway blocked by the *n*-hexylammonium ion pillars.
(2) As the reaction proceeds, the diffusion path length becomes shorter, since the surface concentration of *n*-hexylammonium ion pillars is decreased.
(3) The *n*-hexylammonium cation is cleaved into 1-hexene and ammonium ions, which converts into ammonia leaving a proton on the surface.

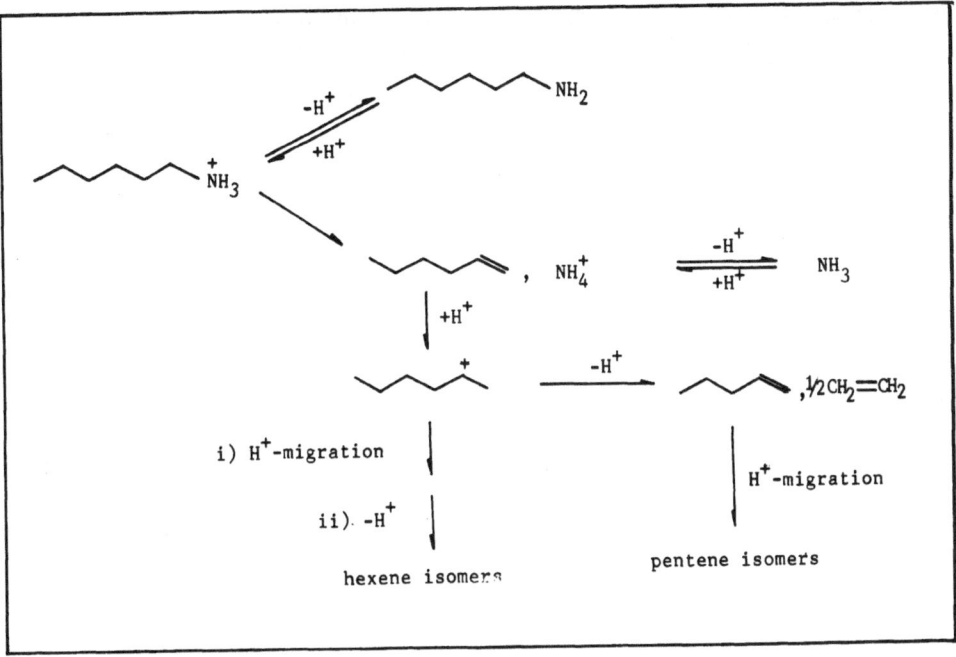

Fig. 5. Proposed mechanism of the catalytic interlayer reaction in the n-hexylammonium silicate complex.

(4) 1-hexene molecules can be protonated on the Brönsted acid sites to form carbocation intermediates on the silicate surface. The isomers of hexene can be produced by proton migration.

(5) The two carbocations produce two 1-pentene and one ethylene molecules leaving two protons on the surface.

(6) The equilibria of steps 1 and 4 become in favor of reactant sites as the temperature increases, because the probabilities of n-hexylamine and 1-hexene contacting the Brönsted acid sites decrease, as the surface concentration of the pillars that control the diffusion pathlength is diminished.

Acknowledgement

This research was supported by the Korean Science and Engineering Foundation (KOSEF).

References

1. R. E. Grim: *Clay Mineralogy*, McGraw-Hill, New York (1968).
2. M. Lipsicas, R. H. Raythatha, T. J. Pinnavaia, I. D. Johnson, R. F. Giese Jr., P. M. Constanzo, and J. L. Robert: *Nature* **309**, 604 (1984).
3. J. M. Adams, S. Evans, and J. M. Thomas: *J. Am. Chem. Soc.* **100**, 3260 (1978).
4. B. Chourabi and J. J. Fripiat: *Clays and Clay Miner.* **29**, 260 (1981).
5. M. S. Whittingham and A. J. Jacobson: *Intercalation Chemistry*, chap. 3, Academic Press, New York (1982).
6. T. J. Pinnavia and F. Farzaneh: *Inorg. Chem.* **22**, 2216 (1983).
7. A. Weiss: *Angew. Chem. Int. Ed. Engl.* **20**, 850 (1981).

8. N. Lahav, U. Shani, and J. Shabtai: *Clays and Clay Miner.* **26**, 107 (1978).
9. S. Yamanaka, T. Doi, S. Sako, and M. Hattori: *Mat. Res. Bull.* **19**, 161 (1984).
10. M. L. Ocelli, F. Hwu, and J. W. Hightower: *Am. Chem. Soc. Div. Pet. Chem.* **26**, 672 (1981).
11. A. Weiss, G. Lagaly, and M. F. Gonzalez: *Clay Miner.* **11**, 173 (1976).
12. C. J. Nicholls, D. S. Urch, and A. N. L. Key: *J. Chem. Soc., Chem. Commun.* 1198 (1972).

Vibrational Spectroscopic Studies of 4,4'-Bipyridyl Metal(II) Tetracyanonickelate Complexes and their Clathrates

ARZU SUNGUR and SEVIM AKYÜZ*
Department of Physics, Hacettepe University, Beytepe, Ankara, Turkey.

J. ERIC D. DAVIES
Department of Chemistry, University of Lancaster, Lancs., LA1 4YA, U.K.

(Received: 2 October 1986; in revised form: 26 January 1987)

Abstract. The infrared spectra of M(4,4'-bipyridyl)Ni(CN)$_4$ complexes (M = Ni or Cd) and their dioxane, benzene, toluene, aniline and N,N-dimethylaniline clathrates are reported. Additional information regarding the structure of the host lattice is obtained from the Raman spectra of the M=Cd complex. It is shown that the structure of the host lattice consists of infinite polymeric layers of $\{M\text{-}Ni(CN)_4\}_\infty$ analogous to those of Hofmann type clathrates that have tetragonal symmetry. Bidentate 4,4'-bipyridyl molecules form bridges between the metal atoms $\{M\}$ in the adjacent $\{M\text{-}Ni(CN)_4\}_\infty$ layers. It is found that the 4,4'-bipyridyl molecules are centrosymmetric in this structure.

Key words: IR and Raman spectra, 4,4'-bipyridyl, tetracyanonickelate clathrates.

1. Introduction

Recently three-dimensional metal complex hosts have been developed from the two-dimensional Hofmann type host lattices, M(NH$_3$)$_2$Ni(CN)$_4$, by replacing the ammonia groups by bidentate ligands, with the aim of enlarging the range of guest molecules which can be accommodated in the host lattices [1–5]. In a previous study Mathey et al. reported the preparation of the Ni(4,4'-bipyridyl)Ni(CN)$_4$ host lattice and its benzene, xylene, naphthalene and anthracene clathrates [5]. We have extended this study and prepared M(4,4'-bipyridyl)Ni(CN)$_4$·2G (M = Ni or Cd; G = dioxane, toluene, aniline or N,N-dimethylaniline) clathrates for the first time. In this study an IR spectroscopic study of the M(4,4'-bipy)Ni(CN)$_4$·nG compounds (where M = Ni or Cd, G = dioxane, benzene, toluene, aniline or N,N-dimethylaniline, n=0–2) (abbreviated henceforth as M-Ni-bipy-G) are reported. Additional information is obtained from the laser-Raman spectrum of the Cd-Ni-bipy complex. We also recorded the powder X-ray diffraction patterns of the M-Ni-bipy complexes.

2. Experimental

The M-Ni-bipy-G samples were prepared by one of the following methods:

(a) Slightly in excess of one mole of 4,4'-bipyridyl was dissolved in a liquid guest (G)

* Author for correspondence. Present address: Department of Physics, Ondokuz Mayis University, Samsun, Turkey.

Presented at the Fourth International Symposium on Inclusion Phenomena and the Third International Symposium on Cyclodextrins, Lancaster, U.K., 20–25 July 1986.

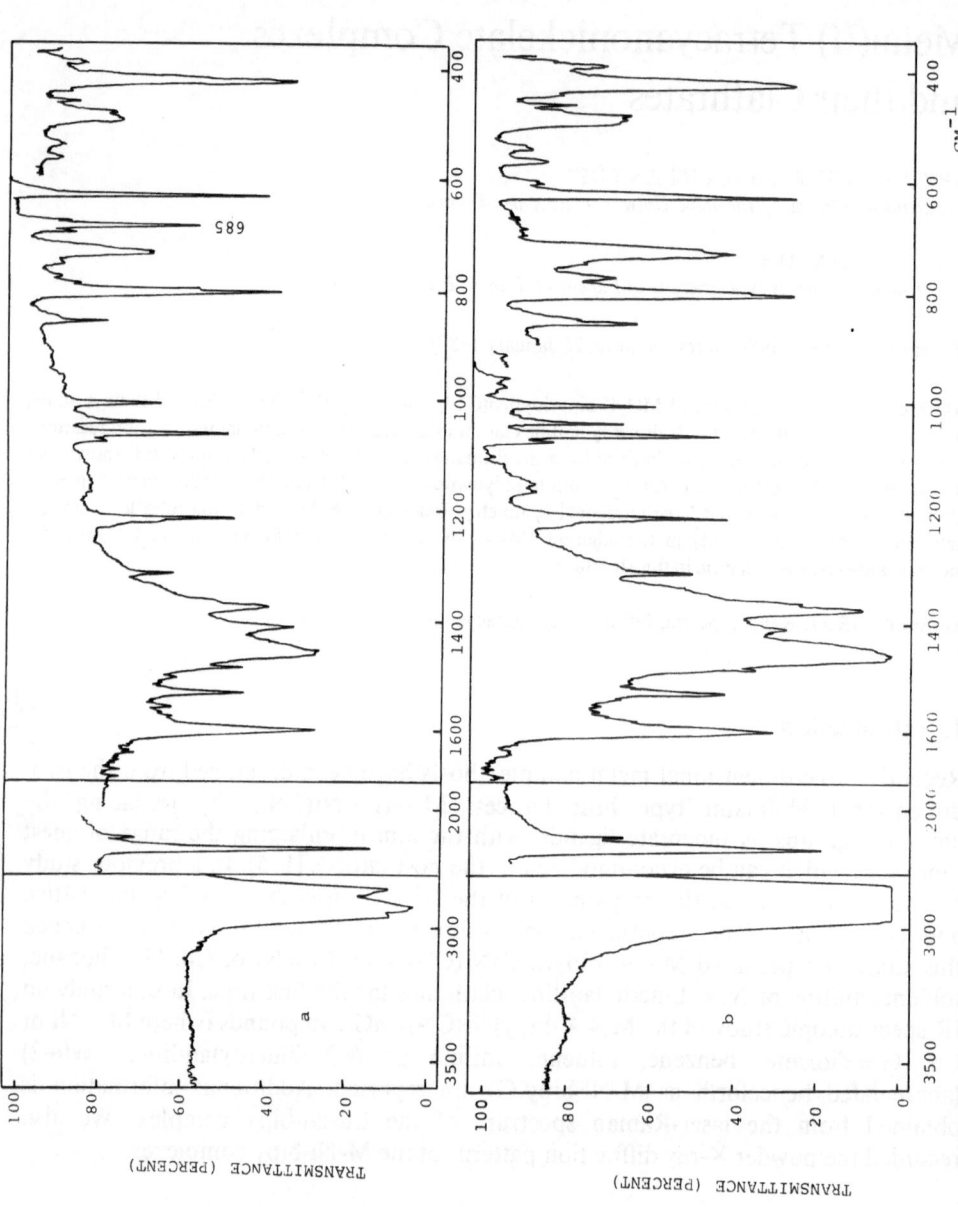

Fig. 1. Infrared spectra of the Cd-Ni-bipy-benzene clathrate (nujol mulls); (a) freshly prepared, (b) after heating at 70 °C for 2.5 hours.

and this solution, together with the aqueous solution of one mole of $K_2Ni(CN)_4$, were added to an aqueous solution of the metal (II) chloride, with constant stirring. The precipitate was filtered and washed with water and acetone.
(b) In this method the M-Ni-bipy complexes were first prepared by decomposing M-Ni-bipy benzene clathrates, then the complexes were immersed in a liquid guest (G) in sealed bottles for a week or more at room temperature.

All the clathrates except G = dioxane were prepared by both methods. However in the case of the dioxane clathrate, the first method yielded $M(dioxane)_2Ni(CN)_4$ complexes. It was also found that M-Ni-bipy-aniline clathrates which were prepared by the first method contained the $M(an)_2Ni(CN)_4$ complex as an impurity.

M-Ni-bipy complexes were prepared by completely decomposing the M-Ni-bipy-benzene clathrates by evaporation or by heating. It was found that almost all the benzene molecules were lost from the Ni-Ni-bipy-benzene and Cd-Ni-bipy-benzene clathrates when kept in an oven at 70 °C for 25 days and 2.5 hours, respectively. Figure 1 shows the IR spectra of the Cd-Ni-bipy-benzene clathrate when freshly prepared and after heating at 70 °C for 2.5 hours.

The IR spectra of the samples were recorded on Perkin-Elmer 621 and Nicolet MX-IE spectrometers which were calibrated using polystyrene and CO_2 bands, respectively.

The Raman spectra were measured with a Cary 81 spectrometer using the 514.5 nm line of a CRL argon ion laser.

3. Results and Discussion

3.1. THE $Ni(CN)_4$ GROUP VIBRATIONS

The IR spectra of the M-Ni-bipy (M = Ni or Cd) complexes are very similar and the X-ray diffraction study showed them to have isomorphous crystal structures with small changes in the unit cell parameters.

Table I. The vibrational wavenumbers (cm^{-1}) of the $Ni(CN)_4$ group of the M-Ni-bipy complexes and M-Ni-bipy-G (G = toluene, aniline or N,N-dimethylaniline) clathrates[a]

Assignment	M-Ni-bipy		M-Ni-bipy-G						Relative intensity
			Toluene		Aniline		N,N-dimethylaniline		
	Cd	Ni	Cd	Ni	Cd	Ni	Cd	Ni	
$A_{1g}\ \nu(CN)$	(2166)								(vs)
$B_{1g}\ \nu(CN)$	(2153)								(s)
$E_u\ \nu(CN)$	2147	2164	2144	2162	2148	2166	2148	2166	vs
$\nu(^{13}CN)$	2105	2132	2103	2130	2106	2133	2106	2133	vw
$E_u\ \nu(NiC)$	540	552	538	552	540	–	540	552	vw
$A_{2u}\ \pi(NiCN)$	444	451	443	451	444	452	444	452	vw
$E_u\ \delta(NiCN)$	426	438	424	438	426	439	427	439	vs
Interplanar distance (c) Å	11.7	10.4							

[a] The bands observed in the infrared spectra are given without parantheses; the bands observed in the Raman spectra are given in parantheses.

The vibrational wavenumbers of the Ni(CN)$_4$ group vibrations of the M-Ni-bipy complexes and M-Ni-bipy-G clathrates [G=Toluene, aniline or N,N-dimethylaniline] are given in Table I. The ν(CN) and δ(NiCN) vibrational wavenumbers are found to be similar to those of Hofmann type clathrates [6] and the pyridine [7] complex, showing that the {M-Ni(CN)$_4$}$_\infty$ layers have been preserved. Since we observed only one ν(CN) (E_u) band in the IR spectrum and the other two ν(CN) (A_{1g} and B_{1g}) bands in the Raman spectrum of the Cd-Ni-bipy complex, we propose a square planar environment around the tetracyanonickelate ion.

We did not observe any differences in the Ni(CN)$_4$ group vibrational modes of the M-Ni-bipy-G clathrates (G=benzene or dioxane) in comparison to those of the corresponding M-Ni-bipy complexes. However, we observed slight shifts of the ν(CN) (E_u) band in the cases of G=toluene, aniline and N,N-dimethylaniline in comparison to those of the M-Ni-bipy complexes (see Table I). The slight shift of the ν(CN) (E_u) band may be attributed to the small change in the host-guest interaction.

The distance (c) between the adjacent {M-Ni(CN)$_4$} networks is calculated from the X-ray diffraction patterns of the M-Ni-bipy complexes and is given in Table I. The c values closely match those calculated from the known interatomic distances, assuming that the principal axis of the ligand is perpendicular to the cyanide planes. This result supports our assumption that 4,4'-bipyridyl molecules form bridges between the metal atoms (M) in the adjacent {M—Ni(CN)$_4$} layers.

4,4'-BIPYRIDYL VIBRATIONS

The IR and Raman spectra of the Cd-Ni-bipy complex are given in Figures 1 and 2, respectively. Since we did not observe any coincidences between the IR and Raman frequencies of the 4,4'-bipyridyl molecule, we propose that the molecule is centrosymmetric and planar in the structure. 4,4'-Bipyridyl therefore belongs to the D_{2h} point group and hence its 54 fundamental vibrations are divided among the symmetry species as follows; 5 B_{1u}, 9 B_{2u}, and 9 B_{3u} IR active; 10 A_g, 9 B_{1g}, 5 B_{2g} and 3 B_{3g} Raman active; and 4 A_u inactive.

It has been noted that IR and Raman data on 4,4'-bipyridyl and its complexes are not plentiful in the literature. In most of the spectroscopic studies on 4,4'-bipyridyl complexes, bipyridyl bands are assigned by comparison with pyridine [8–10]. The only detailed IR and Raman data on crystalline 4,4'-bipyridyl was reported by Gupta [11] who gave approximate interpretations of all of the vibrational modes of the molecule and supported his assignment by a normal coordinate analysis using a valence-force scheme. However he did not apply the theoretical calculations to the deuterium derivatives of 4,4'-bipyridyl. On the other hand Kihara and Gondo [12] carried out a normal coordinate analysis for the in-plane vibrations of 4,4'-bipyridyl only and reported some of the IR and Raman vibrational wavenumbers. We assigned the vibrations of 4,4'-bipyridyl by comparison with the assignment for biphenyl [13, 14], methylviologen [15, 16], pyridine [7] and 4,4'-bipyridyl [11, 12].

The 4,4'-bipyridyl wavenumbers observed in the IR spectra of the M-Ni-bipy complexes, together with our measurement on crystalline 4,4'-bipyridyl are given in Table II. The assignments of biphenyl [14] and 4,4'-bipyridyl taken from Gupta [11] and Kihara and Gondo's [12] papers are included for comparison. Our assignment of the IR spectrum of crystalline 4,4'-bipyridyl agrees with that of Kihara and Gondo. It is also in agreement with that of Gupta with the following exceptions: Gupta assigned

VIBRATIONAL SPECTROSCOPY OF BIPYRIDYL METAL CLATHRATES

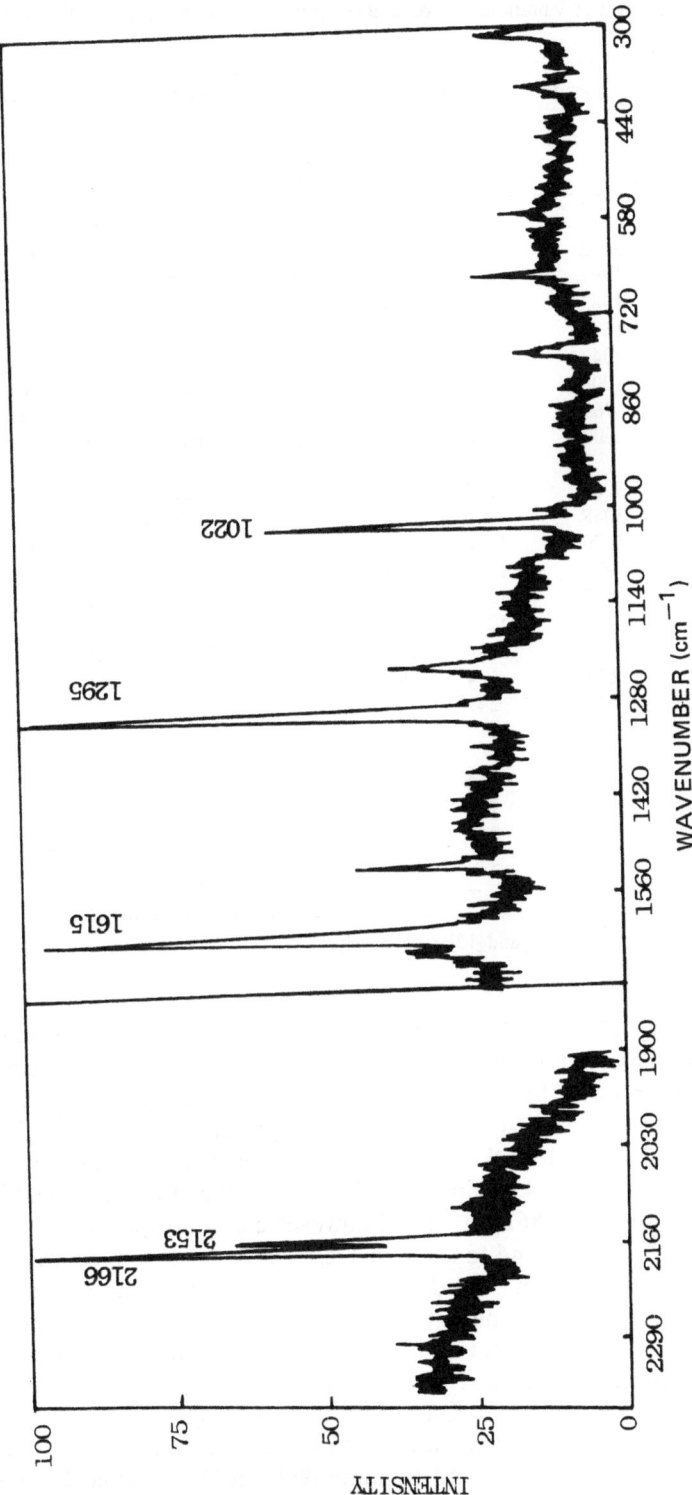

Fig. 2. Raman spectrum of the Cd-Ni-bipy complex.

Table II. The fundamental vibrational wavenumbers (cm^{-1}) of 4,4'-bipyridyl in the M-Ni-bipy complexes

Symmetry and description of mode		Biphenyl[a]	4,4'-bipy			M-Ni-bipy		Relative intensity
			G[b]	KG[c]	This study	Cd	Ni	
B_{3u}	1 ν(CH)	3080	3070	–	3055	3069	3080	w
	2 ν(CH)	3072	3050	–	3047	3053	–	w
	3 ν_{ring}	1597	1590	1585	1598	1605	1610	vs
	4 ν_{ring}	1482	1485	1485	1481	1488	1490	m
	5 δ(CH)	1176	1130	1217	1215	1216	1219	m
	6 δ(CH)	1040	1040	1036	1044	1064	1066	m
	7 $\nu_{ring}+\delta_{ring}$	1008	851	965	967	976	985	w
	8 ν_{ring}	965	1287[d]	990	994	1011	1012	m
	9 δ_{ring}	609	610	607	615	632	635	vs
B_{2u}	10 ν(CH)	3069	3070	–	3078	3080	3084	m
	11 ν(CH)	3068	3030	–	3029	3036	3034	m
	12 ν_{ring}	1570	1540	1528	1532	1534	1536	s
	13 ν_{ring}	1432	1406	1405	1413	1414	1413	s
	14 ν_{ring}	1272	1273[d]	–	1324	1315	1318	m
	15 δ(CH)	1156	1218	–	1223	1219	1222	sh
	16 δ(CH)+ν_{ring}	1074	1074	1073	1074	1078	1078	w
	17 δ_{ring}	628	634[d]	670	677	668	665	w
	18 inter-ring bend. i.p.	116	241[d]	–	–	–	–	
B_{1u}	19 γ(CH)	903	880	–	862	859	854	w
	20 γ(CH)	736	806	–	810	803	806	vs
	21 γ_{ring}	698	733	–	737	729	730	s
	22 γ_{ring}	484	460	–	507	488	–	m
	23 γ_{ring}	174	–	–	367	386	–	m

[a,b,c] Taken from references [14], [11] and [12], respectively.
[d] Calculated values.

the IR band at 1130 cm^{-1} as a fundamental vibrational mode of B_{3u} symmetry species. However we observed a very weak band at 1125 cm^{-1} and preferred to assign the medium intense band at 1215 cm^{-1} to the ν_5 mode in agreement with Kihara and Gondo. Our assignment is compatible with those of pyridine (1217 cm^{-1}) [7] and methylviologen (1181 cm^{-1}) [16]. We observed two weak bands at 862 cm^{-1} and 967 cm^{-1}, in the IR spectrum of 4,4'-bipyridyl. The 967 cm^{-1} band shifted to a higher frequency on coordination; the 862 cm^{-1} band however did not display any shift. As discussed below the B_{3u} modes tend to show increases in frequency on coordination. Therefore we assigned the 967 cm^{-1} band to ν_7 in agreement with the assignment of Kihara and Gondo. The ν_8 mode (the out-of-phase component of the ring breathing mode) is assigned to a strong IR band at 994 cm^{-1}. This is because we observed the corresponding in-phase component at 1002 cm^{-1} as a very strong band in the Raman spectrum of crystalline 4,4'-bipyridyl. Our assignment is in agreement with that of Kihara and Gondo [12] and is also compatible with those for the biphenyl (965 cm^{-1}) [14], methylviologen (993 cm^{-1}) [16] and pyridine (990 cm^{-1}) [7] molecules. Gupta [11] also observed a band at 990 cm^{-1} in the IR spectrum of 4,4'-bipy which was assigned to an A_{1g} mode. This in-phase ring-breathing mode must however be IR inactive in D_{2h} symmetry. In Gupta's paper the out-of-phase ring-breathing mode was calculated as 1287 cm^{-1} (B_{3u}) but the observed value was not given.

The v_{22} (γ_{ring}) mode was assigned to a band at 460 cm^{-1} by Gupta; however, we did not observe any band at this wavenumber value in the IR spectrum of crystalline 4,4'-bipyridyl. Therefore we assigned the strong IR band observed at 507 cm^{-1} to v_{22}.

The vibrational modes of 4,4'-bipyridyl observed in the IR spectra of the M-Ni-bipy complexes and of the M-Ni-bipy-G clathrates (G=benzene or dioxane) are found to be the same in intensity and frequency. However in the cases of the M-Ni-bipy-G clathrates (G=toluene, aniline or N,N-dimethylaniline) we observed slight frequency shifts (up to 2 cm^{-1}) and intensity changes in some 4,4'-bipyridyl vibrational modes in comparison to those of the M-Ni-bipy complexes. These changes are thought to be due to the interaction between the host lattice and the guest molecules.

The 4,4'-bipyridyl vibrational modes observed in the IR spectra of the compounds studied show all the characteristics of a coordinated ligand, e.g. several modes of the coordinated ligand have upward shifts in frequency compared to those in the free molecule and the shifts are metal dependent. As is clear from Table II, considerable shifts to higher frequency occur for the modes of B_{3u} symmetry species. Similar shifts are observed in the pyridine complexes [7] and explained by coupling with low frequency vibrations, particularly the M-N stretching frequency.

Acknowledgement

One of us (Dr S. Akyüz) thanks the British Council for financial support.

References

1. T. Iwamoto: *Isr. J. Chem.* **18**, 240 (1979).
2. T. Hasegawa, S. Nishikiori, and T. Iwamoto: *J. Incl. Phenom.* **1**, 365 (1984).
3. J. E. D. Davies and A. M. Maver: *J. Mol. Struct.* **102**, 203 (1983).
4. S. Nishikiori and T. Iwamoto: *Inorg. Chem.* **25**, 788 (1986).
5. Y. Mathey, C. Mazieres and R. Setton: *Inorg. Nucl. Chem. Lett.* **13**, 1 (1977).
6. S. Akyüz, A. B. Dempster, and R. L. Morehouse: *Spectrochim. Acta* **30A**, 1989 (1974).
7. S. Akyüz, A. B. Dempster, R. L. Morehouse, and S. Suzuki: *J. Mol. Struct.* **17**, 105 (1973).
8. S. Farquharson, P. A. Lay, and M. J. Weaver: *Spectrochim. Acta.* **40A**, 907 (1984).
9. A. I. Popov, J. C. Marshall, F. B. Stute and W. B. Person: *J. Am. Chem. Soc.* **83**, 3586 (1961).
10. I. S. Ahuja, R. Singh, and C. P. Rai: *J. Inorg. Nucl. Chem.* **40**, 924 (1978).
11. V. P. Gupta: *Indian J. Pure App. Phys.* **11**, 775 (1973).
12. H. Kihara and Y. Gondo: *J. Raman Spectrosc.* **17**, 263 (1986).
13. G. Zerbi and S. Sandroni: *Spectrochim. Acta* **24A**, 483 (1968).
14. G. Zerbi and S. Sandroni: *Spectrochim Acta* **24A**, 511 (1968).
15. M. Forster, R. B. Girling, and R. E. Hester: *J. Raman Spectrosc.* **12**, 36 (1982).
16. R. Hester and S. Suzuki: *J. Phys. Chem.* **86**, 4626 (1982).

The Synthesis of Metallocene Calix[4]arenes

PAUL D. BEER* and ANTHONY D. KEEFE
Department of Chemistry, University of Birmingham, P.O. Box 363, Birmingham B15 2TT, U.K.

(Received: 2 October 1986; in final form: 31 January 1987)

Abstract. The condensation of 1,1'-bis(chlorocarbonyl)metallocenes and *p-tert*-butylcalix[4]arene in toluene leads to novel metallocene calix[4]arenes in which the metallocene subunit bridges the opposite hydroxy groups of the parent calixarene.

Key words: Metallocene calix[4]arenes, synthesis, variable temperature NMR studies, electrochemistry.

The calixarenes are phenol-formaldehyde cyclic oligomers which possess hydrophobic cavities capable of forming inclusion complexes with aromatic guest molecules in the solid state [1, 2]. They also have the ability to function as ion and molecular carriers as well as enzyme mimics [1, 3, 4].

Calix[4]arenes in solution are conformationally mobile and can exist in four discrete forms, the 'cone', 'partial cone', '1,2-alternate' and '1,3 alternate' conformations [1]. Above room temperature these conformations are rapidly interconverting as shown by dynamic ^{1}H NMR investigations [5]. In an effort to synthesise 'rigid' calix[4]arenes in the cone conformation a number of workers have introduced bulky substituents on the hydroxy groups with some success, although the partial cone[6] or 1,3-alternate conformations [7] may occasionally be obtained instead. Recently the fixed cone conformation has been achieved by the connection of two *para*-positions by an aliphatic chain of appropriate length [8]. This communication reports the first example of a metallocene calix[4]arene in which the metallocene subunit bridges opposite hydroxy groups of *p-tert*-butylcalix[4]arene.

The condensation of respective 1,1'-bis(chlorocarbonyl)-metallocenes (**1**) and (**2**) [9, 10] with *p-tert*-butylcalix[4]arene [11] (**3**) in the presence of triethylamine gave, after column chromatography (alumina/CH$_2$Cl$_2$), the ferrocene calix[4]arene (**4**) (64% yield, orange crystals) and the ruthenocene analogue (**5**) (47% yield, pale yellow crystals) (Scheme 1). The structures of both new air-stable metallocene calix[4]arenes were verified by elemental analysis and mass spectrometry. The experimental details for their synthesis and characterisation will be reported elsewhere.

The respective 400 MHz ^{1}H NMR spectra of (**4**) and (**5**) in CDCl$_3$ at room temperature (298 K) display two broad doublets for the methylenes of (**4**) and (**5**) typical of an *AB* system, two singlets for the *t*-butyl groups, two metallocene signals and two aromatic absorptions. On warming to 55 °C (328 K) the two doublets sharpen and at 100 °C (373 K) in toluene the respective *AB* systems are still observed. These NMR data are similar to those shown by the dimethyl derivative of *p-tert*-butylcalix[4]arene [12] and by the crowned *p-tert*-butylcalix[4]arenes [3, 13]. These compounds are all

---

* Author for correspondence.

Presented at the Fourth International Symposium on Inclusion Phenomena and the Third International Symposium on Cyclodextrins, Lancaster, U.K., 20–25 July 1986.

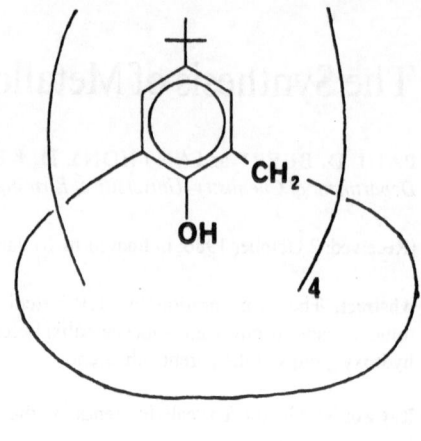

Scheme 1.

THE SYNTHESIS OF METALLOCENE CALIX[4]ARENES

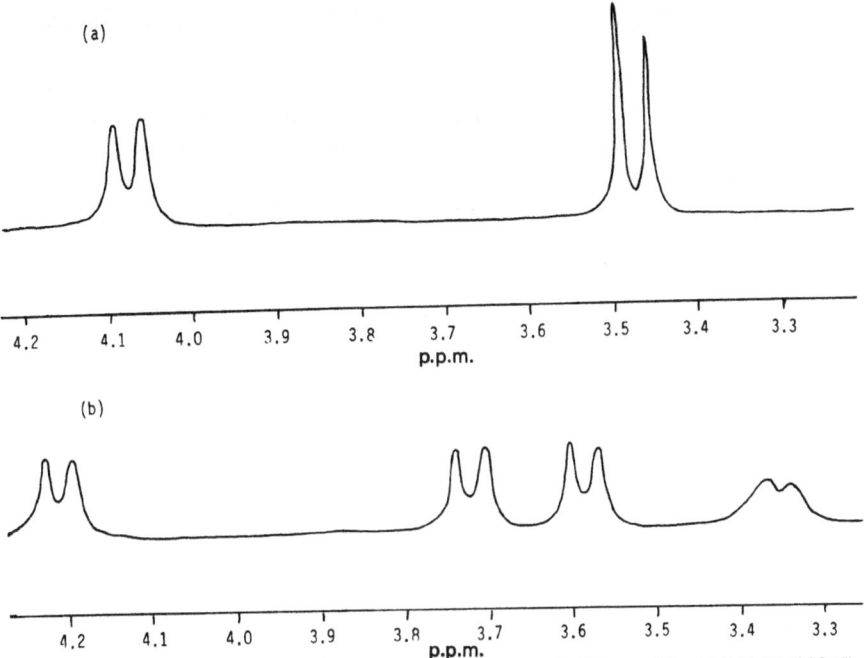

Fig. 1. ^{1}H NMR spectrum of the methylene protons of ferrocene calix[4]arene (**4**) at (a) 25 °C (298 K) and (b) −100 °C (173 K) (CD$_2$Cl$_2$)

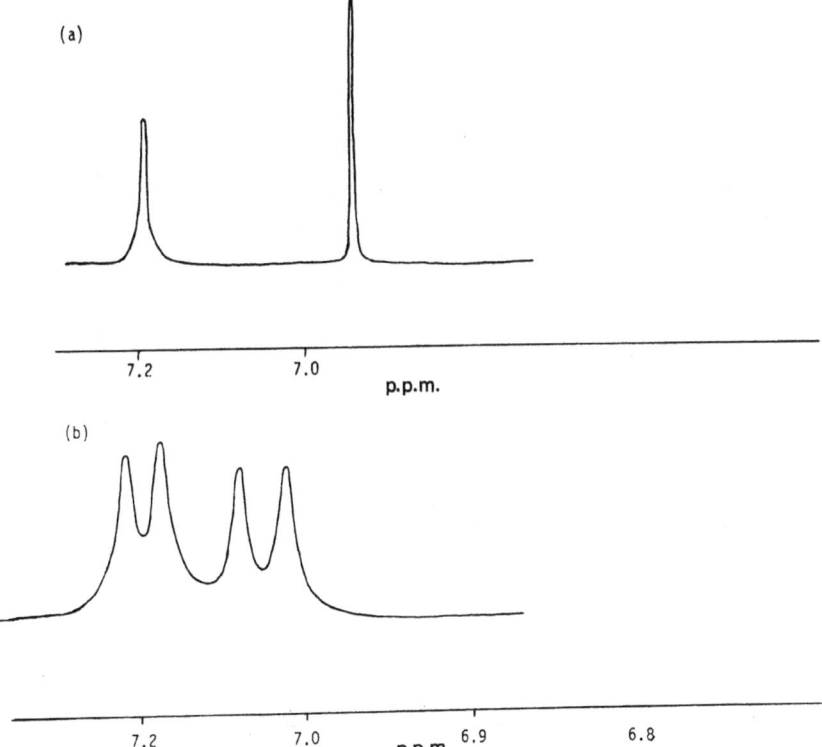

Fig. 2. ^{1}H NMR spectrum of the aromatic protons of ferrocene calix[4]arene (**4**) at (a) 25 °C (298 K) and (b) −100 °C (173 K) (CD$_2$Cl$_2$).

(1,3)-calix[4] arene derivatives and have been proved to exist in a "flattened" cone conformation.

However, unexpectedly on cooling the respective samples (CDCl$_3$) to -50 °C (223 K) and -100 °C (173 K) (CD$_2$Cl$_2$) another pair of *AB* doublets appear (Figure 1) along with four singlets for the metallocene protons, four singlets for aromatic protons (Figure 2) and two hydroxyl signals. The *t-butyl* protons now consist of at least three singlets, one of which remains boad even at -100 °C (Figure 3). Two possible ratio-

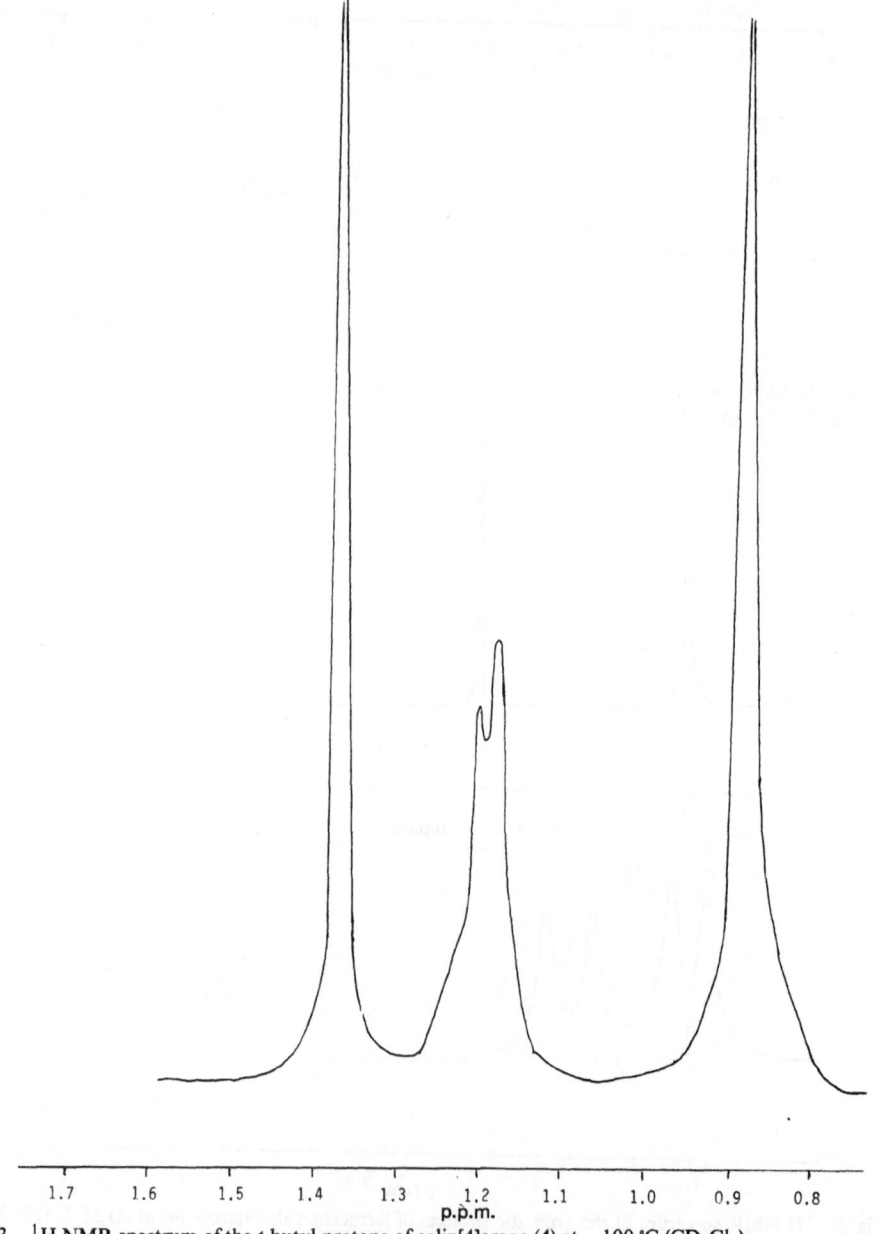

Fig. 3. ^{1}H NMR spectrum of the *t*-butyl protons of calix[4]arene (**4**) at -100 °C (CD$_2$Cl$_2$).

THE SYNTHESIS OF METALLOCENE CALIX[4]ARENES

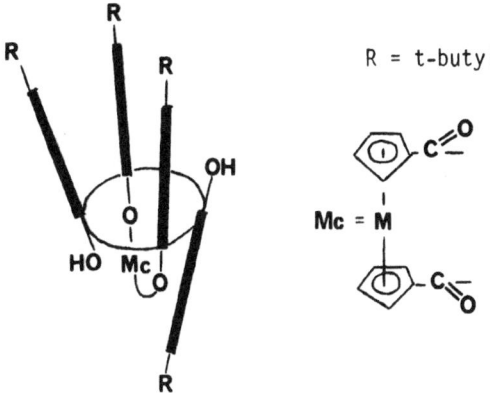

Fig. 4. Partial cone conformation of the metallocene calix[4]arenes.

nalisations may account for this variable temperature NMR behaviour. At low temperatures (−100 °C) the respective ^{1}H NMR spectra of (4) and (5) are consistent with a partial cone conformation [7]. See Figure 4.

Alternatively the ^{1}H NMR spectra may be explained by considering only the intramolecular rotation of the respective metallocene carbonyl groups. At room temperature and above the carbonyl group is rotating fast on the NMR timescale about the metallocene cyclopentadienyl—carbonyl carbon bond. On cooling, this intramolecular fluxional process slows down and more complicated spectra result. The fact that two hydroxyl protons are observed at the lowest temperatures suggests the absence of a C_2 axis in the respective molecules. This result coupled with molecular model considerations imply that at low temperatures the respective carbonyl groups attached to the metallocene subunits are *cis* to one another. See Figure 5.

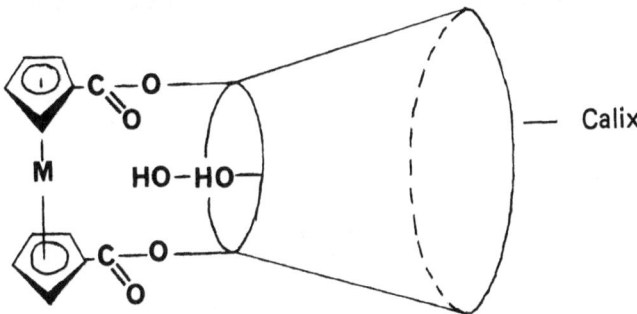

Fig. 5. Metallocene carbonyl groups shown in the *cis* configuration.

Acyl groups in simple acylferrocenes are known to exhibit this type of fluxional behaviour [14]. Unfortunately corroborative evidence from ^{13}C NMR spectra could not be obtained because of poor solubility of both (4) and (5) in organic solvents. The results of preliminary electrochemical investigations on (4) and on a model

Table I. Electrochemical data

Compound	(4)	(6)
$E_{1/2}$ Volts[a]	+1.2	+1.2
ΔE_p^b (mV)	80	100

[a] Obtained in CH_2Cl_2 solution containing 0.2 M $Bu_4\overset{+}{N}\overline{B}\overline{F}_4$ as supporting electrolyte. Solutions were ca. 5×10^{-3} M in compound and potentials were determined with reference to ferrocene as internal standard but are quoted relative to Ag wire reference electrode.

[b] Separation between anodic and cathodic peak potentials of cyclic voltammograms at a scan rate of 200 mv sec^{-1}.

compound (6) are reported in the Table I. Large anodic $E_{1/2}$ values are found for both compounds.

Solution complexation experiments of the metallocene calix[4]arenes with organic guests are currently under investigation.

Acknowledgements

We thank the SERC for an 'earmarked' studentship to A.D.K, for use of the high field NMR service at the University of Warwick and The Research Corporation Trust for additional financial support.

References

1. C. D. Gutsche: *Top. Curr. Chem.* **123**, 1, (1984).
2. G. D. Andreetti, R. Ungaro, and A. Pochini: *J. Chem. Soc. Chem. Commun.* 1005, (1979).
3. R. Ungaro, A. Pochini, and G. D. Andreetti: *J. Incl. Phenom.* **2**, 199, (1984).
4. S. R. Izatt, R. T. Hawkins, J. J. Christensen, and R. M. Izatt: *J. Am. Chem. Soc.* **107**, 63, (1985).
5. G. Happel, B. Mathiasch, and H. Kammerer: *Makromol. Chem.* **176**, 3317, (1975).
6. C. Rizzoli, G. D. Andreetti, R. Ungaro, and A. Pochini: *J. Mol. Struct.* **88**, 133, (1982).
7. C. D. Gutsche, B. Dhawam, J. A. Levine, K. H. No, and L. J. Baner: *Tetrahedron* **39**, 409, (1983).
8. V. Bohmer, H. Goldmann, and W. Vogt: *J. Chem. Soc. Chem. Commun.* 667, (1985).
9. P. D. Beer: *J. Chem. Soc. Chem. Commun.* 1115, (1985).
10. H. J. Lorkowski, R. Pannier, and A. Wende: *J. Prakt. Chem.* **35**, 149, (1967).
11. C. D. Gutsche, M. Iqbal, and D. Stewart: *J. Org. Chem.* **51**, 742, (1986).
12. C. D. Gutsche, B. Dhawan, J. A. Levine, K. H. No, and L. J. Baner: *Tetrahedron* **39**, 409, (1983).
13. C. Alfieri, E. Dradi, A. Pochini, R. Ungaro, and G. D. Andreetti: *J. Chem. Soc. Chem. Commun.* 1075, (1983).
14. J. Sanstrom and J. Seita: *J. Organometal. Chem.* **108**, 371, (1976).

Journal of Inclusion Phenomena 5 (1987), 505–513.
© 1987 by D. Reidel Publishing Company.

Solid State and Solution Structures of the Lanthanide Complexes with Cryptand (2.2.1): Crystallographic and NMR Studies of a Dimeric Praseodymium (2.2.1) Cryptate Containing Two μ-Hydroxo Bridges

J. REBIZANT
Commission of the European Communities, JRC Karlsruhe, Postfach 2266, D-7500 Karlsruhe, Federal Republic of Germany

M. R. SPIRLET*
Physique Expérimentale (B 5), Université de Liège au Sart Tilman, B-4000 Liège, Belgium

P. P. BARTHÉLEMY and J. F. DESREUX*
Laboratoire de Chimie Analytique et de Radiochimie (B6), Université de Liège au Sart Tilman, B-4000 Liège, Belgium

(Received: 7 October 1986; in final form: 8 January 1987)

Abstract. A dimeric lanthanide cryptate was obtained by the addition of an excess of cryptand (2.2.1) to a slightly hydrated solution of the monomeric praseodymium (2.2.1) perchlorate complex in acetonitrile. This new lanthanide compound is centrosymmetric and displays the space group $P2_1/n$. The encryptated metal ions are nine-coordinated, they are bonded to all the heteroatoms of a (2.2.1) ligand and they are linked to each other by two μ-hydroxo bridges. The hydroxyl groups are relegating the cryptands to both end of the dimer and the praseodymium ions are less effectively accomodated in the macrocyclic internal cavities than in the case of the monomeric Pr(2.2.1) complex. The formation of both the monomeric and the dimeric lanthanide complexes is readily observed by proton NMR.

Key words: Crystal structure, lanthanides, praseodymium, dimer, macrocycle, (2.2.1), cryptand, NMR, paramagnetic ion.

Supplementary data relevant to this article are deposited with the British Library Lending Division as Supplementary Publication No. 82050 (24 pages).

1. Introduction

Among the lanthanide macrocyclic complexes [1], the (2.2.1) cryptates (where (2.2.1) stands for pentaoxa-4,7,13,16,21-diaza-1,10-bicyclo[8.8.5]tricosane) are of special interest because they display several unusual properties such as high thermodynamic stability [2–3], kinetic inertness in water [4] and stabilization of the +2 oxidation state [4]. Up to the present, studies of the lanthanide (2.2.1) cryptates have been restricted to solution studies and no crystallographic investigation has yet been reported. The present paper is part of our efforts directed at elucidating the solid-state and solution conformation of these complexes. We report herein the

---

* Authors for correspondence.

Presented at the Fourth International Symposium on Inclusion Phenomena and the Third International Symposium on Cyclodextrins, Lancaster, U.K., 20–25 July 1986.

structure and the proton NMR spectrum of a new dimeric praseodymium complex [$Pr_2(2.2.1)_2(OH)_2$]$4[ClO_4] \cdot 2[CH_3CN]$ whose structure is quite unusual because of the two μ-hydroxo bridges that link the two metal ions. The formation of this dimeric species could account for the unexpected affinity of the lanthanide (2.2.1) cryptates for the hydroxyl ion [4].

2. Experimental

An anhydrous acetonitrile solution of praseodymium perchlorate was obtained by dissolving the dehydrated lanthanide salt [5] in dried acetonitrile. All solutions and reagents were stored and handled in a glove box filled with an inert atmosphere. Mixing equimolar amounts of cryptand (2.2.1) and of the anhydrous praseodymium salt enabled us to obtain crystals of a monomeric complex [$Pr(2.2.1)(CH_3CN)_2$]$\cdot 3[ClO_4]$ whose structure will be reported in detail elsewhere [6]. Adding a 100% excess or more of cryptand (2.2.1) to an acetonitrile solution of the monomeric praseodymium complex brings about drastic modifications of the NMR spectra as well as crystallization of a new compound. Pale green prismatic crystals suitable for X-ray analysis were selected in a glove box and mounted in capillaries. In our experimental conditions, the direct addition of a large excess of cryptand to praseodymium perchlorate leads to the partial precipitation of praseodymium hydroxide.

The proton NMR spectra were recorded on a Brucker AM spectrometer at 300 MHz.

3. Crystal Data

[$Pr_2(C_{16}H_{32}N_2O_5)_2(OH)_2$]$\cdot 4[ClO_4] \cdot 2[CH_3CN]$, Formula weight = 1460.61. Monoclinic, $a = 10.706(3)$, $b = 20.048(5)$, $c = 13.297(2)$ Å, $\beta = 94.12(1)°$, $V = 2847(2)$ Å^3, $Z = 2$, d(calcd) = 1.702 g cm^{-3}, μ(MoKα) = 19.731 cm^{-1}.

4. Structure Determination

Investigations on a selected single crystal with approximate dimensions $0.15 \times 0.15 \times 0.30$ mm were carried out with an Enraf-Nonius CAD-4 X-ray diffractometer at 293 ± 1 K using MoKα radiation ($\lambda = 0.71073$ Å). The computer programs used were part of the Enraf Nonius SDP Programs [7]. The scattering factors and anomalous dispersion corrections were taken from [8]. Unit cell parameters were determined by least-squares refinement of diffractometer setting angles for 25 diffraction maxima. The space group was unequivocally deduced from the systematic absences. A total of 4843 reflections representing 4603 symmetry-independent reflections were measured in the θ-2θ scan mode (collection range $3 \leqslant 2\theta \leqslant 48°$, h 0/+13, k 0/−23, l +16/−16). During data collection (4 days long), three reflections were monitored at 30 min intervals as a check of the stability of the diffractometer and of the crystal. The maximum anisotropic decay correction applied to the data was 6.3%. Intensities were corrected for Lorentz-polarization effects. Absorption corrections were applied by an empirical method based on a set of scans of reflexions having χ values near 90(1)°. The transmission factors ranged from 85.2 to 93.5%.

The structure was solved by direct methods. The E-map calculated with the set of phases presenting the highest combined figure of merit given by the program MULTAN [9] revealed the position of the Pr and of the Cl atoms. Further Fourier and difference Fourier maps in alternance with full-matrix least-squares refinement cycles allowed the location of the remaining non-hydrogen atoms. However, it was not possible to avoid difficulties during the treatment of the data because some parts of the macrocycle underwent high thermal motions. As expected [10], the oxygen atoms belonging to the perchlorate ions also exhibited large temperature factors. Finally, some C—C distances within the macrocycle became abnormally short or long indicating disorder problems. However, when these atoms were omitted in the refinement, they were all recovered at their initial positions in a subsequent difference Fourier calculation. The use of anisotropic thermal parameters resulted in non-positive values for many atoms and only the thermal motions of the Pr, O (hydroxyl) and Cl (perchlorate) atoms were treated anisotropically. The structure was then refined alternatively in two groups of atoms: on the one hand, the atoms belonging to the Pr cryptate, on the other hand, the perchlorate and acetonitrile atoms. Moreover, in order to maintain reasonable interatomic distances, restraints were imposed on some carbon atoms which were forced to ride on one attached neighbour at the distance found in the Fourier map. Finally, the hydrogen atoms were not included in the data treatment. A secondary extinction coefficient applied near the end of the computations refined to $g = 2.71 \times 10^{-7}$ ($F_c = F_c/(1 + g(F_c)^2 Lp$ where L is the Lorentz factor and where p is the polarization factor). All parameters were allowed to vary only in the last two cycles of refinement. The final agreement factors $R_1 = \Sigma(|F_0|-|F_c|)/\Sigma|F_0|$ and $R_2 = [\Sigma w(|F_0|-|F_c|)^2/\Sigma w F_0^2]^{1/2}$ were 0.072 and 0.084 respectively. The largest parameter shift Δ/σ was 0.12. The function minimized was $\Sigma w \, [|F_c|]^2$ with unit weights for 2936 reflections for which $I > 3\sigma(I)$. The standard deviation of an observation of unit weight esd $= [\Sigma w(|F_0|-|F_c|)^2/(m-n)]^{1/2}$ was 3.09. A difference map calculated after the last cycle of refinement showed maximum and minimum excursion densities of +1.8 and −1.5 e Å^{-3}, respectively, around the Pr ion.

The coordinates and the isotropic thermal parameters from the final least-square cycle are given in Table I. Anisotropic thermal parameters and calculated structure factors have been deposited with the British Library Lending Division under SUP 82050.

5. Results and Discussion

5.1. STRUCTURE OF THE DIMERIC Pr CRYPTATE

The structure of the dimeric complex is depicted in Figure 1 together with the atom-labeling scheme. The complex is centrosymmetric and contains two identical nine-coordinated praseodymium ions linked together by two μ-hydroxo bridges. The idealized symmetry of the dimer is $C_2\hat{h}$ with a C_2 axis passing through the two praseodymium ions and the two O(21) atoms and a mirror plane passing through the two hydroxyl oxygen atoms O(B). The dihedral angle between the mean plane of the N(1), N(10), O(21) and Pr atoms and the plane of the two O(B) and Pr atoms is 36(1)°. The coordination polyhedron around each praseodymium ion is reproduced in Figure 2. As expected, the mean Pr—O(hydroxyl) bond length (2.36 Å) is shorter than the mean Pr—O(cryptand) (2.60 Å) or Pr—N(cryptand) (2.76 Å) bond lengths. The

Table I. Fractional atomic co-ordinates and isotropic thermal parameters with standard deviations in the least significant digits in parentheses.

	x	y	z	$B(Å^2)^A$		x	y	z	$B(Å^2)$
Pr	0.99833(7)	0.08687(4)	0.93168(6)	2.89(1)[B]	C(19)	0.805(3)	0.233(2)	0.896(3)	10.3(7)
N(1)	0.779(2)	0.154(1)	0.926(2)	9.7(6)	C(20)	0.908(3)	0.245(2)	0.8283(2)	14.5(7)
C(2)	0.690(3)	0.134(2)	0.840(2)	9.3(7)	O(21)	1.008(1)	0.2023(5)	0.8479(8)	4.5(2)
C(3)	0.763(3)	0.116(2)	0.755(3)	10.4(8)	C(22)	1.101(3)	0.223(2)	0.784(2)	12(1)
O(4)	0.851(2)	0.073(1)	0.767(2)	10.7(6)	C(23)	1.193(2)	0.168(2)	0.774(3)	10.3(6)
C(5)	0.889(2)	0.026(1)	0.684(2)	7.5(6)	O(B)	0.8848(9)	-0.0103(5)	0.9634(8)	3.8(2)[B]
C(6)	1.036(3)	0.012(2)	0.690(3)	13.7(7)	Cl(1)	0.9547(5)	0.4345(3)	0.7172(5)	7.5(2)[B]
O(7)	1.074(2)	0.0176(9)	0.788(1)	11.6(6)	O(11')	0.843(1)	0.408(1)	0.746(1)	14.6(5)
C(8)	1.191(3)	0.018(2)	0.760(2)	12(1)	O(12')	0.929(2)	0.467(2)	0.626(1)	15.6(8)
C(9)	1.280(3)	0.055(2)	0.813(3)	14(1)	O(13')	1.001(1)	0.482(1)	0.792(1)	13.3(5)
N(10)	1.230(2)	0.1238(9)	0.862(1)	11.9(8)	O(14')	1.026(2)	0.380(2)	0.700(2)	15.9(6)
C(11)	1.314(3)	0.141(2)	0.966(2)	10.4(8)	Cl(2)	1.0028(6)	0.1668(3)	0.4728(5)	8.7(2)[B]
C(12)	1.240(3)	0.191(2)	1.009(3)	12(1)	O(21')	0.994(2)	0.209(2)	0.397(2)	15.8(9)
O(13)	1.149(2)	0.159(1)	1.055(2)	10.7(6)	O(22')	1.016(2)	0.200(2)	0.557(2)	15.9(9)
C(14)	1.105(2)	0.182(1)	1.144(2)	7.3(6)	O(23')	1.063(2)	0.114(2)	0.468(2)	15.6(8)
C(15)	0.996(2)	0.138(2)	1.187(2)	14.7(5)	O(24')	0.888(3)	0.129(3)	0.461(3)	15.8(9)
O(16)	0.921(2)	0.132(1)	1.100(2)	11.7(6)	N(A)	0.463(2)	0.152(1)	0.607(2)	15.8(7)
C(17)	0.800(3)	0.163(2)	1.105(2)	10.4(9)	C(A1)	0.414(2)	0.109(1)	0.561(2)	11.0(9)
C(18)	0.723(3)	0.144(2)	1.011(3)	11.1(9)	C(A2)	0.345(3)	0.057(2)	0.506(2)	9.5(7)

[A] All atoms refined isotropically unless noted otherwise.
[B] Atoms refined anisotropically. The isotropic equivalent thermal parameter is defined as $(4/3) [a^2 B(1,1) + b^2 B(2,2) + c^2 B(3,3) + ac \cos\beta B(1,3)]$.
[C] Cl(1) and Cl(2) and the oxygen atoms labelled O(nn') are atoms belonging to perchlorate ions that are omitted in Figure 1.
[D] N(A)—C(A1)—C(A2) is an acetonitrile molecule omitted in Figure 1.

BRIDGED PRASEODYMIUM (2.2.1) CRYPTATE

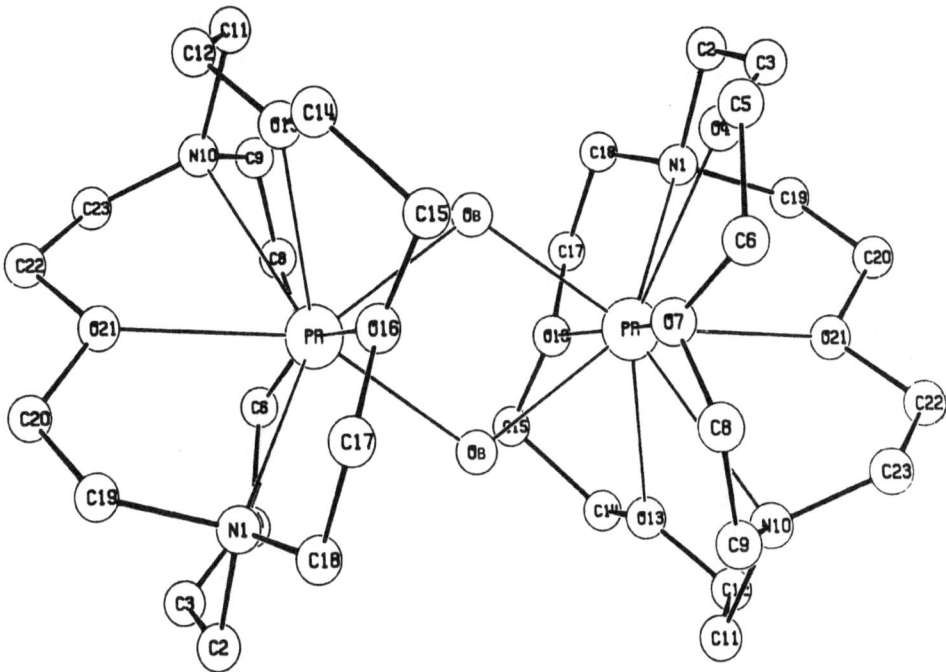

Fig. 1. General view illustrating the molecular structure of the dimeric Pr(2.2.1) cryptate with the two μ-hydroxo bridges.

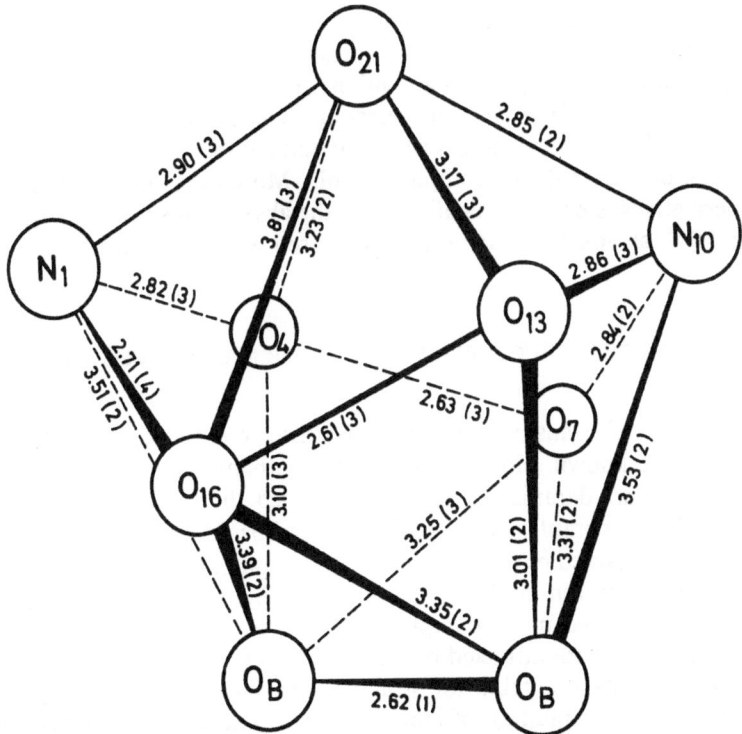

Fig. 2. Coordination polyhedron around each praseodymium ion in the dimeric Pr(2.2.1) complex.

geometry of the coordination sphere of the metal ions is much distorted with respect to the usual nine-coordinate polyhedra [11] but still exhibits an approximate C_2 axis. Distortions were expected because of the steric requirements of the (2.2.1) ligands and because of the constraints that the hydroxyl ions impart on the dimeric complex. Indeed, the (2.2.1) cryptand molecules have to fold in such a way as to accommodate a Pr^{3+} ion in their internal cavity and are relegated to one hemisphere around each metal ion by the strongly coordinating OH^- groups.

Table II. Selected distances (Å) and mean distances and valency angles (°), with standard deviations in parentheses

Pr—N(1)	2.70(2)	Pr—O(13)	2.65(2)
Pr—N(10)	2.81(2)	Pr—O(16)	2.60(3)
Pr—O(4)	2.62(3)	Pr—O(21)	2.57(1)
Pr—O(7)	2.54(2)	Pr—O(B)	2.35(1)
		Pr—O(B)	2.37(1)[a]
Mean C—N distance: 1.54(4)		Mean C—C distance: 1.49(5)	
Mean C—O distance: 1.38(4)			
Mean N—C—C angle: 113(3)		Mean C—N—C angle: 111(2)	
Mean O—C—C angle: 111(2)		Mean C—O—C angle: 110(2)	

[a] Refers to atom at $-x+2, -y, -z+2$.

Selected distances and mean distances and angles are listed in Table II. The apparent shortenings or lengthenings of some bonds as well as the abnormal values of some bond angles are mainly spurious effects believed to arise from inadequate treatment of curvilinear molecular thermal motions in the crystallographic analysis. Structural investigations of macrocyclic complexes are often thwarted by similar difficulties [12]. Disorder of the perchlorate anions [10] also limits the refinement of the structure reported here. However, the selection of non-coordinating anions is a prerequisite to reliable comparisons between the solid state and the solution conformations of labile lanthanide complexes [6]. Moreover, the formation of hydroxylated derivatives is probably facilitated by the absence of strongly complexing anions such as NO_3^-, an ion that has been widely used with lanthanide marcocyclic complexes [1].

5.2. COMPARISON WITH THE STRUCTURE OF SOME OTHER MACROCYCLIC LANTHANIDE COMPLEXES

Numerous examples of transition metal polymeric complexes involving two μ-hydroxyl bridges are known. However, only one lanthanide complex of this type has been reported so far. Baraniak et al. [13] described the structure of a dimeric di-μ-hydroxo ytterbium complex with pyridine-2-carboxaldehyde-2'-pyridylhydrazone. This dimer was obtained by the addition of a solution of the ligand in ethanol to an equimolar aqueous solution of $YbCl_3$. The ligand being a strong base, the formation of a partially hydrolysed metal complex is not surprising. However, polymeric lanthanide complexes have not attracted much interest so far probably because of difficulties in isolating crystals suitable for structure analysis and because hydrolysis most often leads to the precipitation of lanthanide hydroxides instead of the formation of soluble well-defined complexes.

Traces of water in our solutions of monomeric praseodymium cryptate are most probably responsible for the formation of the dimeric complex reported here. Partial hydrolysis of this complex takes place because the excess of (2.2.1) cryptand brings about a pH increase. Incomplete hydrolysis of a lanthanide macrocyclic complex has also been noted by Bünzli et al. [14] who prepared a dimeric praseodymium complex with 1,4,7,10,13-pentaoxacyclododecane (15-crown-5) by dehydrating in vacuo a monomeric species. The metal ions in this dimer are bridged by only one hydroxyl group and by three trifluoroacetate anions. The distance between the two praseodymium ions in the (2.2.1) cryptate reported here is 3.927(1) Å; this value compares very well with the values reported for the two other dinuclear lanthanide complexes mentioned above [13–14].

The solid state structure of the monomeric praseodymium (2.2.1) cryptate [6] synthesised in acetonitrile is presented schematically in Figure 3. The metal ion achieves the same coordination number as in the dimeric cryptate, two hydroxyl groups being replaced by two acetonitrile molecules. The macrocyclic ligand again displays an approximate twofold symmetry and its conformational arrangement is similar to the geometry adopted by the (2.2.1) cryptand in the dimeric compound.

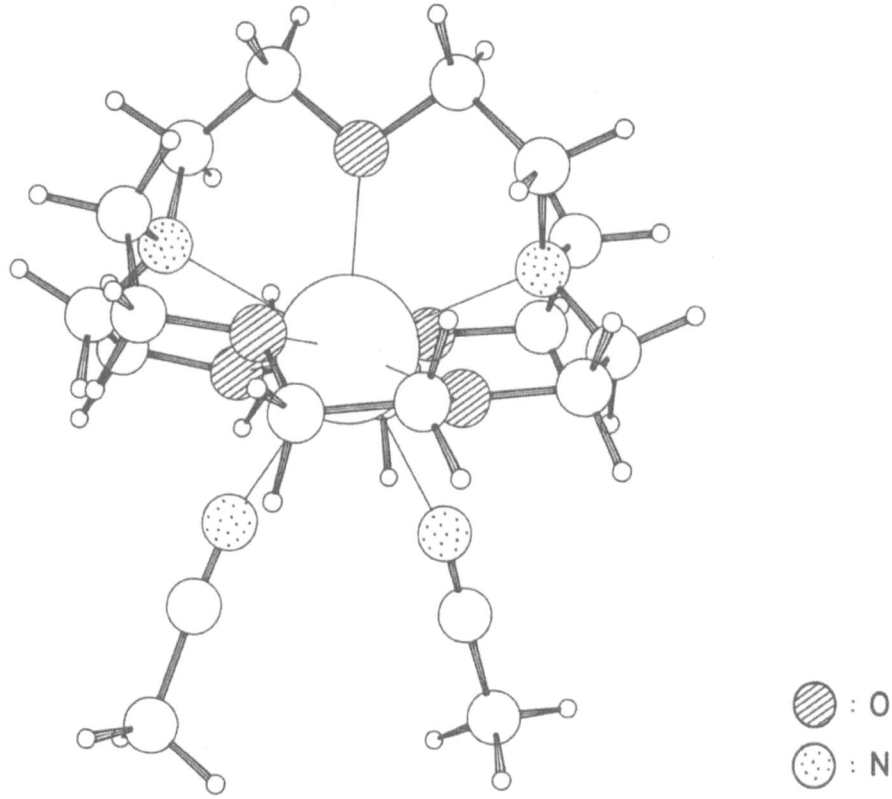

Fig. 3. Schematic presentation of the molecular structure of the monomeric praseodymium (2.2.1) cryptate solvated by two acetonitrile molecules [6]. The approximate C_2 axis is in the plane of the Figure and passes through the praseodymium ion and the oxygen atom of the monooxygen strand.

However, there is a significant difference between the two species: the metal cations are pulled out of the cryptate internal cavity by the two hydroxyl groups. In the monomeric complex, the lanthanide is located 0.13 Å above the mean plane defined by the four oxygen atoms of the 18-membered cycle of the cryptand while it lies 0.18 Å below this plane in the dimeric complex.

Finally, it is worth mentioning here that our structural analysis could account for the unusual affinity of the lanthanide cryptates for the hydroxide ion. According to Yee et al. [4], both the Eu(2.2.1)$^{3+}$ cryptate and the uncomplexed Eu^{3+} ion associate very strongly with the hydroxide ion. The crystallographic structures shown in Figures 1 and 3 clearly indicate that there is ample space available for the coordination of additional ligands around the encapsulated lanthanides and that hydroxylated complexes can be stabilized by the formation of polymers.

5.3. NMR STUDIES

The formation of at least two lanthanide (2.2.1) species in acetonitrile is easily demonstrated by proton NMR spectroscopy. The proton NMR spectrum of monomeric Pr(2.2.1) cryptate displays eight peaks that are shifted from their diamagnetic position by the paramagnetism of the metal ion. The assignment of these peaks required [6] the synthesis of several partially deuterated derivatives of cryptand

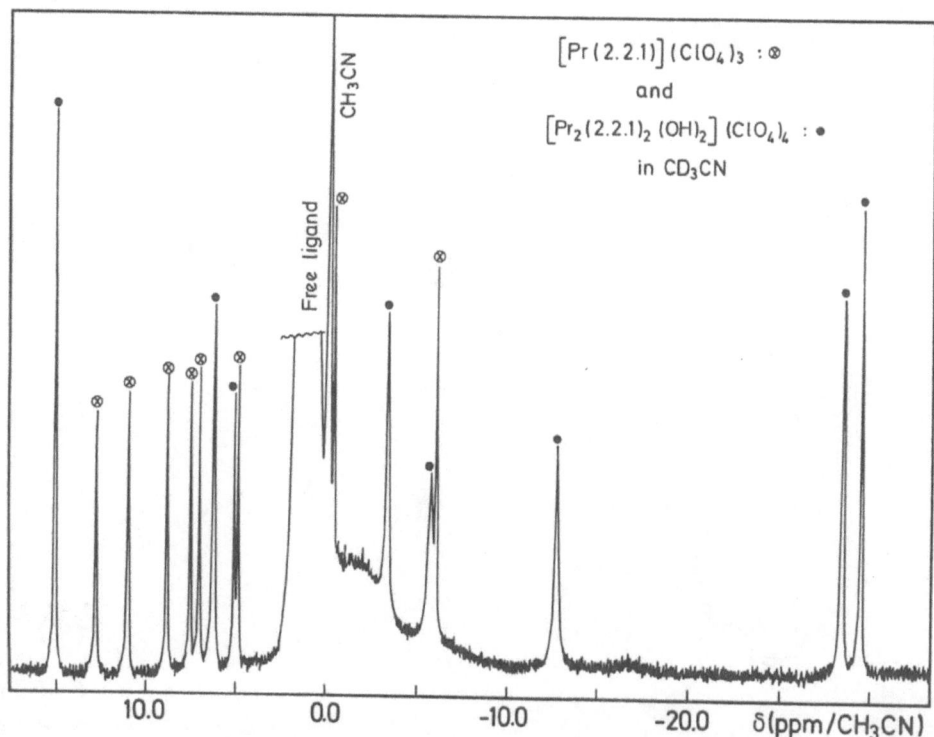

Fig. 4. 300 MHz proton NMR spectrum of an acetonitrile solution of the monomeric Pr(2.2.1) complex containing an excess of cryptand (2.2.1).

(2.2.1). A quantitative analysis of the induced shifts indicated that the NMR spectrum of monomeric Pr(2.2.1)$^{3+}$ arises from metal-centered rearrangements that lead to eight distinct proton environments. The conformational changes occur without alteration of the geometrical configuration of the complex. As shown in Figure 4, the NMR spectrum shows 16 resonances if an excess of cryptand (2.2.1) is added to an acetonitrile solution of monomeric Pr(2.2.1) perchlorate. The new peaks are tentatively attributed to the formation of the dimeric compound because the relative areas of the two sets of peaks depend upon the excess of ligand and because the crystals of the dimer were obtained under the same experimental conditions as used for recording the NMR spectra. However, a more quantitative analysis [15] of the solution conformation of the dimer was hampered by difficulties in interpreting the relative magnitude of the induced shifts. Indeed, the dimeric complex features two paramagnetic centers and the ligands seem to undergo intricate conformational changes.

Acknowledgements

J.F.D. and P.P.B. gratefully acknowledge financial support from the Fonds National de la Recherche Scientifique of Belgium. J.F.D. is Chercheur Qualifié and P.P.B. is Aspirant at this institution. The NMR spectra were obtained at the CREMAN Facility funded by a grant from the FNRS of Belgium.

References

1. J. C. Bünzli and D. Wessner: *Coord. Chem. Rev.* **60**, 191 (1984).
2. M. C. Almasio, F. Arnaud-Neu, and M. J. Schwing-Weill: *Helv. Chim. Acta* **66**, 1296 (1983).
3. J. H. Burns and C. F. Baes: *Inorg. Chem.* **20**, 616 (1981).
4. E. L. Yee, O. A. Gansow, and M. J. Weaver: *J. Am. Chem. Soc.* **102**, 2278 (1980).
5. J. H. Forsberg and T. Moeller: *Inorg. Chem.* **8**, 883 (1969).
6. P. P. Barthélemy, M. R. Spirlet, J. Rebizant, and J. F. Desreux, to be submitted.
7. *Enraf-Nonius Structure Determination Package*: Enraf-Nonius, Delft, Holland, 1981.
8. *International Tables for X-ray Crystallography*, Kynoch Press, Birmingham, England, 1974, Distr. D. Reidel Publ. Co., Dordrecht, Holland.
9. G. Germain, P. Main, and M. M. Woolfson: *Acta Crystallogr.* **18**, 104 (1965).
10. A. Navaza, F. Villian, and P. Charpin: *Polyhedron* **3**, 143 (1984).
11. D. L. Kepert: *Inorganic Chemistry Concepts* v. 6. Springer-Verlag (1982).
12. J.-C. G. Bünzli, G. A. Leonard, and D. Plancherel: *Helv. Chim. Acta* **69**, 288 (1986) and references cited therein.
13. E. Baraniak, R. St. L. Bruce, H. C. Freeman, N. J. Hair, and J. James: *Inorg. Chem.* **15**, 2226 (1976).
14. D. Harrison, A. Giorgetti, and J.-C. G. Bünzli: *J. Chem. Soc. Dalton Trans.* 885 (1985).
15. M.-R. Spirlet, J. Rebizant, J. F. Desreux, and M.-F. Loncin: *Inorg. Chem.* **23**, 359 (1984).

Synthesis and Alkali Metal Cation Complexation of N-Aryl [3.2.2] Cryptands

RICHARD A. BARTSCH*, DAVID A. BABB, and BRIAN E. KNUDSEN
Department of Chemistry and Biochemistry, Texas Tech University, Lubbock, Texas 79409-4260, U.S.A.

(Received: 7 October 1986)

Abstract. Two novel three-nitrogen cryptands with N-aryl substituents are prepared by cyclization of 1,10-diaza-18-crown-6 with mixed anhydrides of 6-arylaza-3,9-dioxadecanedioic acid followed by borane reduction of the resultant bicyclic diamides. For the alkali metal cations, the N-phenyl [3.2.2] cryptand exhibits strongest complexation for Rb^+ in picrate extractions.

Key words: Cryptand, alkali metal cation complexation, picrate extraction.

1. Introduction

Since the introduction of cryptands by Lehn and coworkers [1], the synthesis and cation complexation behavior of these macropolycyclic ligands have been objects of intense research interest in many laboratories [2, 3]. Most common are the bicyclic cryptands with two bridgehead nitrogens and with only oxygen heteroatoms in the bridging arms. The cation complexing properties of such [2]-cryptands is influenced by the replacement of one or more of the oxygen atoms with nitrogen or sulfur [3]. Although replacement of oxygens by —N(H)— and —N(Me)— units has been described, [2]-cryptands with two bridgehead nitrogens and a —N(aryl)— replacement unit for one oxygen atom are unknown. Such a —N(aryl)— unit could serve as a site for the type of functionalization that has been performed with N-phenyl monoaza-18-crown-6 [4]. We now report the synthesis of two N-aryl [3.2.2] cryptands and the alkali metal cation complexing abilities of one of them.

2. Experimental

2.1. PREPARATION OF DIESTERS 1a AND 1b

The appropriate N-aryl diethanolamine (0.105 mol) was dissolved in 500 ml of hot t-BuOH and 58.8 g (0.53 mol) of t-BuOK was added. The mixture was refluxed for 1 h and 19.88 g (0.21 mol) of chloroacetic acid in 150 ml of t-BuOH was added dropwise during 1 h. After refluxing for 48 h and solvent removal by distillation under reduced pressure, 200 ml of 3N HCl was added and the mixture was evaporated in vacuo. The residue was suspended in 200 ml of EtOH, filtered, and 400 ml of benzene was added to the filtrate. The solution was refluxed for 24 h with water removal by anhydrous sodium sulfate in a Soxhlet extractor. After filtration and evaporation in

---

* Author for correspondence.

vacuo, the residue was dissolved in dichloromethane and washed with 1M sodium bicarbonate and then water. After drying over magnesium sulfate and evaporation in vacuo, the residue was distilled under vacuum.

1a: Yield 49%, bp 197–198 °C/0.13 torr. ^{1}H-NMR (CDCl$_3$): $\delta = 1.13$ (t, 6H), 3.59 (s, 8H), 3.9–4.5 (m, 8H), 6.5–7.4 (m, 5H). IR (neat): cm^{-1} = 1753 (C=O), 1138 (C—O). Anal. Calcd: C, 61.17; H, 7.70. Found: C, 61.53; H, 7.85%.

2a: Yield 25%, bp 207–210 °C/0.40 torr. ^{1}H-NMR (CDCl$_3$): $\delta = 1.27$ (t, 6H), 2.25 (s, 3H), 3.68 ($br\ s$, 8H), 3.9–4.5 (m, 8H), 6.89 (ABq, 4H). IR (neat): cm^{-1} = 1755 (C=O), 1140 (C—O). Anal. Calcd: C, 62.10; H, 7.96. Found: C, 62.17; H, 8.12%.

2.2. PREPARATION OF DIACIDS 2a and 2b

The appropriate diethyl ester (22 mmol) was dissolved in 40 ml of 6N HCl and 10 ml of water and refluxed for 4 h. Evaporation in vacuo and addition of benzene (200 ml) followed by refluxing and water removal with a Dean-Stark trap gave a suspension which was filtered and dried under vacuum to produce the hygroscopic hydrochloride salt.

2a: Yield 100%, mp 157–158 °C. ^{1}H-NMR (DMSO-d_6): $\delta = 3.26$ ($br\ s$, 12H), 7.0–8.0 (m, 5H), 10.9–11.7 ($br\ s$, 3H). IR (KBr): cm^{-1} = 3182 (COOH), 2606 (R$_3$NH), 1753 (C=O), 1140 (C—O). Anal. Calcd: C, 50.38; H. 6.04. Found: C, 50.15; H, 6.08%.

2b: Yield 98%, glassy, low-melting solid. ^{1}H-NMR (DMSO-d_6): $\delta = 2.32$ (s, 3H), 3.2–4.6 (m, 12H), 7.40 (ABq, 4H), 8.84 ($br\ s$, 3H). IR (neat): cm^{-1} = 3700–2340 (COOH + R$_3$NH$^+$), 1737 (C=O), 1138 (C—O). Anal. Calcd: C, 51.80; H, 6.38. Found: C, 52.04; H, 6.67%.

2.3. PREPARATION OF CRYPTAND DIAMIDES 4a AND 4b

Finely powdered **2a** or **2b** (7.62 mmol) was suspended in 15 ml of dichloromethane and diluted to 50 ml with toluene. Triethylamine (2.40 g, 24 mmol) was added and the mixture was stirred for 15–20 min until only a light white solid was present in suspension. The mixture was cooled to 0–5 °C and isobutyl chloroformate (2.10 g, 15.4 mmol) diluted in 10 ml of cold toluene was added dropwise. The mixture was stirred at 0–5 °C for 30 min and filtered. The filtrate was diluted to 70 ml with cold toluene to make Solution A. Solution B was prepared by dissolving 2.00 g (7.60 mmol) of 1,10-diaza-18-crown-6 in 10 ml of toluene and diluting to 70 ml with toluene. Solutions A and B were placed in water-jacketed addition funnels maintained at 0–5 °C and were added simultaneously over 8 h to 300 ml of vigorously-stirred toluene at 0–5 °C under nitrogen. The solution was stirred overnight at room temperature and evaporated in vacuo. The residue was purified by chromatography on alumina with dichloromethane-chloroform (3:1) as eluent.

4a: Yield 36%, mp 118–119 °C. ^{1}H-NMR (CDCl$_3$): $\delta = 2.6$–4.5 (m, 36H), 6.5–7.4 (m, 5H). IR (neat): cm^{-1} = 1647 (C=O), 1116 (C—O). MS: 523.5 (M$^+$). Anal. Calcd: C, 59.64; H, 7.89. Found: C, 59.42; H, 7.69%.

4b: Yield 32%, mp 129–130 °C. ^{1}H-NMR (CDCl$_3$): $\delta = 2.20$ (s, 3H), 2.6–4.6 (m, 36H), 6.82 (ABq, 4H). IR (neat): cm^{-1} = 1651 (C=O), 1116 (C—O). Anal. Calcd: C, 60.31; H, 8.06. Found: C, 60.21; H, 8.06%.

2.4. PREPARATION OF CRYPTANDS 5a and 5b

To a solution of the cryptand diamide (2.31 mmol) in THF (20 ml) was added borane-dimethyl sulfide (1.0 ml, 10 mmol) in 19 ml of THF and the solution was refluxed for 9 h. After slow addition of water (5 ml), the suspension was evaporated in vacuo. The residue was refluxed in 25 ml of 3.6N HCl for 12 h. The solution was cooled, made strongly basic with ammonium hydroxide, and evaporated in vacuo. The solid was washed with MeOH (2×25 ml) and the washings were filtered. Diethyl ether was added to the methanolic solution and the mixture was filtered. The filtrate was evaporated in vacuo and the residue was purified by chromatography on alumina with chloroform-dichloromethane (2:1) as eluent.

5a: Yield 74%, light yellow oil. ^{1}H-NMR (CDCl$_3$): $\delta = 2.83$ (t, 12H), 3.4–4.0 (m, 28H), 6.5–7.5 (m, 5H). IR (neat): cm^{-1} = 1126 (C—O). ^{1}H-NMR (CDCl$_3$): $\delta = 2.83$ (t, 12H), 3.4–4.0 (m, 28H), 6.5–7.5 (m, 5H). MS: 495.5 (M$^+$). Anal. Calcd: C, 60.91; H, 9.24. Found: C, 60.93; H, 9.05%.

5b: Yield 74%, colorless oil. ^{1}H-NMR (CDCl$_3$): $\delta = 2.22$ (s, 3H), 2.73 (t, 12H), 3.62 (m, 28H), 6.77 (ABq, 4H). IR (neat): cm^{-1} = 1122 (C—O). Anal. Calcd: C, 63.62; H, 9.30. Found: C, 63.84; H, 9.10%.

2.5. PICRATE EXTRACTIONS

Picrate extractions into chloroform were performed by the reported technique [5]. Extraction constants (K_{ex}) and association constants (K_a) were calculated by the literature methods [6, 7].

3. Results and Discussion

3.1. SYNTHESIS OF N-ARYL [3.2.2] CRYPTANDS

The synthetic route to N-aryl [3.2.2] cryptands **5a** and **5b** is shown in the Scheme. Starting from the N-aryl diethanolamine, reaction with two equivalents of chloroacetic and t-BuOK in t-BuOH followed by conversion of the diacid to the diester for purification gave **1a** and **1b** in yields of 49 and 25%, respectively. Subsequent acid-catalyzed hydrolysis produced the dicarboxylic acid amine hydrochlorides **2a** and **2b** quantitatively. Attempts to form the diacid chloride of **2a** resulted in an unreactive deep blue-colored substance of unknown identity. Ring closure was effected by adaptation of a method from peptide synthesis [8]. Mixed anhydrides **3a** and **3b** formed by reaction with two equivalents of isobutyl chloroformate in the presence of triethylamine were cyclized with 1,10-diaza-18-crown-6 in toluene to produce the corresponding cryptand diamides **4a** and **4b** in 36 and 32% yields, respectively. Reduction with borane-dimethyl sulfide in THF provided 74% yields of **5a** and **5b**.

3.2. ALKALI METAL CATION COMPLEXATION

The alkali metal cation complexing ability of cryptand **5a** was assessed by the picrate extraction method with deuteriochloroform as the organic solvent [5]. Calculated extraction constants (K_{ex}) and association constants (K_a) are presented in

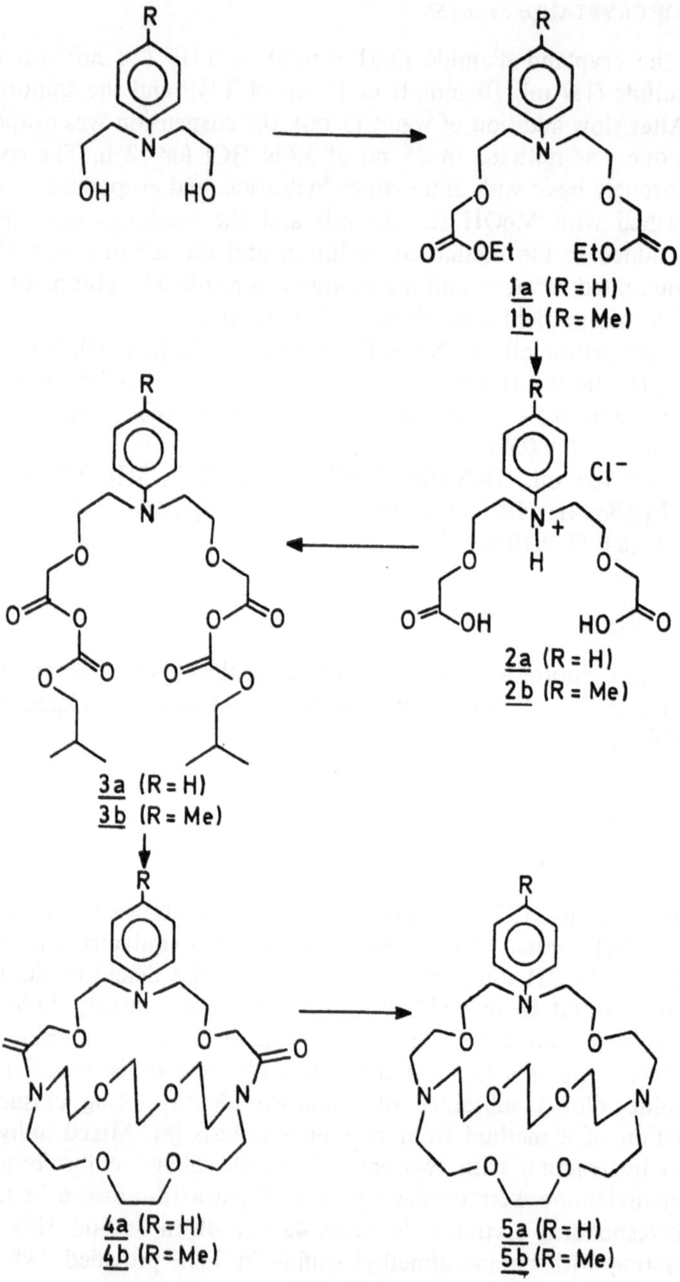

Scheme 1.

Table I. The data reveal that cryptand **5a** exhibits strongest complexation for Rb⁺ in agreement with complexation studies conducted with [3.2.2] cryptand itself [6].

Acknowledgement

This research was supported by Grant D-775 from the Robert A. Welch Foundation.

Table I. Alkali metal picrate extraction into deuteriochloroform by cryptand **5a**.

M^+	$\log K_{ex}$	$\log K_a$	$-\Delta G$(kcal/mole)
Li^+	3.10	5.96	8.10
Na^+	3.95	6.70	9.10
K^+	5.19	7.79	10.58
Rb^+	5.87	8.20	11.15
Cs^+	5.83	8.10	11.00

References

1. B. Dietrich, J. M. Lehn, and J. P. Sauvage: *Tetrahedron Lett.* 2885 (1969).
2. R. Hilgenfeld and W. Saenger: *Top. Curr. Chem.* **101**, 49 (1982).
3. B. Dietrich in *Inclusion Compounds*, Vol 2, (Eds. J. L. Atwood, J. E. D. Davies and D. D. MacNicol), pp. 337–405, Academic Press, New York (1984).
4. J. P. Dix and F. Vögtle: *Chem. Ber.* **113**, 457 (1980).
5. B. P. Czech, D. A. Babb, B. Son, and R. A. Bartsch: *J. Org. Chem.* **49**, 4805 (1984).
6. A. Sadakane, T. Iwachido, and K. Toei: *Bull. Chem. Soc. Jpn.* **48**, 60 (1975).
7. S. S. Moore, T. L. Tarnowski, M. Newcomb and D. J. Cram: *J. Am. Chem. Soc.* **99**, 6398 (1977).
8. J. R. Vaughan, Jr.: *J. Am. Chem. Soc.* **73**, 3547 (1951).
9. J. M. Lehn and J. P. Sauvage: *J. Am. Chem. Soc.* **97**, 6700 (1975).

Redox Responsive Metal Complexes Containing Cation Binding Sites

PAUL D. BEER, CHRISTOPHER J. JONES*, JON A. McCLEVERTY and SITHY S. SALAM
Department of Chemistry, University of Birmingham, PO Box 363, Birmingham, B15 2TT, U.K.

(Received: 9 October 1986; in final form: 16 February 1987)

Abstract. A series of cyclic polyethers, which incorporate the redox active $\{Mo(NO)\}^{3+}$ moiety, have been prepared and characterised. The electrochemistry of these complexes has been investigated and their reduction potentials found to undergo anodic shifts upon the binding of alkali metal cations to the cyclic polyether moiety. The magnitude of the shift appears relatively insensitive to the size of the cyclic polyether ring, but is substantially reduced when K^+ is used in place of Na^+.

Key words: Molybdenum, nitrosyl, pyrazolylborate, electrochemistry, cyclic polyether, complexation, sodium, potassium.

The sterically demanding tripodal ligand $\{HB(3,5-Me_2C_3N_2H)_3\}^-$, L*, restricts the coordination spheres of metals to which it is bound so that only complexes having essentially octahedral or tetrahedral geometries may form. In the case of the $\{Mo(NO)\}^{4+}$ core this results in the formation of formally 16-electron complexes of formula [Mo(NO)L*XY] in which X and/or Y is halide, amide, alkoxide or thiolate. These electron deficient compounds undergo one-electron reduction reactions at potentials which vary substantially depending on the nature of X and Y [1]. Thus in the complexes [Mo(NO)L*Cl{NHC$_6$H$_4$Z}], in which Z is one of a variety of *para* substituents, the reduction potential of the $\{Mo(NO)\}^{3+}$ core is linearly related to the substituent constant, σ_p, with a reaction constant of 6.7 [2, 3].

We have extended these studies by investigating the effects of non-covalently bonded charged moieties located in close proximity to the redox centre. Our initial experiments have involved a study of the effects of cation binding on the electrochemical properties of complexes which contain cyclic polyether cation binding sites. The synthesis and electrochemistry of the redox active compounds I, II ($n=m=1$) and III ($a=0, 1, 2, 3$) has been reported previously [4]. It was found that the binding of Na^+ to the cyclic polyether moiety in these complexes could produce anodic shifts in their reduction potentials. In the case of complexes of type III, the magnitude of these shifts varied from 180 to 320 mV depending upon the size of the polyether ring. We have now synthesised further complexes of type II so that the effect of polyether ring size on anodic shift could also be assessed in these complexes. In addition we have investigated the effect of changing the nature of the bound cation on the anodic shifts produced. Finally, to provide an example of a

---

* Author for correspondence.

Presented at the Fourth International Symposium on Inclusion Phenomena and the Third International Symposium on Cyclodextrins, Lancaster, U.K., 20–25 July 1986.

Mo* = L*Mo(NO)

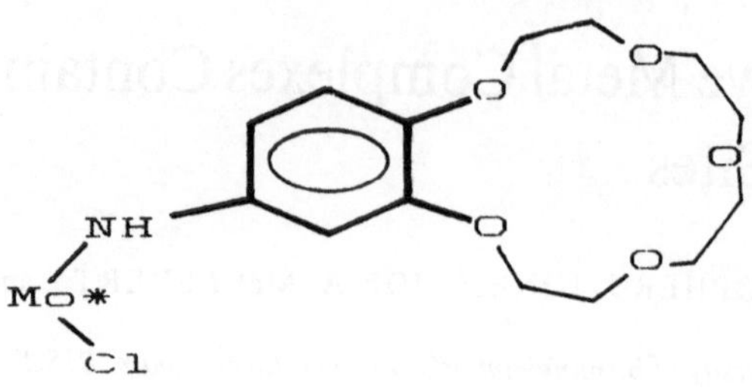

I

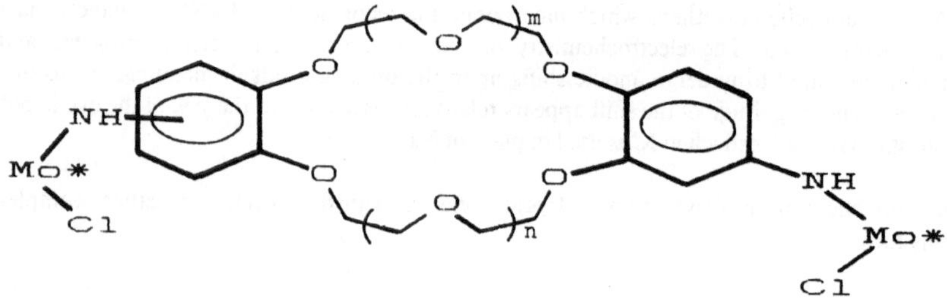

II

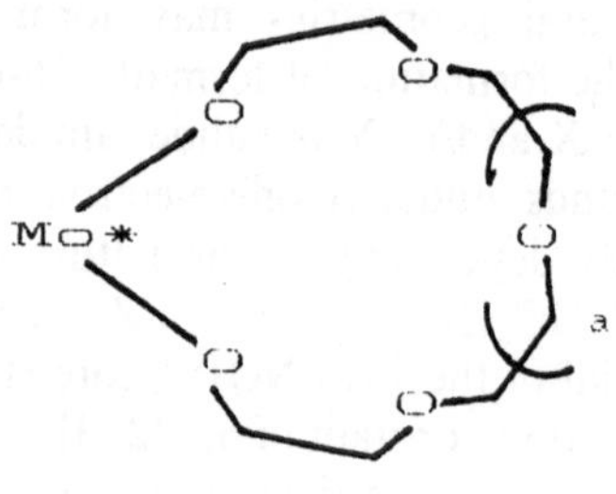

III

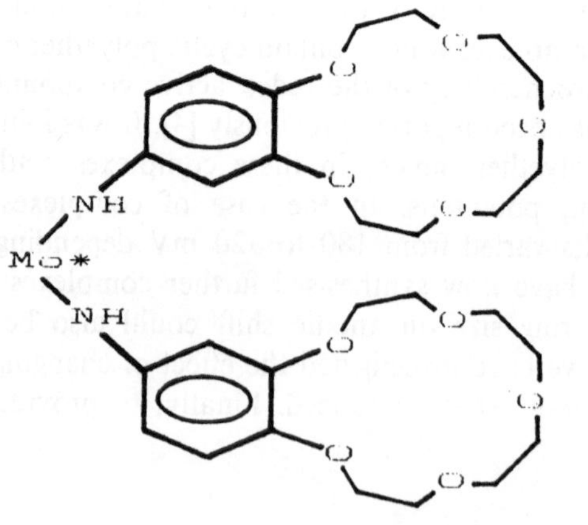

IV

bis-crown ether system, we have also prepared a complex containing two cyclic polyether rings and one redox centre.

The new complexes **II** ($n = 1$, $m = 2$; $n = m = 2$) have been prepared from the reaction between two moles of [Mo(NO)L*Cl$_2$] and the appropriate diaminodibenzo-crown ether. The bis complex **IV** was also prepared from the reaction between the monoiodide derivative, [Mo(NO)L*I{NHC$_6$H$_3$OCH$_2$(CH$_2$OCH$_2$)$_3$CH$_2$O}], and {NH$_2$C$_6$H$_3$OCH$_2$(CH$_2$OCH$_2$)$_3$CH$_2$O} in the presence of sodium naphthalenide. These new complexes were characterised by elemental analysis and by infrared, ^{1}H-NMR and mass spectroscopy. In the cases where $n = m = 1$ and $n = 1$, $m = 2$ it proved possible to separate the *cis* and *trans* isomers of the diaminodibenzo-crown ether precursors and in these cases only one isomer of the bimetallic complex was synthesised and used in the subsequent studies. However, in the case where $n = m = 2$ separation of the *cis* and *trans* isomers was not possible and the complex characterised was a mixture of the two isomeric forms. Since the single isomers and the mixture exhibited similar spectral properties and behaved in a similar manner in the electrochemical experiments it would appear that there are no electrochemically detectable differences between the *cis* and *trans* isomers.

The electrochemical properties of the new compounds have been investigated using cyclic voltammetry and differential pulse techniques. The results obtained, along with previously reported results for **I** and **II** ($n = m = 1$), are summarised in the Table. The observed reduction potentials are in accord with the formulations of **I** and **II** as monoarylamide derivatives, [L*Mo(NO)Cl{NHAryl}], and of **IV** as a bis-arylamide derivative, [L*Mo(NO){NHAryl}$_2$] [6]. In the case of the complexes **II**, coulometric studies have confirmed that a two electron reduction reaction occurs. This is in accord with their formulation as bimetallic complexes containing non-interacting redox centres [5]. Cyclic voltammogram and differential pulse polaragram traces showing the effect of adding up to 2 molar equivalents of NaBPh$_4$ or

Table I.

Complex	I	II			IV
		$n = m = 1$	$n = 1, m = 2$	$n = m = 2$	
E_f (V)[a]	−0.95	−0.96	−0.94	−0.92	−1.36
ΔE_p (mV)[b]	110	80	80	80	120
ΔE(Na) (mV)[c]	60	70	85	85	90
ΔE(K) (mV)[c]	—	40	40	40	40

[a] Obtained in MeCN solution containing 0.2M [Bu$_4$N]BF$_4$ as supporting electrolyte. Solutions were ca. 2×10^{-3} M in complex and potentials were determined with reference to ferrocene as internal standard but are quoted relative to the S.C.E..

[b] Separation between anodic and cathodic peak potentials of cyclic voltammograms.

[c] Shift in reduction potential produced by the presence of Na$^+$ or K$^+$ added as their BPh$_4^-$ salts, in solution, and in aliquots to provide up to 2 molar equivalents with less than 5% volume change in the solution. Broadening of the cyclic voltammogram trace was observed when between 0.2 and 1.0 equivalents of Na$^+$ or K$^+$ had been added. After ca. 1.2 equivalents had been added the trace returned to its original shape and no further changes were observed on adding an additional 0.8 equivalents to give a cation/complex molar ratio of 2. These shifts may be compared with the effects of adding aliquots of NaBPh$_4$ to solutions containing [Mo(NO)L*Cl{NHC$_6$H$_4$-3, 4-(OMe)$_2$}] or [Mo(NO)L*{NHC$_6$H$_4$-3, 4-(OMe)$_2$}$_2$] for which shifts in reduction potential of less than 10 mV were found under similar conditions.

KBPh$_4$ to the electrochemical cell were also obtained. The results of these experiments are also summarised in the Table.

In the case of the complexes **II**, the anodic shifts produced by cation addition appear relatively insensitive to the size of the cyclic polyether ring. A more notable effect is the consistent difference in the ΔE value on going from Na$^+$ to K$^+$. The larger potassium ion, having the smaller charge:radius ratio produces about half the shift of the Na$^+$ ion. A similar effect is apparent with the complex **IV** but, although the electrochemical titrations suggest a 1:1 stoichiometry for cation:complex binding with both Na$^+$ and K$^+$, further studies are needed before an unequivocal interpretation of the electrochemical data can be given for **IV**. The results obtained with the type **II** complexes suggest that the polarising power of the cation is of great importance in determining the magnitude of the anodic shift in E_f produced by cation binding.

Acknowledgement

We are grateful to the Commonwealth Scholarships Commission (U.K.) for support (to S.S.S.).

References

1. J. A. McCleverty: *Chem. Soc. rev.* **12**, 331 (1983).
2. G. Denti, C. J. Jones, J. A. McCleverty, B. D. Neaves, and S. J. Reynolds: *J. Chem. Soc., Chem. Commun.* 474 (1983).
3. N. Al Obaidi, S. M. Charsley, W. Hussain, C. J. Jones, J. A. McCleverty, B. D. Neaves, and S. J. Reynolds: *Trans. Met. Chem.*, in press, Paper 1622, (1987).
4. N. Al Obaidi, P. D. Beer, J. P. Bright, C. J. Jones, J. A. McCleverty, and S. S. Salam: *J. Chem. Soc., Chem. Commun.* 239 (1986).
5. S. M. Charsley, C. J. Jones, J. A. McCleverty, B. D. Neaves and S. J. Reynolds: *Trans. Met. Chem.*, **11**, 329 (1986).
6. N. Al Obaidi, T. A. Hamor, C. J. Jones, J. A. McCleverty and K. Paxton: *J. Chem. Soc., Dalton Trans.* 1525 (1986).

Journal of Inclusion Phenomena 5 (1987), 525–534.
© 1987 *by D. Reidel Publishing Company.*

Binding Properties of Alginic Acid and Chitin

YASUKO TAKAHASHI
Japan Women's University, Department of Chemistry, 2-8-1 Mejirodai, Bunkyo-ku, Tokyo 112, Japan

(Received: 15 October 1986; in final form: 28 February 1987)

Abstract. The binding properties of granular alginic acid(H-alg.) and chitin to iodine, bromine, cadmium ion, calcium ion and cholesterol were investigated. Chitin-iodine and chitin-bromine compounds closely resemble those of H-alg. The amount of iodine included by these polysaccharides increased with a decrease in the concentration of potassium iodide (KI). The number of sugar residues bound to one iodine molecule extrapolated to 0 g of KI was around 6.0 for H-alg. and 6.4 for chitin. These saccharides, which do not form a gel in water, were found to also absorb KI and radioactive iodide or iodine in aqueous solution. H-alg. did not show as much affinity to cadmium ion and calcium ion and cholesterol in iso-propyl alcohol as metalalginates.

Key words: Alginic acid, chitin, iodine, bromine.

1. Introduction

During the course of an investigation of the binding properties of sodium alginate to metal ions and metal alginates to iodine, cholesterol and dyes [1, 2], it became desirable to have experimental values for several measurements of granular alginic acid (H-alg.) obtained from the treatment with hydrochloric acid. Sodium alginate is a binary heterogeneous copolymer of D-mannuronate (M) and L-guluronate (G) residues arranged in a blockwise pattern along the linear chain. There are three types of blocks, M, G and MG blocks in which these two uronic acids occur in some sort of alternation [3]. The ability of alginates to form gels in the presence of calcium ions depends mainly upon auto-cooperatively formed junctions which were interpreted in terms of an "egg-box" model between chain regions enriched in G-sequences [4]. The size of beads or films of metal alginate becomes shorter due to the shrinkage of the gel upon chelation.

By treating sodium alginate with hydrochloric acid, the polysaccharide acquires the form of alginic acid with the appearance of helicoidal structures which are associated supplementarily [5]. X-ray diffraction analysis [6, 7] gave the fibre repeating distance for sodium alginate, poly G-acid and poly M-acid as 15 Å, 8.7 Å and 10.35 Å, respectively. Such differences, reflecting packing differences, are typical of polysaccharide chains. Among crystalline A-, B- and V-amyloses which have a packing behavior based on the helical molecular structure [8], only V-amylose forms an iodine-inclusion compound because of its rather short helix pitch, that is, its rather wide cavity [9].

The metal alginate fibre is very sensitive to acid. Electron microscopic evidence [10] indicated that under acid condition even on treatment with 2% osmic acid solution at the time fixation, the shrinkage of the fibre occurred. The formation of a metal alginate iodine compound at about pH 1, the removal of cholesterol from acetic acid by the addition of metal alginate and the packing of some blocks of dye powder enclosed in a metal alginate net without dissolving in alcohol after acid addition which were reported in the previous paper [1] could be explained by the shrinkage of the metal alginate fibre under the acidic condition.

On the other hand, chitin obtained from the carapace is a popular adsorbent [11] or a starting material for the production of adsorbing media like the amino-acid substituted glucans [12], and is insoluble in water and does not form a gel in water. Being different from the structure of alginic acid, chitin is poly-β-(1,4)-N-acetyl-D-glucosamine. Three polymorphic forms have been identified by X-ray methods. Among α-, β- and γ-chitin, α-chitin appears to be the most stable form and consists of a series of sheets of twofold chains [13].

The present investigation was undertaken to study the binding properties of H-alg. and chitin to iodine, bromine, cadmium ion, calcium ion and cholesterol.

2. Materials and Methods

2.1. MATERIALS

Sodium alginate ... Sodium alginate was obtained from Wako Pure Chemical Industries, Ltd. G/M of this sample is ca. 0.91 and $\overline{DP}$ is 468–473.

Granular alginic acid(H-alg.) ... H-alg. was prepared from the treatment with hydrochloric acid of sodium alginate by the same procedure described in the previous paper [1].

The acidity of the solutions in which 50 mg and 500 mg of H-alg. prepared using 1N HCl was immersed in 50 ml of water were pH 3.3 and pH 2.7, respectively, but for fresh H-alg. the pH was 4.2–4.5. The granular sample was washed with water until the rinsing no longer gave a chloride ion reaction in the preparation, although it showed some evidence of the presence of Cl^-.

Chitin ... Chitin was obtained from Wako Pure Chemical Industries, Ltd. This sample obtained from carapace contains 5–7% of nitrogen and shows less than 10% loss on drying over sulphuric acid. This sample was used without further purification.

The acidity of the solutions in which 50 mg and 500 mg of chitin was immersed in 50 ml of water were pH 4.1 and pH 3.5, respectively, but for the sample washed with water the solution showed pH 6. Purification of chitin requires treatment with dilute acid to remove carbonate, although this technical grade chitin was negative to the Cl^- reaction.

Cd or Ca-alginate beads ... In a typical experiment, 25 ml of 0.4% sodium alginate was added dropwise to 50 ml of 0.01M cadmium nitrate or calcium nitrate with stirring. After standing for 24 h at 30°, alginate beads were separated by a nylon cloth.

Cholesterol detectable solution ... In the previous experiment the cholesterol content of the solution was determined by the ferric chloride reaction. The method lacks reproducibility. The most versatile method of cholesterol detection is with an enzyme. In this paper a one step enzymatic method called the COD·DAOS color former method was used. (Cholesterol esters are hydrolyzed to free cholesterol by cholesterol ester hydrolase. The free cholesterol produced is oxidized by cholesterol oxidase to cholest-4-en-3-one with the simultaneous production of H_2O_2, which oxidatively couple with 4-aminoantipyrine and PhOH in the presence of peroxidase to yield a chromogen with max. absorption at 600 nm.) This cholesterol test kit named the Cholesterol E-Test was obtained from Wako Pure Chemical Industries, Ltd.

All other commercially available chemicals used were of the highest available purity.

2.2. METHOD

Iodine... The weighed amount (0.3–3.0 g) of sample in a glass-stoppered conical flask was mixed with 18 ml of a 0.15M iodine-potassium iodide solution (KI 5, 10 and 15 g/100 ml) and 2 ml of 1N HCl. In one part of the iodine experiment, 30 ml of 0.15M I_2 solution was used. After storing for 24 h, an aliquot of the supernatant liquid was taken out with a pipette and the amount of iodine consumed by the sample was determined by absorbance at 285 nm or by titration with 0.1N $Na_2S_2O_3$.

For the experiment with radioactive iodine, three kinds of radioactive solutions were prepared by spiking with radioactive sodium iodide-^{125}I solution (aqua, KI (5 g/100 ml) and 0.15M I_2). The specific radioactivity of the radioiodine added was so high that its weight was negligible. The activities of the solutions were adjusted to about 2×10^4 cpm/ml, each 15 ml of radioactive solution was mixed with the sample and after standing for 1 h, the activity of the supernatant liquid was measured with a Aloka Auto Well ARC-500.

After removal of the solvent, the solid phase was dried in air until a constant weight was recorded. For X-ray analysis the solid phase sample ground into a powder or a standard material (I_2, KI, H-alg. and chitin) powder was spread on a Cellophane tape and the tape was attached to the sample window. The mesurement and the scanning electron microscopy (SEM) investigation were also carried following the same procedure described in the previous paper [1] respectively.

Bromine... The method used here was virtually identical with that described for iodine, but the latter was replaced with 30 ml of 0.029M Br_2 (chitin) or the mixture of 27 ml of 0.05M Br_2 and 3 ml of 1N HCl (H-alg.). After storing for 5 h (chitin) or 6 days (H-alg.) in the dark, the amount of bromine consumed by the sample was examined both by absorbance at 394 nm and titration with 0.1N $Na_2S_2O_3$. Ca-alg. beads prepared with 30 ml of 0.02M $Ca(NO_3)_2$ and 20 ml of 0.4% Na-alg. were immersed into the mixture of 20 ml of 0.025M Br_2 and 2 ml of 1N HCl. Several parallel experiments using different concentrations of Br_2 in the solution were carried out.

Metal ion... To investigate the pH effect on the binding ratio, 25 ml of 0.2% Na-alg. or the mixture of 25 ml of water and 50 mg of H-alg. was added dropwise with stirring to the mixture of 25 ml of 0.01M $Cd(NO_3)_2$ and 25 ml of buffer solution (pH 4 and pH 7... standard buffer solution for a pH meter obtained from Wako Pure Chemical Industries, Ltd. pH 8.3... 0.1M NH_4Cl 16:0.1N NH_4OH 1. pH 9... 0.1M NH_4Cl 4:0.1N NH_4OH 1. pH 10... 0.1M NH_4Cl 1:0.1N NH_4OH 4. pH 11... 0.1M NH_4Cl 1:0.1N NH_4OH 32) in a beaker. The pH of the mixture was measured and the electrode of the pH meter spiked in the solution was washed with 10 ml of water. The mixture was left to stand for 20 h at 40° and filtered. After rinsing the precipitate, the filtrate and the wash liquid were mixed and the amount of metal ion in the solution was determined by titration with 0.01M EDTA. To investigate the metal concentration effect on the binding ratio, metal alginate-beads were immersed in 50 ml of 0.01–0.14M $Cd(NO_3)_2$ or $Ca(NO_3)_2$ for 24 h, and then the solution was adjusted to pH 1 with 1N HCl. After standing for 24 h, a 10 ml portion of the supernatant liquid was taken out and the amount of metal ion consumed by the sample was determined by titration.

Cholesterol ... Each weighed sample (0.05–0.5 g chitin and H-alg.) was mixed with 25 ml of cholesterol standard solution (150 mg/isopropyl alcohol 1000 ml). After shaking for 10 min and standing for 1 h, a 1 ml portion of the supernatant liquid was taken out and transferred into 3.0 ml of the cholesterol E-Test solution. The mixture was shaken vigorously and left to stand for 5 min at 37°, and then the extinction was measured at 600 nm. For every run of absorbance, titrating and radioactivity experiments a blank solution was made and measured simultaneously.

3. Results and Discussion

Iodine ... Both Chitin and H-alg. react with iodine to give dark purple stained adducts. In this paper results for chitin are presented. Figure 1 shows the relation between the amount of chitin added and the concentration of iodine. The numerical value of the absorbance or the titration decreased with an increase in the amount of chitin. The amount of chitin corresponding to the iodine used was calculated

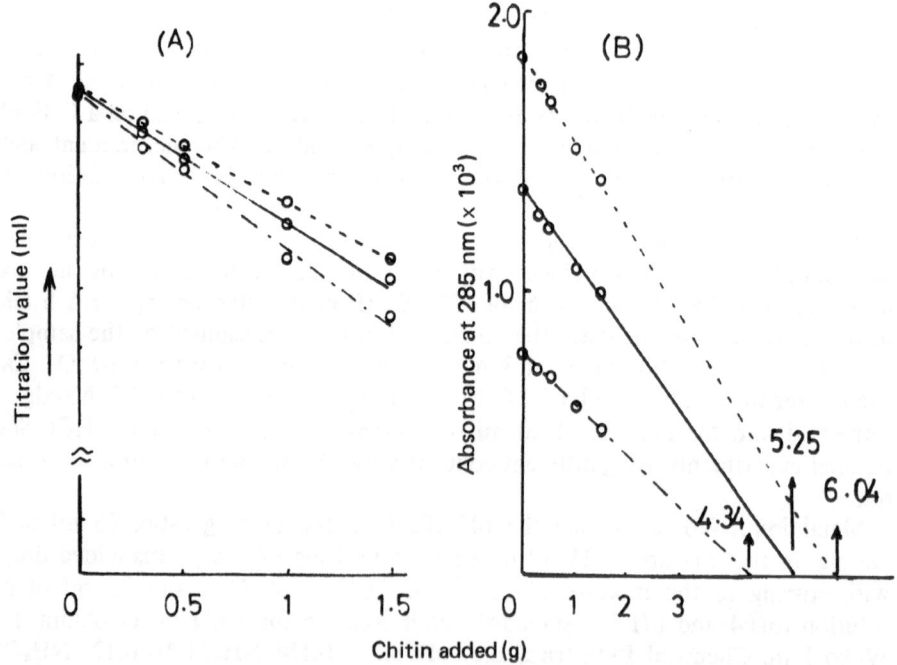

Fig. 1. Effect of the amount of chitin added on the concentration of iodine (A) by titration with $Na_2S_2O_3$ and (B) by UV absorbance. KI content in I_2 solution —·—0.05, ——0.10 and ··· 0.15 (g/ml).

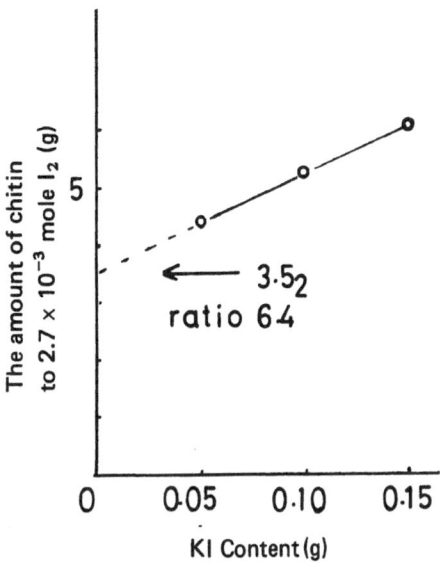

Fig. 2. The amount of chitin bound to iodine.

from the slope obtained from the least squares fit of the plots. The value increased with an increase in the content of potassium iodide as shown in Figure 1(B). In Figure 2 the values obtained were plotted against the KI content, and the number extrapolated to zero g of KI was also indicated. It shows that 6.4 residues of chitin correspond to one iodine molecule in the chitin-iodine compound. Taking into account the impurity of the chitin sample the value is in good agreement with the value (6) reported before for the H-alg.-iodine compound.

The proposed model for the helicoidal structure of alginic acid by Simionescu et al. [5] showed the same binding ratio for the H-alg.-iodine complex as ours. On the other hand, the adsorption of iodine and bromine on chitosan, but not on chitin has been reported [14]. The binding ratio for chitosan-halogen complexes showed 0.5 molar halogen per one hexosaminyl residue in the proposed structure, in which a halogen molecule links to two amino groups of chitosan to form a bridged ladder like structure. The structure differs from the well-known amylose-iodine complex in which iodine atoms are aligned in channels of the polysaccharide.

Table I. Weight of chitin before and after experiment

Chitin taken (g)	0.500	1.000	3.000
Chitin after experiment (g)	0.680	1.310	3.759
Iodine content of the chitin $-I_2$ adduct (g)	0.084	0.149	0.350

In Figure 3 X-ray powder diffractograms are shown. The presence of potassium iodide was shown by the peak at 25°. The characteristic peak observed in the chitin-iodine sample corresponds to 3.16 Å. This value differs from the value of 3.06 Å observed for the I-I distance of iodine packed in the helical cylinder of the

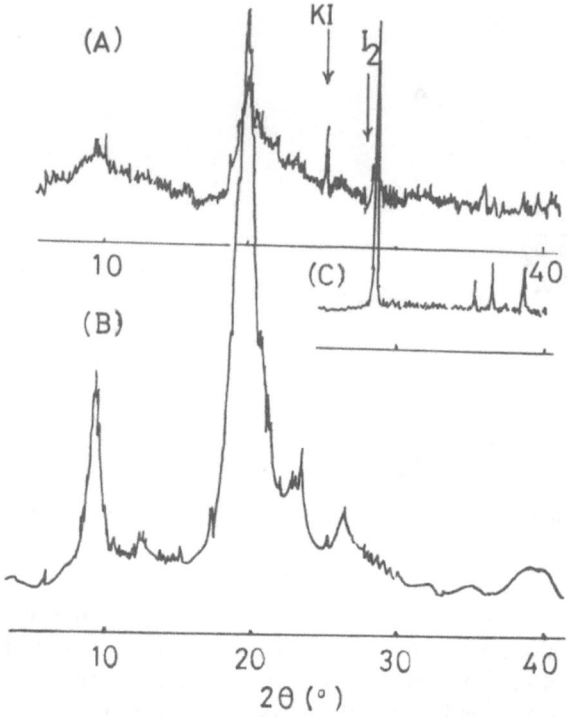

Fig. 3. X-ray diffractograms of (A) chitin-iodine (B) chitin and (C) iodine.

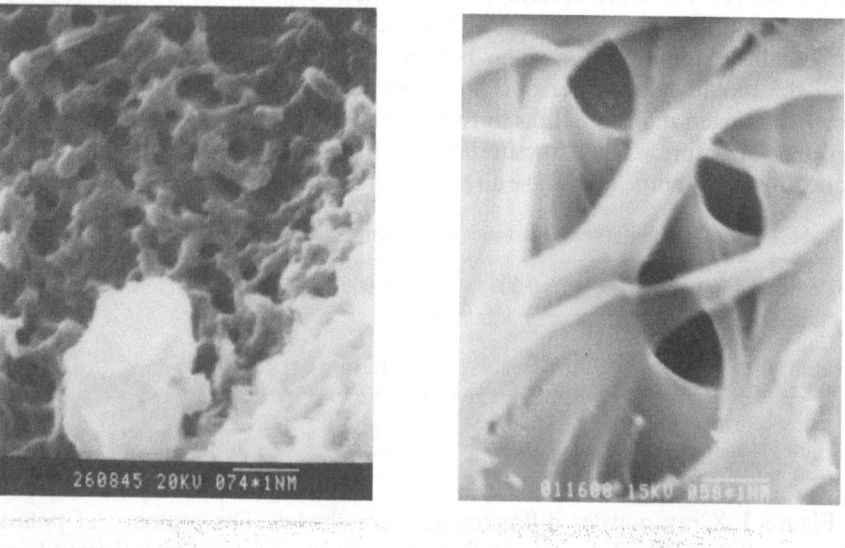

Fig. 4. Scanning electron micrographs of iodine-poly saccharides (A) H-alg., (B) chitin.

α-cyclodextrin-iodine complex [15] slightly but resembles those in the Ca-alg.-iodine compound (3.13 Å) and the Cd-alg.-iodine compound (3.17 Å) prepared at pH 1 [1]. The results by X-ray diffraction [16] for the chitin sample used here agree with an orthorhombic unit cell of α-chitin with dimensions $a = 4.76$ Å, $b = 10.28$ Å, and $c = 18.85$ Å. This fibre repeating distance of about 10.3 Å is the same as that for poly M-acid. H-alg. does not possess as high a stereoregularity as chitin, although the fibre repeat distance of 8.7 Å showed clearly [16]. Therefore, it is very likely that the fiber axis of poly G-acid contains two D-guluronic acid residues per turn, which is the repeat obtained when two α-(1,4)-linked residues in the stable 1 C conformation are arranged in a 2_1 helix [17].

Figure 4 shows the porous structure of iodine-poly saccharide. The microscopic structure of chitin is very similar to that of granular alginic acid.

The results of the experiments for radioactive iodine are shown in Figure 5. In a radioisotopic exchange reaction, the activities are divided in portion to weight among exchangeable species at equilibrium, so the activities reduced by these saccharides are affected by various factors. As shown in Figure 5, the reduction of the plotted line is quite reasonably attributable to the reversible equilibrium between the solid phase and the liquid phase. Chitin and H-alg. also absorb radioactive iodide and radioactive iodine. Chitin absorbs them faster than H-alg.

Bromine. In this experiment with bromine the absorption equilibrium between the solid phase and the liquid phase was attained after 1 hr for chitin but after 5 days for H-alg. Although the limited solubility of bromine in water precludes measurements at high concentrations, as shown in Figure 6, these saccharides absorb bromine (A) and bead form Ca-alg. has a high binding ratio corresponding to the concentration of bromine (B). Being different from chitin, chitosan absorbs bromine in water up to 1.12 molar bromine per two hexosaminyl residues [14] and neither bromine nor iodine is adsorbed on N-acetylchitosan [14]. The latter conflicts with our results for the chitin-iodine and chitin-bromide adducts.

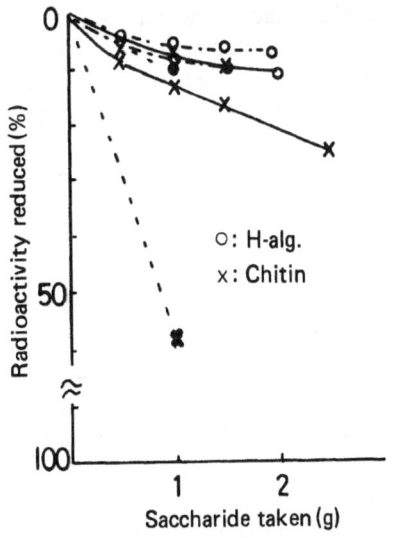

Fig. 5. Effect of poly saccharides on ^{125}I-solution ··· aqua, ——I_2 and —·—KI.

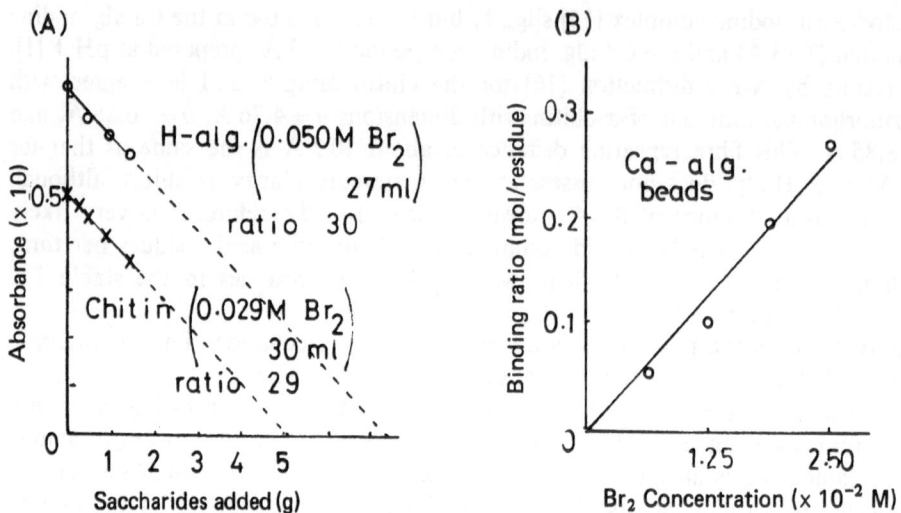

Fig. 6. Effect of sample form on the binding ratio of (A) H-alg. or chitin to Br_2 and (B) Br_2 to bead form Ca-alg.

Metal. In Figure 7(A) results for Na-alg. and H-alg. are presented. Chitin scarcely reacted with 0.1M and 1.0M $Cd(NO_3)_2$, respectively. The binding ratios are similar in the range pH 3–8 and then increase markedly up to pH 10. Taking into account the amount of sample used (H-alg. 2.8×10^{-3} mole, Na-alg. 2.5×10^{-3} mole) the two values are the same. According to Rees [18] alginates gel reversibly and are therefore suspected of an ability to form an ordered conformation when stabilized by cohesion with other chains. They are formed by the controlled mixing of the polysaccharide with the salt of a suitable cation and liquified by ion exchange with an alkali-metal cation such as Na^+ or by sequestering the gelling cation. The ring conformations are known to be as shown for both the solid state and solution. The chains are ribbon-like and extended, and they pack like planks in a timber-yard. The chains have twofold screw symmetry. This conformation for poly-G [19], appears to persist [20] in all of the salt forms so far studied. Mackie et al. [21], reported that the conformation of polyguluronate remains unchanged in the condensed and solution phases.

Figure 7(B) shows the different reactivities of metal alginates, H-alg. and Na-alg. with Cd^{2+} or Ca^{2+}. The high affinity of metal alginate to metal ion may be caused by dispersion in the gel net. These bindings by the "egg box" model conformation and the gel net of metal alginates suggest nothing about the formation of a dark blue iodine-H-alg. adduct or a dark purple iodine-chitin one. With regard to the binding sites of Cu^{2+} in chitin, examination of the crystal structure of α-chitin suggests that a site consisting of two N and two O ligands is more probable [22]. The size of a chitin particle effects the adsorption of dyes [23] which was explained as intraparticle diffusion [24]. These reports indicate that the binding mechanism of chitin to iodine is different from those.

Figure 7(A) shows no evidence of the binding of H-alg. to Cd^{2+} at pH values below 3. pH 2.85 is the acidity at which a sample of poly-G acid is prepared from the treatment with hydrochloric acid by Haug's method [25]. X-ray diffraction

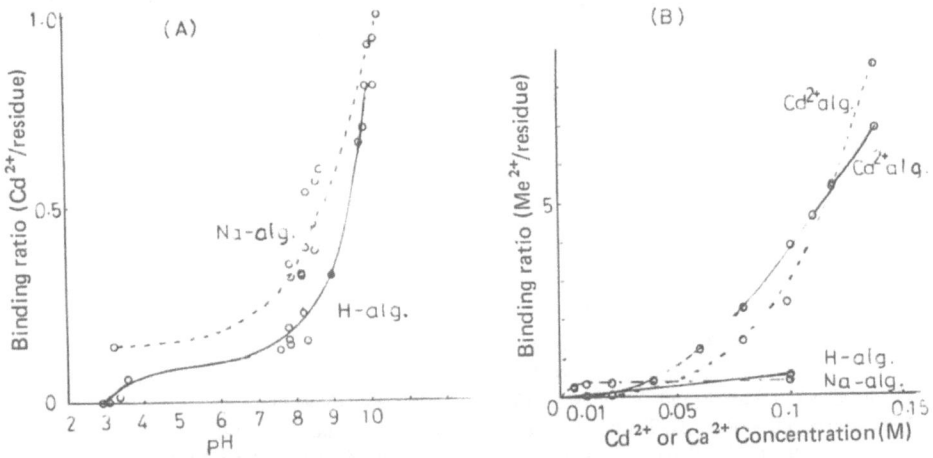

Fig. 7. Changes of binding ratio (A) pH-induced (B) M^{2+} concentration-induced.

points to the high degree of crystallinity of alginic acid, obtained after gel dehydration (achieved by hydrochloric acid addition) as against the altogether amorphous sodium alginate [5]. On the other hand α-chitin may be obtained by treating β-chitin with 6N HCl [17]. No method is known to reverse this change. These data suggest that the formation of metal alginate-iodine observed at pH 1, H-alg.-iodine or chitin-iodine compounds depends on the fiber chain conformation of these polysaccharides at the acidity below pH 2.85.

Cholesterol... None of chitin was obtained pure, so the sample of chitin obtained from carapace was positive to the cholesterol E-Test and the extinction to standard cholesterol solution increased with an increase in the amount of chitin, but any reduction of cholesterol by chitin was not observed at all. In the previous experiment the removal of cholesterol from acetic acid solution by the addition of metal alginate beads was observed. Such reduction of cholesterol from iso-propyl alcohol by the addition of metal alginate beads (unpublished), H-alg. or chitin is not observed in this experiment, so it indicates that the removal of cholesterol from the acidic solution depends on the shrinkage of the metal alginate beads, but not on the fiber chain packing.

These results suggest that the binding capacity of H-alg. or chitin is fixed, so the amount of iodine enclosed in H-alg. or chitin decreases with an increase in the amount of KI.

4. Conclusions

It may be concluded from the results of this investigation that crystalline alginic acid and chitin have almost the same binding ability to iodine, bromine, cadmium and cholesterol. Chitin-iodine and chitin-bromine compounds closely resemble those of crystalline alginic acid. The amount of iodine included by these polysaccharides decreases with an increase in the amount of potassium iodide. These polysaccharides which do not form a gel in water also absorb potassium iodide and radioactive iodide or radioactive iodine, but do not have an affinity to cholesterol. H-alg. does not have as high a binding ratio to cadmium ion and calcium ion corresponding to the concentration of the nitrate metal ions as metal alginate shows after the treatment with hydrochloric acid. The affinity of H-alg. to 0.01M Cd^{2+} is zero around pH 3.

Acknowledgements

The author wishes to thank Prof. N. Ikeda and Dr R. Seki of the University of Tsukuba for carrying out this radioactive iodine experiment.

References

1. Y. Takahashi: *J. Incl. Phenom.* **2** (1984) 399.
2. Y. Takahashi and K. Tsuji: *Eisei Kagaku* **27** (1981) 30.
3. H. Grasdalen, B. Larsen, and O. Smidsrod: *Carbohydr. Res.* **68**, (1979) 23.
4. E. R. Morris, D. A. Rees, D. Thom, and J. Boyd: *Carbohydr. Res.* **66** (1978) 145.
5. CR. I. Simionescu, V. I. Popa, V. Rusan, and A. Liga: *Cellulose Chem. Technol.* **10** (1976) 587.
6. W. T. Astbury: *Nature* **155** (1945) 667.
7. E. D. T. Atkins, W. Mackie, and E. E. Smolko: *Nature* **225** (1970) 626.
8. H.-C. H. Wu and A. Sarko: *Carbohydr. Res.* **61** (1978) 7 and 27.
9. F. Horii, A. Hirai, R. Kitamaru, and H. Yamamoto: *Polym. Preprints, Jpn.* **34** (1985) 2469.
10. Y. Takahashi and A. Yamada: *Jpn. Women's Univ. J.* **33** (1986) 113.
11. G. Mckay, H. S. Blair, and J. Gardner: *J. Appl. Polym. Sci.* **27** (1982) 4251.
12. R. A. A. Muzzarelli, F. Tanfani, M. Emanuelli, and L. Bolognini: *Biotech. Bioeng.* **27** (1985) 1115.
13. D. Carlstron: *J. Biophys. Biochem. Cytol.* **3** (1957) 669.
14. S. Hirano, Y. Kondo, M. Fuketa, and A. Yamashita: *Chitin Chitosan Proceedings Int. Conf.* **2** (1982) 57.
15. F. Cramer: *Chem. Ber.* **84** (1951) 855.
16. Y. Takahashi and M. Takahashi: *Jpn. Women's Univ. J.* **34** (in press).
17. A. G. Walton and J. Blackwell: *Biopolymers*, Academic Press (1973).
18. D. A. Rees: *Biochem. J.* **126** (1972) 257.
19. E. D. T. Atkins, I. A. Nieduszynski, W. Mackie, K. D. Parker and E. E. Smolko: *Biopolymers* **12** (1973) 1879.
20. W. Mackie: *Biochem. J.* **125** (1971) 85.
21. W. Mackie, S. Perez, R. Rizzo, F. Taravel, and M. Vignon: *Int. J. Biol. Macromol.* **5** (1983) 329.
22. S. Schlick: *Macromol. Chem.* **19** (1986) 192.
23. G. Mckay, H. S. Blair, and J. Gardner: *J. Appl. Polym. Sci.* **27** (1982) 3043.
24. G. Mckay, H. S. Blair, and J. Gardner: *J. Appl. Polym. Sci.* **28** (1983) 1767.
25. A. Haug, B. Larsen, and O. Smidsrød: *Acta Chem. Scand.* **20** (1966) 183.

# The Molecular Anvil Model of an Enzyme Taking into Consideration the Flexibility of Enzyme Molecules

KAZUO AMAYA
National Chemical Laboratory for Industry, 1-1 Azuma, Yatabe, Tsukuba, Ibaraki, 305, Japan

(Received: 24 October 1986; in final form: 19 May 1987)

Abstract. The concept of a molecular anvil model of an enzyme, assuming a rigid enzyme molecule, is introduced. Two distinct features of enzymes, high catalytic power and high specificity, are reasonably and consistently explained. The dynamic nature of molecular anvil action is stressed. The origin of the high catalytic power is the spontaneous creation of a high energy state at the anvil site. The origin of the high specificity is a high sensitivity of the maximum accessible potential energy to the relatively extruded distance of the molecular anvil. The flexible model is developed by assuming a flexible enzyme molecule. It is deduced from this flexible model that enzyme activity shows a maximum with a wide range of monotonous change of the configuration of the enzyme molecule. This is the origin of the general property of enzymes that enzyme activity shows a maximum with monotonous variation of environmental parameters such as pH, temperature, pressure or some times concentration of chemical substances. The induced fit theory of Koshland is reasonably explained. The relation and differences between individual theories of enzymes are discussed. The enzymological basis of the complex regulation of biological organisms is discussed. The inversion of the sign of control of effectors is predicted when environmental parameters are varied. This concept may be useful in designing artificial enzymes or high specificity catalysts.

Key words: Molecular anvil, high catalytic power, high specificity, spontaneous creation of a high energy state, relatively extruded distance, semi-inclusion phenomenon, slightly imperfect fitting, antibody, hapten, enzyme, enzyme activity, regulation of biological system, Boltzman factor.

1. Introduction

The concept of the molecular anvil has been proposed by us in a previous paper [1] in order to explain the high specificity of enzymes for discriminating optical isomers. In this paper we will describe this molecular anvil in a more general way and make clear its features and the importance of its dynamic nature. Then the flexibility of the enzyme molecule is introduced and we will discuss how the efficiency of the molecular anvil will vary with a change of configuration of the enzyme molecule. By assuming a linear relationship between the configuration of the enzyme molecule and the environmental parameters, the general properties of enzyme, *viz.* that enzyme activity shows a maximum with a monotonous change of environmental parameters such as pH, temperature etc. is derived. The importance of this general property of enzymes in the regulation of biological organisms is discussed.

2. The Rigid Molecular Anvil Model

2.1. THE MOLECULAR ANVIL AND THE CONDITIONS OF ITS FORMATION

An anvil is a mechanical instrument to produce very high pressures from low pressure sources using pistons. Such a mechanism can be realized at the molecular level if the following two conditions are satisfied. The first is that two molecules can contact simultaneously at multiple points. This may be called the simultaneous multi-point contact. The first condition is satisfied for a pair of substrate and enzyme molecules because it is confirmed by X-ray experiments that an enzyme molecule has a cleft over its surface into which the substrate molecule can fit. It is generally realized that when the concave surface of one molecule comes into contact with the convex surface of the other molecule. This is called the semi-inclusion phenomenon. The second condition may be satisfied because it is not probable that different molecules can always fit strictly exactly. Various types of molecular anvils are schematically shown in Fig. 1. In these figures the number of simultaneous contacts, n, is chosen as 5. The relatively extruded point is called an anvil site and the $n-1$ contact points in the enzyme molecule distinct from the anvil site are called the contact sites. We do not exclude a case where a small molecule like water is sandwiched between the two molecules as the molecular anvil as shown in Fig. 1 (e).

Let us explain how the molecular anvil operates. When the two molecules are separated at distances greater than the equilibrium distance, an attractive force exists between every pair of molecular contacts. However when they come closer to each other, a repulsive force begins to exert only at the anvil site and the other contact sites remain to be attractive. In such a range of distances attractive forces at the contact sites are focused at the anvil site and high pressure is produced at this point and a high energy state is spontaneously created at this anvil point. This is molecular anvil action.

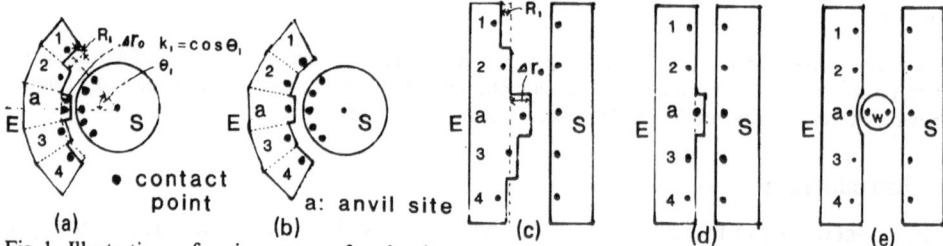

Fig. 1. Illustrations of various types of molecular anvil for $n=5$. a: $k_i s \neq 1$, $R_i s \neq 0$. b: $k_i s \neq 1$, $R_i s = 0$. c: $k_i s = 1$, $R_i s \neq 0$. d: $k_i s = 1$, $R_i s = 0$. e: a small molecule like water is sandwiched. S – substrate; E – enzyme; w – water.

2.2. POTENTIAL ENERGY AT THE ANVIL SITE

Let us discuss molecular anvil action more quantitatively. The total potential energy of the molecular anvil $E_t(r)$ is expressed by assuming pair approximation, that is the total energy is the sum of the energy of each molecular contact pair,

$$E_t(r) = f_a(r - \Delta r_0) + \sum_{i=1}^{i=n-1} f_{ci}(k_i r + R_i) \tag{1}$$

where r is the intermolecular distance, r_0 is the relatively extruded distance, $f_a(x)$ is a potential energy function of the anvil site pair, $f_{ci}(x)$ is the one for the i-th contact pair in the $n-1$ contact pairs, k_i is a ratio of the intermolecular distance change at the i-th contact site to the one at the anvil site, and R_i is a constant relating to the

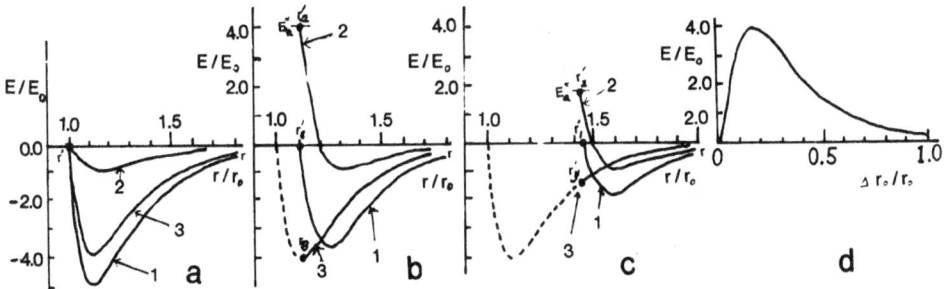

Fig. 2. Potential energy curves of the molecular anvil for $n=5$ with different values of the relatively extruded distance Δr_0 and its relationship between maximum peak potential energy at the anvil site, E_a^*. Figures 2(a)–(c) are curves for values of $r_0/r_0 = 0$, 0.2 and 0.5 respectively. Curves 1, 2 and 3 in each figure indicate the anvil site, the contact sites and the potential energy respectively. Figure 2(d) is the dependence of E_a^* on Δr_0.

intermolecular surface distance of the i-th contact pair in the contact sites. For simplicity, we can assume without loss of intrinsicity of the molecular anvil that all $k_i s = 1$ and all $R_i s = 0$ and further $f_a(x) = f_{c1}(x) = f_{c2} = \cdots = f(x)$, then Equation (1) is reduced to

$$E_t(r) = f(r - \Delta r_0) + (n-1) \cdot f(r) \qquad (2)$$

If we assume a proper function for $f(x)$ we can calculate the numerical value of each term of Equation (2). By assuming a Lenard-Jones 12-6 type potential function

$$f(r) = 4E_0\{(r_0/r)^{12} - (r_0/r)^6\} \qquad (3)$$

where E_0 is the minimum value of the potential energy and r_0 is the distance at which the potential energy crossed zero. For $f(r)$, we calculated each term of Equation (2) for the molecular anvil $n = 5$ with $\Delta r_0/r_0 = 0$, 0.2 and 0.5 as functions of r and are shown in Figure 2. The potential energy of the state when the two molecules are separated at infinite distance at rest is taken to be zero. When the two molecules begin to approach, they are accelerated by their attractive force and finally they stop at the distance where the total potential energy $E_t(r) = 0$, due to the principle of conservation of energy, and then reverse again. During these processes $E_t(r)$, $f(r - \Delta r_0)$, and $(n-1) \cdot f(r)$ values change along curves 1, 2 and 3 respectively and each molecular contact pair moves from r_1 to $r_{1'}$, r_2 to $r_{2'}$, and r_3 to $r_{3'}$ respectively. Positive values of potential energy appear only at the anvil site except when $\Delta r_0 = 0$.

The peak positive potential energy values at the anvil site for a particular value of the relatively extruded distance Δr_0 varies very sharply with Δr_0 and is shown in Figure 2(d) for $n = 5$. The maximum value in the curve of this figure E_m^* is a function of n and is equal to $(n-1) \cdot E_0$. Values of $\Delta r_0/r_0$ corresponding to E_m^* vary from 0.153 for $n = 2$, 0.200 for $n = 5$ and 0.237 for $n = 10$ and are nearly near 0.2. Since r_0 is an order of a few Å, the molecular anvil has its maximum efficiency for Δr_0 of about 1 Å. This means that the efficiency of the molecular anvil is very sensitive to the relative size and shape of the substrate and enzyme molecules. This is the origin of the high specificity of enzymes.

2.3. TIME VARIATION OF POTENTIAL ENERGY AT THE ANVIL SITE

In an ideal case when the two molecules are isolated from the surroundings only one collision may occur if they had kinetic energy initially. In an actual case, each molecule has kinetic energy and is confined to a small space by collisions with surrounding molecules. In such case we can reasonably assume that the two molecules forming the molecular anvil complex will continue to oscillate in the intermolecular potential well. The range of distance of oscillation may be approximated between r_{min} at which $E_t(r) = 0$ and r_{max} at which $E_t(r) = -kT$, the mean kinetic energy. Then the potential energy at the anvil site may vary periodically with time as shown in Figure 3. A high energy state is spontaneously and periodically created at the anvil site. Since chemical reactions are triggered by instantaneously created high energy states, the peak values govern the enhancement factor of an enzymatic reaction. The peak and average values of potential energy at the anvil site of the molecular anvil of maximum efficiency are calculated for various values of n and are shown in Figure 4. The average value is approximated to the one when the enzyme-substrate complex is at rest at equilibrium distance where $E_t(r)$ is the minimum. It is seen from this figure that the peak values are significantly higher than the average values. The peak potential energy is produced by the dynamic nature of molecular anvil action. It is stressed that the high catalytic power of an enzyme is produced by its dynamic nature. During repeated appearance of a high energy state at the anvil site, enzymatic reaction may be triggered.

Preliminary calculations of the peak pressures at the anvil site give values of several hundred thousand Mpa. for $n = 5$ assuming E_0 corresponding to a vaporization energy of 10 Kcal/mol and r_0 corresponding to a molar volume of 100 ml.

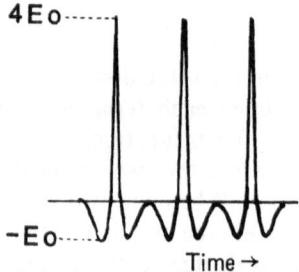

Fig. 3. Schematic illustration of periodic variation of potential energy values at the anvil site.

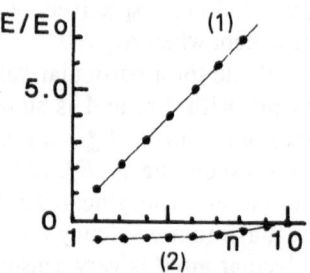

Fig. 4. Peak (1) and average (2) values of potential energy at the anvil site of the molecular anvil for various values of n with maximum efficiency.

2.4. THE ORIGIN OF HIGH CATALYTIC POWER AND HIGH SPECIFITY OF ENZYMES

The total reactivity of an enzymatic reaction is governed by two factors. The first one is the probability of formation of an enzyme-substrate complex. The second one is the enhancement factor of lowering the activation energy in a chemical process. The complexation energy E_c relates to the probability of formation of the complex by the factor $\exp(E_c/kT)$. The value of E_c is equal to the minimum value of the total potential energy of the molecular anvil $E_t(r)$. We can calculate E_c as a function of Δr_0 of the extruded distance for a particular value of n. The relationship between E_c and $\Delta r_0/r_0$ for $n=5$ is shown by curve 1 in Figure 5. It is seen that the complexation energy decreases with increasing Δr_0 due to the increase of the repulsive force at the anvil site. We can assume that the local high energy state at the anvil site is effectively utilized to enhance enzymatic reaction. High pressure at the anvil site may help chemically bonded enzyme-substrate intermediate formation. The enhancement factor is proportional to $\exp(E^*/kT)$, where E^* is the peak positive potential energy value at the anvil site. E^* is reproduced as curve 2 in Figure 5 for $n=5$. The total reactivity is the product of the above mentioned two factors and is shown as curve 3 in Figure 5.

For the molecular anvil of maximum efficiency the ratio of the peak potential energy at the anvil site $(n-1) \cdot E_0$ to E_c is nearly 1 ranging from 0.810, 0.946, 1.036 to 1.243 for $n=3, 4, 5$ and 10 respectively. It may be said that in enzymatic reactions specificity is greatly elevated by utilization of the complexation energy twice, while catalytic power is enhanced by the factor of $\exp(E_c/kT)$, corresponding to utilization of E_c once. This is the origin of the high catalytic power and high specificity of enzymes.

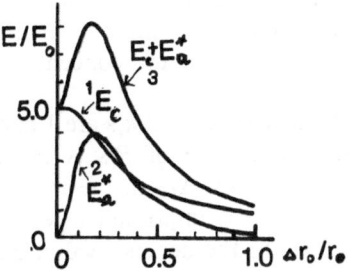

Fig. 5. Dependence of complexation energy, E_c, peak energy at the anvil site, E_a^*, and their sum on the relatively extruded distance, Δr_0, for the molecular anvil with $n=5$.

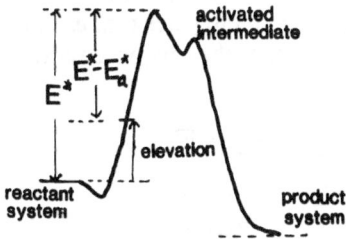

Fig. 6. Energy diagram of an enzymatic reaction showing elevation of starting level of a reactant system by molecular anvil action.

It may be noted that enhancement of a chemical reaction by the molecular anvil action is due to elevation of the starting level of the reactant system but not the lowering of activation energy at the top of the reaction intermediate state as shown in Figure 6. It also may be noted that molecular anvil action is useful in discriminating a substrate smaller than the size of the cleft of the enzyme.

3. The Flexible Molecular Anvil Model

3.1. ENZYME ACTIVITY AND CONFIGURATION OF ENZYME MOLECULE

We assume that a flexible molecular anvil is composed of at least one rigid molecular anvil with one anvil site and $n_a - 1$ contact sites and n_c pieces of other rigid contact sites with n_j contact points and each rigid body is connected by a flexible chain and the relative positions are changeable. Various types of flexible molecular anvil are shown in Figure 7. We do not exclude the case of having two rigid molecular anvils by which the substrate molecule is sandwiched as shown in Figure 7-c. The relative position of each rigid body is supposed to be changed by various environmental parameters such as temperature, pressure, pH and also concentration of chemical substances in the medium in which the enzyme molecule exists.

The total potential energy of the flexible molecular anvil is generally expressed as a function of various environmental parameters such as

$$E_t(r) = f_a(r - \Delta r_0) + (n_a - 1) f_{ac}(r) + \sum_{i=1}^{i=n_c} n_j f_c k_i(T, P, C_i) r + R_i(T, P, C_i) \quad (4)$$

where $f(x)$s are intermolecular potential functions and the suffixes a and c indicate anvil site and contact site respectively and ac indicates contact site in the anvil, k_i and R_i are the same as Equation (1) but now they are not constant but functions of environmental parameters such as temperature T, pressure P, and concentration of chemical substances c.

For simplicity we can assume without loss of intrinsicity that all k_is are 1 and only R_is are varied with environmental parameters. The change of R_is is essentially equivalent to the change of the relatively extruded distance Δr_0 in the rigid model since the change of R_is relative to the anvil site surface is important in expressing molecular anvil action. Then this problem is reduced to the dependence of enzyme activity on the relatively extruded distance except that it is caused environmentally but not structurally. Thus it is derived that enzyme activity varies with values of environmental parameters. If one of the environmental parameters is varied over a wide range, enzyme activity should show a maximum at some optimum value. This conclusion is in accordance with the general properties of enzymes that each enzyme has its optimum pH or optimum temperature. When the major environmental

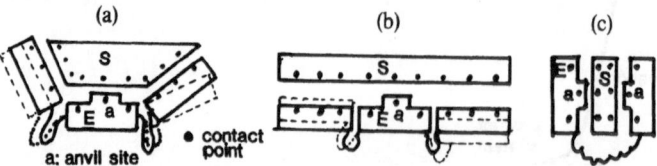

Fig. 7. Illustration of various types of flexible molecular anvil for $n_c = 3$. a: $k_i \neq 1$ and variable. b: $k_i s = 1$ and constant. c: Two anvil site. Solid line indicates original position and dotted one after shifting.

Fig. 8. Enzyme activity vs. environmental parameter curve showing maximum. Arrows indicate direction of change of environment to both sides.

Fig. 9. Schematic illustration of gradual change of peak energy at the anvil site with time during a period of induced fitting.

parameters are fixed and only a minor environmental parameter is changed monotonously, the change of enzyme activity is either a monotonous decrease or a slight increase followed by a decrease, since the original point is not always under conditions showing maximum activity. This is illustrated in Figure 8. The induced fit proposed by Koshland is a case where the configuration of an enzyme molecule is induced by approaching the substrate molecule so as to increase its activity. A gradual increase of peak potential energy at the anvil site may occur and is illustrated in Figure 9.

It is predicted theoretically that some special chemical substances may cause a bell shaped change of enzyme activity with a maximum even over a very small concentration range. This seems important in that if this substance is an effector controlling a certain enzyme in a biological organism, inversion of sign of control will occur when its concentration exceeds some threshold value or other environmental parameters are changed. It seems closely related to the enzymological basis of inversion from an unstable positive feedback state to a stable negative feedback one or vice versa in a biological regulation system.

In actual biological organisms many enzymes exist not in free solution but attached to cell membranes or the skeleton. In such cases changes of configuration of the enzyme molecule caused by a morphological change of the cell are much greater than in free enzymes in solutions. This model offers the principle of activity change of enzyme in cells playing an important role in the regulation of biological systems. This concept seems to be useful in understanding complex biological phenomena from an enzymological point of view.

4. Discussions

The most distinct feature of the molecular anvil is the spontaneous creation of a local high energy spot at the anvil site by accumulating energy from the surroundings even though instantaneously and periodically. The probability of appearance of a high energy state at the anvil site is higher than that expected from the Boltzman factor. The usual thermodynamic laws seem to be invalidated even though instantaneously and locally. The high catalytic power produced by this molecular mechanism of pumping energy from the surroundings by molecular anvil action may be the thermodynamical basis of producing the most far-from equilibrium state such as a biological organism.

The famous key and lock or template theory proposed by E. Fischer [2] implicitly assumes a perfect fitting without an extruded part in the template and in such cases a high energy state is never created and cannot explain the important properties of enzymes of high catalytic power. A slightly imperfect fitting of the molecular anvil is essential to produce high catalytic power.

Recently it was found that an antibody for a bapten works like an enzyme, though weak, for a substrate molecule which has a structure similar to the bapten molecule [3, 4]. This fact indicates that the bapten molecule fits perfectly with the antibody molecule and does not form a molecular anvil, but a certain molecule slightly different from the bapten may form a molecular anvil with the antibody molecule since these molecules fit slightly imperfectly. This seems strong support for the concept of molecular anvil and encourages artificial enzyme design.

According to the molecular anvil model it is predicted that if a certain enzyme has several substrate molecules of different size, relative enzyme activities for these substrates may vary with their molecular volume showing a maximum value for the proper size. Preliminary experiments on urease for substrates such as urea, formamide, methylurea, ethylurea, urethane, show that relative enzyme activities, vary with molecular volume of the substrate molecules, as shown in Figure 10.

There are many individual theories explaining the high catalytic power of enzymes based on particular experimental results on a particular enzyme such as a charge relay system [5], or facilitated proton transfer [6], polyfunctional catalysis [7] etc. Our model does not contradict with these theories because these models explain the lowering of activation energy at the metastable intermediate state while our model is to explain the elevation of the starting level of the reactant system resulting in equivalent lowering of the activation energy. We do not deny that the ease of an enzymatic

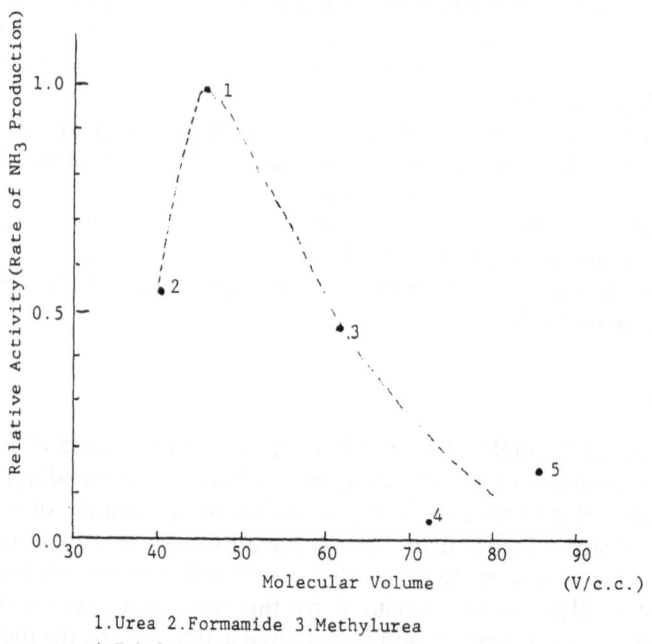

1. Urea 2. Formamide 3. Methylurea
4. Ethylurea 5. Urethane

Fig. 10. Relationship between relative activity of urease and the molecular volume of various substrates.

reaction depends on the kind of group in contact at the anvil site, that is the enzymatic reaction is not wholly determined by the potential energy value at the anvil site but there exist many factors to affect it.

Of all enzymes, about 30% are metal containing ones and the metal atoms play important roles in enzymatic reactions. The electronic properties of these metal atoms may contribute greatly to enhance related enzymatic reactions. But it may be pointed out that molecular anvil action may to some extent contribute to enhancing these enzymatic reactions, because metal atoms are usually greater in size than atoms such as H, C, N, and O which constitute an organic compound surrounding metal atoms and the metal part may be relatively extruded. In some metal porphyrin containing enzymes, the position of the metal atoms are affected by environmental parameters such as temperature, or concentration of chemical substances or substrates and metal atoms come in and out of the plane of the porphyrin molecule. Unusual properties of this group of enzymes may be interesting if examined from the stand point of molecular anvil. The peculiar properties of hemoglobin is also related to this subject, as mentioned by Perutz [8].

According to the flexible molecular anvil model it is predicted that enzyme activity may show a maximum with change of concentration of some effectors on chemicals.

Such a prediction is actually observed, in the L-Lactate-Debydrogenase system [9].

A monotonous decrease of enzyme activity was observed by the addition of 10 to 40 p.p.m. of soap to the amylase-starch system. Whereas in the case of dodecyl benzene sulfonate a slight increase followed by a decrease was observed for the some concentration range [10].

The conclusion of the flexible molecular anvil model that enzyme activity is varied by the change of configuration of the enzyme molecule offers an enzymological basis of regulation of complex biological systems. The inversion of sign of control of effectors may be related to the enzymological basis of cytodifferentiation as well as carcinogenesis [11].

The concept of the molecular anvil may relate to some extent to catalytic power and also to the specificity of non-enzymatic catalysis. This concept may be useful in designing high specific catalysts and artificial enzymes.

5. Conclusions

Two distinct features of enzymes, high catalytic power and high specificity are reasonably and consistently explained by the rigid molecular anvil model. The flexible model similarly explains some general properties of enzymes such as showing maximum activity at the optimum values of environmental parameters such as pH, temperature, concentration of chemical substances. The relation between these fundamental properties of enzymes and biological phenomena is discussed.

References

1. K. Amaya: *J. Incl. Phenom.* **2**, 675 (1984).
2. E. Fischer: *Ber. Dtsch. Chem. Ges.* **27**, 2985 (1894).
3. A. Tramontano, K. D. Janda, and B. A. Lerner: *Science* **234**, 1566 (1986).
4. S. J. Pollack, J. W. Jacobs, and P. G. Schultz: *Science* **234**, 1570 (1986).

5. D. M. Blow, J. J. Birktoff, and B. S. Hartley: *Nature* **221**, 337 (1969).
6. J. H. Wang: *Science* **161**, 328 (1968).
7. C. G. Swain. and J. F. Brown, Jr.: *J. Am. Chem. Soc.* **74**, 2538 (1952).
8. M. F. Perutz: *Nature* **228**, 726 (1970).
9. H. Taguchi, M. Machida, H. Matsuzawa, and T. Ohta: *Agric. Biol. Chem.* **49**, 359 (1985).
10. K. Amaya: *Annual Report of the Japanese Research Society for Synthetic Detergents* **8**, 13 (1985).
11. K. Amaya: *Man and Environment* (in Japanese) **12**, 2 (1986).

Journal of Inclusion Phenomena 5 (1987), 545–549.
© 1987 by D. Reidel Publishing Company.

Display of Cross-Sections Showing Packing in Inclusion Compounds

JANUSZ LIPKOWSKI
Institute of Physical Chemistry, Polish Academy of Sciences, ul. Kasprzaka 44/52, 01-224 Warsaw, Poland

TOSCHITAKE IWAMOTO*
Department of Chemistry, College of Arts and Sciences, The University of Tokyo, Komaba, Meguro, Tokyo 153, Japan

(Received: 30 October 1986; in final form: 27 March 1987)

Abstract. A system of BASIC programs for a personal computer allowing the display of plane sections through crystal structures has been developed in order to analyse intermolecular contacts as well as the shape and size of intra- or intermolecular voids in the structure of inclusion compounds. The data consisting of lattice parameters, type of Bravais lattice, symmetry operations, and positional parameters of atoms in an asymmetric unit are stored in a data file, reusable and correctable. The structure is considered as an assembly of spherical atoms of given van der Waals radii. The section plane may be defined by either coordinates of three points or the crystallographic index. A series of sections parallel to the original one at a specified interval of distance may be calculated and displayed automatically, and memorized for further reviewing, modifying graphic features, etc. Hard copies may be prepared using a dot-matrix printer or plotting devices. The program may be applicable to any crystal structure with an arbitrary choice of atomic radii.

Key words. Crystal packing, personal computer display.

1. Introduction

A crystal structure can be understood in itself through the given set of crystal data and atomic coordinates described in the literature. However, it is often difficult, even for crystallographers, to make an image of the inclusion circumstances in a crystal of complicated structure. In such a case a simple program system for a personal computer is desirable to display the crystal packing. The program can also be used to design or estimate novel inclusion structures of selected host and guest species. The program system we have developed allows the display of plane sections through crystal structures, and is particularly useful in analysing intermolecular contacts between the host and the guest as well as the shape and size of intra- or intermolecular voids in the structures. The program written in BASIC was originally loaded on one of the most economic personal computers with an 8-bit CPU, the NEC-PC8001-MkII [1]; a revised version applicable to a 16-bit CPU computer the NEC-PC9801 is now commercially available [2].

---

* Author for correspondence.

Presented at the Fourth International Symposium on Inclusion Phenomena and the Third International Symposium on Cyclodextrins, Lancaster, U.K., 20–25 July 1986.

2. Program

2.1. OUTLINE

The principle of this program system WANDA is illustrated in the flow chart in Figure 1. The system is comprised of three main programs RAZ, DWA, and SAN. A data file is created by RAZ to memorize the file name, the file title, the lattice parameters, the type of Bravais lattice, the symmetry operations if necessary, the atomic coordinates, the atomic radii, and the colours in the display. By defining a section plane in DWA a work file is created to memorize the atoms to be displayed along with the plane equation, and the successive display of the section on a colour screen is executed by SAN. If one wants to have more sections parallel to the originally-defined plane at a given distance, the cycle between DWA and SAN is run until the required displays have been completed; each work file defined for each section is memorized in the disk. A complimentary program SHI is added to the system for review or printout of the contents of a data file or work file, or for reviewing the contents of a work file.

For the execution of the whole program system a dual floppy disk unit is necessary: $5\frac{1}{4}''$ in the original 8-bit CPU version and $8''$ in the revised 16-bit CPU version.

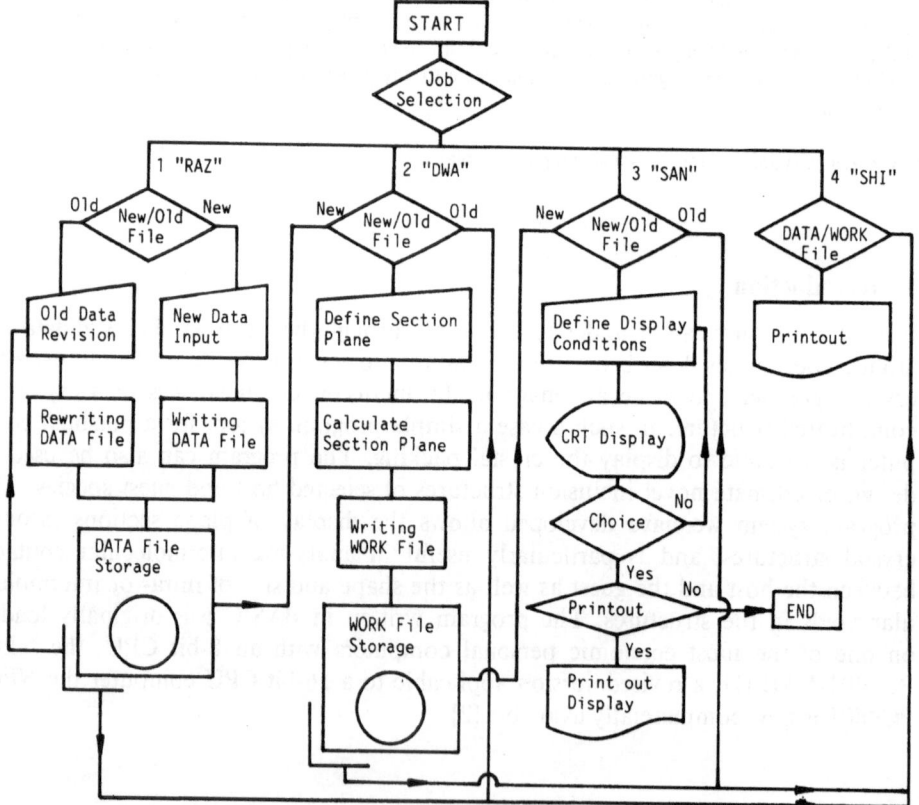

Fig. 1. Construction of the program system WANDA.

2.2. ALGORITHM AND FUNCTIONS

2.2.1. RAZ

Its function is to prepare a data file for a given crystal structure with a file name, a file title noting down information of the crystal within 256 letters, lattice parameters, type of Bravais lattice, symmetry operations, and atomic coordinates, van der Waals radius, and colour code for each atom. The space group emerges in the data file from the data inputted for lattice parameters, Bravais lattice type, presence or absence of inversion centre, and symmetry operations. If all the atoms to be displayed are on the special positions in the given space group, input of the operations related to the special ones is enough for the following calculations.

At each stage of input a question "data correct?" and y/n answer system is prepared. By calling the data file again the contents can be modified and/or corrected according to the enquiry system. After the correction or modification a new file name can be given independently of the previous file name but giving the same file name to the previous one cancels the previous file.

2.2.2. DWA

It first recalls the data file already prepared by RAZ from the disk. Then it defines the section plane by its crystallographic index or three points in the unit cell. The program calculates the orthogonalization matrix and the plane equation in orthogonal coordinates. Taking each atom from the input list, the program checks successively whether its sphere and the spheres of those generated by the symmetry operations are intersected by the defined plane. The approximate computing time estimated roughly from the volume of the unit cell is displayed along with the order number of the atom currently being considered by the program on the screen. Up to 278 atoms within the range of the unit cell ± 2 Å intersected by the plane are memorized into a work file with the name of the data file followed by two digits. The program automatically runs the third part SAN.

2.2.3. SAN

As for DWA, a job selection is first done either by recalling the work file written previously or a review of already existing work files by referring to the data file name.

The scale and the mode of drawing are preset in the program: three scale factors, 15.5×9.5, 31×19, and 62×38 Å^2, and four modes, to paint circles and print out, to paint circles but not to print out, open circles and print out, and open circles but not to print out. The most appropriate selection of the scale factor and the mode of display is ascertained by successive trials. Successive displays of the sections parallel to the original one with input of the number of planes and the interval in Å, or display of non-parallel sections with input of the plane definition are available. In these processes, DWA-SAN cycles run automatically and the work file for each section is memorized into the disk.

In order to review already defined work files, the data file name and the number (last two digits) of the work file is called. After the selection of the scale factor and display mode, one can define the origin of the drawing on the screen.

3. Results

Examples of displays executed by this program system are shown for four types of clathrates, (i) quinol-host hydrogen chloride, **I**, [3], (ii) (deutero)hydrate of ethylene oxide, **II**, [4], (iii) 'Werner-type' bis(isothiocyanato)tetrakis(4-methylpyridine)-nickel(II)-p-xylene(1/1), **III**, [5], and (iv) 'Hofmann-dahxn-type' 1,6-diaminohexane-cadmium(II) tetracyanonickelate(II)-o-toluidine(1/1), **IV**, [6] in Figure 2; the data files were written according to the crystalographic data in the relevant references. Although Figure 2 shows the monochrome output from the printer, the original

Fig. 2. Examples of display. Left upper to lower; a: (001), b: (003), c: (002), and d: (110) section of **I**; middle upper to lower; e: (111), f: (400), and g: (110) section of **II**, and h: (001) section of **III**; right upper to lower; i: (010), j: (020), and k: (200) section of **IV**, and l: (040) section of **III**.

colour display on the VDU is far better than the monochrome one for understanding the chemical structure. However, as the knowledge of the crystal structure should be presupposed to the users, enclathration of each guest molecule is recognizable on the monochrome output when the section is appropriately selected. For example, the $(00l)$ sections of I, a–c, on the printer-output show the positions of guest hydrogen chloride at each corner of the rhombus and at $\frac{2}{3}, \frac{1}{3}, z$ and $\frac{1}{3}, \frac{2}{3}, z$, respectively. The (400) section of II, f, gives the image of guest ethylene oxide molecule in disorder centered at 0.25, 0, 0.5 clearly but it is difficult to judge the image at 0.25, 0, 0 either as that of the guest or of the host moiety, unless one knows the crystal structure in detail.

4. Concluding Remark

As the four programs comprising WANDA can run independently with the respectively independent data and work files, it is possible to display an inclusion structure for both the cases in the presence and absence of the guest molecule. The section once calculated is reviewable as occasion demands with variable size and mode of display.

The program system WANDA is applicable to display arrangement of atoms and/or molecules on a cross section of any crystal including ionic and metallic ones, if the atomic radii are chosen for the relevant types of chemical bonding. Since the ionic and metallic radii are related to the distance of contact between two atomic species, a packing structure should be given by the same procedure as that applied to the inclusion compounds. However for a covalent crystal the display of the packing structure should give rise to too much vacancy if the covalent radii are applied.

Acknowledgements

This work was supported by Grant-in-Aid for the Special Project Research on the Properties of Molecular Assemblies (No. 60104003) from the Ministry of Education, Science and Culture, Japan, and by the Joint Research Program between the Japan Society for Promotion of Sciences and the Polish Academy of Sciences.

References

1. Printout of the program set written in N80 DISC-BASIC will be sent on request to one (T.I.) of the authors.
2. Vendor: Softscience Inc., Shinjuku NS Building, 2-4-1 Nishi-shinjuku, Shinjuku-ku, Tokyo 163, Japan.
3. J. C. A. Boeyens and J. A. Pretorius, *Acta Crystallogr.*, Sect. B **33**, 2120 (1977).
4. F. Hollander and G. A. Jeffrey: *J. Chem. Phys.* **66**, 4699 (1977).
5. J. Lipkowski, K. Suwińska, G. D. Andreetti, and K. Stadnicka: *J. Mol. Struct.* **75**, 101 (1981).
6. T. Hasegawa, S. Nishikiori, and T. Iwamoto: *J. Incl. Phenom.* **2**, 351 (1984).

Erratum

S. F. Lincoln, J. H. Coates, B. G. Doddridge, and A. M. Hounslow: 'The Inclusion of the Drug Diflunisal by Alpha- and Beta- Cyclodextrins. A Nuclear Magnetic Resonance and Ultraviolet Spectroscopic Study', *J. Incl. Phenom.* **5**, 49–53 (1987).

In the structural formula for diflunisal (2-hydroxy-5-(2,4-difluorophenyl)benzoic acid), the –COOH and –OH substituents should be shown in the 1 and 2 positions, respectively.

AUTHOR INDEX

Abou-Hamdan, A., I. M. Brereton, A. M. Hounslow, S. F. Lincoln and T. M. Spotswood - An equilibrium and kinetic study of the complexation of lithium and sodium ions by the cryptand C21C5   137
Akyuz, S. and T. Akyuz - IR investigations of adsorption and oxidation of N,N-dimethylaniline by sepiolite, loughlinite and diatomite   259
Akyuz, S. - See Sungur, A. et al.   383
Akyuz, T. - See Akyuz, S.   259
Aladko, E. Ya. - See Dyadin, Yu. A. et al.   203
Amaya, K. - The molecular anvil model of an enzyme taking into consideration the flexibility of enzyme molecules   427
Andera, L. and E. Smolkova-Keulemansova - The effect of water vapour on the cyclodextrin-solute interaction in gas-solid chromatography   289
Andreetti, G. D., G. Calestani, F. Ugozzoli, A. Arduini, E. Ghidini, A. Pochini and R. Ungaro - Solid state studies on p-t-butyl-calix[6]arene derivatives   123
Andreetti, G. D. - See Calestani, G. et al.   269
Arduini, A. - See Andreetti, G. D. et al.   123
Babb, D. A. - See Bartsch, R. A. et al.   407
Bacca, G. - See Calestani, G. et al.   269
Barthelemy, P. P. - See Rebizant, J. et al.   397
Bartsch, R. A., D. A. Babb and B. E. Knudsen - Synthesis and alkali metal cation complexation of N-aryl [3.2.2.] cryptands   407
Beer, P. D. and A. D. Keefe - The synthesis of metallocene calix[4]arenes   391
Beer, P. D., C. J. Jones, J. A. McCleverty and S. S. Salam - Redox responsive metal complexes containing cation binding sites   413
Bell, M. N., A. J. Blake, R. O. Gould, A. J. Holder, T. I. Hyde, A. J. Lavery, G. Reid and M. Schroder - Transition metal complexes of homoleptic polythia crowns   169
Bell, T. W., A. Firestone, F. Guzzo and L-Y. Hu - Torands: Planar polyazamacrocyclic ligands for metal ions   149
Belosludov, V. R. - See Dyadin, Yu. A. et al.   195
Benetollo, F., G. Bombieri and M. R. Truter - Crystal

structures of 1:1 complexes between urea and two
  crown ether derivatives of phthalic acid   165
Bishop, R., I. G. Dance, S. C. Hawkins and M. L.
  Scudder - Molecular determinants of a new family of
  helical tubuland host diols   229
Blake, A. J. - See Bell, M. N. et al.   169
Bombieri, G. - See Benetollo, F. et al.   165
Bottino, F. - See Pappalardo, S. et al.   153
Brereton, I. M. - See Abou-Hamdan, A. et al.   137
Bujtas, K., T. Cserhati and J. Szejtli - Reduction of
  phytotoxicity of nonionic tensides by cyclodextrins   313
Burrows, C. J. and R. A. Sauter - Synthesis and con-
  formational studies of a new host system based on
  cholic acid   117
Calestani, G. - See Andreetti, G. D. et al.   123
Calestani, G., V. Sangermano, C. Rizzoli, G. Bacca and
  G. D. Andreetti - Generation and management of 3D
  structural diagrams for zeolites on an IBM PC   269
Carilla, J. - See Veciana, J. et al.   241
Celebi, N., O. Shirakura, Y. Machida and T. Nagai -
  The inclusion complex of piromidic acid with dimethyl
  beta cyclodextrin in aqueous solution and in the
  solid state   299
Chau, L. T. - See Szente, L. et al.   331
Chekhova, G. N. - See Dyadin, Yu. A. et al.   187
Chekhova, G. N. - See Dyadin, Yu. A. et al.   195
Choy, J. H., C. E. Kim, K. W. Hyung and J. C. Park -
  Correlation between layer charge and activation
  energy of thermally induced deintercalation in organo
  layer silicates   253
Choy, J. H., Y. J. Shin, G. Demazeau and P. Hagenmuller
  Isomorphous substitution effects on the thermally
  induced interlayer reaction in N-hexylammonium lay-
  ered aluminosilicates   375
Chujo, R. - See Inoue, Y. et al.   55
Clayden, N. J., C. M. Dobson, S. J. Heyes and P. J.
  Wiseman - ^{2}H NMR studies of metallocenes in host
  lattices   65
Coates, J. H. - See Lincoln, S. F. et al.   49
Coates, J. H. - See Schiller, R. L. et al.   59
Cserhati, T. - See Bujtas, K. et al.   313
Cserhati, T. - See Szogyi, M. et al.   325
Dance, I. G. - See Bishop, R. et al.   229
Dauphin, G. - See Prudhomme, M. et al.   99
Davidson, D. W., M. A. Desando, S. R. Gough, Y. P.
  Handa, C. I. Ratcliffe, J. A. Ripmeester and J. S. Tse
  Some physical and thermophysical properties of clath-
  rate hydrates   219
Davies, J. E. D. - See Sungur, A. et al.   383
Demazeau, G. - See Choy, J. H. et al.   375
Dengler, L. - See Wiener, H. L. et al.   215

AUTHOR INDEX

Depmeier, W. - Aluminate sodalites: a family of inclusion compounds with strong host-guest interactions — 279
Desando, M. A. - See Davidson, D. W. et al. — 219
Desreux, J. F. - See Rebizant, J. et al. — 397
Dobson, C. M. - See Clayden, N. J. et al. — 65
Doddridge, B. G. - See Lincoln, S. F. et al. — 49
Duran, A. - See Veciana, J. — 173
Dyadin, Yu. A., G. N. Chekhova and N. P. Sokolova - Solid clathrate solutions — 187
Dyadin, Yu. A., V. R. Belosludov, G. N. Chekhova and M. Yu. Lavrentiev - Clathrate thermodynamics for the unstable host framework — 195
Dyadin, Yu. A., F. V. Zhurko, E. Ya. Aladko, Yu. M. Zelenin and L. A. Gaponenko - Clathrate formation in water - tetra alkylammonium iodide systems at high pressure — 203
Fastrez, J. - See Lepropre, G. — 157
Federov, V. E. - See Mischenko, A. V. et al. — 263
Fenton, D. E., B. P. Murphy, R. Price, P. A. Tasker and D. J. Winter - Metal-free macrocycles via template method: a starting point for selective complexation studies — 143
Fenyvesi, E. - See Szeman, J. et al. — 319
Fenyvesi, E. - See Suzuki, M. et al. — 351
Finocchiaro, P. - See Pappalardo, S. et al. — 153
Firestone, A. - See Bell, T. W. et al. — 149
Fonagy, A. - See Gerloczy, A. et al. — 307
Fronczek, F. R. - See Pappalardo, S. et al. — 153
Fukushima, Y. and S. Inagaki - Synthesis of an intercalated compound of montmorillonite and 6-polyamide — 365
Gadd, K. F. - Metal-containing cellulose: some novel materials — 265
Gaponenko, E. Ya. - See Dyadin, Yu. A. et al. — 203
Gerloczy, A., L. Szente, J. Szejtli and A. Fonagy - Improvement of fat digestion in rats by dimethyl beta cyclodextrin — 307
Ghidini, E. - See Andreetti, G. D. et al. — 123
Gies, H. - Synthesis, crystallographic and thermal properties of a new porous silica — 283
Goldberg, I., H. Shinar, G. Navon and W. Klaui - Organometallic ionophore for alkali metal cations — 181
Gough, S. R. - See Davidson, D. W. et al. — 219
Gould, R. O. - See Bell, M. N. et al. — 169
Gresh, N. - See Prudhomme, M. et al. — 99
Grey, H. - See Lockhart, J. C. — 113
Guyot, J. - See Prudhomme, M. et al. — 99
Guzzo, F. - See Bell, T. W. et al. — 149
Hagenmuller, P. - See Choy, J. H. et al. — 375
Hamilton, A. D. - See Pant, N. et al. — 109

Handa, Y. P. - See Davidson, D. W. et al. 219
Harris, K. D. M. - See Hollingsworth, M. D. et al. 273
Hasegawa, T. - See Iwamoto, T. et al. 225
Hattori, K. and K. Takahashi - Novel HPLC adsorbents by immobilization of modified cyclodextrins 73
Hawkins, S. C. - See Bishop, R. et al. 229
Herbstein, F. H., M. Kapon and G. M. Reisner - Catenated and non-catenated inclusion complexes of trimesic acid 211
Holder, A. J. - See Bell, M. N. et al. 169
Hollingsworth, M. D., K. D. M. Harris, W. Jones and J. M. Thomas - ESR and X-ray diffraction studies of diacyl-peroxides in urea and aluminosilicate hosts 273
Hoshi, H. - See Inoue, Y. et al. 55
Hounslow, A. M. - See Lincoln, S. F. et al. 49
Hounslow, A. M. - See Abou-Hamdan, A. et al. 137
Heyes, S. J. - See Clayden, N. J. et al. 65
Hu, L-Y. - See Bell, T. W. et al. 149
Huang, N-J. - See Zhang, D-D. et al. 335
Huang, Y-M. - See Zhang, D-D. et al. 335
Hyde, T. I. - See Bell, M. N. et al. 169
Hyung, K. W. - See Choy, J. H. et al. 253
Iijima, M. - See Ikeda, T. et al. 93
Ikeda, H. - See Yoon, C-J. et al. 85
Ikeda, H. - See Ikeda, T. et al. 93
Ikeda, T., R. Kojin, C-J. Yoon, H. Ikeda, M. Iijima and F. Toda - Catalytic activity of beta cyclodextrin - histamine 93
Ikeda, T. - See Yoon, C-J. et al. 85
Ilardi, L. - See Wiener, H. L. et al. 215
Inagaki, S. - See Fukushima, Y. 365
Inoue, Y., M. Kitagawa, H. Hoshi, M. Sakurai and R. Chujo - The geometry of the alpha cyclodextrin inclusion complex with m-nitrophenol deduced from quantum chemical analysis and ^{13}C chemical shifts 55
Ishizaki, H. - See Sakuraba, H. et al. 341
Iwamoto, T., S-I. Nishikiori and T. Hasegawa - Hofmann diaminoalkane-type clathrates 225
Iwamoto. T. - See Lipkowski, J. 437
Jeffas, S. A. - See Wiener, H. L. et al. 215
Jeminet, G. - See Prudhomme, M. et al. 99
Jones, C. J. - See Beer, P. D. et al. 413
Jones, S. P. and G. D. Parr - Ability of the acetotoluides to form cyclodextrin inclusion complexes 45
Jones, W. - See Hollingsworth, M. D. et al. 273
Jurczak, J. - See Koscielski, T. et al. 69
Kapon, M. - See Herbstein, F. H. et al. 211
Kata, M. and B. Selmeczi - Increasing the solubility of drugs through cyclodextrin complexation 39
Keefe, A. D - See Beer, P. D. 391
Kijima, T., S. Takenouchi and Y. Matsui - Complexes of

AUTHOR INDEX

Na, Ca and Zn-montmorillonites with an aminated cyclodextrin	361
Kim, C. K. - See Choy, J. H. et al.	253
Kitagawa, M. - See Inoue, Y. et al.	55
Klaui, W. - See Goldberg, I. et al.	181
Knudsen, B. E. - See Bartsch, R. A. et al.	407
Kojin, R. - See Yoon, C-J. et al.	85
Kojin, R. - See Ikeda, T. et al.	93
Koscielski, T., D. Sybilska and J. Jurczak - Separation processes in gas liquid chromatography based on the formation of alpha cyclodextrin - chiral hydrocarbon inclusion complexes	69
Lavery, A. J. - See Bell, M. N. et al.	169
Lavrentiev, M. Yu. - See Dyadin, Yu. A. et al.	195
Le Bas, G. - See Tsoucaris, G. et al.	77
Lepropre, G and J. Fastrez - Size and charge dependence of binding by azacyclophanes	157
Liberati, P. - See Wiener, H. L. et al.	215
Lincoln, S. F., J. H. Coates, B. G. Doddridge and A. M. Hounslow - The inclusion of the drug Diflunisal by alpha and beta cyclodextrins. An NMR and UV spectroscopic study	49
Lincoln, S. F. - See Schiller, R. L. et al.	59
Lincoln, S. F. - See Abou-Hamdan, A. et al.	137
Lipkowski, J and T. Iwamoto - Display of cross-sections showing packing in inclusion compounds	437
Lockhart, J. C. and H. Grey - Molecular graphics in the study of the calcium-binding sites of carp parvalbumin and other proteins	113
Lucken, E. A. C. - See Pang, L.	245
Machida, Y. - See Celebi, N. et al.	299
Machida, Y. - See Szeman, J. et al.	319
MacNicol, D. D., P. R. Mallinson, A. Murphy and C. D. Robertson - Synthesis and structure of hexakis-(p-hydroxyphenyloxy)benzene	233
Mallinson, P. R. - See MacNicol, D. D. et al.	233
Mamo, A. - See Pappalardo, S. et al.	153
Mann, M. - See Pant, N. et al.	109
Matsui, Y. - See Kijima, T. et al.	361
McCleverty, J. A. - See Beer, P. D. et al.	413
Miravitlles, C. - See Veciana, J. et al.	241
Mischenko, A. V., Yu. V. Moronov, P. P. Samojlov and V. E. Fedorov - Lithium intercalation cluster compounds	263
Miyata, M., F. Noma, K. Okanishi, H. Tsutsumi and K. Takemoto - Inclusion polymerisation of diene and diacetylene monomers in DCA and ACA canals	249
Molins, E. - See Veciana, J. et al.	241
Moronov, Yu. V. - See Mischenko, A. V. et al.	263
Murphy, A. - See MacNicol, D. D. et al.	233

Murphy, B. P. - See Fenton, D. E. et al. 143
Nagai, T. - Developments in cyclodextrin applications in drug formulations 29
Nagai, T. - See Celebi, N. et al. 299
Nagai, T. - See Szeman, J. et al. 319
Navon, G. - See Goldberg, I. et al. 181
Nishikiori, S-I. - See Iwamoto, T. et al. 225
Noma, F. - See Miyata, M. et al. 249
Okanishi, K. - See Miyata, M. et al. 249
Pang, L. and E. A. C. Lucken - ^{35}Cl nuclear quadrupole resonance studies of CCl_4 as a guest in various clathrates 245
Pant, N., M. Mann and A. D. Hamilton - Synthetic analogs of peptide-binding antibiotics 109
Pappalardo, S., F. Bottino, P. Finocchiaro, A. Mamo and F. R. Fronczek - Synthesis of symmetrical N-tosyldiazamacrocycles and complexation properties of their derivatives 153
Park, J. C. - See Choy, J. H. et al. 253
Parr, G. D. - See Jones, S. P. 45
Pietraszkiewicz, M. - Synthesis and complexing properties of a chiral macrocyclic molecular receptor with convergent binding sites 177
Pochini, A. - See Andreetti, G. D. et al. 123
Price, R. - See Fenton, D. E. et al. 143
Prudhomme, M., G. Dauphin, J. Guyot, G. Jeminet and N. Gresh - Modifications of benzoxazole ring substituents in A.23187 (Calcimycin). Effect on cation carrier properties. 99
Ratcliffe, C. I. - See Davidson, D. W. et al. 219
Rebizant, J., M. R. Spirlet, P. P. Barthelemy and J. F. Desreux - Solid state and solution structures of the lanthanide complexes with cryptand (2.2.1) 397
Reisner, G. M. - See Herbstein, F. H. et al. 211
Ripmeester, J. A. - See Davidson, D. W. et al. 219
Rizzoli, C. - See Calestani, G. et al. 269
Robertson, C. D. - See MacNicol, D. D. et al. 233
Rysanek, N. - See Tsoucaris, G. et al. 77
Saba, S. - See Wiener, H. L. et al. 215
Sakuraba, H., H. Ishizaki, Y. Tanaka and T. Shimizu - Asymmetric halogenation and hydrohalogenation of styrene in crystalline cyclodextrin complexes 341
Sakurai, M. - See Inoue, Y. et al. 55
Salam, S. S. - See Beer, P. D. et al. 413
Samojlov, P. P. - See Mischenko, A. V. et al. 263
Sangermano, V. - See Calestani, G. et al. 269
Sasaki, Y. - See Suzuki, M. et al. 351
Sauter, R. A. - See Burrows, C. J. 117
Scudder, M. L. - See Bishop, R. et al. 229
Selmeczi, B. - See Kata, M. 39
Severcan, F. - Topology of N-ethylmaleimide in normal

AUTHOR INDEX

human erythrocyte membrane   127
Schiller, R. L., S. F. Lincoln and J. H. Coates - The inclusion of Pyronine B and Pyronine Y by beta and gamma cyclodextrins. A kinetic and equilibrium study   59
Schmidtchen, F. P. - Multiple recognition in polytopic anion hosts   161
Schroder, M. - See Bell, M. N. et al.   169
Shimizu, T., Y. Tanaka and K. Tsuda - Ion transport, ion extraction and ion binding by a synthetic cyclic octapeptide   103
Shimizu, T. - See Sakuraba, H et al.   341
Shin, Y. J. - See Choy, J. H. et al.   375
Shinar, H. - See Goldberg, I. et al.   181
Shirakura, O. - See Celebi, N. et al.   299
Smith, N. O. - See Wiener, H. L. et al.   215
Smolkova-Keulemansova, E. - See Andera, L.   289
Sokolova, N. P. - See Dyadin, Yu. A. et al.   187
Spirlet, M. R. - See Rebizant, J. et al.   397
Spotswood, T. M. - See Abou-Hamdan, A. et al.   137
Sungur, A., S. Akyuz and J. E. D. Davies - Vibrational studies of 4,4´-bipyridyl metal tetracyanonickelate complexes and their clathrates   383
Suzuki, M., Y. Sasaki, J. Szejtli and E. Fenyvesi - ^{13}C NMR spectra of cyclodextrin monomers, derivatives and their complexes with methyl orange   351
Sybilska, D. - See Koscielski, T. et al.   69
Szejtli, J. - See Gerloczy, A. et al.   307
Szejtli, J. - See Bujtas, K. et al.   313
Szejtli, J. - See Szeman, J. et al.   319
Szejtli, J. - See Szogyi, M. et al.   325
Szejtli, J. - See Szente, L. et al.   331
Szejtli, J. - See Suzuki, M. et al.   351
Szeman, J., E. Fenyvesi, J. Szejtli, H. Ueda, Y. Machida and T. Nagai - Water soluble cyclodextrin polymers: their interaction with drugs   319
Szente, L. - See Gerloczy, A. et al.   307
Szente, L., J. Szejtli and L. T. Chau - Effect of cyclodextrin complexation on the reduction of menthone and iso-menthone   331
Szogyi, M., T. Cserhati and J. Szejtli - Cyclodextrins lessen the membrane damaging effect of nonionic tensides   325
Takahashi, K. - See Hattori, K.   73
Takahashi, Y. - Binding properties of alginic acid and chitin   417
Takemoto, K. - See Miyata, M. et al.   249
Takenouchi, S. - See Kijima, T. et al.   361
Tanaka, Y. - See Shimizu, T. et al.   103
Tanaka, Y. - See Sakuraba, H. et al.   341
Tasker, P. A. - See Fenton, D. E. et al.   143

Thomas, J. M. - See Hollingsworth, M. D. et al. 273
Toda, F. - See Yoon, C-J. et al. 85
Toda, F. - See Ikeda, T. et al. 93
Truter, M. R. - See Benetollo, F. et al. 165
Tse, J. S. - See Davidson, D. W. et al. 219
Tsoucaris, G., G. Le Bas, N. Rysanek and F. Villain - Conformational and enantiomeric discrimination in cyclodextrin inclusion compounds 77
Tsuda, K. - See Shimizu, T. et al. 103
Tsutsumi, H. - See Miyata, M. et al. 249
Ueda, H. - See Szeman, J. et al. 319
Ugozzoli, F. - See Andreetti, G. D. et al. 123
Ungaro, R. - See Andreetti, G. D. et al. 123
Veciana, J. and A. Duran - Isolation, properties and association phenomena of alkaline salts and their crown complexes with a radical anion 173
Veciana, J., J. Carilla, C. Miravitlles and E. Molins - Free radicals as host molecules 241
Villain, F. - See Tsoucaris, G. et al. 77
Wiener, H. L., L. Ilardi, P. Liberati, L. Dengler, S. A. Jeffas, S. Saba and N. O. Smith - Selectivity of the host $Ni(4-Mepy)_4(NCS)_2$ towards aromatic guests 215
Winter, D. J. - See Fenton, D. E. et al. 143
Wiseman, P. J. - See Clayden, N. J. et al. 65
Xue, L. - See Zhang, D-D. et al. 335
Yoon, C-J., H. Ikeda, R. Kojin, T. Ikeda and F. Toda - Reaction of cyclodextrin-nicotinamide as a NADH coenzyme model 85
Yoon, C-J. - See Ikeda, T. et al. 93
Zelenin, Yu. M. - See Dyadin, Yu. A. et al. 203
Zhang, D-D., N-J. Huang, L. Xue and Y-M. Huang - Beta cyclodextrin catalysed effects on the hydrolysis of esters of aromatic acids 335
Zhurko, Yu. M. - See Dyadin, Yu. A. et al. 203

SUBJECT INDEX

Alginic acid	417
Apocholic acid host	249
Azacyclophanes	157
Azamacrocycles	149,153
Calcimycin	99
Calixarenes	123
metallocene type	391
Catalytic effects	93,335
Cellulose	265
Chitin	417
Cholic acid hosts	117
Chromatography	
gas liquid	69
gas solid	289
HPLC	73
Clathrate hydrates	219
Clathrates	
at high pressure	203
solid solutions of	187
thermodynamics of	195
Coenzyme models	85
Conformational	
discrimination	77
Crown ethers	
complexes with	
free radical	173
transition metals	169
urea	165
polythiacrowns	169
redox properties	413
Cryptands	137,397
aryl	407
Cyclic octapeptide	103
Cyclodextrin polymers	319
Cyclodextrins	
complexes with	
acetotoluides	45
diflunisal	49

iso-menthone	331
menthone	331
pyronine B	59
pyronine Y	59
conformational	
discrimination	77
enantiomeric	
discrimination	77
NMR spectra	351
use in	
chromatography	69,289
drug formulations	29,319
drug solubility	39
reduction of	
phytotoxicity	313
Cyclodextrin - *alpha*	
complex with	
m-nitrophenol	5
^{13}C NMR of	55
quantum analysis	55
use in GLC	69
Cyclodextrin - *beta*	
catalytic effect	335
complexes with	
esters	335
histamine	93
nicotinamide	85
use in HPLC	73
Cyclodextrin - *dimethylbeta*	
complex with	
piromidic acid	299
use in	
improvement of fat	
digestion	307
Cyclophanes	
aza type	157

SUBJECT INDEX

Deoxycholic acid host	249	of trimesic acid	211
Diatomite	259	of urea	273
Diol hosts	229	polymerisation	249

Discrimination
  conformational 77
  enantiomeric 77
  isomer 215
  isotopomer 215
  product 331,341
Drug formulations 29,319
Drug solubility 39

Enantiomeric
    discrimination 77
Enzyme models 93,427
Equilibrium study 59,137
ESR spectra 273

Fat digestion 307
Free radical guests 273
Free radical hosts 241

GLC separations 69
GSC separations 289

Hexa hosts 233
Hofmann type hosts 225,383
Host lattice
  aluminate sodalite 279
  apocholic acid 249
  calixarene 123
  cholic acid 117
  deoxycholic acid 249
  diol 229
  free radical 241
  hexa host 233
  Hofmann type 225,383
  polytopic anion 161
  porous silica 283
  selectivity 215
  trimesic acid 211
  urea 273
  Werner 215
HPLC separations 73

Immobilised CDs 73
Inclusion
  compounds
    catenated 211
    non catenated 211

Intercalates
  aluminosilicate 375
  chalcogenide 263
  cluster compound 263
  interlayer reaction 375
  montmorillonite 361,365
  silicate 253,273
Ion binding 103
Ion extraction 103
Ion transport 99,103
Ionophores 99,103
  organometallic 181
IR spectra 259,383

Kinetic study 59,137

Lanthanide complexes 397
Loughlinite 259

Macrocycles
  aza 149,153
  chiral 177
  metal free 143
Metallocene guests 65
Molecular
  anvil model 427
  graphics 113,269,437
  receptors 177

N-Ethylmaleimide 127

NMR spectra
  ^{13}C spectra of
    CD derivatives 351
    CD/m-nitrophenol 55
  ^{2}H spectra of
    metallocene guests 65
  ^{19}F spectra of
    CD/diflunisal 49

Nonionic tensides 313,325

NQR spectra
  ^{35}Cl spectra 245

PC displays 269,437
Peptide binding 109

SUBJECT INDEX

Polytopic hosts	161
Porosil	283
Raman spectra	383
Redox properties	413
Selectivity	
complexational	143
conformational	77
enantiomeric	77
isomer	217
isotopomer	217
product	331,341
Sepiolite	259
Silicate intercalates	253
Sodalites	279
Synthetic	
antibiotics	109
receptors	109
Torands	149
UV spectra of	
CD/diflunisal	49
CD/pyronine B and Y	59
Zeolites	269

CYCLODEXTRIN NEWS

AN INTERNATIONAL NEWS LETTER

Editors: J. Szejtli (Hungary), J. Pagington (U.K.)

Published monthly in Britain and Hungary

CYCLODEXTRIN NEWS offers a comprehensive abstracting service with over 530 papers on cyclodextrins abstracted in Volume 1 (10 issues). Each monthly issue also includes a feature on some aspect of CDs or their complexes, as well as legislation, approvals, applications, etc.

Extra services for subscribers include the sale of small quantities of various cyclodextrins, and an abstracting service providing detailed literature searches tailored to specific requirements.

Twelve month subscription (Sept '87 to Aug '88)   150$

Limited number of Volume 1 copies still available   75$

Prices include airmail delivery and a binder

For a complimentary copy
and
further details contact:

FDS Publications, P.O. Box 41, Trowbridge, England BA14 8UE.

Science on Form

Proceedings of the First International Symposium held at the University of Tsukuba, Japan, November 26–30, 1985

edited by
S. ISHIZAKA, Y. KATO, R. TAKAKI and J. TORIWAKI

1987, 690 pp. ISBN 90-277-2390-7
Hardbound Dfl. 310.00/£109.00/US$129.00
D. Reidel Publishing Company/KTK, Japan

This work is a collection of papers presented at The First International Symposium for Science on Form held at the University of Tsukuba in 1985. The scope of the Symposium was very wide and covered all kinds of fields concerned with 'Form', e.g. pattern recognition, stereology, computer graphics, mechanism of pattern formation, development of organisms or a social system, etc. The Symposium had no parallel sessions, and participants from a variety of fields were able to have exciting discussions in an interdisciplinary atmosphere. This volume, which contains the Proceedings of the Symposium, will be of interest to physicists, geoscientists, biologists, mathematicians, computer scientists, and engineers.

Contents
Preface. Pattern Formation and Morphogenesis I – Fractal Phenomena – Pattern in Systems of Active Elements. Morphometry and Stereology I – General Theory and Statistical Method. – Micromechanics of Inhomogeneous Materials. – Morphological Analysis of Materials. Pattern Recognition and Image Processing I – Three-dimensional Shape Features and Image Understanding. – Image Generation and Graphics. Pattern Formation and Morphogenesis II – Mechanism of Form and Pattern Creation in Living Systems I – Mechanism of Form and Pattern Creation in Living Systems II. Pattern Recognition and Image Processing II – Two-dimensional Shape Features and Textures. – Pattern Formation and Morphogenesis III. Morphometry and Stereology II – The Form in Living Structures-Quantification and Functional Correlation. – Pattern Formation and Morphogenesis IV. – Quasicrystals. Indexes.

No sales rights in Japan

Kluwer academic publishers group

P.O. Box 989
3300 AZ Dordrecht
The Netherlands

101 Philip Drive
Norwell,
MA 02061, U.S.A.

Falcon House
Queen Square
Lancaster, LA1 1RN, U.K.

NEW

Lanthanide and Actinide Research

A Multidisciplinary International Journal covering all Aspects of f-Block Elements

Editor
Prof. **Shyama P. Sinha,** *Dept. of Chemistry, University of Dayton, OH, USA*

Lanthanide and Actinide Research is an international journal covering *all aspects* of theoretical and experimental studies on the lanthanides, actinides, and related elements (Y, Sc). It is a *rapid* communication journal, established for the purpose of increasing *multidisciplinary interaction* by reporting original research in the form of short articles, notes, and communications. Communications will normally deal with comments on articles published in *Lanthanide and Actinide Research* or elsewhere. Book reviews will be published occasionally.

All contributions submitted will be refereed, and the editor and editorial board members reserve the right to reject any manuscript deemed unsuitable for publication in *Lanthanide and Actinide Research* and to manuscripts when necessary. Only articles written in the English language will be accepted. Only original work, not published previously or scheduled for publication elsewhere, should be submitted.

Subscription Information   **ISSN** 8755-5301
1987, Volume 2 (6 issues)
Institutional rate: Dfl. 290.00/US$125.00 incl. postage/handling
Private rate: Dfl. 135.00/US$59.00 incl. postage/handling

Private subscriptions should be sent direct to the publishers

Kluwer academic publishers group

P.O. Box 989	101 Philip Drive	Falcon House
3300 AZ Dordrecht	Norwell,	Queen Square
The Netherlands	MA 02061, U.S.A.	Lancaster, LA1 1RN, U.K.